Historical Encyclopedia of Natural and Mathematical Sciences

Ari Ben-Menahem

Historical Encyclopedia of Natural and Mathematical Sciences

Volume 6

Springer

Professor Ari Ben-Menahem
Weizmann Institute of Science
Rehovot 76100
Israel

Library of Congress Control Number: 2008944294

ISBN: 978-3-540-68831-0
This publication is available also as:
Electronic publication under ISBN 978-3-540-68832-7 and
Print and electronic bundle under ISBN 978-3-540-68834-1

Springer is part of Springer Science+Business Media

springer.com

Editor: Christian Witschel, Heidelberg, Germany
Development Editor: Sylvia Blago, Heidelberg, Germany
Production: VTEX, Vilnius, Lithuania
Cover Design: Frido Steinen-Broo, Girona, Spain

Printed on acid-free paper SPIN: 12250899 2109 — 5 4 3 2 1 0

Editorial Assistant and Scientific Advisor

Shahar Ben-Menahem M.Sc., Ph.D.
Department of Physics, Stanford University, USA

To my beloved wife Batia for her support, forbearance and counsel during the hardship years of creating this treatise; without her help this feat could not have been accomplished

* * *

To the memory of my dear parents:
Sara-Haya and Moshe

* *
*

"Whoever is ignorant of the past remains forever a child. For what is the worth of human life, unless it is woven into the life of our ancestors by the records of history?"

Marcus Tullius Cicero (106–43 BCE)

* *
*

"He lives doubly, who also lives the past".

Marcus Valerius Martial (ca 100 CE)

* *
*

"Mathematics was born and nurtured in a cultural environment. Without the perspective which the cultural background affords, a proper appreciation of the content and state of present-day mathematics is hardly possible".

Leonhard Euler (1707–1783)

* *
*

"History is the essence of innumerable biographies".

Thomas Carlyle (1795–1881)

* *
*

"To understand a science it is necessary to know its history".

Auguste Comte (1799–1857)

* *
*

"The history of science is science itself".

Johann Wolfgang von Goethe, 1825

* *
*

"No subject loses more than mathematics by any attempt to dissociate it from its history".

J.W.L. Glaisher (1848–1928)

* *
*

"Take three hundred men *out of history and we should still be living in the stone age".*

Arthur Keith (1866–1955)

* *
*

"The history of science is very largely the history of great men".

George Sarton, 1927

* *
*

"Without the concepts, methods and results found and developed by previous generations right down to Greek antiquity, one cannot understand the aims or the achievements of mathematics in the last fifty years".

Hermann Weyl, 1950 (1885–1955)

* *
*

"There is nowhere else to look for the future but in the past".

James Burke, 1978

* *
*

"The genesis of the majority of mathematical theories is obscure and difficult. Often, today's presentation of a classical topic, will be much more accessible and concise than it could even have been when it was developed. Any scientist's work, can only be understood within its contemporary scientific framework".

Walter K. Bühler

Preface

Many books are published every year on the history of science, but I know of no comprehensive treatise that blends the essential historical data (chronology, biographies, major background political and economical events, etc.) *together* with science proper (principles, laws, experiments, observations, theories, equations, etc.). The present encyclopedic treatise does just that; it tells the reader not only *who* did it and *when* it was done but also precisely *what* was done.

The saga of this history of ideas, discovery and invention in the natural and mathematical sciences – spanning about 100 generations of great thinkers from Thales to Feynman – unfolds in all its grandeur before the eyes and mind of the reader. Whether to professional scientists, students, or unassuming curious laymen, the doors of this shrine are open, inviting them to browse, linger and study whatever suits them. I believe that every intelligent person can understand the development of science when properly presented from its beginnings; The historical method is the best for introducing scientific facts and ideas to unprepared minds in a thoroughly understandable manner.

The history of science is more than the arithmetical sum of the histories of all sciences, for it also explains the interrelations of them all. Indeed, our division of science into many branches is largely artificial. Like the branches of a living tree which have no separate existence, but grow together – the progress of each science is dependent upon the progress of all the others.

The main postulate of science is the unity of nature: *nature is one*; and therefore, *science is one*. Finally, the fact that simultaneous discoveries have been made by different groups of workers, in different settings, organizations and nations, demonstrate that *mankind is one*: one mankind through one science is unfolding the mysteries of one nature.

It follows that the only rational way to subdivide this history is not according to the sciences or countries involved, but only according to time; for each period of time we have to consider at once the *whole* of science's historical and intellectual development. This calls for the marshaling of all scientific facts, activities and ideas in a definite order; which means that we must try to assign to each of them a date as precise as possible – not just the date of their birth or their publication, but also that of their *actual incorporation into our knowledge* – often a very difficult thing to do, as the reader will not fail to appreciate. Such work of erudition is the bedrock upon which this history is built.

We have also considered some other departments of life which have bearing on the evolution of science. These are:

- General natural and human history, especially the history of civilization.
- The history of technology.
- The history of philosophy and religions.

To make this general synthesis possible, we found it expedient to write a large number of monographs on various subjects which emphasize the interrelations between environmental, economic, social, cultural, political and scientific 'events' (e.g. the history of epidemics is needed to correctly estimate the evolution of medical ideas).

The history of science is a field of endless complexity and incredible scope. There are many ways to study it and many points of view, none of which is exclusive of the others. The chronological order of discoveries is often very different from their logical sequence. What some people call the logic of scientific discovery is largely a retrospective construction; it is nevertheless useful to bring it out. Discoveries are not always made in logical order but it is worthwhile and helpful to attempt to explain them in such an order: the actual path of progress is not straight but very crooked, although the general direction is clear enough.

We have interspersed our history with a narration of general intellectual climate and of major social, cultural, political, economic and environmental events: science does not develop in a social vacuum and every man of science needs a modicum of food and other amenities in order to do his work; if called to arms and killed in battle his activities come to an end; if he is an empirical scientist, his opportunities will depend upon the laboratory or observatory to which he has been admitted or which he was able to fashion, and his freedom to pursue his work will be limited by the good or bad will of administrators or fellow workers. Yet nobody can completely control his spirit; he may be helped or hindered, but his scientific ideas are not determined by social factors. Honest men of science and mathematics have often continued activities detrimental to their material interests.

In this treatise, I have tried to draw a map of science, technology and great ideas that would be as accurate and complete as possible, yet sufficiently free from unnecessary details and sufficiently condensed so as not to obstruct the general view.

The book was composed through intensive work during 1991–2008. I had planned to present it to my readers just at the turn of the millennium, but unexpected difficulties prevented me from this symbolic gesture.

The diverse sources used in my work are listed in the bibliography and sometimes in the text itself. Whenever possible, the data was cross-checked between different sources.

January 2009 Ari Ben-Menahem

About the Author

Ari Ben-Menahem (Schlanger) was born in Berlin, Germany and came to Israel in 1934.

He earned his M.Sc. degree in physics from the Hebrew University of Jerusalem (1954) and his Ph.D. degree from the California Institute of Technology (1961). He has been a Professor of Mathematics and Geophysics at the Weizmann Institute of Science in Israel since 1964 and a visiting Professor at MIT, Stanford University and the University of Paris. During 1970–1995 he was the incumbent of the Ayala and Sam Zacks Chair of Geophysics at the Weizmann Institute.

Prof. Ben-Menahem also served as Chief Seismologist of the Government of Israel, member of the European Seismological Commission, US National Research Council Fellow and Director of Adolpho Bloch Geophysical Observatory.

Between 1958 and 2008 he published over 150 papers including graduate textbooks (Springer 1981, Dover 2000). He pioneered the birth of modern seismic source elastodynamics based on his observation of wave radiation from finite rupturing faults with subshear velocity (1960). He introduced the fundamental concepts of Directivity (1959), and Potency (1965) from which the Moment Tensor was derived. He unraveled (1975) the mystery of the Tunguska bolide explosion of 1908.

In this comprehensive tour de force, Prof. Ben-Menahem – a polymath equally versed in pure and applied science, and the humanities – rises to the challenge of tracing the tapestry of human thought and action throughout history and up to our own age, placing it in the context of our niche in the universe.

Alternating between fine details and bird's-eye vistas, this Encyclopedia encompasses man's science and artifice, illuminating his connections with his environmental, economic, sociopolitical and cultural development.

Acknowledgments

I would like to express heartfelt thanks to our family members: Dr. Shahar Ben-Menahem, assisted me in the formulation of essays dealing with modern theoretical physics, contemporary pure mathematics and the nature and evolution of scientific discovery and invention. He is also the text editor of chapters 4 and 5 of this encyclopedia. I am deeply indebted to him for dedicating long time and great efforts to the completeness, accuracy and streamlining of the book.

Dr. Gali Oren-Amit helped me with the formulation of the medical essays. Savyon Amit (LL.B, MBA) assisted me in all legal matters, and helped me keep on track with the publication process. Their children Erel and Ella provided a warm atmosphere of friendly cooperation and fun.

My secretary, Sarah Fliegelman, word-processed the manuscript, until her untimely death in 2002. The camera-ready text was then prepared by Denis Simakov, Ph.D. and Olga Shomron, M.Sc.

The librarians of the Weizmann Institute Wix Library: Miriam Gordon, Libeta Chernobrov and Anna Ilionski rendered constant assistance in search of bibliographic material.

I convey my deep appreciation to professor Max Jammer of Bar-Illan University who heeded the publishers invitation to review the manuscript.

Special thanks go to my editor Dr. Christian Witschel, his assistant Marion Schneider and the Development editor Dr. Sylvia Blago.

My teachers throughout my student years deserve my gratitude for opening to me the doors of the world of science; At the Hebrew University I owe tribute to Profs. G. Racah, A.H. Fraenkel, Y. Levitski, M. Schiffer and C.L. Pekeris; At the California Institute of Technology I had the privilege of being taugth by Profs. F. Press, H. Benioff, O. Taussky, A. Erdelyi and R.P. Feynman.

At the Weizmann Institute I benefited from the friendship of Profs. D. Vofsy, J. Gillis and S. Freier.

But, first and foremost, I am honored to pay my homage to my father Moshe Ben-Menahem (Schlanger): poet, writer and educator, who set me on the course of my scientific career through his imperative maxim: "Always think like a philosopher, see like an artist, and feel like a poet."

Contents

Part I

Part II

Science 'Progress' Reports

Worldviews

List of Tables

List of Figures

List of Photographs

Acknowledgements

Dr Shahar Ben-Menahem, of Stanford University, wrote the following sections for chapter 6: *The Trees of Knowledge and Life*, *Science vs. Technology*, *Discovery vs. Invention*, *The Ultimate Machine*, *The Iterative Nature of Scientific Knowledge* and *Front-Lines of Science 1950–2008.* I am extremely grateful to him for lending me his expertise in various fields of pure mathematics and theoretical physics.

Professor Carlos Gitler, of the department of Chemical Biology of our Institute, kindly read the section "From Cell to Biosphere..." of this chapter and suggested useful changes which were consequently implemented. His professional advice and encouragement is highly appreciated.

Overview

The roots of the evolutionary progress of mathematics, physics and technologies based upon applied science during the second half of the 20th century are anchored in the early days of the Industrial Revolution.

Science interacted strongly with political, economic and other social changes in Western society since the Industrial Revolution in the middle of the 18th century. Major events such as the advent of modern industrialization in England (1749–1760) and the consequent rise of the first British Empire, the nascence and rise of the United States from British North America (18th century on), the French Revolution (1793), the Unification of Germany (1871), WWI (1914–1918) and WWII (1939–1945), impacted science and were impacted by it in return.

In 1700, England was still chiefly a rural land – there were no big towns except London, and the economy was primarily based on agriculture. Previously, the revocation (1685) of the edict of Nantes (1598) caused extensive emigration of French Huguenots and their dispersal in various countries, enriching these with their talents. The ensuing industrialization was partly a result of the oppression of the Puritans. Until 1800, science in England was the business of gentlemen and noblemen, mostly occupied with Evolution and Geology. Science and philosophy were empirical. At the turn of the 19th century British science as a whole became more theoretical again: ***Thomas Young*** *(1773–1829) revived the wave theory of light (1801) and* ***John Dalton*** *(1803) introduced atomic theory into chemistry.*

In France, during 1740–1819, philosophers sowed the seeds of a rationalist revolt which, inter alia began the French Revolution. In a complementary way, the French then became more empirical, spurred on to experimental and applied science by the needs of the Napoleonic Wars.

German science, with its metaphysical roots, influenced by the philosophy of ***Kant***[1] *(1781) and* ***G.H.F. Hegel*** *(1817), began to blossom in the first half of the 19th century and reached its peak during 1870–1930.*

The industrial revolution had changed the status of science and scientists: in the first place it caused scientists to be considered as representatives of radicalism (i.e. progressive). Then it shifted the emphasis towards the applied side of science: The rapid growth of applied mathematics in the wake

[1] **Berkeley** and **Kant** in turn were influenced by **D. Hume** while Berkeley influenced **Mach** and **Einstein**.

of the industrial revolution called for the establishment of a discipline of approximations through which algorithms could be systematized and developed methodically to answer the growing needs of the exact sciences.

Above all, since **J.C. Maxwell** *and* **Hertz**, *science began to become a dominant force in Western society, in times of peace as well as war.*

The 19th century was rushing toward its close, propelled by steam and electricity. The railroad, the steamship, the telegraph, the telephone, the phonograph, photography, radio and the internal combustion engine took their places in human life in bewildering succession.

It had been a magnificent century for mathematics; most of the mathematical tools later needed for GTR and quantum mechanics were forged by 19th century mathematicians. In fact, 20th century physics is essentially based on 19th century mathematics! Thus GTR is based on Riemannian geometry and tensor analysis; Quantum mechanics and quantum field theory are based upon abstract algebra and functional analysis – and specifically the theories of matrices, Lie algebras and Lie groups, differential geometry, differential and algebraic topology, as well as Hilbert spaces and c^*–algebras. Many of these mathematical fields also supported Nonlinear Dynamics.[2]

The 20th century was an epoch of Revolutions in Physics and Biology. Physics was the first of the natural sciences to become fully modern and highly mathematical. Chemistry followed in the wake of physics, but biology, the retarded child, lagged far behind. Even in the time of **Newton** *and* **Galileo**, *men knew more about the moon and other heavenly bodies than they did about their own.*

It was not until the late 1940's that this situation changed. The postwar period ushered in a new era of biological research.

World-War I made chemistry respectable (even **Ernest Rutherford** *often presented himself as a chemist). World-War II made physics respectable (Atomic bomb, Radar, Rockets, Jet planes, Computers).*

During 1916–1928 there emerged the two most revolutionary physical theories ever – the General Theory of Relativity and Quantum Mechanics, neither

---

[2] *Nonlinear Dynamics* – a field of applied mathematics that started in the 19th century by the advent of computer simulations in the second half of the 20th century. And all of these fields, of course, benefited from 19th century work on differential and integral equations, integral transforms and complex analysis. The needs of 20th century developments in technology and in the natural and social sciences also spurred many extensions and ramifications of pre-existing mathematical fields – e.g. the theory of stochastic processes (from probability theory), algorithms and computational complexity, *index theorems*, etc.

of which had been adumbrated by any scientist, philosopher or even science-fiction writer. In terms of breadth of applicability and accuracy of predictions, Quantum Mechanics (augmented by STR in the 1930's and 1940's to become QFT) proved itself the most fruitful theoretical framework ever developed by man.

After WWII the USA emerged as a political, military, economic, technological and scientific superpower. It supported the largest scientific establishments in the history of mankind (accelerators, Manhattan Project, NASA). New discoveries are constantly being made, and many of these discoveries have important political, economical and social overtones.

But the 20th century left in its wake an overcrowded, polluted and dangerous world – disillusioned by global wars and conflicts, and with the very concepts of progress and individual responsibility that gave rise to it; a morally rudderless world stricken by major environmental problems, social disorders and the rise of a new barbarism.

Perhaps, no words are more becoming then those of **Leonard Bernstein** *in his succint 'Requiem-like' summation of the 20th century (1973):*

"Ours is the century of death: the end of faith. Why is our century so uniquely death-ridden? Couldn't we say this of other centuries as well? Yes, true: all human histories have been a long record of the struggle to survive, to deal with the problem of mortality. Yes; but never before has mankind been confronted by the problem of surviving global death, total death, the extinction of the whole human race. In fact, all the truly great works of art, music and literature have been born of despair and protest or refuge from death: think of Tolstoy, Rilke, Kafka, Sibelius, Mahler, Picasso, Sartre and Camus.

The 20th century has been a badly written drama from the beginning: Greed and hypocracy leading to genocidal world – wars, totaliterianism, post-war hysteria, existentialism, galloping technology, the flight into outer space, the doubting of reality, sub cultures and counter – cultures, new religions movements – all under the aegis of planetary death."

FRONT-LINES OF SCIENCE 1950–2001

Mathematics

- CONTINUATION OF ABSTRACTION AND UNIFICATION OF PURE MATHEMATICS:
  - Combinatorial Geometry;
  - Pure topology and graph theory (e.g. 4-color Theorem);
  - Abstract Algebra and Group Theory; Affine Lie Algebras; classification of finite groups;
  - Functional Analysis; Categories, Functors;
  - Fusion of Differential Geometry, Algebraic Topology and Algebraic Geometry; K-theory; Homotopies, Homologies and Cohomologies; Knot Theory, Topology of Smooth 4-manifolds; Fiber Bundles and Non-abelian Gauge Theories; Complex and almost-complex structures on manifolds; Spinors on manifolds and spin structures; sheaves; Cobordism; characteristic classes and index theorems;
  - Modular Functions and analytic Number Theory; Elliptic Curves;
  - Logic, Set Theory, Foundations of Mathematics and the genesis of Computer Science (Halting problem, foundations of mathematics, meta-mathematics proofs, Axiom of Choice, Continuum Hypothesis);
  - p-adic Analysis;
  - Non-standard Analysis: 'emancipation' from infinitesimals.
- DISCRETE MATHEMATICS AND THE UBIQUITOUS ALGORITHM.
  - Digital Signal Processing.
- COMPUTATION BEYOND ALGORITHMS:
  - Soft Decisions and Fuzzy Logic;
  - Bayesian Inference;
  - Artificial Neural Networks;
  - Genetic Algorithms;
  - Simulated Annealing.

- Nonlinearity, Stochasticity, Optimization, Multiple Scales and the Science of Complexity:
  - Nonlinear Dynamics, Chaos and Fractals, Self-Organizing Criticality and Cellular Automata;
  - Algorithmic Complexity;
  - Optimization and Control Theories;
  - Integrable Systems, Invariant Tori and the KAM Theorem;
  - Nonlinear Diffusion and Wave Propagation; Turbulence; Reaction-Diffusion-Advection equations;
  - Queuing Theory;
  - Homogenization, Multiscale Analysis and Singular Perturbation Theory;
  - Theory of Stochastic Processes and Measure Theory;
  - Ito Calculus and Stochastic Differential Equations.
- Mathematization of the Biological, Environmental and Social Sciences:
  - Game Theory and Microeconomics;
  - Macroeconomics and Stochastic Processes;
  - Population Dynamics and Population Genetics;
  - Morphogenesis; Epidemiology; Neuronal and Brain Modeling;
  - Modeling of Protein Folding and Molecular Biology.
- Abstract Algebra and group theory in modern Theoretical Physics:
  - Groups (Discrete, Continuous, finite, infinite) in Quantum Mechanics, Chemistry, Crystal lattices, Quantum Field Theories, Particle Physics; Classification of all Finite Groups;
  - Global vs. Local (gauge) symmetry groups;
  - Spontaneous symmetry breaking;
  - C^*-Algebras applied to QFT, condensed matter and string theory;
  - Affine Lie Algebras;
  - 2-D Conformal field theory;
  - Current Algebras;
  - Quantum Groups and Braid Groups;
  - Operator-product expansions.

Physics

- Particle physics and the Coming of Gauge:
  - Weak and Strong Nuclear Forces;
  - Quantum Field Theory beyond QED;
  - The 8-Fold way, the Quark model and Current Algebra;
  - Non-Abelian (Yang-Mills) Gauge Theories and Differential Geometry of Fiber Bundles;
  - Path Integrals, Fiber Bundles and the Quantization of Gauge Theories;
  - Partons, Quarks, Gluons, Quantum Chromodynamics, Scaling, Confinement and Asymptotic Freedom;
  - Heavy-Ion Collisions and the Quark-Gluon Plasma;
  - Phase Transitions and Spontaneous Symmetry Breaking, Nambu-Goldstone Modes and the Higgs Mechanism;
  - The Standard Model of Particle Physics;
  - Solitons and Instantons in Quantum Field Theories;
  - Supersymmetry, Supergravity and Grand Unified Theories;
  - Quantum Effects in Background Gravitational Fields;
  - Extra Dimensions, Kaluza–Klein models and String Theories;
  - Attempts at a Quantum Theory of Gravity;
  - Non-Accelerator Experiments and Indirect Evidence.
- The New Cosmology: Whence, Whither and Why?
  - The Big Bang Theory and Hubble's Constant;
  - Cosmic Microwave Background Radiation, Large Scale structure, and "Precision Cosmology";
  - Dark Matter, Dark Energy and the Cosmological Constant;
  - The Early Universe as a Hot Quanta Soup;
  - CP violation and Baryogenesis in the Early Universe;
  - The "First Three Minutes" and Nucleosynthesis;
  - The Recombination Era;
  - The "Dark Ages", re-ionization and first Stars and Galaxies;
  - The Planck Era and the Question of Initial Conditions;

  - Stellar and Galactic Models, Pulsars, Neutron Stars, Black Holes, Quasars, Active Galactic Nuclei and Gamma Ray Bursters, Brown Dwarves, Dark Matter Surveys and Candidates, the Intergalactic Medium;
  - New Windows to the Heavens: Infrared, Ultraviolet, X-ray, Gamma Ray, Neutrino and Gravitational Wave, Gravitational Lensing of Light Astronomies.

- General Relativity Matures and Meets Particle Physics, Astrophysics and Quantum Mechanics:
  - The Unmanageable Infinities of Quantum Gravity;
  - Singularity Theorems and self-limitations of GTR;
  - Quantum Field Theories in Curved Background Metrics;
  - Radiation from Black Holes;
  - Supergravity and Models of Hidden Dimensions;
  - Strings, Branes and "M-Theory";
  - Gravitational Lensing and Advent of "Applied GTR";
  - The evolution of theories of fundamental strings from Dual Resonance Models to Theories Of Everything;
  - Effective (non fundamental) strings in QFT, Astrophysics and Cosmology;
  - New Tests of GTR: Pound-Rebka experiment, Shapiro's radar-ranging, atomic clocks, the binary pulsar, orbiting gyroscope, LISA, and LIGO;
  - Gravitational Waves from Astrophysical Cataclysms and from Early Universe Phase Transitions.

Biochemistry

- Biochemistry of Nature and Man:
  - Cell Structure and Signaling;
  - Chemical and Morphological Logic of Life;
  - Molecular Biology;
  - Non-Equilibrium Thermodynamics in Physical Chemistry and the Biosphere;

- Neuroscience and functioning of the Mind;
- The Human Information and Storage systems: Genome, Immune System and Brain.

Technology and Engineering

- QUANTUM TECHNOLOGY:
  - Semiconductors; superconductors, superfluids, lasers and holography; advances in Magnetics, Ferroelectricity, Quantum Optics and Electrooptics;
  - Quantum Electronics, Optronics and Photonics;
  - Quantum Metrology (precision physical measurements using quantum effects);
  - Quantum Computing, cryptography and Teleportation;
  - Reversible Computing;
  - Ultra-low Temperatures; Magneto-Optical Traps and the Bose-Einstein Condensate (BEC) technology, and the slowing and freezing of light waves;
  - Electromagnetics of a single atom or particle in an RF cavity, cavity QED, observation of single-atom decays and the Quantum Zeno effect;
  - Microdevices and Nanotechnology;
  - Designer's Atoms and Materials: Quantum Dots, Wires and Wells, and Quantum Hall Effect, Superlattices and heterostructures;
  - Advances in Atomic Clocks;
  - Few-Quanta experiments, Quantum Entanglement and empirical tests of the foundations of Quantum Mechanics;
  - Electromagnetically Induced Transparency, "Dark States" and Lasing without population inversion;
  - Carbon nanotubes and Buckeyballs;
  - Optical ponderomotive forces: optical tweezers and ratchets; Light Lattices and Laser Molasses;
  - Role Reversal: Control of Material Object with Light Structures.

- RF Engineering: Electron guns, Klystrons, Waveguides and Junctions and Particle-accelerator Technology; Charged Particle Traps; Magnetic focusing and guidance of Particles beams.
- Worldwide Sensing, Communication and Information Technology:
  - Recording and Reproduction of Light and Sound; Television and Video Technology;
  - Sonar, Radar and Laser Ranging, Satellite Telemetry and Reconnaissance, Remote Sensing;
  - Optronics: Diode Lasers and detectors, Electro-optical chips, Optical Fibers and Optical Amplifiers, Optical Isolators and Polarization Dependence, Interferometers, AO (Acousto-Optic, devices) and SAW filters, Digital Optical Fiber and HFC (Hybrid Fiber Coax) Networks;
  - Client-Servers, Routers; Cable, optical, twisted – wire and wireless Networks; GPS;
  - Computer Based Communication Systems and the Rise of the Internet;
  - RFID (Radio Frequency ID) tags for tracking goods and national security;
  - Geosatellites in service of Telecommunication.
- Modern Microscopy and Telescopy:
  - ***Microscopes***: Field Emission; Scanning Electron; Transmission Electron; Scanning Tunneling; Atomic Force; confocal; FRET (Fluorescent Resonant Energy Transfer); Near-field Optical (to resolve distances much smaller than a wavelength); TPM (Two-Photon Microscopy);
  - ***Telescopes***: Spaceborne; computerized; Terrestrial and Spaceborne Interferometers; Detection of planets of other stars; with adaptive optics; with charge coupled devices; Hubble; Observing the Universe via Infrared, ultraviolet, X-ray, Gamma-ray, radio and Neutrino radiations; Gravitational-wave telescopes (LIGO and others).
- The Klystron and Particle Accelerator Technology:
  - Application to particle-physics experiments, X-ray sources, material science, medical imaging, time-resolved spectroscopy in physical chemistry.

- Material Science:
  - Metals, ceramics, plastics and composite materials;
  - Research on Smart Materials;
  - Doping in Electronics and photonics;
  - Rare-Earth Magnets;
  - Designer's Atoms and Superlattices.
- Computer-Related and Computer-Enabled Technology:
  - The Microchip;
  - Electro-Optical and Microwave chips; MEMS (Micro Electro Mechanical Systems);
  - Robotics and Control, Machine Vision, Machine Learning and AI;
  - Applications of Abstract Algebra, Number Theory, Topology and other branches of mathematics: Error Correction Codes and Cryptography; Wavelet Transform and Multiresolution Compression, Signal Processing and Pattern Recognition; Dynamic Topology of Computer Networks; Computer design.
- Computer Simulation – the Third Mode of Science:
  - Monte Carlo Simulations;
  - Finite Element, Finite Difference and Particle-In-Cell Simulations;
  - Computerized Symbol Manipulation;
  - Special-Purpose Computer architectures;
  - Computers as an integrative tool of Science Research.
- 'Big Science':
  - The Age of Accelerators and the Particle Physics Frontier;
  - Observatories and Detectors for Exotic Particles from Space, Exotic Particle Decay Modes, and Gravitational Waves;
  - NASA and the Manned and Unmanned exploration and Observation of the Solar System and Beyond;
  - Satellites, Rockets and Missiles;
  - Mapping the Human Genome;
- Biomedical Informatics, Radiology, Imaging, Molecular Medicine and Biotechnology:
  - MRI (Magnetic Resonance Imaging) and fMRI (*functional* MRI);

- PET (Positron Emission Tomography) scans;
- CAT (Computer Assisted Tomography) scans;
- Nuclear medicine;
- Particle-beam radiology and therapy;
- Ultrasound Imaging and Therapy;
- Stem-cell research, gene therapy, telomerase research, cloning, and genetically modified crops;
- DNA Markers and Retinal Scans for identification; individual genome sequencing for health care;
- Prions and abnormal protein folding;
- Bioinformatics, genomics and *proteomics* (the study and quantification of gene expression);
- Remote and sensor-assisted surgery;
- Analysis of microscopic tissue samples using optics, microfluidics, gene chips, tagging, and digital processing.

The Trees of Knowledge and Life – New Trends in Contemporary Science

Certain trends in mathematics, the natural sciences and technology are apparent:

1. Inward-bound progress

Great progress in the "inward" direction. In Physics this is manifested as ever deeper, more unified and more aesthetic mathematical principles underlying physical reality – matter, energy, space and time. This "inward" direction points both to the very small (in spatial and temporal extents) and the large (large energy concentrations at the high temperatures achieved in particle accelerators and inferred for the early epochs after the Big Bang, and large scales of time and space in Cosmology).

In material science and the biological sciences this "inward" trend is always towards smaller scales of space and mass (and sometimes of time and/or of temperature as well), and is manifested by dramatic improvements in small-scale imaging (of cells and their constituents, microscopic lifeforms, molecules, and individual atoms). It is also manifested by indirect elucidation of microscopic structures, as in the sequencing of genomes, identification of proteins, and molecular-dynamics computer simulations of physical, chemical and biological processes at the smallest scales.

In material science and the biological sciences, the "inward" journey involves no new fundamental principles of nature, but merely newly-discovered interplays of known physical principles.

All these feats of analysis, imaging, simulation and deduction usually involve computerization, in addition to analog scientific hardware and mathematical theories and models. Here 'computerization' refers to various interfaces with digital sensing, actuating, memory and computation circuits and devices – whether embedded within the hardware, controlling it, interfacing externally with it, or used for offline analysis of data.

2. Outward-bound progress – the Aegis of 'Complexity'

Impressive progress also in the "outward" direction – the study of complex phenomena and structures in all the natural sciences (including population dynamics, evolution, ecology, the new fields of mathematical biology and bioinformatics, neuroscience and neuronal modeling, epidemiology, continuum mechanics, material science, meteorology and geophysics) as well as in traditionally humanistic fields (economics, sociology, etc.); and the study of complexity itself as a new branch of applied mathematics.

This "outward" progress is largely driven by the computing and information revolution, but also involves experimentation, observations and mathematical modeling, as do all other scientific endeavors. The aegis of "complexity" covers such phenomena as chaos and fractals that were found to be ubiquitous in complex systems.

New mathematical tools, as well as older ones, were pressed into service to help ferret out, model and understand complex structures; examples of such tools are nonlinear dynamics, multiscale analysis, wavelets, artificial neural networks, reaction-diffusion partial differential equations and stochastic differential equations.

3. Bridges and Unifications

The forging of a robust web of bridges between different structural and functional levels of description in different sciences, resulted in a powerful trend towards convergence of our disparate "islands of knowledge" of nature.

Thus, for instance, modern research in biology tends to gradually reduce biotic structures and functions to physical and chemical ones, organized at all scales (including the so-called nanoscale); all this new knowledge is placed in computerized, online databases which are then mined via sophisticated algorithms (the new field of bioinformatics) for the purpose of developing new drugs and treatments, identifying new genes and their locations, sequences and functions, identifying new proteins, elaborating the "tree of life" (i.e., maps of where each organism that ever lived is placed in the tree of evolving DNA), et cetera.

We see that science at the start of the 21st century is on an inexorable trend toward unification of the growing "tree of knowledge" with the "tree of *life" – a trend with incalculable potential effects on human lifespan, health, cognitive powers, and material and mental well being.*

4. New Mathematical Structures

The demarcations between our chapters 4, 5 and 6 were largely driven by developments in the natural sciences and technology. Pure mathematics has its own intrinsic dynamics; although it undergoes dramatic new developments in response to impetus from the sciences, it then typically continues to hone and develop the resulting structures for very long periods of time – often generating, in the process, new tools that are then adopted for modeling physical reality.

As an example, the revolutionary theories of 20th century physics – quantum mechanics and the theories of relativity – used pre-existing mathematical structures that are properly classified as belonging to the "Abstraction and Unification" phase of mathematics, even if the discovery of these structures is listed by us in both chapters 5 and 6. And in turn, work by physicists on relativity and quantum theory led to new types of mathematical structures.

There are many other instances that illustrate this type of two-way influence between mathematicians and scientists working on the two sides of the "Math-Science divide".

In view of the above, it should now be clear that the mathematics of the second half of the 20th century can be roughly divided into two endeavors (with, however, interactions and hybrids abounding):

(A) The continuation *of the "Abstraction and Unification" themes of chapters 4 and 5 in such fields as analytic number theory, differential geometry, abstract algebra, topology, algebraic geometry, algebraic topology, functional analysis and the foundations of mathematics.*

(B) The development of new, computer-driven branches of mathematics, including studies of complexity (as outlined above) and of computer science itself (algorithms for data processing within and by computers). These computer-driven branches are just as rigorous as the older ones, but they tend to deal with discrete objects and processes (a representative paradigm being that of "cellular automata").

5. Man outdoing Nature

In addition to the ongoing and novel trends listed above, there appears a new feature of modern science: the growing capability of human technology to outdo nature itself in certain respects.

We do not, as yet, have the technology to shift stars from their courses, create new planets and stars, or create galaxies and quasars; nor to create life from non-life, or sentience from non-sentience. We simply cannot compete with inanimate nature at extremes of high energy, large size or long duration, at least not in the foreseeable future. Nor are we anywhere near to computing, as engineers, with Life's devices in terms of density, adaptivity, plasticity, miniaturization, 3-D packing and multiscale structure and function. However, we are already able, *on some specific fronts, to outdo both life and inanimate nature.*

The philosophical conclusion drawn from this observation can be stated as follows: In the sense of the Strong Anthropic Principle, sentient beings (which themselves presumably arose spontaneously from inanimate matter via molecular and biological evolution) are, in a sense, nature's vehicle for developing in certain directions that are extremely unlikely to occur without the agency of such beings. Nature's way of enabling this option was achieved by endowing man with intelligence and consciousness *which he acquired through the process of evolution.*

Let us now specify those directions along which human technology outdid nature:

(A) *Low Temperatures: Using Magneto-Optical Traps (MOT) and Gravito-Magnetic Traps – a combination of magnetic fields, gravitation, laser beams and radio waves – physicists (since the 1990's) routinely cool small samples (several million atoms each) of various species of atoms down to temperatures as low as several nanodegrees kelvin or even below a nano-kelvin. At such low temperatures, the de Broglie wavelengths at the atoms' quantum wave-functions become larger than inter-atomic separation, in effect turning the entire mesoscopic sample into a single, coherent quantum probability wave.*

Quite apart from the rich scientific and technological implications of the attainment of these so-called 'Bose-Einstein Condensates' (BEC), we note that as far as is known, such low temperatures and state of matter have never before existed in nature *– neither in the depths of intergalactic space, nor on any planet, nor in the remote reaches of our universe or in its early history – save, perhaps, in the laboratories of some other, non-human civilization of sentient beings.*

(B) *Symbolic representation, recording, transmission and processing:* since the dawn of human culture, people have been using physical manifestations (cuneiform etchings, ink marks, et cetera) to represent ideas and convey information in a symbolic manner. As far as is known to science, inanimate nature never does that.

Now, this particular direction in physical parameter space does not require hi-tech or even intelligence – bees, ants and many other low orders of animals and plants (and even individual cells) convey symbolic information, by means chemical, electric, optical and acoustical; and indeed, any form of life whatever on earth does so by means of the *genetic code.* Nonetheless, we note that the *information technology revolution* of the late 20th century has introduced on our planet such complex mazes of symbolic information storage, retrieval, transmission and processing (mainly electromagnetic) as to rival – and, by many measures, to surpass – the main three information processing/storage facilities in the human body (genome, immune system, brain – in order of increasing storage capacity).

The latter three systems, while 'natural' (i.e. not devised by humans), are once again hallmarks of *animate* matter. Inanimate matter does not seem to have any uses for storing and manipulating information in a symbolic manner.

Granted, the *physical laws* which govern all matter and energy are, arguably, a sort of highly condensed logico-mathematical *symbolic code;* but this code is not physically stored anywhere (except again, in the writing and minds of human beings – and presumably in those of other advanced sentient beings in the cosmos, if they exist).

(C) *Efficient information processing:* granted that symbolic information processing/storage/transmission does not seem to exist outside the realm of life and its products and artifacts (such as genomes, brains, cuneiform tablets, abaci and computer hard-discs), the trends in modern computer technology also point to ever increasing *efficiency* of such symbolic manipulations.

This enhanced efficiency either already exceeds, or will before too long, what animate, yet pre-technological, nature has wrought – in several directions in physical parameter space. These directions are *high density* (bits per volume), *high speed* (access and processing speeds in bits per second), *low energy cost* (Joules per bit processed), and *efficient particle usage* (low number of elementary particles whose quantum state is affected by each bit change). We note, though, that the entire human genome (several Gigabytes of digital data, not counting partially-analog epigenetic information) is packed within almost each cell of the human *body – indeed, most of it within the cell's nucleus; human technology cannot yet achieve such information storage densities.*

(D) *Design optimization: At its most fundamental, we know that nature's laws may invariably be cast as* optimum principles *(principles of least time, least action, etc). At the biological level, nature optimizes at a different level – efficient use of wax in honeycombs; optimizing paths of ants and birds; survival strategies of species at both the genomic and population levels; and so forth.*

This latter type of optimization is opportunistic and a rigorous 'global optimum' is almost certainly never achieved. Humans, however, are capable of modeling problems mathematically, and the resulting optimizations – especially when computers are utilized – may well be far more comprehensive than pre-technological nature is able to achieve. And even without advanced mathematical models, various engineering solutions to practical problems (such as streamlined design of vehicles and furniture) can easily exceed the efficiency found in animate, pre-technological nature. (However, nature seems at present to be better at optimizing at the 'systems level' – witness the self-inflicted ecological problems humanity is grappling with!)

(E) *Composite and smart materials:* Advances in material and computational sciences – and even more importantly, the convergence of the two (the emerging branches of engineering sometimes referred to as *nanotechnology*) – are enabling another form of optimization: at the level of the organization of atoms into macroscopic matter.

Thus, "designers' materials" can be planned and fabricated, taking the process of technological optimization [discussed in section (D)] to new heights. While human technology has already outstripped *inanimate nature* along these directions in physical parameter space, it is still lagging behind what *animate* nature has accomplished (even a single living cell represents a far more sophisticated feat of 'natural nanotechnology' than any fashioned in a human lab).

However, there is every reason to believe we will surpass even biology in these directions. This is even more apparent when one considers that molecular genetics (and the burgeoning biotechnology based upon it) is enabling humanity to 'co-opt' various naturally occurring biochemical and biophysical mechanisms, rearranging them almost at will – either with one another, or in conjunction with inanimate human technology (such as in the so-called "gene chips").

(F) *Quantum Computing:* *Almost all of today's computers are based on simple Turing Theory and employ Boolean logic based on binary mathematics. Even "parallel" computers are really complex Turing engines employing multiple computing modules which deal with pieces of incoming and internally-generated data (digitized acquired data, instructions, etc). There has been some research into biological computing using enzymes or large-molecule systems as memory, shift registers, etc, but this has not yet proven to be very practical.*

Quantum Computing is based on a different physics than ordinary digital computing. Instead of having two (or three) states per element: off, on, or in between ('hung'), quantum computers consist of elements each of which may at any given moment be in a superposition of "on" and "off" states. An eight bit digital computer can exist in only one of 256 states at a time while an eight bit quantum computer can exist in all 256 states at a time and theoretically, work on 256 calculations at once (quantum parallelism). Each of the 256 numbers in this 8-bit example has an equal probability of being measured, so that a quantum processor functions, in effect, as a random number generator. The actual register represents all these combinations of bit-values at once, but a single 8-bit value output only occurs at measurement.

While a classical digital computer would have to operate on each number from 0 to 255, a quantum computer requires only one pass through the "processor" — radically reducing calculation time. Of course, the larger the register size, the larger the reduction factor – even a simple 10-bit quantum computer could make a supercomputer pale in comparison.

Where the digital computer uses binary digits (bits), the quantum computer uses qubits ("quantum bits"), but qubits are extremely difficult to generate. A quantum switch may not be disturbed by anything – a single photon of light, a single impinging molecule, or ambient fields – for the proper operation of a quantum computer depends on the interaction of the various qubits without any outside influence (measurements included). When disturbed, the qubit temporarily "collapses" into a conventional 1-bit register.

A quantum computer can perform an arbitrary reversible classical computation on all its qubit data registers simultaneously, and also has some ability to produce interference, constructive or destructive, between various different computational paths. By doing a computation on many different input data-sets at once, then interfering the results to get a single answer, a quantum computer has the potential to be much more powerful than a classical computer of the same size.

The most famous example of the extra power of a quantum computer is an algorithm for factoring large numbers. Factoring is an important problem in cryptography; for instance, the security of public key cryptography depends on factoring being a hard problem. Despite much research, no efficient classical factoring algorithm is known.

There are many proposals for how to build a quantum computer, with more being made all the time. The 0 and 1 of a qubit might be represented by the ground and excited states of an atom in a linear ion trap[3]*; they might be represented by polarizations of photons that interact in an optical cavity; they might even be represented by the excess of one nuclear spin state over another in a liquid sample in an NMR machine. As long as it admits of a way to put the system in a quantum superposition and there is a way to store, couple and occasionally measure multiple qubits, a physical system can potentially be used as a quantum computer.*

In order for a system to be a good choice, it is also important that many operations may be performed before losing quantum coherence. It may not ultimately be possible to make a quantum computer that can do a useful calculation before decohering, but if we can get the error rate low enough, we can use a quantum error-correcting code *to protect the data even when a certain fraction of the individual qubits in the computer decohere.*

In conclusion: As of the 1990's, quantum and computer technologies reached the ability to produce states of matter and energy that (barring other

---

[3] Here, the charges, voltages and magnetized domains which represent data bits in conventional digital computers, are replaced with photons and ions trapped by electromagnetic fields. Because they are encased in these fields, the ions are in a coherent state for fractions of a second. The obstacle in using ions is that they may lose their quantum coherence too quickly to be useful as a computing resource. Los Alamos researchers have been able to apply a single laser pulse to a single ion in an electromagnetic trap.
From their demonstration, the researchers said that as many as 100,000 logic operations could be applied to registers that consist of up to 50 trapped ions. It would take but a few microseconds for a register to complete a single operation. Although it is smaller and faster than any silicon-based device, quantum computers are not expected to replace desktop PCs or supercomputers, at least not at first. Instead, these machines would be dedicated to specialized tasks such as generating keys for strong cryptography, an operation that requires a computer to factor very large numbers.

intelligent species) could almost certainly never be produced in nature outside of human laboratories. These frontier states can be grouped into three classes:

- *Low-temperature frontier (BEC)*
- *Mesoscopic quantum coherence frontier (e.g. Quantum Computing)*

Nature finds it hard, if not impossible, to reach these niches in physical parameter-space (as well as the niches described in (B), (C), (D) and (E) above) except through the agency of sentience. By the anthropic principle, we exist, as part of nature, to enable nature to accomplish such tasks.

6. "Lest he take also of the tree of life" (*Gen*: **3**, 22)

The current state of development of the theory of stochastic processes, as well as our ability to realistically simulate the dynamics of complex biotic macromolecules, is too primitive to compute with any reliability the probability of the spontaneous arousal of living from non-living molecules.

But we have HINTS such as:

- ***I. Prigogine**'s theory (1947) of far-from-equilibrium chemical processes,*
- *The Jacob-Monod theory (1961) of the operon mechanism (genes controlling other genes),*
- *Knowledge about various chemical pathways and point mutations in biology,*

which indicate that naive probability models may be completely useless in estimating the probabilities and likely timescales involved in a spontaneous genesis of life.

Thus, Prigogine's models show that complex spatiotemporal structures can arise from homogeneous mixtures in open thermodynamic systems far from equilibrium.

The evidence from molecular biology (such as the operon mechanism) illustrates that, since the proverbial typewriting monkeys can sometimes accidentally learn to program a sequence of keystrokes into a single keypunch, and since this process is hierarchical, the expected time for a homogeneous mixture of pre-biotic molecules (under primordial-earth conditions) to produce living organisms – might be drastically shorter *than what the simplest (***Hoyle***-like) combinatorial arguments would suggest.*

No single example better illustrates the primitive state of our bio-stochastic modeling and simulation tools than the humbling fact that present-day (2008) optimization algorithms (whether stochastic or deterministic) and current computing power, are unable even to predict the correct shapes of folded proteins of known amino-acid sequences in aqueous solutions – a problem that nature herself "integrates empirically" (as **Albert Einstein** *would say) in a matter of minutes!*

Science vs. Technology

Science may be loosely defined as the well-ordered, systematic and programmatic gathering of positive knowledge concerning the universe and ourselves. The history of science, therefore, is concerned with the story of gradual unveiling of objective truths and the conquest of matter, energy, space and time by mind; it also describes an age-long and endless struggle for freedom of thought – freedom from violence, intolerance, error and superstition. This process of gathering and explanation of systematized positive knowledge is the only human activity which is truly cumulative and progressive.

Science is also the systematic attempt to understand and comprehend the deep principles, hidden order, complex structures and beauty of operation that lies behind all natural phenomena. It aims to discover the true facts about, and the rational relationships between, observable phenomena in nature, and to establish theories that serve to organize these facts and relationships in the language of *mathematical symbolism*.

The scope of science is not just the external world, but ourselves as well – our bodies, perceptions, thoughts, imaginings, emotions and actions. Science deals with ideas and is a curiosity-driven, abstract, cultural activity, but unlike philosophy – it always concerns itself concrete phenomena, within which the general, abstract principles and patterns are reified. It is motivated primarily by intrinsic interest, not by utility.

Science flourishes best under special conditions of society, under which there is freedom to exercise to the full the two aspects of the *scientific method*: on the one hand, creative imagination aided by rigorous logic must seek to build and examine *hypotheses* extending beyond existing knowledge; on the other hand, *experimental* investigation must subject these hypotheses to the most rigorous empirical testing, employing the most elaborate scientific instruments available to the scientist. Yet, in the end, no matter how complex the apparatus is, the information that it delivers has to be observed by a scientist, who must examine it critically in relation to his hypotheses, and then, perhaps, reformulate his hypotheses.

Technology (or *Applied Science*, as it is sometimes called) is an effort to apply empirical scientific knowledge to some useful purpose. It is science applied to the business of life. It deals with tools, techniques and procedures that people use for utilizing the findings of science.

There are technologists in many countries where there are no scientists. In fact, all countries have technologists; even counties in the Stone Age, for example, had experts in the manufacture of stone axes.

Technologists (who, however, are sometimes also scientists) are responsible for all the marvelous inventions that have transformed the conditions of our life. Think of the spectrum of revolutions in the means of communication, in the materials for all purposes, and in electronics; think of all the new inventions relating to medical practice, to agriculture and food, to the chemical industry, and to computers. This human activity is quite different from that of science, although it utilizes and exploits discoveries of science and, in turns, aids scientific research. Of course, a technologist has to have wide-ranging knowledge, imagination, and high intelligence, just as does a scientist.

Faraday's laws of electricity and magnetism are science. Marconi invented wireless technology. Clausius contributed to the science of thermodynamics. Watt invented the modern steam engine. It was science that clarified the nature of nuclear binding energy; but it was technology that, in an astonishingly short time, converted it to a weapon of unimaginable power.

Whereas the first clear records of scientific concerns date back to about 600 BCE, the history of technology is much older. There is evidence that toolmaking goes back as far as one million years. Invention that produced technology did not require scientific reasoning until relatively modern times.

Progress in technology was an important component of natural selection as our human ancestors learned to cope with recurrent ice ages and other conditions hardly conducive to creative contemplation. By the time of the apparently abrupt appearance of scientific thought, a considerably sophisticated technology was at hand: fire; wheels; metalworking; agriculture; weights and measures; elements of arithmetic, algebra and geometry; an astronomical data base; navigation; land surveying; medicine and surgery; calendars.

The history of science is constantly interwoven with the history of technology and it is impossible to fully separate one from the other, especially in recent centuries: industrial requirements are always putting new questions to science while the progress of science continually gives birth to new industries or brings new life to old ones. Let us review some examples:

Science has been progressing at breakneck speed during the last century, allowing us to regard the universe from a much loftier vantage point than had ever been possible before. Thus viewed, it is infinitely greater, infinitely more complex and yet amenable to a more unified description; infinitely more harmonious, more beautiful, and yet much more mysterious. Although not all veils have been lifted, many have been during the past decades.

On the other hand, the very progress of science created new riddles, stirred up new contradictions, which seem more difficult than ever to explain. One revolutionary discovery after another seemed to put everything into question.

The natural sciences have made enormous strides in discerning and intuiting the fundamental laws of nature, as well as in explaining a myriad of complex structural and operational mechanisms (both artificial and naturally-occurring) predicated upon these laws.

The riddles cracked, or in the process of being cracked, by science – a relatively new human activity – are extremely diverse, ranging over some many tens of orders of magnitude in terms of space, time, mass and energy. Thus, physics – the most fundamental of sciences – directly explain atoms, stars, and galaxies, and provide us – through a sub-discipline dubbed cosmology – with an embryonic theory of the universe on the largest possible scale.

Furthermore, physics also explains (either on its own or through successive layers of emergent *disciplines, such as chemistry and the life sciences) a good deal of what we observe on scales intermediate between those of elementary particles and the universe as a whole.*

Microbes and DNA, salt crystals, quasicrystals and critical opalescence, flowers and hurricanes, brains and ecosystems have all become subjects of meticulous examination under the mental microscope of the scientific enterprise. Perhaps surprisingly, some of the most complex (and thus most difficult of disentangling) systems known to us lie at comfortably human scales of distance and time – well away from the extremes of the observable universe on the one hand (about 10^{28} cm), or the Planck length (10^{-33} cm) on the other.

Indeed, the human brain – a marvel of complexity and function – occupies rather 'mundane' dimensions (circa 10 cm). The consideration of complex systems could thus be a compensation of sorts for a humanity that has been rudely evicted from any semblance of centrality in the cosmic scheme of things, by successive scientific discoveries. But beyond such comforts, it points to an essential duality in the nature of science: it seeks to analyze *all objects and systems down to their basic components and underlying principles. And having analyzed them, it then strives to assemble the pieces back again, and explain how composite systems may arise (or be engineered) from its building blocks.*

The scope, activity and growth of a scientific field can be evaluated from the size of the population of scientists who work in it or from the number of scientific publications that describe their research results.

In 1910 all the German and British physicists and chemists put together amounted to perhaps 8000 people. By 1990 the number of scientists and engineers actually engaged in research and development in the world was estimated at 10 million, of whom almost 2 million were in the USA, a slightly larger number in the states of Europe, and about 3 million in Russia. Statistics shows that, by the above measures, science as a whole now doubles every 20 years.

It follows that at the end of a professional career of about 40 years, an aging scientist finds that there are 4 times as many scientist and scientific books and journals as when he was a student. In subjects such as molecular biology, particle physics, oceanography, and few other fields, there is a doubling every 4 years! During a professional career there are thus 10 doublings, which is to say, a 1000-fold expansion.

The scientist nowadays is always surfing up the swelling wavefront, just trying to keep his head above a flood of scientific literature. This enormous expansion cannot continue forever, but it does not yet appear to be slackening. This is because it is constantly refueled by revolutionary scientific discoveries and by the need for additional information as an exploding human population exploits nature.

Discovery vs. Invention

An invention is the design and creation of a teleonomic (i.e., purpose-oriented) device or procedure:

The purpose of an invented device or procedure can be anything desired by the inventor, including e.g. labor-saving, entertainment, edification, or assistance in achieving some other human goal. The terms "design" and "creation" refer to a key characteristics of an invention that distinguished it from a discovery – namely, that an invention is a pattern of energy, matter, information and/or organization that cannot be naturally thought of as having existed "out there" before it was brought into being by its inventors.

Thus, a statue hewn of rock cannot be said to have existed within the original slab of rock; nor is it reasonable to regard a newly-fabricated Lexus automobile as having been somehow "discovered" in the patterns of the plastics, metals, chemicals and composite materials of which it consists. (In fact, many of those components are themselves quite elaborate inventions – plastic polymers, new materials, electronics, hydrocarbon fuels, etc.) Even something as abstract as a computer algorithm qualifies as an invention, provided it is complex enough, arbitrary (contingent) enough, and sufficiently oriented towards and informed by a specific purpose.

If it doesn't fulfill these conditions, the algorithm might be regarded as a 'discovered fact' of mathematics (an example is the Euclidean algorithm ubiquitous in number theory and abstract algebra). And this brings us to the concept of discovery – often confused with invention.

A discovery is the elucidation of a physical pattern, structure or effect in nature, or of a mathematical pattern, structure or fact (theorem) within some set of axioms. Unlike an invented device or procedure, a discovered fact or effect is neutral vis-a-vis any possible application – i.e., it is not inherently teleonomical.

Furthermore, a "discovery" has the essential hallmark suggested by the word itself: to wit, it emerges from the basic fabric of reality (or from a given set of mathematical axioms) in a manner sufficiently natural and non-arbitrary that it can be regarded as being in some sense "out there" (covered, as it were) in the universe before its discovery. (Thus, the Pythagorean theorem, the irrationality of the number $\sqrt{2}$, or any of the laws of mechanics, electromagnetism, or the rest of physics and chemistry, cannot be reasonably viewed as having been invented by the persons who happened to first realize or demonstrate them.)

With regard to strictly mathematical discoveries, one must add the caveat that the theory (a set of hopefully-consistent axioms) within which the discovery is made, is itself the arbitrary creation of the human mind – and this circumstance seems prima facia to preclude the assignment of any a priori existence to a discovered mathematical theorem.

But anyone who has had sufficient experience with pure and applied mathematics would agree that "interesting" mathematical theories are those that keep coming up and interconnecting in novel and unexpected ways; these interconnections are "inside" and "amongst" the mathematical theories themselves, and also between these theories and physical reality. Branches of "invented" mathematics that lack this kind of connectivity do not seem to last very long or draw the interest of many practitioners.

Likewise, discovered laws of nature – if they withstand the test of time – are invariably seen to be gradually refined and forged into general principles of great aesthetic appeal, which are also deeply interrelated with those mathematical theories having the above-mentioned enhanced-connectivity property.

Finally, we note that inventions are often thought of as not having been realized in the natural world before humans conceived of and reified them; discoveries, on the other hand, are commonly thought of as being realized in nature quite apart from humanity's (or other sentient civilizations) endeavors.

But this is not always the case! For instance, life – with its intricate, teleonomic mechanisms, patterns and structures – clearly qualifies, by our definition above, as a vast trove of "inventions" – although modern science realizes that there were likely no inventors of these inventions; rather, they arose through random interactions of molecular assemblies, incessantly pruned by the mathematical patterns of nonlinear reaction-diffusion-advection partial differential equations.

Thus, life is a striking example of natural laws and patterns of physics, chemistry and mathematics – "discovered" facts – giving rise to spontaneous, "inventorless inventions", including human beings and their brains.

And just as blind natural forces can "invent" (thus forms of heavier-than-air flight and symbolic processing of information were instantiated in biology long before humans and their technology) so is the converse true: many physical effects that qualify as having been "discovered" (in the sense defined above) are thought never to have actually been instantiated during the universe's lifetime, outside of human-devised laboratories (or those of other sentient beings elsewhere).

Examples of discoveries in this category are legion: the Mössbauer effect and the Bose-Einstein condensation of weakly-interaction clouds of Alkaline atoms are two likely examples. Such effects or laws of nature were, of course, always in existence – but only as unrealized potential phenomena; it took sentient, biotic beings ("invented" by inanimate nature!) to reify these effects and principles into actual, material reality.

Invention characterizes the living being; it attests to an effort of adaptation to the environment. Therefore it is found in the animal kingdom as well as in humankind. In Africa monkeys have been seen to use a stick to get at food which is out of reach: they have invented a tool. This ability is not restricted to mammals: birds can be watched dropping thick shells on rocks in order to break them and then eating the mollusks inside. Perhaps the Californian mosquito shows proof of invention too, as it increases by 200 times its production of the gene which enables it to synthesize the enzyme antagonistic to an insecticide.

Animal inventions are limited, however, and traditionally the history of inventions starts with the emergence of Homo sapiens sapiens. Hardly had he appeared but Homo sapiens sapiens not only adapted to the environment but also adapted the environment to himself.

Early humans were at the mercy of hostile predators and the climate. Without their ability to invent, it is doubtful whether our relatively weak and slow ancestors would have survived for long. Spears with fire-hardened points served first of all as hunting weapons and then for building fences.

Some eight thousand years BCE human populations were no longer content with hunting, fishing and gathering. They captured wild animals which they domesticated: horses for draught and for riding, dogs for protection, sheep for meat and poultry for eggs and meat; they also selected crops to cultivate, and therefore founded settled communities. So agriculture and farming began.

The countryside changed drastically, through the burning of the woodlands and then by repeated cultivation. Man invented pottery, affixed shaped flints onto shafts of wood, and made different tools according to his needs, such as the scraper, axe, pruning knife and adze.

At approximately the same time the settlers in Anatolia and on the banks of the Danube invented a rational organization of their shelters which were to become permanent. The food stores were wisely situated in the middle of the communities. Such was the origin of town planning.

During the Copper, Bronze and finally Iron ages, in the 5th, 3rd *and* 1st *millennia BCE respectively, the techniques and consequently the inventions began to diversify.*

One or more unknown persons invented successively the technique for working iron, then iron with carbon to make steel, then how to make needles, then metal wheels, armour, and cooking utensils which could withstand *high temperatures. The historical age, very roughly taken as starting in the 5th century BCE, took over from the prehistoric ages.*

Then great waves of invasion and commercial trading played a role in the spreading of inventions, comparable to that of swarms of insects in the pollination of plants. The conqueror and the tradesman placed instruments and products from one end to the other of the ancient continents, Africa and Eurasia.

Tables 1–3 summarize man's major inventions up to the 9th century CE. It was around this time that northwestern Europe began to climb to an ascendancy in technology that it has held to for centuries since. The poorer climate of this region, combined with the need to develop a new form of agriculture, was responsible for the emergence, around the 8th century, of the crop rotation methods still used today.

The technologies which have had the most profound effects on human life are usually simple. A good example of a simple technology with profound historical consequences is hay. Nobody knows who invented hay, the idea of cutting grass in the autumn and storing it in large enough quantities to keep horses and cows alive through the winter. This technology was unknown to the Romans but was known to every village of medieval Europe. It was a decisive event which moved the center of gravity of urban civilization from the Mediterranean basin to Northern and Western Europe. The Roman Empire did not need hay because in a Mediterranean climate the grass grows well enough in winter for animals to graze. North of the Alps, great cities dependent on horses and oxen for motive power could not exist without hay. So it was hay that allowed populations to grow and civilizations to flourish among the forests of Northern Europe.

In Northwestern Europe, the wind was put to use in both sea-going vessels and land-based mills. The region grew more populous and, by the 11th century, Northern Europeans were extending their influence into the Mediterranean and Middle East.

With few exceptions, most inventions made until the 16th century are anonymous. But even later, despite the work of historians and documenters, it is impossible to compile a complete history of inventions dating from ancient times to the end of the 19th century, for the following three reasons:

- *No large-scale control of production existed until well into the 19th century; the work of each inventor was carried out to fill his own needs and wishes and so did not resemble that of anyone else. (Thus, hundreds of types of lifting instruments have existed throughout the centuries, and one just gets lost in speculation as to the one which was used in the* *building of the fabulous Colossus of Rhodes in the 3rd century BCE, which was 35 m high.*

- *Many inventions which were designed on paper, were never realized, so the principle of selection cannot be applied to unrealized inventions.*
- *Many inventions are undoubtedly lost forever and other have been lost and found again, or reinvented. For example, not all the works of the inventors of the famous Alexandrian School are known. Before printing, the only documents describing inventions were manuscripts, of which often only a very few copies existed.*

The Ultimate Machine

At first all machines were mechanical, whatever their motive power; mechanical design requires strict specifications of parts' dimensions, material properties, relative positions, and mechanical contact, rolling, sliding etc. between parts. This can be thought of as the initial stage in the evolution of engineering – a stage in which machine functionalities and structure are very tightly correlated with the physical properties and configuration of the machine's parts.

Later, when electromagnetic theory was put on a sure footing, electrical circuitry and their standardized electronic components were developed (capacitors, resistors, inductances, batteries, vacuum tubes, transistors, and a variety of thermal, mechanical and optical transduction devices). This allowed the modular design and hookup of devices in which functionality and structure was partially liberated from the tyranny of geometry and mechanics; instead, function and structure tended to inhere to a large extent in circuit connectivity.

In the middle of the 20th century, analog electronic circuits were for the first time used to perform digital *(i.e. discrete logical) functions; thus was born the field of* digital design. *The ubiquitous* digital computer *is an extreme case of a machine whose entire function (except actuators and transducers interfacing with peripheral equipment) is to manipulate digital information; but digital components have since been interwoven into many other machines types as well (automobiles, wristwatches, camcorders, stereo systems, televisions, telescopes, microscopes...).*

Digital design is even more flexible than analog-electric one, because in addition to being liberated from geometric and physical constraints, it is also tolerant (up to a point) of major distortions of the electrical wave-forms and signals themselves (a signal representing the bit "1" can be recognized as such even after undergoing significant distortion from its original shape – much as human handwriting can be recognized across a fairly wide spectrum of individual penmanship, orthographic styles, font types, script size, orientations, etc).

Digital design need not be implemented electronically, of course – or even electrically; thus, an old fashioned, Pascal-type calculating machine is a digital computer implemented mechanically or electro-mechanically, while a pre-digital telephone exchange performs rudimentary digital computations via analog electric circuitry and electromechanical switches.

In the second half of the 20th *century, electronic circuits became increasingly miniaturized – especially digital computation circuits, but also to a certain extent analog circuits, and even some electromechanical devices). In such miniaturized devices, individual components are crammed together in dense configurations – which demands close attention to issues of heat dissipation, distributed analog electromagnetic effects and, of course, mechanical design and material science. This trend culminated (thus far!) in the VLSI silicon chip, and extended to electro-optical and optronic devices ("photonics"), Field Programmable Gate Arrays (FPGAs), Acousto-Optical (AO) devices, Micro Electro-Mechanical devices (MEMs), a variety of high-density digital memory devices, miniaturized RF devices, etc.*

Despite the fact that shrinking chip sizes re-introduce physical and geometric constraints, once a batch of chips is successfully manufactured – and assuming chip operation is reasonably resistant to thermal, chemical, mechanical and external electromagnetic disturbances and stresses – the chips can be treated as if they were simple electronic components with known logic functionality and assembled onto motherboards to render machines whose overall (digital and analog) functionality is, once again, flexible and modular.

Looking ahead into the coming decades, advances in photonics, electronics, material science and nanotechnology promise to both continue the miniaturization trend, and allow machine functionalities to be controlled by light alone, or by a combination of photons and electrons – in some cases, only one photon or electron at a time – as opposed to the millions of electrons whose motion states must change to flip a single bit from '0' to '1' in present-day computer chips.

This raises the further issue of how robust can a machine be when controlled by a single quantum particle! Even contemporary chips are susceptible to random bit-flips due to stray cosmic rays, natural radioactivity or external electromagnetic interference; this problem is customarily resolved via error correction codes *(the modern mathematical theory of which involves abstract algebra and Galois fields!).*

There are further challenges ahead to digital robustness, stemming from quantum uncertainties and superpositions of states; *researchers have already begun to tackle those, extending standard computer science into the new uncharted waters of* quantum computing.

Nanoscopic, and even quantum-coherent, machines now being contemplated – in which the "gears" are single quanta, atoms or molecules – mark man's final efforts (so far) in his long way of emancipating structure and functionality from physical and geometrical constraints of matter and energy.

The Iterative Nature of Scientific Knowledge

The manner in which **Kepler** *used the empirical astronomical data available to him (consisting of antiquity's accumulated lore plus* **Tycho Brahe***'s observations) is an instructive case study in how science's knowledge of laws of nature is actually abstracted from experience – through sequences of* interactive, iterative *and* convergent *interplays between empirical investigations on the one hand, and theoretical speculations and modeling on the other.*

Various (sometimes conflicting) misapprehensions about how this process operates are rampant among the general lay public and certain types of historians and sociologists of science: on the one hand, one often hears that laws of nature are discovered by painstaking accumulation of empirical observations, followed by suitable generalizations; then further experiments are carried out to verify new predictions of the generalized laws; and so on. (This is the view of the "Scientific Method" whose earliest formal champion was **Francis Bacon***.)*

But on the other hand, some modern philosophers and academics in the social sciences are of the opinion that general laws of nature are "underdetermined" by empirical data, and are thus (in part or even wholly) cultural constructs – which these scholars then, of course, gleefully rush to deconstruct *(a term enjoying unfortunate ubiquity in today's best universities).*

But both these schools of thought are simply wrong; the actual progress of our understanding of how the universe works is a far more interesting – and more robust – process than "dreamed of in their philosophies". And the resulting laws of nature – though always subject to eventual re-interpretation and modification from the vantage point of the next, deeper-level round of understanding – are real, not culturally-relative, and are never "repealed" or "go out of style" as, say, do legislations or dress fashions or styles of musical composition and architecture.

Thus, **Newton***'s theory of gravitation was not repealed by the newer, deeper understanding furnished by* **Einstein***'s theory of General Relativity – only its range of validity was circumscribed; after all, the Newtonian theory is still the coin of the realm in NASA's computations of satellite orbits, deep-space probe trajectories and the like.*

Nor do we cease to apply Newtonian mechanics to the modeling of aircraft flight, mechanical gears, engines, water spouts, golf-ball physics and all other manner of macroscopic machinery and motion, just because this theory breaks down for very small or very fast (or large or massive) objects – in which regimes, we now know, it must be replaced by quantum mechanics and (special and general) relativistic physics, respectively.

There is an iterative, back-and-forth interplay between theory and experiment in science which always seems to converge sooner or later! This is an amazing fact, given the ever-increasing length of the logical chains used by science in deciphering Nature.

These long chains of reasoning, deductions and modeling used to "interpolate" between experimental data points are necessitated by the limitations imposed upon humans by many constraints: their size, duration, particular location in the universe; by the limited spatial and temporal resolution of their sense-perception apparata; and by their limited ability to process and interpret concepts and data.

These limitation are, of course, vastly compensated for by human-invented technology, *which itself benefits greatly from scientific discovery – thus introducing another element into the dynamics of theory/experiment iterative interplay.*

Not least among humanity's technological aids to scientific research are those enabling it to summarize, record and propagate to subsequent generations its ever-increasing storehouse of data, theories, models and conclusions. Among such technologies are writing, printing, and (of late) electronic data storage; even the university *may be viewed as such an invention – for it perpetually trains new cadres of scientists who can process, add to, and (where necessary) modify the accumulated knowledge and understanding of the ages).*

Chronological Lists of Inventions

Table INV-1: Major inventions up to 3000 BCE

Invention	*Inventor(s)*	*Date*	*Use*
Tools made from sharp-edged stones; creating fire	Homo Habilis Homo Erectus	ca 2 million years ago ca 400,000 years ago	Fight off predators; cook and hunt for food. Surviving the second ice-age
• Collecting seeds and plants for 'next-year' planting; cereal crops			
• Distillation			
• Systematic crop-rotation (Europe)			Preserving soil fertility
• Producing wine, beer, mead, and bread, using fermentation			
• Producing yogurt, cheese and vinegar			
• Domestication of millet and rice			
• Domestication of animals			
• The wheel (ca 3400 BCE)	Agrarian valley-civilizations and societies	8000–3000 BCE	Transportation, milling devices
• Smelting metals; glass; firing of bricks and pottery (potters wheel); alloyed bronze			Irrigation systems
• Ploughshares, sickles; spinning wheel, saw, mirrors, balance; *lock*; *ball-bearings*			Organized city-life and agriculture
• River boats; *irrigation systems*			
• Calendar; sundial; hieroglyphics			
• Bridge (China); corbeled arches and domes (Mesopotamia) • Loom (Egypt, 4400 BCE) • Cosmetics (Egypt, 4000 BCE) • Sailing boats (Sumer, 4800 BCE)			

Table INV-2: MAJOR INVENTIONS 3000 BCE–100 CE

Invention	*Inventor(s)*	*Date*	*Use or comment*
• Silk	China	ca 2700 BCE	
• Cartography	Mesopotamia	ca 2300 BCE	Earliest surviving map (city map of Lagash, Mesopotamia created in stone)
• Bitumen	Mesopotamia	ca 2400 BCE	
• Wheeled vehicle on tracks	Babylon	ca 2250 BCE	Parallel lines of grooved stone blocks
• Horses tamed for transport	Near East	2000 BCE	
• Plant medicines	Egypt	2000 BCE	
• Toilet flush	Crete (Minoan)	2000 BCE	
• Pitch	Mesopotamia		Watertight boats
• Rubber (Latex)	Maya	ca 1600 BCE	Rubber human figurines;
• Alphabet (23 letters)	Phoenicians	ca 1700 BCE	
• Waterclock	Egypt	ca 1450 BCE	
• Wrought iron	Mesopotamia	ca 1400 BCE	Warfare and agriculture
• Natural gas	China	ca 900 BCE	Cooking and light
• Arches	Etruscans	ca 900 BCE	Used by Romans for bridges and aqueducts
• Iron ploughshares		ca 900 BCE	
• Stone canal and aqueduct		ca 700 BCE	
• Bridge (Euphrates)		ca 600 BCE	
• Coins	Asia Minor	ca 650 BCE	
• Papyrus	Egypt	ca 650 BCE	Paperlike writing material
• Seasonal harvesting	Mesopotamia	ca 600 BCE	
• Windmills	Persia (Neh)	ca 644 BCE	
• Mining of silver	Greece	ca 580 BCE	At Laureion
• Rails	Greece	ca 470 BCE	Grooved stone wagonway 8.8 km Athens-seaport of Piraeus

Table INV-2: (Cont. [a])

Invention	*Inventor(s)*	*Date*	*Use or comment*
• World map	Herodotos	ca 450 BCE	
• Automaton	Archytas of Tarentum	ca 350 BCE	Consists of hydraulic circuits and counterweight activating levers
• Connecting-rod crank system; propeller; hydraulic pump; levers; screws; pulleys	Archimedes	ca 250 BCE	Warfare
• Water clock	Ctesibios	ca 270 BCE	Measurements of time-intervals
• Chain drive	Philo of Byzantium	ca 250 BCE	Catapults (warfare)
• Astrolabe	Hipparchos	ca 150 BCE	Astronomy
• Concrete		1st century BCE	Road construction network
• Aqueducts	Vitruvius and	1st century CE	Water supply systems
• Urban water supplies	S.J. Frontinus		
• Water mill			Grinding corn
• Coal			Fuel in Northern Europe
• Commercial book-publication			
• Water wheel			Grounding grain and olives
• Arch		19 BCE	Seen in the aqueduct of Pont du Gard (Niem)
• Differential gear	Geminos of Rhodes	80 BCE	
• Production of cogwheel system			Gear box
• Steam engine	Hero of Alexandria	ca 60–80 CE	
• Turbine			Powered by steam or water devices
• Float regulator			Automatic regulation of thermomechanical devices

Table INV-3: MAJOR INVENTIONS IN THE FAR EAST 400 BCE–1400 CE

Invention	*Inventor(s)*	*Date*	*Use or comment*
• Use of coal for iron making; cast iron; fuel oil; fire wells (CH_4)		ca 350 BCE	Cooking, lighting and heating metallurgy
• Animal harness		300 BCE	Animal power agriculture and transportation
• The crossbow		300 BCE	
• Explosives		150 BCE	
• Universal (Cardan) joint		140 BCE	
• Roller bearings (wooden)		100 BCE	
• Suspension bridge		25 CE	Span = 15 m
• Porcelain		50 BCE	
• Paper from pulp		100 CE	
• Abacus	China	300 CE	First mechanical calculator
• Hydraulic piston mill	400 BCE–1400 CE	550 CE	For steam piston engine
• Block book printing		700 CE	
• Gun powder		700 CE	
• Printed newspaper		750 CE	
• Escapement in a mechanical clock		725 CE	Clock driven by hydraulic power; warfare
• Chain driven mechanical clock		970 CE	
• Magnetic compass		1086 CE	Navigation
• Mechanical water-clock		1092 CE	
• Multicolor printing		1107 CE	
• Rockets		1150 CE	Warfare
• Bombs		1221 CE	Warfare
• Small cannon		1288 CE	Warfare
• Chess	Indo-China	ca 450 CE	Modern version of the game "Goose-egg"
• The zero	India	ca 660 CE	Sign for a *number* (not just a placeholder)

Table INV-4: Major inventions in Europe ca 800 CE–1800 CE

Invention	*Inventor(s)*	*Date*	*Use or comment*
• Crop-rotation methods			Increase of food supply
• Hay	Northwestern Europe	Middle Ages	Motive power for horses and oxen
• Blast furnace			Cast-iron production
• Treatment of ores			Metals for housing and coinage and weapons
• Spinning wheel			
• Crossbow		ca 1050 CE	
• Spectacles	Roger Bacon S. Armati A. della Spina	ca 1250–1285 CE	Lenses for improved vision and reading
• Water-powered saw-mill	Villard de Honnecourt	1235 CE	
• Movable-type printing press	Gutenberg	1450 CE	Speeding up transfer of technology and scientific knowledge between people
• Concave lenses	Nicolas of Cusa	1451 CE	Treat nearsightedness
• Parachute • Flying machines • Military engineering • Centrifugal pump • Gas turbine • Hydraulic cooling system	Leonardo da Vinci	1480–1497 CE	Warfare
• Domestic spring-driven clock	Peter Henlein	1504 CE	
• Compound microscope	Zacharias Janssen	1590 CE	
• Thermometer	Galileo Galilei	1593 CE	
• Regular newspaper	Abraham Verhoeven	1605 CE	
• Use of coke	Hugh Plat	1603 CE	
• Telescope	Hans Lippershey	1608 CE	First optical refracting telescope
• Slide rule	William Oughtred	1622 CE	
• Micrometer	W. Gascoigne	1636 CE	
• Mechanical calculator	Blaise Pascal	1642 CE	

Table INV-4: (Cont. [a])

Invention	*Inventor(s)*	*Date*	*Use or comment*
• Barometer	E. Torricelli	1643 CE	
• Pendulum clock	Christiaan Huygens	1656 CE	
• Improved microscope	Robert Hooke	1662	
• Free escarpment		1666	Pendulum clock
• Matches	Robert Boyle	1680	
• Inoculation	Giacomo Pylarini	1701	Advent of immunology
• Dead-beat escapement	George Graham	1715	Mechanical clock
• Hammer klavier	Bartolomeo Christofori	1709	Father of the modern piano
• Achromatic lens	Chester Moor Hall	1733	Refracting telescope
• Marine chronometer	John Harrison	1737	
• Free escapement	Pierre Le Roy	1748	
• Lever escapement	Thomas Mudge	1765	First modern clock
• Balloon flight	Montgolfier brothers	1783	
• Gold-leaf electroscope	Abraham Bennet	1787	
• Vacuum pump	Otto von Guericke	1645	
• Optical refracting telescope	Isaac Newton	1668	
• Clinical thermometer	Santorio Santorio	1615	
• Mercury thermometer	G.D. Fahrenheit	1714	
• Centigrade thermometer	A. Celsius	1742	
• Sextant	John Hadley	1731	
• Machine tools	John Wilkinson Henry Maudsly Marc I. Brunel	1775–1801	
• Appendectomy	Claudius Aymand	1763	First successful
• Pianoforte with pedals	John Broadwood	1783	
• Railway		1681	4 km track in Stourbridge, England
• Railway	Charles Brandlin	1758	First railway for a colliery wagonway
• Solar heating	Antoine Lavoisier	1780	

Table INV-5: (Cont. [b])

Field	*Inventor(s)*	*Date*	*Invention*
Steam-Engine Technology	Robert Stirling	1816	Single-cycle heat engine
Internal Combustion Engine	Jean Lenoir	1860	Fuel-powered engine
	Sigfried Marcus	1875	Internal combustion engine
	Nikolaus Otto	1877	Gas-engine
	Gottlieb Daimler	1888	
	Karl Benz	1885	Gasoline-powered automobile
	Rudolph Diesel	1892–1896	Diesel-engine
Non-Electric Machines and Devices	Coulomb	1777	Torsion balance
	L.S. Lenormand	1783	Parachute
	Joseph Bramah	1795	Hydraulic press
	Alois Senefelder	1798	Lithography
	N.F. Appert	1809	Preservation of food by canning
		1814	Large scale gas lighting (London)
	Karl D. Sauerbronn	1816	Bicycle
	Augustus Siebe	1819	Diving suit which receives air pumped down from the surface
	B. Fourneyron	1827	Water turbine
	John Ericsson	1833–1837	Screw propeller; wind powered generator
	Gaston Plante	1834	Rechargable battery
	Linus Yale	1851	Lock
	J.B.L. Foucault	1852	Gyroscope
	J.E. Lundstrom	1855	Safety matches
	Elisha G. Ottis	1857	Steam-driven elevator
	Christopher Shoals	1867	Typewriter
	Pierre Michaux	1862	Steel ball-bearings
	I.W. McGaffey	1869	Vacuum cleaner
	A. Mouchet	1878	Solar generator to power a printing press
	John Milne	1880	Seismograph

Table INV-5: (Cont. [c])

Field	*Inventor(s)*	*Date*	*Invention*
Non-Electric Machines and Devices	Ottmar Mergenthaler	1886	Linotype machine
	Tolbert Langston	1887	Monotype printing
	W.L. Judson	1891	Zipper
	Benjamin Holt	1900	Tractor
	King C. Gillette	1901	Shaving with double-edged safety razor
	Jacob Perkins	1834	Mechanical compression system for air-conditioning
	James Harrison; A. Twining	1850	Refrigerator
	Ferdinand Caree	1854	Ammonia absorption system for air-conditioning
	Thomas S. Mort	1861	Artificial refrigeration of food
	Carl von Linde	1874	Ammonia compression system
Calculating Machines	Charles Babbage	1822–1834	Principle of programmable computer
	Ada A. Lovelace (Byron)	1842	Programming
	William S. Burroughs	1885	Mechanical calculating machine
	Hermann Hollerith	1890	Punched cards
Chemical Industry	August W. von Hofmann	1845	Aniline dye from benzene and *nitric acid* (HNO_3)
	C.F. Schönbein	1846	Nitrocellulox from cellulose and *citric acid*
	W.H. Perkins	1856	First synthetic dyestuff; quinine

Remarks: HNO_3 used for fertilizers, explosives, dyestuffs.
H_2SO_4 used for fertilizers, explosives, dyestuffs, petrochemicals, dehydration, iron and steel pickling.
HCl used for food preservation, metallurgy, oil industry.
NH_3 used for nylon, fertilizers, explosives, plastics, films.
Na_2SO_4 used for wood pulp, glass, detergents.
Na_2CO_3 used for glass, soap, detergents, textile treatment.

Table INV-5: (Cont. [f])

Field		*Inventor(s)*	*Date*	*Invention*
HEALTH AND MEDICINE		H. Davy	1800	Pioneer of anaesthetics
		F. Serturner	1805	Morphine
		Crawford Long	1842	First surgical operation using anaesthetic
		Charles Pravaz	1853	Hypodermic syringe
		Alexander Wood	1853	
		Joseph Lister	1867	Pioneer of antisepsis
		Louis Pasteur	1868–1870	Pasteurization
			1882	Rabies vaccine
		Augustus Waller	1887	Electrocardiogram
		Paul R. Ehrlich	1891	Pioneer of chemoteraphy
		Charles Gerhardt	1893	Aspirin
		Wilhelm Conrad Röntgen	1895	X-ray photography
ELECTRIC AND ELECTROMAGNETIC DEVICES	POWER	Ewald G. von Kleist	1745	Leyden jar (condenser)
		C. van Mussahenbroek	1746	
		P. van Mussahenbroek		
		Alessandro Volta	1800	Electric battery
		Luigi Brugnatelli	1805	Electroplating
		Moritz H. von Jacobi	1837	
		G.R. Elkington	1896	
		H. Elkington	1896	
		Michael Faraday	1821–1831	Principles of the motor, generator (dynamo) and the transformer (induction coil)
			1834	Electrolysis
		Joseph Henry	1829–1832	Principle of autotransformer
		William Sturgeon	1823	Electromagnet
		James B. Francis	1849	Hydroelectric turbine
		Georges Leclanche	1868	Dry cell
		William Stonely	1885	AC Transformer

Table INV-5: (Cont. [g])

Field		*Inventor(s)*	*Date*	*Invention*
Electric and Electromagnetic Devices	Power	Z. Theophile Gramme	1859–1869	DC and AC electric generator
		Werner von Siemens	1881	Electric railway (Germany)
		Charles R. Parsons	1884	Reaction steam turbine
		Carl G. de Laval	1883	Impulse steam turbine
		George Westinghouse	1885	AC power transmission (long distance)
		Frank J. Sprague	1884	Electric locomotive (US)
		P. Heroult; C. Hall	1886	Extraction of aluminum by electrolysis
		Lester Allen Pelton	1884–1889	The Pelton wheel (water turbine with *curved buckets* fixed to its periphery)
		N. Tesla	1899	AC induction motor
		Thomas A. Edison	1882	First steam electric power plant
		Sebastian de Ferranti	1899	High voltage AC generation and transmission
		Charles G. Curtis (US)	1900	High-power turbine plant (5000 kw)
	Light	J. Starr	1845	Long-lasting light source powered by electricity; carbon filament in vacuum
		Joseph Swan	1878	
		Thomas A. Edison	1879	
		Charles F. Brush	1879	Carbon arc street lamp
	Visual Communication and media	Alexander Bain	1842	Facsimile telegraphy
		F.C. Bakewell	1847	Facsimile with rotating scanning drums
		Leon S. de Martinville	1855	Phonoautograph
		J. Plücker	1859	First gas-discharge tube (CRT)

Table INV-5: (Cont. [h])

Field			*Inventor(s)*	*Date*	*Invention*
ELECTRIC AND ELECTROMAGNETIC DEVICES	VISUAL COMMUNICATION AND MEDIA		Louis May; Willoughby Smith	1873	Photoconductivity of selenium
			Eadweard Muybridge	1872	Cinematography
	COMMUNICATION	RECORDING, REPRODUCTION AND TRANSMISSION OF SIGNALS	Thomas A. Edison	1877	Phonograph
			Charles Cros	1877	Gramophone record of photo-galvanoplastic process
			Constantine Selencq	1878	Facsimile via selenium photo-conductivity
			Oberlin Smith	1878	Magnetic recording and reproduction of sound
			William Crooks	1879	Cathode-ray tube (CRT)
			Julius Elster; Hans Geitel	1880	First practical photocell
			Shelford Bidwelll	1881	Picture transmission via photocell and scanning
			Charles Fritts	1883	Solar cells
			Paul Nipkow •	1884	Mechanical TV scanning system photocell
			Louis de Prince	1886	Camera and projector system for cinematography
			Emile Berliner	1887	Grooved audio disc in a gramophone system
			Heinrich Hertz	1888	Wireless communication over long distance
			The Lumiéré brothers	1895	The cinematograph

Table INV-5: (Cont. [i])

Field			*Inventor(s)*	*Date*	*Invention*
Electric and Electromagnetic Devices	Communication	Recording, Reproduction and Transmission of Signals	Guglielmo Marconi	1895	Radio waves transmission and reception
			Karl F. Braun	1897	Cathode-ray tube oscillograph
			Valdemar Poulsen	1898	Magnetic wire recording and reproduction of sound
		Telegraphy	C.F. Gauss; W.E. Weber	1833	Short-distance communication
			Charles Wheatstone	1837–1858	Long-distance communication
			Samuel Morse	1838	Signal alphabet code
			Lord Kelvin Cyrus W. Field (US)	1866	Transatlantic telegraph cable-line
		Telephony	Charles Wheatstone	1827	Microphone
			Philip Reiss	1855–1861	Microphone
			Alexander G. Bell	1876–1877	Microphone
			David Hughes	1877–1878	Carbon microphone
			Almon B. Stowager	1891–1906	Automatic exchange with dial switches
			Michael Pupin	1894	Long-distance telephony (New York–Chicago)

Table INV-6: MAJOR INVENTIONS IN THE 20th CENTURY

Field		*Inventor(s)*	*Date*	*Invention*
TRANSPORTATION TECHNOLOGY	AUTOMOBILE	Henry Ford	1908	Assembly line method of automobile manufacturing
		Charles F. Kettering	1911	Self-starter; Engine ignition system
		Jean A. Gregoire Pierre Feneille	1926	Front wheel drive
		The Sturtevant brothers	1904	Automatic transmission
		Franz Wankel	1951	Rotary engine
	AVIATION	Samuel P. Langly Clement Ader	1896 1890	Airplane pioneers
		F.W. Lanchester	1897–1908	Wing theory and design
		Ferdinand von Zeppelin	1900	Built first successful dirigible
		The Wright brothers	1903	Self-powered controlled flight of heavier than air vehicles
		Rene Lorin A.A. Griffith Frank Wittle	1909 1926 1930	Turbojet engine
		Max von Opel	1929	Flight powered by rocket engine
			1947	First supersonic jet flight
			1970	Concord supersonic jet reaches Mach 2
		H. Anschutz-Kempfe Elmer Sperry (USA)	1908 1911	The gyrocompass
		Enrico Forlanini	1900	Hidrofoil boat
		Christopher Cockerell	1956	Hovercraft
		Paul Cornu	1907	Built first helicopter
		Juan de la Cierra	1923	Autogyro
		Heinrich Focke Rene Leduc Hans von Ohain Igor I. Sikorsky	1936 1938 1939 1939	Helicopter

Table INV-6: (Cont. [a])

Field		*Inventor(s)*	*Date*	*Invention*
Transportation Technology	Trains		1955	SNCF electric locomotive (France) on the Bordeaux line travels at speed 331 km/hr
			1990	TGV (France) on the Courtalain-Tour line travels at speed 515 km/hr
			1997	*Magnetic levitation* vehicle moves at 550 km/hr
Material Technology		Leo H. Baekland	1907	Bakelite
		Julius A. Nieuwland	1904–1924	Synthetic rubber
		J.E. Brandenburger	1912	Cellophane
		H. Brearly	1913	Stainless steel
			1917	X-ray crystallography for analysis of crystal structure and polymers
		Hermann Staudinger	1920	Birth of polymer science
			1927	PVC for pipes and bottles
		Eugene Freyssinet	1927	Prestressed concrete
		Richard Drew	1929	'Scotch' tape
		William Chalmers	1930	Plexiglass (perspex)
			1930	Polystyrene (styrofoam) for cups, packaging and thermal insulation
		Thomas Midgley	1930	Freon-refrigeration coolant
		Wallace Carothers	1931	Nylon: neoprene
		Roy J. Plunkett	1938	Teflon
		Thomas; W. Sparks	1937	Butyl rubber
			1941	Styrene butadiene rubber to make *tires*
		J.R. Whinfield; J. Dickson	1941	Terylene
			1941	Polyethylene for packing, piping and toys
		W.E. Hanford; D.F. Holmes	1942	Synthetic polyurethanes
			1952	Mylar polyester film
		Karl Biegler; G. Natta	1953	Polyethylene; polypropylene rubber

Table INV-6: (Cont. [b])

Field	*Inventor(s)*	*Date*	*Invention*
	Michael Schwartz	1954	Polyisoprene rubber (imitation natural rubber)
	Edward Schmitt; R. Polistine	1963	Synthetic sutures (polyglyolic acid): tissue engineering
	Bruce Merrifield	1964	Synthetic proteins and peptides
	W.J. Buehler	1967	Shape-memory alloys(*)
	Ronald Rosenzweig	1968	Ferrofluids (magnetic liquid)
	James Fergason Hoffmann La Roche	1969–1970	Liquid crystal display (LCD)
MATERIAL TECHNOLOGY	James Economy	1970	Ekonol (moldable high temperature polymer for electronic devices and aircraft engines)
	Stephanie Louise Kwolek	1971	Kevlar (polymer fiber stronger than steel) for bullet-proof vests and fire-proof garments (up to 300°C)
	A.G. McDiarmid; H.J. Heager	1977	Conductive plastics
	L.E. Lyons; Neville Mott; Hugh McDiarmid	1978	Molecular transistors (carbon based materials)
	Dan Shechtman et al.	1984	Quasicrystals(**)
	R. Smalley; H. Kroto	1985	C_{60} ('Bucky balls')
	G.J. Bendorz; Alex K. Müller	1987	High-critical temperature ceramic superconductors
	H. Naarmann; N. Theopilou	1987	Polyacetylene-Iodine compound: efficient conductor of electricity

(*) The study of shape-memory alloys takes off when **William J. Buehler** of the Naval Ordnance Laboratory, discovered the *shape-memory effect* in equiatomic Nickel-Titanium alloy (NITINOL: NIckel-TItanium Naval Ordnance Lab). The effect is governed by the thermoelastic behavior of the martensite phase.
Shape memory alloy have found applications in medicine (Vascular Stents), orthodontistry (braces), coffeepots, thermostats and other *smart materials*.
The Shape Memory Effect (SMF) was first discovered by the Swedish physicist **Arne Olander**, for a gold-cadmium alloy in 1932.
(**) *Discovered*, but may not exist in nature outside laboratories. Produced via invented annealing technique. May lead to many inventions and products.

Table INV-6: (Cont. [c])

Field		*Inventor(s)*	*Date*	*Invention*
World-Wide Communication and Information Technology	Radio	J.A. Fleming	1904	Diode tube
		Greenleaf W. Pickard	1906	Crystal radio detector
		R.A. Fessenden	1906	Amplitude-modulation; alternator
		Lee de Forest	1907	Triode tube
		Edwin H. Armstrong	1912	Frequency-modulation
		Irving Langmuir	1913	Gas-filled electron tube
		Edwin H. Armstrong	1920	Superheterodyne receiver
		Frank Conrad	1920	First radio broadcast station (Pittsburgh)
		Alan D. Blumlein	1932	Stereo sound system
		Seymour Benz	1942	Use of germanium crystal as radio detector: precursor to the transistor radio
	Television		1900	The word 'television' coined
		Manfred von Ardenne Boris Rosing A. Campbell-Swinton	1901–1923	Early experiments with television: CTR with fluorescent screen and electron scanning to replace Nipkow's mechanical system
		Charles F. Jenkins Philo T. Farnsworth	1923–1930	Many major TV components: power, focusing systems synchronization, contrast and scanning
		Denes von Mihaly	1928	
		Vladimir Zworykin	1928–1935	All-electronic system: iconoscope and kinescope
		E.F.W. Alexanderson	1928	First home TV receiver
		John Logie Baird	1939–1945	Electromechanical TV transmitter
		Peter C. Goldmark	1940	Color TV transmission
		Henri de France	1961	SECAM = Sequenced Color And Memory
		Walter Bruch	1963	PAL = Phase Alternating by Line

Table INV-6: (Cont. [d])

Field		*Inventor(s)*	*Date*	*Invention*
World-Wide Communication and Information Technology	Recording and Reproduction of Sound and Light	Eugene A. Lauste	1906	Early sound motion pictures
		Samuel Waters	1920	Early stereophonic reproduction of sound
		T.W. Case	1922–1926	Sound motion pictures
		Paul M. Ramey	1924	Pulse code modulation (to be used subsequently in computers and audio CD)
		Fritz Pfleumer	1927	Sound recording on plastic magnetic tape
		J.A. O'Neill	1927	Sound recording on coated paper ribbon
		H.C. Harrison; F.V. Hunt A.H. Frederick; J.A. Pierce	1931–1933	Early attempts to produce long-playing records
		W.D. Lewis		
		H.J. von Braumühl Walter Weber	1940	Modern tape recorders: high-frequency paramagnetization and AC bias
		Hedy Lamarre (cinema star actress)	1940–1942	Frequency hopping (spread-spectrum); subsequently used for torpedoes and cellular phones
		Peter C. Goldmark	1947	Long-period high-fidelity records
		Emery Cook	1951	Stereophonic LP recording
		Charles Ginsburg; Ray Dolby	1956	First commercial VCR recorder (videotape)
		Klass Compaan	1969	The idea of digitally encoded disc
		Klass Compaan; P. Kramer	1970	Glass-disc prototype
		Joop Sinjou; Toshitada Doi	1979–1983	Audio compact disc (500 MB) (CD)
			1996	DVD technology (5 GB)
	Photography	Edwin H. Land	1932	Polaroid instant camera
		Chester Carlson	1938	Xerography
		Dennis Gabor	1947	Holography
			1969–1990	Video digital camera
			1969	Charged Coupled Device (CCD)

Table INV-6: (Cont. [e])

Field		Inventor(s)	Date	Invention
World-Wide Communication and Information Technology	Telecommunication	W. Chen	1977	Videophone
			1983	Advent of cellular phone
			1984	Cable TV; coaxial development
			1984	Single-mode optical fiber communication system deployed. Transmitters use multi-longitudinal-mode Fabry-Perot lasers in the 1.3 μm infrared wavelength band.
			1987	Light amplified by light: First practical wideband, high-gain EDFA (Erbium Fiber Amplifiers) demonstrated: 1.5 μm wavelength-band signals amplified by nonlinear quantum photonics, with the energy provided by a shorter-wavelength pump laser.
			1987	Advent of WDM (Wavelength Division Multiplexing) in optical fiber communications.
			1987	HFC (Hybrid Fiber Coax) technology introduced (AM based)
			1988	Cellular phone network developed
			1989	DSL (Digital Subscriber Line) data-communication technology over twisted copper wire is introduced. Originally planned for video transmission, in the late 1990's it became a leading high-speed domestic internet technology.
			1990	Domination of world-wide news coverage (CNN)
			1993	Internet merges with new WWW (World Wide Web)
			1993	US launched first Direct Broadcast Satellite (DBS)
			1994	Advent of Coaxial Cable Modem (for internet)
			1995	World-wide web explodes
			1995	VoIP (Voice over Internet Protocol): internet telephony
			1996	Digital cellular phone system takes off

Table INV-6: (Cont. [f])

Field		*Inventor(s)*	*Date*	*Invention*
World-Wide Communication and Information Technology	Telecommunication		1998	Telecommunications go wireless
				Iridium global satellite phone
				High definition TV
			2000	Military-grade GPS (Global Positioning System) position tracking and time signals become available to civilians, but can still be jammed by U.S. military in war zones or during global alerts.
			2001	Cable internet connections take off
Warfare and Weapons Technology			1901	Britain's first submarine launched
		Isaac Lewis	1911	Lewis gun
		John Browning	1918	Automatic rifle
			1918	Gyroscope-guided missile (US Navy)
		Wernher von Braun	1942	V-2 rockets, powered by liquid oxygen and alcohol, travel 200 km from Peenemünde to London
			1944	Electromagnetic method of "enrichment", separation of uranium isotopes U-235 and U-238, is achieved in Oak Ridge Tennessee. Necessary to produce fissionable material
		Le Prieur	1915	Air-to-air unguided rockets
		Herbert Wagner	1943	Air-to-air guided missile
		Leo Szilard; Enrico Fermi;	1945	US exploded two atomic bombs over Japan
		Hans Bethe, et al.	1952	US hydrogen *fusion* nuclear bomb
		E. Teller; Stanislaw Ulam	1956	US 'suit-case' hydrogen bomb
		Wernher von Braun	1958	US launched *Atlas*, ICBM (range: 15,000 km) powered by liquid oxygen and kerosene
			1959	US launched *Titan 1*, ICBM
			1960	Underwater firing of solid-fuel *Polaris*
			1970	Anti-tank missile (US 'Tow')

Table INV-6: (Cont. [g])

Field	*Inventor(s)*	*Date*	*Invention*
Warfare and Weapons Technology		1972	Laser-guided missile ('smart bomb')
		1977	US Neutron bomb
		1983	US *Tomahawk* cruise missile
		1985	US assault-rifle M16A2
			'Laser-gun', 'Laser-pistol', Laser-targeting
		1990	Stealth fighter aircraft F-17A
		1991	Advanced medium range air-to-air missile
		1993	"Star-wars" initiative (SDI) announced
		1999	SDI missile system
Medical and Bio-Technology	Willem Einthoven	1903	Electro-cardiograph
	Mikhail Tsvett	1906	Chromatography (for separation of organic compounds)
	Paul Gelmo	1908	Sulfa drugs
	Paul R. Ehrlich	1909	Chemoteraphy: salvarsan ('magic bullet')
	Chaim Weizmann	1915	Discovered the microbe C. Acetobutylicum which converts starch into acetone and butanol, and uses the process to produce *cordite* explosive for the British WWI war effort (artillery shells)
		1917–1920	Sewage treatment using bacteria. Germans use acetone produced from plants to make explosives
	Alexander Fleming	1928	Penicilin
	A.G. Weyman	1932	Cardiac pacemaker
	W.B. Kouwenhoven	1932	Cardiac defibrillator
	Arnold O. Beckman	1934	pH-meter
	Arne W.K. Tiselius	1937	Electrophoresis (biochemistry)
	W.H. Florey; E.B. Chain	1940	Antibiotics
	A. Martin; R. Synge	1944	Chromatography (biochemistry)
	Erwin Chargaff	1978	
	R. Burns; William Doering	1944	Synthetic quinine
	Robert Woodward	1951–1972	Synthetic cholesterol, cortisone, chlorophyll, vitamin B12

Table INV-6: (Cont. [h])

Field	*Inventor(s)*	*Date*	*Invention*
Medical and Bio-Technology	Jonas Salk; Albert Sabin	1954	Polio vaccine (AS = safe oral live virus)
	George Pincus; Carl Djerassi	1955	Oral contraceptive
	Wilson Greatbach	1956	Silicon pacemaker
	Clarence W. Lillehei	1957	Pacemaker (internal)
	Willem Kolff	1957	Artificial heart
	Ian McDonald	1958	Ultrasound as diagnostic and therapeutic aid
		1959	Laser assisted surgery
	René Faraloro	1967	First coronary artery bypass operation
	Christiaan Barnard	1967	First human heart transplant
	Arieh Aviram	1974	Human genome project proposed using carbon-based
	Mark Ratner	1974	molecules as computer switches – the *molecular switch*
	Paul Berg; H.W. Bayer; S. Cohen	1977	Recombinant DNA technology (gene transfer)
		1977	First human protein manufactured by a bacteria
		1978	Laparascopy
		1978	First 'test-tube' baby born
		1980	Genetic engineering: production of human insulin
	Kary B. Mullis	1983	Polymerase Chain Reaction (PCR)
	Alec Jeffreys	1984	DNA 'fingerprinting'
	Robert Jarvik	1982	Artificial heart implant
		1986	Genetically engineered human vaccine
	Peter Schultz	1986	Combination of antibodies with enzymes ('abzymes')
		1988	Advent of the 'Human Genome Project'
	Richard Friend	1988	Diode model of electrochemical circuits of living cells
	Steven Rosenberg	1989	Human gene transfer
	Michael Blease	1989	Human gene transfer
	French Anderson	1989	Human gene transfer
		1990	First 'gene teraphy'
		1995	Immune system modulation

Table INV-6: (Cont. [i])

Field	*Inventor(s)*	*Date*	*Invention*
MEDICAL AND BIO-TECHNOLOGY		1997	Artificial human chromosoms
		1999	First cloning of a human embryo
ARTIFICIAL INTELLIGENCE	Karel Capek	1917	Coined the word '*Robot*'
	John von Neumann	1928	*Game Theory*
	Alan A. Turing	1937	'*Turing Machine*'
	John P. Eckert John W. Mauchly	1946	*ENIAC*, the world's first fully electronic programmable electronic computer
	Norbert Wiener	1947	'*Cybernetics*': information and control theory
		1947	The discipline '*Artificial Intelligence*' is born
	Maurice Wilkins	1949	Built 'EDSAC' – the world's first *stored-program* computer
	Alan Turing	1950	'*Turing Test*' for machine intelligence
	Claude Shannon	1950	Proposed a computer *chess program*
	D. Devol; J.F. Engelberger	1956	First industrial robot '*Unimate*'
	John von Neumann	1966	Developed theory of self-replicating automata. Such systems can find applications in *nanotechnology* and planetary explorations and terraforming.
		1974	First *computer-controlled* industrial robot
	Gordon Matthews	1979	Patent filed for *voice-mail* invention
		1984	Wabot-2, a 100 kg robot (in Japan) reads sheet music through its camera eyes and plays an organ with its 10 fingers and two feet
		1986	A robotic ping-pong player wins against humans
		1992	*Smart Weapons* incorporate electronic copilots, pattern-recognition techniques and other advanced technologies for tracking, identification and destruction
		1996	Speech recognition systems enter corporate voice menus, facilitating the handling of customer calls
		1997	First continuous-speech dictation software released

Table INV-6: (Cont. [j])

Field	*Inventor(s)*	*Date*	*Invention*
Artificial Intelligence		1997	*DEEP BLUE* IBM computer wins a chess match against the world's champion
Computer Technology	J. Abraham; E. Bloch	1918	Binary electric calculating machine
	W.H. Eccles; F.W. Jordan	1919	First "flip-flop" circuit design
	Vannevar Bush	1928	Partly electronic analog computer
	Konrad Zuse	1935	Programmable electromechanical computer
	Alan A. Turing	1941	The '*BOMBE*' computer (*Enigma* project)
	George R. Stibitz	1940	Electronical digital computer
	Russell Ohl	1940	Silicon pn junction demonstrated at Bell Labs.
	John von Neumann	1942–1952	Stored programs – advent of computer science
	John Atanasoff; C. Berry	1942	Electronic digital computers
	T.H. Flowers; A.W.B. Coombes	1943	The '*COLOSSUS*' computer (*Enigma* project)
	H.A. Aiken; Grace Hopper	1944	Harvard Mark I – automatic digital computer
	John W. Mauchly; J.P. Eckert	1946	ENIAC I (20,000 vacuum tubes)
	John Bardeen William B. Shockley Walter H. Brattain	1947–1948	Doped semiconductor transistor
	Frederick Williams; T. Kilburn	1948	First stored-program computer, SSEM, built as a test for newly invented tube memory device, Williams storage CRT.
	Yoshiro Nakamata	1950	Floppy disc
	Jay Forester	1951	Development of RAM (Random Access Memory); magnetic core memory at M.I.T.
	G.W.A. Drummer	1952	The idea of integrated circuits
	John Backus	1954	High level programming language
	J.S. Kilby; Robert N. Noyce	1959	Integrated circuit – the "chip"
	Douglas Engelbart	1964–1965	The 'mouse'

Table INV-6: (Cont. [k])

Field		*Inventor(s)*	*Date*	*Invention*
Computer Technology		James T. Russell	1965	Created the digital Compact Disc (CD)
		Lofti Zadeh	1965	Fuzzy logic
		Marcian Hoff; F. Faggin; S. Mazo	1970	The microprocessor
		Alan Shugart	1971	Reintroduced the floppy disc (1.44 MB)
			1972	Magnetic disc; microcomputers based on microchips
		J.S. Kilby; J.D. Merryman	1972	Electronic pocket calculator
		Robert Metcalf	1973	Networking (Ethernet)
		David Ahl	1974	First PC computer
		Edward Roberts	1975	PC, the 'Altair'
		Stephen Wozniak	1976	Apple I, PC
		Steve Jobs	1977	Apple II, PC (first commercial PC)
		Seymor Rubenstein; R. Barnaby	1979	Word processor
		Kenneth H. Olsen	1986	Magnetic core memory
		Tim Berners-Lee	1989	World-Wide Web (WWW)
		Scott Fisher	1989	Virtual Reality (VR)
		Thomas Bein; C.G. Wu	1994	'Molecular wire' nano-electronic device
			1997	PC in 35% of US homes; 27 million US adults regularly use the internet
Imaging Technology	Sonar, Radar and Lidar	Paul Langevin	1917	Sonar echolocation system (tracing U-boats)
		Christian Hulsmeyer	1904	One-mile range electromagnetic echo system
		Albert W. Hull	1917–1921	Magnetron diode; LIDAR-Light Detection and Ranging
		G. Breit and M. Tuve	1925	First practical RADAR
		Robert Watson-Watt	1935	Microwave RADAR used in WWII
		The Varian brothers	1938	*Klystron* microwave generator
		J.T. Randall; Henry Boot	1940	Multicavity *magnetron* generator
			1945	Reflection of RADAR signals from the moon

Table INV-6: (Cont. [l])

Field		*Inventor(s)*	*Date*	*Invention*
IMAGING TECHNOLOGY	MICROSCOPY	Max Knoll; E.A.F. Ruska	1929	Electron microscope
		Frits Zernike	1935	Phase-contrast microscope
		James Hillier	1937	Improved electron microscope
		Erwin W. Mueller	1955	Field-ion microscope ($\times 10^6$)
		Albert Victor Crewe	1970	Scanning electron microscope
		Gerd Binnig; Heinrich Rohrer	1980	Scanning Tunneling Microscope (STM)
				Atomic-force microscope
				Field-emission microscope
				Near-field optical microscope
				Confocal microscope
			1988	Positron transmission microscope
	TOMOGRAPHY	Johann Radon	1917	The Radon transform
		Pierre-Michel Duffieux	1940–1966	Optical imaging
		André Maréchal	1946–1960	Photographic image enhancement
		Felix Bloch; E.M. Purcell	1946	Nuclear Magnetic Resonance (NMR)
		Norbert Wiener	1948	Mathematical basis for magnetic resonance imaging (Cybernetics = computer controlled systems)
		Godfrey Hounsfield; Alan Cormack	1967–1968	Computerized tomography imaging
		Godfrey Hounsfield	1972	Computed Tomography Scanner (CT)
		Raymond Damadian	1971	NMR imaging
		Paul Lauterbur	1973	Magnetic Resonance Imaging (MRI)
		J.E. Greenleaf; S.A. Johnson	1974–1980	Ultrasound computed tomography
		Louis Sokoloff	1978	Emission Computed Tomography (ECT)
		Douglas Boyd	1979	Computed Tomography Scanner

Table INV-6: (Cont. [m])

Field		*Inventor(s)*	*Date*	*Invention*
IMAGING TECHNOLOGY	TOMOGRAPHY			Electron Paramagnetic Resonance (EPR)*
				Ferromagnetic Resonance (FMR)*
				Anti-Ferromagnetic Resonance (AFMR)*
			1979	Single-Photon Emission Computed Tomography (SPECT)
			1985	Positron Emission Tomography (PET)
	EARTH-BASED TELESCOPY		1917	*Mount Wilson* 254 cm Optical Reflecting Telescope began operation (ORT)
		Karl Jansky	1930	Built a 30 m long rotating aerial radio telescope
		Bernhard V. Schmidt	1934	First 35.5 cm Schmidt Optical Reflecting Telescope (SORT)
			1936	*Palomar* 45.7 cm SORT began operation
		Grote Reber	1937	Built a 945 cm Radio Telescope (RT)
		Bernard Lovell	1947	*Jodrell-Bank* 66.5 m Non-Steerable Radio Telescope (NSRT)
			1949	*Palomar* 122 cm SORT began operation
			1949	*Palomar* 508 cm ORT began regular operation
		Robert Hanbury Brown	1949	Correlation interferometer for measuring angular diameters of stars
		Bernard Lovelll	1957	*Jodrell-Bank* 76 m Steerable Radio Telescope (SRT)
			1960	*Owen-Valley* 27 m RT
			1963	*Arecibo* (Puerto Rico) 300 m RT
		Martin Ryle	1964	Cambridge (England) 1.6 km radio interferometer
			1965	*Owen-Valley* 40 m RT
			1967	First VLBI images – 182 km baseline
			1969	*Big-Bear* solar observatory

* These were *discoveries*, but applied to imaging *inventions*.

Table INV-6: (Cont. [n])

Field		*Inventor(s)*	*Date*	*Invention*
IMAGING TECHNOLOGY	EARTH-BASED TELESCOPY		1970	*Cerro-Tololo* (Chile) 4 m ORT
			1970	*Kitt-Peak* (Tucson, AZ) 4 m ORT
			1974	Siding-Springs (Australia) 3.88 m ORT
		G. Smith; F. Landauer J. Janesick	1975	Used CCD to observe Uranus
			1979	*Amado* (Arizona) 4.47 meter Infrared Reflecting Telescope (IRT)
			1979	*Mauna Kea* (Hawaii) 3.81 m IRT
			1880	*Socorro* (NM) VLA
			1992	*W.M. Keck* (Mauna Kea Volcano, Hawaii) 10 m optical and infrared telescope: the position of its 36 hexagonal glass segments are continually aligned by computer controlled actuators (Honeycomb 'fly's-eye' pattern); equipped with *wide-field CCD imagers* and *adaptive optics* techniques to compensate for atmospheric distortion
	SPACE-BASED TELESCOPY		1962	First orbital telescope on GB's *Ariel1*: energy spectrum of cosmic rays
			1970–1973	First X-ray satellite
			1972	UV telescope on *Copernicus* satellite
			1979–1981	Gamma-ray telescope on the *Einstein* satellite
			1992	COBE (Cosmic Background Explorer) satellite discerns anisotropy in Cosmological Microwave Background Radiation; advent of *precision cosmology*
			1993	*Hubble space telescope*: 2.4 m aperture of primary mirror: images of celestial bodies
			1999	Chandra X-ray space observatory: 4.2 m diameter telescope: imaging of cosmic X-ray sources

Table INV-6: (Cont. [o])

Field		*Inventor(s)*	*Date*	*Invention*
NUCLEAR[4] TECHNOLOGY	MEDICINE	George Parthes; Walter B. Cannon Henri A. Danlos; E. Bloch	1798–1905	Early uses of X-rays and radium for diagnosis and treatment
		Frederick Proescher	1913	Intravenous radium injection
		Georg von Hevesy	1923	Radioactive tracing (e.g. Lead radioisotope)
		Herman Blumgart (US)	1927	Radioactive traces for diagnosis of heart diseases
		The Joliot-Curies	1934	Artificial radioactivity
		John H. Lawrence	1936–1939	Phosphoros-32 used for leukemia treatment
		John Livingood; S.M. Seidlin	1936–1946	Iodine-131 and Cobalt-60 used for cancer treatment
		Emilio Segré; Glenn Seaborg	1939	Technetium-99 (half-life = 6h)*
		Glenn Seaborg	1950	Californium (transuranium elements)*
		Benedict Cassen	1951	Radioisotope photo-scanner
		Gordon Bronwell; H.H. Sweet	1954	First positron-emission detector
		Hal Anger	1958	Scintillation camera (viewing an organ as a whole)
			1958	Bone scanning with technetium-99
		William Oldendorf	1960	Brain studies with radioisotopes
		Rosalyn Yalow; Solomon Berson	1961	Radio-Immuno-Assay (RIA)
		David Kuhl	1962	Emission reconstruction tomography (the father of SPECT, PET and CT)
		Henry Wagner	1963	Radiolabeled Albumin for imaging lung
			1973	SSRL (Stanford Synchrotron Radiation Laboratory) produced high-intensity X-rays used for medical and material research
			1975	Cardio-vascular imaging with Thalium-201
		Michel Ter-Pogossian	1975	PET scanner (Positron Emission Tomography)
		John Keyes	1976	Single Photon Emission Computed Tomography (SPECT)

* *discoveries* of elements not present in nature

Table INV-6: (Cont. [p])

Field		*Inventor(s)*	*Date*	*Invention*
NUCLEAR TECHNOLOGY	MEDICINE	R. Damadian; L. Minkoff M. Goldsmith	1977	Magnetic Resonance Imaging (MRI): first examination of a human body
		David Goldenberg; J.P. Mach	1978–1981	Radiolabeled antibodies for tumor imaging
		Steve Larson; J. Carrasquillo	1982	Iodine-131 labeled monoclonal antibodies used to treat cancer
	PARTICLE DETECTORS AND ACCELERATORS	Boltwood	1907	Radioactive dating
		W.F. Libby	1955	Radiocarbon dating
		Charles Wilson	1894–1911	Cloud-chamber for tracking charged particle in supersaturated gases
		William Crookes	1903	Scintillation counter
		Johannes Geiger	1910	Geiger counter
		A.J. Dempster; F.W. Aston	1918–1919	Mass spectrometer
		Ralph Videröe	1928	RF linear accelerator
		R.J. van de Graff	1931	Electrostatic accelerator
		J.D. Cockroft; E.T.S. Walton	1932	Linear proton accelerator
		Ernest O. Lawrence	1934	Cyclotron
		D.W. Kerst	1940	Betatron
		Marcus L.E. Oliphant	1943–1947	Proton-synchroton
		E.M. McMillan; V. Vexler	1945	Synchro-cyclotron
			1944	Scintillation counter (with photomultiplying tube)
		Donald A. Glaser	1952	Bubble chamber
			1952	Cosmotron (at BNL)
			1953	First "breader reactor" (US): U-235 is split to provide energy, while U-238 is changed into plutonium, to be used to fuel further nuclear fission reaction
		S. Fukui; S. Miyamoto	1959	Spark chamber
			1966	Stanford electron LINAC (LINear ACcelerator)

Table INV-6: (Cont. [q])

Field		*Inventor(s)*	*Date*	*Invention*
Nuclear Technology	Particle Detectors and Accelerators	G. Charpak; R. Bouclier	1968	Multiwire proportional particle detection chamber
			1965	Alternating Gradient Synchrotron (at BNL)
		Simon van der Meer	1982	Proton-cooling (at CERN)
			1984	Tevatron (Fermilab)
			1987	Storage ring collider (Fermilab)
			1989	Large Electron-Positron Collider (LEP, at CERN)
			1989	Stanford Linear Collider (SLC): single pass collider
			2000	Relativistic Heavy Ion Collider (RHIC) at BNL
Nuclear and Particle-acceleration Technology	Energy		1951	Nuclear fission produce useful electricity in Idaho
			1954	First nuclear power reactor (US)
			1955	First town (Arco, Idaho) to be powered by nuclear energy
			1967	Project 'Gasbuggy': release of natural gas into a nuclear explosion underground chamber
			1975	Laser separation of uranium isotopes
			1976	Full-Electron Laser (FEL; John Madey)
	Applications		1985	Lead-iron phosphate glass for durable containment of nuclear waste
			1991	Cesium atomic clock (1 second drift in 1.6×10^6 years)
				Free-electron laser
	Weapons	Lise Meitner; Otto Hahn	1938	Uranium fission achieved (Germany)
		Enrico Fermi	1942	First self-sustaining nuclear chain reactor (Univ. of Chicago)

Table INV-6: (Cont. [r])

Field		*Inventor(s)*	*Date*	*Invention*
NUCLEAR AND PARTICLE-ACCELERATION TECHNOLOGY	WEAPONS	Seth Neddermeyer	1945	"Fat Man" implosion atomic bomb (Nagasaki)
				"Little Boy" atomic bomb (Hiroshima)
		Edward Teller, Stanislaw Ulam et al.	1952	Thermonuclear hydrogen bomb
			1961	Largest man-made explosion: USSR H bomb, yield = 57 MT (Novaya Zemlya)
			1973	Miniature nuclear warhead (50 ton TNT); USA
			1977	Neutron bomb (radiation damage); USA
		Hyman G. Rickover	1958	Launching of the first nuclear submarine (USS 'Nautilus')
		Hyman G. Rickover	1961	Launching of the first nuclear aircraft carrier (USN)
GEOSATELLITES AND SPACE PROBES	ROCKETS	Konstantin Tsiolkovsky	1903	Rocket propulsion
		Robert Goddard	1926	Launch of first liquid-fuel rocket
		Herman Oberth	1929	Rocketry and space travel technology
		Charles S. Draper	1939	Inertial guidance system
		Theodore von Karman	1941–1963	Long-distance rockets and space travel
		W. von Braun; W. Dornberger	1942–1969	V-2 rockets; Atlas, Titan and Apollo rockets
	SPACE EXPLORATION	Sergei Korolev (project Head)	1957	Sputnik I – first orbiting satellite(USSR)
			1958	Explorer I – first US satellite
		NASA	1962	Mariner 2 – first mission to *Venus*
		Freeman Dyson	1963	Gravity-assist fly by maneuvers
		NASA	1973–2001	

Table INV-6: (Cont. [s])

Field		*Inventor(s)*	*Date*	*Invention*
GEOSATELLITES AND SPACE PROBES	SPACE EXPLORATION	NASA, USA	1965	Mariner 4 – first clear pictures of *Mars*
		NASA	1966	Luna 10 – first spaceship to orbit the *moon*
		NASA	1969	First manned mission to the moon; Apollo 11 launched by Saturn V rocket
		NASA	1971	Lunar rover
		NASA	1974	Mariner 10 photographs of *Mercury*
		NASA	1975	Venera 9 sent pictures of the surface of *Venus*
		NASA	1976	Viking I, II landed on *Mars*
		NASA	1979	Voyager I, II sent images of *Jupiter* and its system
		NASA	1980–1981	Voyager I, II sent images of *Saturn* and its system
		NASA	1981	First flight of space shuttle
		NASA	1986	Voyager II sent images of *Uranus* and its system
		NASA	1989	Voyager II sent images of *Neptune* and its system
		NASA	1989	COBE satellite launched
		NASA	1990	Hubble space telescope
		NASA	1998	'Deep space' ion-rocket prototype
		NASA	2000	US 'Image' space weather satellite
	COMMUNICATION	Arthur C. Clarke	1945	The idea of communication satellite in geosynchronous orbit
		NASA	1960	Weather satellite *Tirus 1*
		NASA	1962	*Telstar 1* re-transmits TV program watched by 200 million people in US and Europe
		NASA	1964	*Syncom 3* launched into a geostationary orbit and used to broadcast Olympic games from Japan
				INTELSAT also relayed Olympic games to North America
			1965	*EARLY BIRD* – first geostationary commercial satellite over the Atlantic Ocean

Table INV-6: (Cont. [t])

Field		*Inventor(s)*	*Date*	*Invention*
GEOSATELLITES AND SPACE PROBES	COMMUNICATION	NASA	1965–1992	*INTELSAT 2, 3, 4, 5, 6*. The last one weighing 2500 kg and relaying 24,000 telephone lines plus 3 television channels
			1969	First live television broadcast from the *moon*, by US *Apollo 11* moon-landing mission, is transmitted to some 600 million viewers worldwide
			1972–1992	*LandSat 1–6* launched to observe the earth, providing warning of crop disease, flooding, icebergs and monitoring pollution
			1977	*Eutelsat*, linked telecommunication networks of the European countries
			1987	Soviet radar satellite (20 tons) launched for map-making, oceanography, crop predictions, ice monitoring and prospecting for minerals
		NASA	1993	*ACTS* was launched from shuttle *Discovery*. It is a testbed for future communication satellite technology such as multi-beam antennae and advanced signal processing
			1996	*INTELSAT* has 18 satellites in orbit working for the world wide web
			2000	Civilian access to GPS satellite data.
QUANTUM TECHNOLOGY	LASERS	Albert Einstein	1916	Proposed the process of stimulated emission (in addition to absorption and spontaneous emission)
		Alfred Kastler	1950	Optical pumping
		Charles Townes; J.P. Gordon H.J. Ziegler	1953	Built the *Maser* which used ammonia to produce coherent microwave radiation
		A. Schawlow; C. Townes Gordon Gould	1954 1957	Proposed the *Laser*

Table INV-6: (Cont. [u])

Field		*Inventor(s)*	*Date*	*Invention*
QUANTUM[5] TECHNOLOGY	LASERS	Theodore H. Maiman	1960	Built the first laser from synthetic ruby
		P.A. Franken; A.E. Hill C.W. Peters; G. Weinreich	1961	Harmonic generation of light by passing the pulse from a ruby laser through a quartz crystal
		A. Javan; W.R. Bennet D.R. Harriott	1961	First *gas laser* (Helium + Neon near infrared)
		M.I. Nathan	1962	Gallium arsenide (semiconductor) laser
		Kumar Patel	1963	CO_2 laser
		W.B. Bridges	1964	Noble-gas Ion lasers
		J.V. Kaspar; G.C. Pimental	1964	Iodine laser (used in holography)
		E. Snitzer	1964	First *optical fiber amplifier* using Neodynium doped fiber, near wavelength 1.06 micron
		P.P. Sorokin; J.R. Lankard	1966	Organic-dye laser (*molecular* energy levels)
		John M.J. Madey	1971–1976	Free Electron Laser (FEL)
		R.H. Stolen; E.P. Ippen	1973	*Raman amplification* of optical signals via stimulated Raman scattering
			1975	Laser separation of Uranium isotopes
		D.L. Mathews	1985	X-ray laser
		M. Nakazawa	1989	Erbium-Doped Fiber Amplifier (EPFA) pumped by high power laser diode of wavelength 1.48 micron
		R.H. Stolen; H.A. Haus	1989	Raman amplifier in silica-core optical fibers. Beginnings of all-photonic computation and communication
		John Bardeen; William B. Shockley Walter H. Brattain	1947	Semiconductor-based diodes and transistors
		Harold Lyons	1949	Atomic clock (based on the quantum-mechanical vibrations of the Ammonia molecule

Table INV-6: (Cont. [v])

Field	*Inventor(s)*	*Date*	*Invention*
QUANTUM TECHNOLOGY		1933–1952	*Zenner-diode* voltage regulator. Used subsequently for light-amplification devices
	Russell Ohl	1941	Silicon junction solar cell
	W.H. Schottky	1948–1956	Schottky barrier diode
	Gerald L. Pearson	1952	Alloy-junction diode
	Narinder S. Kapany	1955	*Optical fiber* communication
	Leo Esaki	1958	*Tunnel diode*
	George Heilmeier	1964	*Light-Crystal-Display* (LCD)
	Nick Holonyak	1965	*Light-Emitting-Diode* (LED)
	Charles Kao; G. Hockham	1966	Proposed using glass fiber for an optical transmission system
		1975–1977	Transmission of video signals by optical fibers
	C. Goy and D. Kleppner	1983–1987	Single-atom cavity[a] QED
		1988	First trans-Atlantic optical-fiber system
	George J. Bendorz Alex K. Müller	1986	High-temperature superconductor alloys
	Bernard Raveau		*Superconductor chip*
	Bennet; Brassand	1989	The idea of quantum computer
	Gerald Pearman	1991	Low-power silicon solar battery
	Sumio Iijima	1991	Carbon nanotubes and their uses as quantum-wire switches
	Peter Shor; N. Gershenfeld I.L. Chung; M.G. Kubinec	1994–1977	First attempts to build a two-bit quantum-mechanical computer using NMR
	Eric Cornell; Carl Wieman	1995	Bose-Einstein Condensate[b]
	L.K. Grover		

[a] Modification of spontaneous emission properties of single atoms in electromagnetic cavities due to coupling with macroscopic cavity modes (RF or optical).

[b] Using a combination of *laser-cooling* and *magnetic evaporative cooling*, they cooled a gas of about 2000 rubidium atoms to a record-breaking temperature of 20 nanoKelvin, enough for the atoms' de-Broglic wave to overlap – thus producing the first ever weakly interacting BEC. Further refinements of these cooling technique led to even lower temperatures (the record as of 2004 is about 500 picoKelvin).

Table INV-6: (Cont. [w])

Field	*Inventor(s)*	*Date*	*Invention*
	Anton Zeilinger	1997	Quantum teleportation
	W.H. Carrier	1902	Air conditioner
	P.C. Hewitt	1903	Mercury vapor lamp
	Georges Claude	1910	Neon light
	G. Claude	1911	Neon lamp
			Halogen light
	Irving Langmuir	1915	Tungsten filament
	Charles Strite	1919	Automatic electric toaster
	B. von Platen; C. Munters	1923	Electric refrigerator
	C.W. Rice; E.W. Kellog	1924	Loudspeaker
	Clarence Birdseye	1924	Frozen food
GENERAL TECHNOLOGY	Theodor Svedberg	1925	Ultra-centrifuge (for determination of molecular weights)
	Hans Wilsdorf	1927	First waterproof wristwatch (Rolex)
	Joseph Horton; Warren Morrison	1928	Quartz crystal clock
	Charles Beebe	1934	Bathysphere (depth = 924 m)
	Edward Germer	1934	Fluorescent light
	Jacques-Yves Cousteau	1943	Dived with compressed-air aqualung
	Percy L. Spencer	1947	Microwave oven
	Maria Telkes	1948	Solar heated home
	Douglas Ross	1952	Computer control of machine tools
	Felix Trombe	1954	Solar power plant with parabolic mirror
	D.M. Chaplin; C.S. Fuller	1954	Solar cell
	Gerald L. Pearson	1954	Solar cell
		1960	Large-scale desalination of sea water (US)

Table INV-7: NOTABLE CULTURAL STRUCTURES ON EARTH SINCE 5000 BCE

Builder	*Location*	*Project*	*Date*	*Height*	*Details*
	Egypt	Sphinx (Giza)	~5000 BCE	20 m	Length = 73 m
	Mesopotamia	Irrigation canal system (Sumer)	~3000 BCE		
Imhotep	Egypt	Pyramid of Khafre (Giza)	~2700 BCE	60 m	Base = 121 × 109 m^2
Hemon	Egypt	Pyramid of Khufu (Giza)	~2600 BCE	146 m	Base = 230 × 230 m^2
	Egypt	Sad el Kafara Dam	~2500 BCE	11 m	
	England	Stonehenge (Wiltshire)	~1900 BCE		
	Egypt	Abu Simbel	~1250 BCE	20 m	
	Cambodia	Temple of Angkor Wat	~1150 BCE		
	Jerusalem	Temple of King Solomon	960 BCE	40 m	Base = 30 × 30 m^2
	Babylon	Marduk Ziggurate (Etemenanki)	~600 BCE	100 m	
	Babylon	Hanging Gardens	~600 BCE	23 m	Area= 120 × 120 m^2
Cherisphron; Metegenes	Asia Minor	Temple of Artemis (Ephesos)	~550 BCE	12 m	Base = 115 × 55 m^2
	Iran	Takht-e-Yamshid (Persepolis)	~550 BCE		Length=1100 m
	Jordan	Petra (Nabataean city)	~550 BCE		
	Peru	Geoglyphs and lines of Nasca	~550–500 BCE		Area = 450 km^2
Eupalinos	Greece	Rock-cut water supply tunnel (Samos)	~530 BCE		
	Jerusalem	The Second Temple	516 BCE	51 m	Base = 51 × 51 m^2
	Burma	Shwe-dagon Pagoda (Rangoon)	~500 BCE	112 m	
Pheidias	Greece	Statue of Zeus (Olymphia)	457 BCE	12 m	
Ictinus; Callicratos	Greece	Parthenon (Akropolis, Athens)	438 BCE	18 m	Area = 37 × 34 m^2
Satyrnos; Pythios	Greece	Mausoleum (Halicarnassos)	353 BCE	41 m	
Appius Claudius Caecus	Italy	Appian Way (and aqueduct)	312 BCE		Length = 589 km
Chares of Lindos	Greece	The Colossos (Rhodes)	283 BCE	37 m	
Sostratos	Egypt	Pharos Lighthouse (Alexandria)	270 BCE	134 m	
Romans	Spain	Baebolo Silver mining tunnel	250 BCE		Length = 2200 m
	China	Gukow River Dam	250 BCE		Length = 30 m
	China	Great Wall	214 BCE	8 m	Length = 2400 km
	Asia Minor	Closed-pipe water system (Pergamon)	170 BCE		Length = 378 m

Table INV-7: (Cont. [a])

Builder	*Location*	*Project*	*Date*	*Height*	*Details*
	Philippines	Banaue Rice Terraces (Ifuago)	~1 CE		Area = 100 × 100 km^2
	Italy	Narni River Bridge	14 CE		Length = 43 m
	Italy	Lake Fucinus drainage tunnel (Rome)	~45 CE		Length = 5600 m
	Italy	Subiaco River Dam	50 CE		Length = 39 m
	Italy	Colosseum (Rome)	80 CE	49 m	Area = 180 × 150 m^2
	China	Lan Chin Bridge (Yunnan)	65 CE		Length = 76 m
	Syria	Palmyra	~100 CE		
	England	Hadrian Wall	125 CE	6 m	Length = 117 km
	Italy	Pantheon (Rome)	128 CE		
	Mexico	Pyramid of the Sun (Teotihuacan)	300 CE		
	Easter Island	Moai Statues (Rapa Nui)	400 CE		
	Turkey	Hagia Sophia (Istanbul)	537 CE	56 m	
	Mexico	Pyramid and Temple of Inscriptions (Palenque)	~650 CE		
	China	Grand Canal	~650 CE		Length = 1735 km
	Japan	Horyu-Ji Temple (near Kyoto)	~690 CE		
	Spain	Cathedral of Cordoba	786 CE	91.5 m	
	Java	Borobudur Temple	~790 CE	42 m	Area = 123 × 123 m^2
	Japan	Toji Pagoda (near Kyoto)	~796 CE	57 m	
	Guatemala	Maya Temple (Tikal)	~870 CE	145 m	
	France	Mont Saint-Michel (Normandy)	~1050 CE		
	England	Westminster Abbey	1065 CE		
	Italy	St. Mark Cathedral (Venice)	~1071 CE		
	England	Tower of London	1086 CE	30 m	
	France	Reims Cathedral	1211 CE		
	France	Chartres Cathedral	1219 CE	114 m	
	France	Rouen Cathedral	~1220–1876 CE	151 m	
	Germany	Cologne Cathedral	1248–1880 CE	157 m	Length = 142 m (completed 1880)

Table INV-7: (Cont. [b])

Builder	*Location*	*Project*	*Date*	*Height*	*Details*
	France	Amien Cathedral	1270 CE	112 m	
	France	Notre Dame Cathedral (Paris)	1330 CE	69 m	Area = 130 × 55 m^2
	Spain	The Alhambra (Granada)	1354 CE		
	Italy	Leaning Tower of Pisa	1372	55 m	
	Tibet	Chak-Sem Bridge (Brahmaputra)	1420		Length = 137 m
Filippo Brunelleschi	Italy	Santa Maria Del Fiore Cathedral and Duomo (Florence)	1471	113 m	
	Mexico City	Great Aztec Temple	1487		
	Spain	Cathedral of Seville	1519	56 m	
Inca (Manco II)	Peru	Machu Picchu (High Andes: 2430 m)	1537		
	England	Lincoln Cathedral	1548	160 m	
	Mexico	Nochistongo drainage tunnel	1609		Length = 6400 m
	Italy	St. Peter Basilica (Vatican City)	1626	123 m	Size = 210 × 137 m^2 (cross)
Shah Jahan	India	Taj-Mahal (Agra)	1654	41 m	
Christopher Wren	England	St. Paul Cathedral (London)	1661	149 m	
	France	Palace of Versailles	1678		Length = 800 m
	England	Telford Bridge: first cast-iron major structure	1775	30 m	
	Italy	Cathedral of Milan	1813	108 m	Area = 149 × 55 m^2
	USA	Erie Canal	1825		Length = 600 km
	England	Menai Straight Bridge	1826		Length = 177 m
	Sweden	Göta Ship Canal	1832		Length = 382 km
	Italy	Mole Antonelliana (Turin)	1863	167 m	
Ferdinand de Lesseps	Egypt	Suez Canal	1869		Total length = 191 km
	Switzerland	St. Goothard Tunnel	1880		Length = 15 km
	India	Darjeeling Himalayan Railway	1881		
	USA	Brooklyn Bridge (NY)	1883		

Table INV-7: (Cont. [c])

Builder	*Location*	*Project*	*Date*	*Height*	*Details*
	USA	Washington Monument (Washington, DC)	1884	169 m	
	USA	Statue of Liberty (NY Harbor)	1886	46 m	
Gustave Eiffel	France	Eiffel Tower (Paris)	1889	300 m	Base = 101×101 m^2
	USA	Cathedral of St. John the Divine (NY)	1892–2001		Area = 183×98 m^2
Antoni Gaudi	Spain	Casa Mila in Barcelona	1905		
Ferdinand de Lesseps	Italy–Switzerland	Simplon Tunnel	1906		Length = 20 km
					(completed 1922)
G.W. Goethals	Panama	Panama Canal	1914		Length = 81.6 km
	USA	George Washington Bridge (NY)	1931		Length = 1067 m
R. Shreve, T.A. Lamb, F. Harmon	USA	Empire State Building (NY)	1931	381 m	Area = 86×86 m^2
Paul Landowsky	Brazil	Statue of Cristo Redentor (Rio)	1931	30 m	
	Russia	White Sea–Baltic Canal	1933		Length = 227 km
	Italy	Apennine Tunnel	1934		Length = 18.5 km
	USA	Hoover Dam (Arizona–Nevada)	1936	223 m	Capacity = $35,154 \times 10^6$ m^3
	USA	Golden Gate Bridge (San-Francisco)	1937		Length = 1280 m
	USA	Overseas Highway (Florida–Keywest)	1938		Length = 11 km
	Belgium	Albert Ship Canal	1939		Length = 130 km
Gutzon Borglum	USA	Mount Rushmore Memorial	1942	150 m	(above valley)
	USA	Grand Coulee Hydroelectric Plant (Washington)	1942		Capacity = 6494 MW
	USA	The Pentagon Building (Washington, DC)	1943		Area = 344,243 m^2 $\simeq$ (587 m)^2
	USA	Alaskan Highway	1946		Length = 2288 km
	USA	Delaware Aqueduct Tunnel	1944		Length = 169 km

Table INV-7: (Cont. [d])

Builder	*Location*	*Project*	*Date*	*Height*	*Details*
E. Saarinen	USA	Gateway Arch (St. Louis)	1948	192 m	
	Holland	Amsterdam–Rhine Canal	1952		Length = 72 km
	Russia	Volga–Don Canal	1952		Length = 101 km
Frank Lloyd Wright	USA–Canada	St. Lawrence Seaway (Montreal–Duluth)	1959		Total length = 3840 km
	USA	The Guggenheim Museum (NY)	1959		
	Canada	Daniel Johnson Dam	1960	214 m	Capacity = $141,852 \times 10^6$ m^3
Jacque Piccard Donald Walsh	USA	US Navy Bathyscaphe '*Trieste*'	1960		Deepest ocean descent 10,916 m in the Mariana Trench
	Russia	Bratsk Hydroelectric Plant	1961		4500 MW
	USA	Seattle Space Needle	1962	185 m	
	Switzerland	Grand Dixence High Dam (Valais)	1962	284 m	
	USA–Chile	Pan-American Highway	1962		Total length = 47,516 km
	USA	KTHI TV mast (steel) (Fargo, ND)	1963	629 m	
	Russia	Volga–Baltic Canal	1964		Total length = 1100 km
	USA	Verrazano–Narrows Bridge	1964		Length = 1298 m
	Israel	Shalom Tower (Tel-Aviv)	1966	140 m	
	England	The "Queen Elizabeth 2" Passenger Ship	1967	294 m	Speed = 53 km/hr; Passengers and crew = 2600
	Russia	Krasnoyarsk Hydroelectric Plant (Yenisei)	1968		6000 MW
	Egypt	Aswan High Dam	1970		
	Canada	Mica Dam	1970	243 m	Capacity = $24,670 \times 10^6$ m^3
	USA	World Trade Center Twin Towers (NY)	1972	417 m	(destroyed by terrorists Sept. 11, 2001)
	USA	New Cornelia Tailings Dam	1973		Capacity = $209,500 \times 10^3$ m^3
Joern Utzon	Australia	Sydney Opera House	1973	67 m	Area = 185×120 m^2
	USA	Sears Tower (Chicago)	1974	442 m	
	Canada	CN Tower (Toronto)	1975	533 m	

Table INV-7: (Cont. [e])

Builder	*Location*	*Project*	*Date*	*Height*	*Details*
	Pakistan	Tarbela Dam	1976		Capacity $= 121,720 \times 10^3$ m^3
	Tajikistan	Nurek High Dam	1980	300 m	Capacity $= 10,500 \times 10^6$ m^3
	England	Humber Bridge	1981		Length = 1410 m
	Brazil–Paraguay	Itaipu Hydroelectric Plant (Parana River)	1983		14,000 MW
	Tajikistan	Rogun High Dam	1985	335 m	Capacity $= 11,600 \times 10^6$ m^3
	Japan	Seikan Tunnel	1988		Length = 54 km (23 km underwater)
	Russia	Sayano–Shushensk Hydroelectric Plant and Dam (Yenisei)	1989		6400 MW; Capacity $= 31,300 \times 10^6$ m^3
	England–France	Channel Tunnel	1994		Length = 50 km (38 km underwater)
	Hong Kong	Tsing Me Bridge	1997		Length = 1377 m
	Sweden	Höga Kusten Bridge	1997		Length = 1210 m
	China	Civic Plaza Building (Guangzhou)	1996	391 m	
	Japan	Akashi Kaikyo Bridge (Hyogo)	1998	283 m	Length = 1990 m
Cessar Pelli	Kuala-Lumpur	Petronas Twin Towers	1998	452 m	
	Denmark	Storebaelt Bridge	1998		Length = 1624 m
	China	Jiangyin Bridge (Yangtze River)	1999		Length = 1385 m
	Shanghai	Jin Mao Building	1999	421 m	
	Norway	Laerdal Tunnel	2000		Length = 24.5 km
	Turkey	Izmit Bay Bridge	2001		Length = 1668 m
	Denmark–Sweden	Oresband Bridge	2001		
	Canada	Syncrude Tailing Dam			Capacity $= 540,000 \times 10$ m^3
	Argentina	Chapeton Dam			Capacity $= 296,000 \times 10^3$ m^3
	Argentina	Pati Dam			Capacity $= 238,000 \times 10^3$ m^3
	Argentina–Chile	Tunnel through the Andes	2001		Length = 25 km

Footnotes in tables

4 Hans Dehmelt developed the *Penning trap*, with which electrons and ions – and (later) positrons, baryons and antibaryons, and even *anti-hydrogen* atoms – are confined by DC and AC electromagnetic fields to a small apace for long periods of time, during which the properties of the trapped particles can be studied. In 1973 Dehmelt succeeded in isolating a *single* electron in such a device. This feat enabled high-precision measurements of electron attributes such as the mass magnetic moment and the fine structure constant. The electron (and positron) gyromagnetic ratio was measured to an accuracy of a few parts in a trillion. Its comparison with QED theoretical predictions, within experimental and theoretical uncertainties, represents the *the best quantitative success yet of any scientific theory.* The Penning trap was also used to measure mass ratios and EM multipole moments of the positron, muons, protons and antiprotons and neutrons, and the neutron lifetime. It was used to measure atomic spectral frequencies and observe individual quantum jumps. A similar electromagnetic trap, the *Pauli trap*, was developed by Wolfgang Pauli in the 1950s.

5 In 1995 J. Prestage et al., at CalTech, by comparing precise Hg^+ (positive ion of mercury) ion-trap atomic clock with hydrogen-maser clock, derived an empirical upper bound of 0.07 parts per trillion annual variation of the dimensionless fine structure constant. Less stringent bounds were obtained from studying Samarium isotope ratios in a natural 2×10^9 (2 billion) year old fission reactor in Oklo mine, Gabon, and from spectroscopy of high-redshift cosmological objects.
In 2004, scientists at NIST (National Institute of Standards and Technology, formerly NBS) have unveiled a miniature cesium atomic clock. The heart of the clock is about the size of a grain of rice, capable of being mass-produced on semiconductor wafers using existing technology. The entire package, including external electronics, can be as small as a cubic centimeter. The device consumes less than 75 mW (and thus can be battery-powered). In this chip-scale clock, the difference between two infrared-laser frequencies is resonantly tuned to a cesium RF transition frequency, producing a "dark state" in which the cesium atoms cease absorbing and emitting light. This defines the atomic RF frequency standard, relative to which an external oscillator is then stabilized. This miniature atomic clock is not yet as accurate as standard atomic clocks, which are much larger (of order 1 meter) as well as expensive and power-guzzling. Chip-scale atomic clocks potentially offer a 1000 fold improvement in long-term timing precision as compared with quartz crystal oscillators.

Timelines of Contemporary Science and Modern Technologies

The Emergence of Worldwide Communications and Associated Technologies

Introduction

In 1844, four years after Samuel F.B. Morse proved that he could send coded messages along electrical wires, he managed to persuade the U.S. congress to appropriate $30,000 to build a telegraph line between Washington and Baltimore. It was on the opening of that earliest line that Morse sent his historic telegram[6] *– "What hath God wrought" (Numbers* ***23****, 23). With that Morse opened the age of telecommunications and triggered one of the most dramatic commercial developments of the* 19th *century. He started a powerful process that is still unfolding in our time.*

Because so much business now depends on getting and sending information, companies around the world have been rushing to link their employees through electronic networks. These networks – which today are digital and often have computers as nodes – form a key infrastructure of the 21th *century, as critical to business and national economic development as the railroads were in Morses' era.*

Some of these networks are "local area networks" (LAN), which merely hook up computers in a single building or complex. Other are globe-girdling nets that connect people the world over. For example, IBM alone connects 400,000 terminals around the world through a system called VNET, which in 1997 handled an estimated 20 trillion characters of data. By itself, a single part of that system, called PROFS, saved IBM the purchase of 10 million envelopes, and IBM estimated that without PROFS it would need nearly 50,000 employees to perform the same work. Networking has spread to the smallest businesses. With some 150 million PCs in use in the United States, many companies now advertise over the Internet.

Companies grow more dependent by the day on their electronic nets – for billing, ordering, tracking, and trading; for the exchange of design specification, engineering drawings, and schedules.

Regarded as purely administrative tools, networked information systems are increasingly seen as strategic assets, helping companies protect established markets and attack new ones.

The race to build these networks has taken on some of the urgency that accompanied the age of railroad construction in the 19th *century, when nations*

[6] The first telegraph message, sent on May 24, 1844 from the Supreme Court Room of the U.S. Capitol to the Mount Claire Station of the Baltimore and Ohio Railroad, Baltimore, Maryland.

became aware that their fates might be tied to the extensiveness of their rail systems.

The networks of Morse (telegraph), Bell (telephone) and others were unintelligent. Common sense taught that a straight line is the shortest distance between two points. So engineers sought this straight line, and messages sent from one city to another were always sent over this pathway. As this first-stage network *expanded, it was discovered that in the world of networks this is not necessarily the best way to get a message from one place to the other.*

Thus, when a network began to monitor its own performance *it could inject "intelligence" into the system and become, as it were,* "self-aware". *Crisscrossing the entire planet with wires running into hundreds of millions of homes, and incorporating complex switching systems and transmission, these* second-stage networks, *constantly improved, and endowed with ever more intelligence, were among the true marvels of the industrial age.*

Nowadays, as millions of computers, from giant Crays to tiny laptops, are linked to form a dense interconnected mesh, a still higher level of intelligence or "self-awareness" is needed to process the incredibly vast volumes of information pulsing through them. As a result, researchers are racing to make network even more intelligent. Their goal is the third-stage neural networks. *These will not only route and reroute massages, but actually* learn from their own past experience, *forecast where and when heavy loads will be, and then automatically expand or contract sections of the networks to match the requirements.*

Yet even before this major effort comes to fruition, another, even more gigantic leap is being taken. We are moving not into a fourth–stage system but to another kind of intelligence altogether.

Until now, even the smartest networks, including the hoped-for neural networks, had only what might be called "intra-intelligence" *– with all its smartness* aimed inward. *It is akin to the intelligence embedded in our own autonomic nervous system, which regulates the involuntary operations of the body, such as heartbeat and hormonal secretion.*

One could, however, reach beyond intra-intelligence toward networks that do not just transfer data, but analyze, combine, repackage, or otherwise alter messages, sometimes creating new information along the way.

Thus enhanced, what comes out the other end in such a network would be different from what is fed in – changed by software embedded in the network. These are termed "Value Added Networks" *(VAN); they are* "extra-intelligent".

Combined with the third-stage neural-networks capacity, the advent of VANs would give communication networks not only self-awareness and the ability to change themselves, but also the ability to intervene directly in our lives, beginning first with our businesses. If this comes to pass, networks will take on revolutionary new roles in business and society.

I. The electromagnetic telegraph (1800–1902)

During the 16th, 17th and 18th centuries, steady but slow progress in the study of magnetic and electrical phenomena was recorded. Already **William Gilbert** *(1600) suggested a link between electricity and magnetism. However what progress occurred between 1600 and 1800 is nevertheless negligible when compared with the strides made between 1800 and 1840.*

The rapid advance which then occurred was not precipitated by any manufacturing process in which electricity or magnetism played a direct role. It was chiefly fostered by a rapidly growing interest in theoretical chemistry during the rise of chemical manufacture and thus benefited form the outlook of new industrial leaders eager to exploit new discoveries.

For example, the chemist **Stephan Gray**, *while experimenting with insulators and conductors (1729), transmitted static electrical charges (generated by an electrostatic friction-type generator) along a brass wire 100 meter long.*

On the other hand, experiments with frogs (1780) led the biologist-physician **Luigi Galvani** *to observations which quickly culminated with the discovery of the cell or battery, which later became the electrical energy source powering the telegraph.*

1800 **Alessandro Volta** *(Italy) invented the voltaic pile (primitive battery).*

1816 **Joseph Henry** *(USA) proposed a single-wire telegraph.* **Francis Ronalds** *(1788–1873, England) demonstrated such a device but the verdict of the British Admiralty was (1832): "Telegraphs of any kind are wholly unnecessary".*

1820–1821 *Hans C. Oersted (Denmark),* **Ampère** *(France) and* **Michael Faraday** *(England) created the science of electromagnetism.*

[Oersted discovered that an electric current creates a magnetic field; Faraday discovered that a changing magnetic field creates an electric current; and Ampère formulated some of the basic laws of electrodynamics.]

1825 **William Sturgeon** *(England) built the first electromagnet for practical technological use.*

1830–1833 **Gauss** *and* **Weber** *(Germany) developed a small scale working telegraph system in Göttingen (3-km line). In the U.S.,* **Joseph Henry** *experimented with a closed circuit in which a source of current activates an electromagnet which in turn activates a bell at a distance.*

1836–1837 **John Daniell** *and* **Charles Wheatstone** *(England) improved the voltaic cell, creating a stable current source.*

1837–1844 **Samuel Morse** *(USA) invented the practical telegraph. He set up a 60 km telegraph line between Washington DC, and Baltimore (1844).*

1838 **Carl August von Steinheil** *(1801–1870; Germany) discovered the possibility of using the* earth *for a return conductor in telegraphy (grounding).*

1851 *First telegraph cable laid across the English Channel.*

1855 **David Edward Hughes** *(England and USA) invented a keyboard telegraph with rotating type-wheel printer that grew into the modern* telex *industry.*

1865 *A telegraph system established between India and England. It took on average 6 days to telegraph a message overland between the two countries.*

1866 *First successful transatlantic telegraph cable laid. Promoted by the financier* **Cyrus West Field** *(1819–1892, USA) and* **Lord Kelvin** *(1824–1907, England).*

1884 **Emile Baudot** *(1845–1903, France),* **J.B. Stearns** *(USA) and* **Michael Pupin** *(USA), independently invented a* multiplex system *– a time sharing device to increase transmission speeds along the telegraph cable.*

1902 *First pacific telegraph cable between Canada and New-Zealand.*

With the advent of railroads, the telegraph became one of the causes and one of the consequences of the industrial revolution. It changed the face of international commerce, opening the door to worldwide communications and accelerating the flow of information.

II. The Telephone – adventure of mass communication (1854–1991)

Mark Twain had a weakness for new inventions; he was fascinated by them, and over the years he lost more than half a million dollars investing in various contraptions.
Once, after a series of bad investments had temporarily tempered his enthusiasm for technology, he was approached by a tall young man with a mysterious device under his arm.
Mark Twain listened politely to what the young man had to say, but explained that he had been burnt once too often and was not interested.

"But I'm not asking you to invest a fortune", said the young man.
"You can have as large a share as you want to for $500."

The author shook his head and the tall, stooped figure started away. Mark Twain, saddened by the sight of his pathetic young man, called after him. "What did you say your name was again?" "Bell", was the reply. "Alexander Graham Bell."

Telephone changed almost everything about business. It permitted operations over a greater geographical area. Top executives could now speak directly with branch managers or salesmen at distant regional offices to find out, in details, what was going on. Voice communication conveyed far more information, through intonations, inflection, an accent, than the emotionless dash-dots of Morse Code ever could.

The telephone made big companies bigger. It made centralized bureaucracies more efficient. Phones helped integrate the industrialized economy.

Capital markets became more fluid; commerce easier. Deals could be struck swiftly, with a confirming letter as a follow-up. Phones accelerated the pace of business activity, which, in turns, stepped up the rate of economic development in the more technically advances nations.

People did not lie awake through the centuries dreaming of making a call. Telephone history did not proceed in a straight line; it was a series of events – mostly technological, some accidental – that made the telephone possible.

As the electrical telegraph was making its way around the world, especially with the laying of the first transatlantic cable in 1866, a new technology was being developed. It was first called the "talking telegraph". Many scientists have conceived the telephone:

1854 **Charles Bourseul** *(France), a Belgian telegraph agent, discovered that the vibration of the human voice could be transmitted: a flexible plate would vibrate in response to a varying pressure of the air and used to open or close an electric circuit; a similar plate at the receiving station would be acted on electromagnetically, and thus produce as many pulsation as there are breaks in the current. His idea met with skepticism, and he never built a telephone.*

1855–1861 **Philip Reiss** *(Germany) Constructed a talking machine based on the ideas of Bourseul and called it* das Telephon. *He could not, however, reproduce the human voice with sufficient clarity. It was the first non-talking telephone.*

1876–1877 **Alexander Graham Bell** *(USA) and* **Elisha Gray** *(USA) independently invented the telephone. The system was based on the principle of the electromagnetic induction [Faraday 1831]: human speech caused a membrane to vibrate, which modulated the magnetic flux threading through an electric circuit and generated by a magnet situated in front of the membrane; the modulated magnetic flux in turn, generated alternating electric currents. At the receiving end, the sound was reproduced by a reverse process. Bell accidentally discovered the principle of the telephone when trying to improve telegraphy.*

Bell's telephone represented a powerful threat to the vested interest of the telegraph companies. This led Western Union *to call in Edison to develop an alternative instrument. Since Edison's microphone turned out to be a more efficient transmitter of sound while Bell had the more efficient receiver, the situation was resolved when* Western

Union assigned their rights on Edison's device to the Bell Company.

1877–1878 **David Edward Hughes** *(England and USA) invented the carbon microphone [the word microphone was coined by Charles Wheatstone in 1827].* **Edison** *(USA) and* **Emile Berliner** *(Germany and USA) hit upon the same idea at about the same time. All three arrived at the invention by way of their efforts to improve the telephone transmitter. Indeed, the device played a critical role in the development of the telephone by increasing its transmission capacity.*

1878 *The first commercial telephone exchange opened in New Haven, CT.*

1880 *The first UK national phone directory.*

1891–1892 **Almon B. Stowager** *(USA) patented the automatic exchange dial system, the first electromechanical switching system, which made possible the extension of the telephone network. By 1892, the Bell Telephone Company had 240,000 subscribers in the US.*

1893 **Oliver Heaviside** *(England) solved the problem of the electrical transmission line, enabling engineers to construct transoceanic cables and realize long-distance telephony.*

1894 **Michael Pupin** *(USA) invented the 'Pupin coil': he devised a means of greatly extending the range of distant telephone communication by placing loading coils at predetermined intervals along the transmission wire. It made long-distance telephony practical by amplifying the signal at intervals along the line without distortion.*

1903 *3,278,000 telephones in the United Sates. By the 20th century, the telephone had become the symbol of modern society. It was ubiquitous in most businesses and had impacted both urban and rural life.*

1904–1912 *The advent of vacuum tubes: rectifiers (***Fleming***; 1904), amplifiers (***Lee de Forest***, 1906) and feedback oscillators (***Edwin H. Armstrong***, 1912).*

Tubes improved telephone and radio communications and led to US national phone services.

1914 *9.7 telephone per 100 people in US.*

1915 *First transcontinental telephone line opened between New York and San Francisco.*

1916–1931 *The first teleprinter (which made it possible to send messages over the telephone line) was invented in the US. A teleprinter system becomes operational in 1928 and was extended nationally by Bell Labs (1931) under the name telex (= teleprinter exchange).*

1924–1941 *Pulse code modulation (PCM), invented and completely worked out by* **Paul M. Ramey**. *It is the basis of digital audio and used in voice transmission and reproduction. It incorporates the three stages of Sampling, quantization and coding. But this groundbreaking work was then apparently forgotten. The idea was reinvented in 1939 by* **A.H. Reeves**, *forgotten again and finally resurrected during WWII by Bell Labs during research into methods of encoding phone conversation.*

1929 **H.A. Affel** *and* **L. Espenschied** *of the AT&T/Bell Laboratories (US) patented a coaxial telephony system: the wide bandwidth enabled large number of telephone channels to be assembled in frequency division multiplexing – important for meeting the ever increasing telephone traffic. As an added bonus, the outer conductor of the cable screened the signals from interference of radio stations and powerlines.*

1947 *First microwave relay station for long-distance telephone communication was adopted between Boston and New York. It eliminated the need for expensive trunk lines.*

1948 **Claude Shannon** *created Information Theory, containing the basis for data compression (source encoding), error detection and correction (channel encoding), and estimation of data channel capacity.*

1950 **Richard Hamming**'*s work on error detection and correction codes.*

1950 *First terrestrial microwave telecommunication system, installed to support 2400 telephone circuits.*

1956 *First transatlantic telephone cable laid. It consisted of 7242 km of coaxial cable, laid in waters up to 4 km deep. The complete system provided 35 high-quality telephone circuits from London to New York (29) and Montreal (6).*

1963 *First push-button telephones were introduced.*

1965 *First commercial electronic telephone exchange brought into service by AT&T. It embodied for the first time the principle of stored program control, which gave greatly increased flexibility in the services provided by the exchange and better maintenance.*

1966 *First successful transatlantic direct-dial phone calls are made.*

1971 *Regular direct-dial phone call began from New York to Paris and London.*

1976 *High capacity transatlantic cable went into service; it carried 4000 conversations simultaneously.*

1975–1985 *First cellular telephone system launched in Sweden (1975) by the Ericsson Company. Within five years, some 200,000 Scandinavians were using these mobile phones. A cellular (cell) phone is essentially a portable analog or digital radio transmitter and receiver; cell phones are linked via microwave radio to base transmitter and receiver stations that connect the user to conventional telephone networks. They operate in the 829–949 MHz frequency band (wavelength at 900 MHz is 33 cm) or the 1850–1990 MHz band. The geographic region served by a cellular system is subdivided into areas called cells, each cell using frequencies different from those used by its surrounding cells. When the phone carrier moves from one cell to another, the telephone call is transferred from one base station (and its frequency) to the next using a computerized switching system. Cellular phone networks in Japan (1980) and the US (1983). Cellular phone went into cars (1985).*

1983 *565 million telephones in the world.*

1985 *A single optical fiber carries the equivalent of 300,000 simultaneous phone calls in a Bell Labs test.*

1987 *AT&T completed digitization of all its long-distance facilities. Voice transmission signals are converted into compressed digital computer codes for more efficient transmission trough lines, then 'decompressed' at the receiving end.*

1991 *The total annual volume of international telephone traffic is estimated at 35 billion telephone minutes, globally averaging about 6 telephone-minutes per person per year.*

III. Wireless communication and the birth of Radio (1864–1961)

Maxwell's research into electrodynamics began a few weeks after his graduation from Cambridge (1854), and ended just before his death (1879). By 1864 he had established both a dynamical theory of the electromagnetic field and the electromagnetic theory of light, accounting for the phenomenon of electromagnetic waves, propagating in vacuo or in a medium at the velocity of light. With this discovery he opened the way for a future technological civilization in which electromagnetic effects play cardinal roles – including the propagation of electromagnetic waves which enables radio, cell phones, television, radar, and optical fiber communications. The first experimental confirmation of Maxwell's theory was rendered by **Heinrich Hertz** *(1887) who produced and detected the first man-made radio wave signals*[7].

[7] Certain amateurs allegedly transmitted wireless signal before Hertz or even transmitted voice signals before Marconi.
Their results, however, are not true Hertzian radiation effects but rather *induction–field* phenomena. One such person was **Malcolm Loomis** (1826–1886, USA), a dentist by trade; At the close of the Civil War in 1865, he flew two kites, carrying wires, from mountain tops 23 km apart. The wire from one kite was attached to ground through a telegraph key; the other kite wire was grounded trough a galvanometer that could measure very small currents. When he operated

1874–1906 *Evolution of wave detectors:* ***Karl F. Braun*** *(1874, Germany) discovered 'one way conduction' properties in metal sulfide crystals. [In 1901 he introduced the use of a crystal detector as part a wireless receiver.]*

History of the Crystal Radio

The crystal radio receiver is a very simple kind of radio receiver. It needs no battery or power source except the power received from radio waves by a long outdoor wire antenna.

Simple crystal radios are often made with a few hand made parts, like an antenna wire, tuning coil of copper wire, crystal detector and earphones. Because crystal radios are passive radio receivers, they are technically distinct in many respects from ordinary radios containing active powered amplifiers. This is because they must receive and preserve as much electrical power as possible from the antenna and convert it to sound power whereas ordinary radios amplify the weak electrical energy "signal" from the radio wave.

A crystal radio receives programs broadcast from radio stations. Radio stations convert sound into radio waves and send out the waves everywhere. Radio waves travel across the crystal radio antenna all the time. Radio waves make radio wave electricity flow between the antenna wire and the ground wire. This electricity is connected to the crystal radio by the antenna and ground wire. The crystal radio uses a tuner to tune the electricity to receive just one station. Then it uses a crystal detector to convert this radio

---

the key, detectable changes of current occurred in the other kite wire. He was granted a patent on his system in 1872, but no known attempt was made to make commercial use of this phenomenon.

It is believed that Loomis merely interrupted current in the antenna resulting from flying an antenna into a cloud, transmitting information between two points by conduction or induction but not via far-field electromagnetic waves.

Another, equally obscure experimenter lurking in the shadows of the early history of radio was **Nathan Stubblefield** (1858–1928, USA) a mendicant Kentucky melon farmer and telephone repairman who could allegedly send messages on a wireless telephone over a distance 800 m (1892–1902). It is assumed that he, too, was relying on the electromagnetic induction field *wave electricity back to sound electricity. It uses earphones to convert the sound electricity to sound you can hear.*

Crystal radio was invented by a long, partly obscure chain of discoveries in the late 1800s that gradually evolved into more and more practical radio receivers in the early 1900s; and constitutes the origin of the field of electronics. The earliest practical use of crystal radio was to receive dot and dash coded radio signals transmitted by early amateur radio experimenters using very powerful spark-gap transmitters. As electronics evolved, the ability to send voice signals by radio caused a technological explosion in the years around 1920 that evolved into today's radio broadcasting industry.

Early radio telegraphy used spark gap and arc transmitters as well as high-frequency alternators running at radio frequencies. At first a primitive detector called a Branley Coherer was used to indicate the presence (or absence) of a radio signal. However, these lacked the sensitivity to convert weak signals.

Around 1906, researchers discovered that certain metallic minerals, such as galena, could be used to detect signals. These devices were called "crystal detectors". ***Greenleaf Whittier Pickard*** *on August 30, 1906 filed a patent for a silicon crystal detector, which was granted on November 20, 1906. Pickard's detector was revolutionary in that he found that a fine pointed wire known as a "cat's whisker", in delicate contact with a mineral produced the best semiconductor effect. A crystal detector includes a crystal, a special thin wire that contacts the crystal and the stand that holds the components in place. The most common crystal used is a small piece of galena. Several other minerals also performed well as detectors. Another benefit of crystals was that they could demodulate amplitude modulated signals. This mode was used in radiotelephones and to broadcast voice and music for a public audience. Crystal sets represented an inexpensive and technologically simple method of receiving these signals at a time when the embryonic radio broadcasting industry was beginning to grow.*

In 1922 the (then named) U.S. Bureau of Standards released a publication entitled, Construction and Operation of a Simple Homemade Radio Receiving Outfit. This article showed how almost any family having a member handy with simple tools could make a radio and tune in to weather, crop prices, time, news and the opera. More than any other system, the design contain therein, was responsible for bringing radio to the general public.

While there were a number of earlier experiments with radio broadcasts to the general public, some historians consider the Autumn of 1920 to be the beginning of radio broadcasting for entertainment purposes. Pittsburgh, PA, station KDKA, owned by Westinghouse, received its license from the United States Department of Commerce just in time to broadcast the Harding-Cox presidential election returns. In addition to reporting on special events, broadcasts to farmers of crop price reports were an important public service, in the early days of radio.

In 1921, factory-made radios were very expensive. Many of them cost more than $2,000 USD (in year 2005 equivalent dollars), and less affluent families could not afford to have one. Newspapers and magazines in many countries urged readers interested in radio to acquire one of the inexpensive crystal sets or build their own. To minimize the cost, many of the plans suggested winding the tuning coil on an empty cylindrical oatmeal box. For years afterwards, home experiments used oatmeal boxes as coil forms for homemade radios. Even the crystal itself could be made by mixing powdered sulfur into molten lead to form the lead sulfide "crystal". The crystal radio did not require batteries, but it did require the user to purchase a commercially made set of headphones (or telephone receivers as they were called in those days), since that accessory was not suitable for home construction.

"Carbon amplifier" consisting of a carbon microphone and an electromagnetic earpiece sharing a common membrane and case. This was used in the telephone industry and in hearing aids nearly since the invention of both components and long before vacuum tubes. This could be readily bought or handcrafted from surplus telephone parts for use with a crystal radio. Unlike vacuum tubes, it could run with only a flashlight or car battery and had an almost infinite lifetime.

In the early 1920s Russia, devastated by civil war, young scientist **Oleg Losev** *was experimenting with applying voltage biases to various kinds of crystals, with purpose to refine the reception. The result was astonishing – with* *a zincyte (zinc oxide) crystal he gained amplification. This was negative resistance phenomenon, decades before the tunnel diode. After the first experiments, he built regenerative and superheterodyne receivers, and even transmitters. However, this discovery was not supported by authorities and soon forgotten and no device was produced in mass quantity beyond a few examples for research. This was partly due to the low education and overall ignorance of leadership, and partly due to the totalitarian nature of the USSR regime.*

The USSR opposed freedom of information, and registered all radio receivers until 1962, typewriters and copy machines until its demise. Crystadine was produced in primitive conditions; it can be made in a rural forge – unlike vacuum tubes and modern semiconductor devices. It was an unwanted discovery to the authorities, and was consigned to obscurity. Oleg Losev died 1943 in besieged Leningrad, abandoned and nearly forgotten.

When Allied troops were halted near Anzio, Italy, during the spring of 1944, personal portable radios were strictly prohibited, as the Germans had radio detecting equipment that could detect the local oscillator signal of superheterodyne receivers. Some resourceful GIs found that a crude crystal set could be made from a coil made of salvaged wire, a rusty razor blade and a pencil lead for a diode. By lightly touching the pencil lead to spots of blue on the blade, or to spots of rust, they formed what is called a point contact diode and the rectified signal could be heard on headphones or crystal ear pieces. The idea spread across the beachhead, to other parts of the war, and to popular civilian culture. The sets were dubbed "foxhole receivers" by the popular press, and they became part of the folklore of World War II.

In some Nazi occupied countries there were widespread confiscations of radio sets from the civilian population. This led to particularly determined listeners building their own "clandestine receivers" which frequently amounted to little more than a basic crystal set. However anyone doing so risked imprisonment or even death if caught and in most parts of Europe the signals from the BBC (or other allied stations) were not strong enough to be received on such a set. However there were places such as the Channel Islands where it was possible.

While it never regained the popularity and general use that it enjoyed at its beginnings, the circuit is still used. The Boy Scouts (who emerged as the unofficial custodians of crystal radio lore) kept construction of a set in their program since the 1920s. A large number of prefabricated novelty items and simple kits could be found through the '50s and '60s, and many children with an interest in electronics built one.

Building crystal radios was a craze in the 1920s, and again in the 1950s. Recently, hobbyists have started designing and building sophisticated examples of the instruments. As much effort goes into the visual appearance of these sets as well as their performance, and some outstanding examples can be found. Annual crystal radio DX contests and building contests allow these sets to compete with each other and help form a community of interest in the subject.

The long wire type antennas often used with crystal radios are monopoles. To receive signals from this type of antenna, a ground reference is needed to provide a place for the antenna signal electricity to flow into and out of. Because crystal radios have no other source of power than the electrical power they receive from the antenna, the grounds for crystal radios must be much better than those used by amplified radios. Amplified radios use energy detectors and as such do not need to take much raw power from the antenna and need little or no physical ground. Crystal radios rely on power detection and need to encourage as much antenna current as possible to flow. This requires effective grounding.

A crystal set *is the simplest radio receiver. There are a variety of circuit designs available. A common design consists of a long-wire antenna, a variable inductor and a variable capacitor forming a tuner or tank circuit to select the desired radio signal frequency, and a detector consisting of a diode demodulator usually consisting of a sharp wire called a cat's whisker pressing against a sensitive point on a mineral crystal in a holder.*

A semiconducting mineral crystal, typically lead sulfide (galena) is fixed inside a brass cup and the radio operator finds the loudest signal by touching the cat's whisker to various points on the surface of the crystal. Alternately, a discrete semiconductor diode can replace a makeshift cat's whisker diode. The most expensive part can be the length of antenna wire.

The detector extracts the amplitude modulation from the radio signal by rectifying it, and provides an audio output in proportion to the strength of the signal coming from the antenna. The entire set is passive, requiring no external power. Because no electrical amplification is used, sensitive earphones are required. These sets have no way to control the audio volume.

Temistocle Calzechi Onesti *(1853–1922, Italy) observed that the electric resistance of a container packed with metal granules is decreased upon the passage of electromagnetic waves. This so called 'coherer' acts like a kind of a macroscopic semiconductor; the device was later developed into a detector of radio waves by* **Edouard Eugene Desire Branly** *(1844–1940, France) in 1890 and by* **Oliver Joseph Lodge** *(1851–1940, England) in 1894. [Lodge was also the first to suggest in 1894 that the sun might be a source o radio waves; this was confirmed in 1942.]*

Greenleaf Whittier Pickard *(1877–1956, USA) discovered (1899) that a contact between a fine metallic wire ("cat whisker") and the surface of a certain crystalline material (notably silicon) rectifies and demodulates high-frequency radio waves. He patented such a device in 1906 and it became an essential component of the crystal radio set. The point-contact rectifier was the forerunner of the transistor (1948).* **E.G Acheson** *first produced carborundum (silicon carbide) in 1891. In 1906* **N.C. Dunwoody** *observed that carborundum could be used as a radio detector.*

1897 *Improvement of the Hertz oscillator by* **Augusto Righi** *(1850–1920, Italy) and* **Adolf Slaby** *(1849–1913, Germany). Righi designed the ball discharger (spark gap) which could generate stable electromagnetic waves with wavelengths as short as 2.5 cm (12,000 MHz). In Germany Slaby transmitted radio signals to a distance of 21 km using as a transmitter a spark coil connected to an antenna wire. At the receiving end the signal was picked by another antenna, passing through a coherer which in turn activated a bell. Thus, Morse-key pressed at the transmitting end was* *'heard' at the receiver. Slaby was assisted by* **George Wilhelm von Arco** *(1869–1940). The first antenna was used by* **A.S. Popov** *who devised it but never used it himself for radio communications.*

1901 *Marconi developed the first practical wireless telegraph system. He was the last in the long line of contributors during 1884–1901. He combined the* **Ruhmkorff** *induction-coil, the 3-spark* **Augusto Righi** *oscillator, the* **Onseti-Branly-Lodge** *coherer and the* **Popov** *antenna into a workable system that could transmit coded Morse signals over great distances. Finally in Dec 1901 he succeeded in broadcasting a signal over the Atlantic from Poldhu (England) to Glace-Bay Newfoundland, over a distance of some 3400 km.*

Human speech was first transmitted via radio waves by **R.A. Fessenden** *(US).*

1902 **Wilhelm Schlömilch** *(1870–1969, Germany) invented an electrolytic detector.*

Discovery of a radio-wave reflection layer in the upper atmosphere.

First radio chess-match: passengers on the American liner Philadelphia and the Cunard liner Campania, 70 miles away in the Atlantic, played the first match by radio, transmitting their moves via wireless operators aboard the ships.

Valdemar Poulsen *(Denmark) transmitted human voice via radio waves over a distance of 200 m.*

1904–1911 *Advent of the 'crystal set' receiver based on the crystal detector of* **Karl F. Braun** *(1901) and* **Pickard** *(1906). Amateurs could from now on build their own wireless receivers and hear early radio broadcasts. It was popular until the crystal detector was superseded by the vacuum-tube radio.*

1904–1914 *High frequency alternators, vacuum tubes, amplitude modulation, feedback and heterodyne systems ushered in modern radio and improved reception:* **J.A. Fleming** *(England) and* **Arthur Wehnelt**[8] *(1871–1944, Germany) introduced the thermionic rectifier valve (1904).*

---

[8] **Wehnelt** invented the oxide cathode; it consists of a metal wire or sheet which is coated with a mixture of metal oxides and is heated to incandescence directly or indirectly. Its thermal electron emission is by many orders of magnitude larger

***R.A. Fessenden** (Canada and US) made the first AM radio broadcast (Dec 24, 1906): sound waves of speech modulate the amplitude of a transmitted radio-frequency carrier wave, creating a* band *of transmitted radio frequencies. The modulated wave is then demodulated by the receiver to recover the original sound wave*[9]*.*

***Lee de Forest** (1906, USA) and **Robert von Leiben** (1878–1913, Germany) invented the triode amplifier tube. De Forest (1907) began regular radio music broadcasts.*

***Edwin H. Armstrong** designed (1912, USA) the feedback oscillator vacuum-tube. It was used (1913) by **Alexander Meissner** (1883–1958, Germany) to generate a radio-frequency signal carrying a spoken conversation between Berlin and Nauen.*

***R.A. Fessenden** had suggested (1900) that an* alternator *[a device that converts direct current into alternating current capable of producing continuous radio-frequency waves] could generate electromagnetic waves capable of carrying sound and music. He uses a spark generator to send a human voice to a distance of about 1600 m. In his 1906 broadcast he used the alternator invented by **E.F.W. Alexanderson** (1906)*

1910 *First successful radio communication from airplane to ground station.*

1911 *Method for locating a radio source by direction finding was developed.*

1912 *The 63 kW spark-transmitter distress call of the* Titanic *is received by the liner* Carpathia *93km away.*

1913 *The first trans-Atlantic two-way radio-telegraph service is lunched by Marconi between Nova Scotia and Ireland.*

---

than that of pure metals of the same temperature. We know today that this empirical fact agrees well with *semiconductor physics*, But this was unknown to Wehnelt at the time.

9 Already in 1899, **G.W. Pickard** had transmitted spoken message from the Blue Hills observatory (Milton, MA) over a distance of 15 km using this "cat whisker" receiver to receive audible signals that had been impressed upon (modulated) a radio-frequency wave.

William David Coolidge *(US) invented a hot-cathode X-ray tube and introduced the tungsten filament into the incandescent lamp.*

Irving Langmuir *(US) made the first gas-filled lamp at atmospheric pressure.*

1915 *Physicist* ***Manson Benedicks*** *(USA) discovered that a germanium crystal can rectify ac current.*

First major demonstration of long-distance voice communication (radio telephony) from Arlington, VA to Paris.

1918 *Development of the superheterodyne radio receiver.*

1919 *Development of the short-wave radio.*

1920 *First commercial radio station (KDKA) in Pittsburgh, PA began broadcasting.*

1921–1925 *Experiment revealing advantage of short-wave radio transmission for long–distance voice communication.*

1922–1926 ***Oleg Vladimirovich Losev*** *(1903–1942, Russia) developed novel kinds of crystal radio sets (with new crystals he fabricated himself) and was the first to study the effects of bias voltage upon the functioning of crystal diodes in circuits, essentially discovering "negative resistance" before the tunneling diode, as well as a pre-ATT version of the transistor and associated amplifiers. He then constructed completely solid-state radios that function up to 5 MHz, a quarter of a century before the transistor.*

1922 ***Edwin H. Armstrong*** *built the first portable radio, known as 'Operadio'.*

U.S. President Harding had a radio installed in the White House. The Ford T-Model car was equipped with a radio.

1922–1924 ***E.A. Appleton*** *(1892–1965, England) confirmed the existence of the 'ionosphere', a region of partially ionized air surrounding the earth at a height of 130–320 km. This acts like a great mirror, reflecting radio waves back to the earth, which also acts as a reflector. In consequence, medium and short-wave radio signals bounce back and forth between sky and earth in a two dimensional "channel" thousands of kilometers in lateral extent, enabling long-distance radio telecommunication.*

AM band was assigned; it spans 550–1550 kHz. More than 1000 radio stations operating in the US

1925 *Advent of short–wave radio broadcasts in the US.*

1927 *The Pentode, a vacuum tube with 5 electrodes, was introduced by* **H.S. Black** *of Bell Laboratories (US) who conceived the idea of 'negative feedback'. It is later found to be one of the most significant inventions in electronics and communications. It took 10 years for the patent to be approved.*

First commercial radio-telephone service operated between the US and Britain.

1928 *Crystal radio sets were being gradually replaced by home radio sets with vacuum tubes, loudspeakers and connection to main. A combined radio-gramophone appeared on the market.*

1929 **Edwin H. Armstrong**, *in collaboration with* **Michael Pupin**, *invented the Frequency Modulation (FM) method of radio broadcasting. According to this method, the transmitted signal is made to modulate the frequency of the carrier wave. This means that FM is static-free and capable of high-frequently sound reproduction. Despite its advantages, FM did not get off the ground until after WWII.*

1932 *Automatic Volume Control (AVC) was introduced.*

Abraham Esau *(1844–1955) increased the radio band to include ultra short waves.*

Karl Jansky *invented the radio telescope.*

1937 **Grote Reber** *invented the parabolic disc antenna*

1944 *57 million radio sets in the US.*

1945 *Arthur C. Clarke suggested using satellites to relay radio broadcasts.*

1952 *U.S. President Harry S Truman created National Security Agency (NSA), a part of the Department of Defense. It is an organization that monitors the telephone, radio and other communications of both friends and adversaries of the United States. Surreptitiously, it reads the world's mail.*

1952 *The first appearance of the Sony pocket sized radio transistor.*

1958 *First monolithic integrated circuit ('microchip') was demonstrated by* **Jack St.Clair Kilby** *and co-inventor* **Robert N. Noyce** *(1927–1990, USA). It revolutionized the design and manufacture of electronic components in radio and television.*

The first integrated circuit consisted of a slice of Germanium on which were formed a transistor, a capacitor and 3 resistors, constituting a simple phase-shift oscillator. The components where linked by fine gold wires and the connections embedded as a part of a manufacturing process. The integrated circuit put entire systems of tiny transistor switches, capacities, resistors, diodes and other electronic devices on one tiny microchip. Made chiefly of silicon, aluminum and oxygen – the three most common substances in the earth's crust – the microchip eventually reduced the price of electronic circuits by a factor of order one million.

1961 *FM stereo broadcasting was authorized in the US. Stereo radio systems became available.*

1962 *The first commercial satellite goes into orbit.*

1963 **James B. Gunn** *discovered the Gunn effect: a nonlinear deviation from Ohm's law in gallium arsenide. The presence of a negative differential conductivity region on the current-voltage characteristics of the GaAs crystal makes it possible to devise ultra–high frequency oscillator, knows as Gunn diodes. In 1966 a first commercial UHF generator, working at a frequency of 2–3 GHz with power output of approximately 100 W in pulsed operation, was produced. It was used (1968) in radar technology to measure the speeds of moving objects. These radars were small enough to be carried by hand.*

1985 *Sony built a radio the size of a credit card.*

1987 **Leo Esaki** *invented the tunneling diode, based on a quantum mechanical effect whereby electrons can travel through a region of electrostatic potential that they would be unable to penetrate classically. Consequently electrons are able to 'tunnel' from one region of a semiconductor to another (by their Schrödinger waves passing through a classical barrier),* *causing resistance to* decrease *rather then* increase *with increasing current. This results in a negative differential resistance which permits oscillations. The tunnel diode is thus used as a very high frequency oscillator at low voltage and power (ca. 100 MHz).*

IV Facsimile (1842–1980)

Smoke and drum signals are believed to have been the earliest form of rapid, long-range communications by humans.

We owe the development of the fax to a Scottish inventor, **Alexander Bain** *(1842). Even now, after the advent of electronics and computerized digital communications, Bain's original concept is still the basis of modern facsimile machines.*

Facsimile (fax) is a method of encoding data, transmitting it over telephone lines or radio broadcast channels, and receiving hardcopy text, line drawings, or even photographs.

A modern, digital fax machine scans an image, whether it be text, pictures or mixed, by reading a very small area of the image at a time. The fax machine decides whether the area it is reading is light or dark and assigns the area a number such as "0" for white and "1" for dark. Then the fax transmits the number to a remote facsimile receiver (usually via telephone lines). The receiver makes a mark on paper corresponding to the area on the original image.

This process continues as the transmitting machine scans a series of small areas horizontally across the image, and transmits that information to the remote receiver. The transmitting fax then scans the next lower line and so on until the entire image has been scanned, digitized, and transmitted. (In today's computerized fax machines, the scanned texts are often stored and queued for later transmission.) Facsimile telegraph is one of the oldest telegraph techniques.

Bain's invention was improved by six generations of scientists and engineers, the most important of which are:

1847 **Frederick Collier Bakewell** *(England). Physicist. The (analog) images were transmitted and received on cylinders that rotated at a uniform rate by means of a clock mechanism. A demonstration took place in 1851 at the World's Fair in London.*

1862 ***Giovanni Caselli*** *(1815–1891, Italy) Physicist. Improved on Bakewell's version. His pantelegraph used two extremely accurate clocks and made the synchronization timers independent of the current relayed by the telegraphic line itself. He could send handwritten messages as well as photographs. The first commercial facsimile service run between Paris and Lyon. He transmitted nearly 5,000 faxes in 1865.*

Giovanni Caselli was born in Siena in 1815; he studied literature and science. From 1841 to 1849 he lived in Modena as tutor of the sons of Marquis of San Vitale, but as he took part in the riots for the annexation of the Duchy of Modena to Piedmont, he was expelled from the Duchy. He spent all the money he had saved during his Modenese period in experiments which eventually led to his Pantelegraph.

1877–1880 ***Constantin Senlecq de Ardres*** *(1842–1934, France). Physicist. Made use of the photoconductive properties of selenium. The transmitted image was first focused onto a glass plate of a Camera Obscura. Then, the image was traced line by line with a selenium stylus which converted the amount of light at each spot on the image to an electric current. The electrical signals were sent by wires to the receiver where they magnetically controlled a pencil that re-recorded the document. In 1880 de Ardres published a book entitled "The Telectroscope", and in 1881 he outlined photo-telegraphy.*

1881 ***Shelford Bidwell*** *(England). Demonstrated a device that transmitted silhouettes using both selenium and a scanning system. It was called phototelegraph.*

1895 ***Ernest A. Hummel*** *(USA). Watchmaker in St. Paul, Minnesota.*

His system used synchronized rotating 8-inch drums, with a platinum stylus used as an electrode in the transmitter. The original image was drawn on tin foil using a non-conducting ink made from shellac mixed with alcohol. The image was received on carbon paper wrapped between two sheets of blank paper. When the electrode touched the tin foil in the transmitter the circuit was closed; when it touched the shellac the circuit was opened.

The system was known as the Telediagraph, one of several fax-line devices sending pictures via telegraph lines.

The first machines were installed in the office of the New York Herald in 1898. By 1899, Hummel had improved the machines, and they were in use in the offices of the Chicago Times Herald, the St. Louis Republic, the Boston Herald, and the Philadelphia Inquirer.

1902–1907 **Arthur Korn** *(1870–1945, Germany). Physicist. Father of telephotography (1902): sent photos on telephone lines over a distance of 1000 km from Munich to Nuremberg. In 1907 he sent the first wire photos from the continent to England. In 1922, May 06, he wired a picture from near Rome to Berlin, whence it was radioed across the Atlantic to a Navy radio station in Maine in about forty minutes.*

Korn put a sheet of photographic film on a revolving glass drum. Light scattered off the picture, traversed both film and glass, went through a prism and was projected on a selenium cell connected with a battery. The ensuing current were then transmitted by wire to the distant receiver, where it was decoded.

Korn was born in Germany and was appointed professor of Physics at the university of Munich (1903–1908). From 1914–1936 he was professor of Electro-Physics at the Berlin Institute of Technology. Emigrated to the United States in 1939. Korn's idea of scanning the object and transforming light signals into electronic signals, preceded the first meaningful research into television by just two years.

1921 **Eduard Belin** *(1876–1963, France). Engineer. First transatlantic transmission of photos by radio. His system, known as belinograph, is based on Korn's idea, but is automatic while the previous transmissions were manual.*

1922 *RCA provided the first transatlantic facsimile service.*

1927 *First commercially available equipment for phototelegraphy goes into operation. Radio transmission of pictures becomes an important tool for press news reporting and weather services around the world.*

Photo 1: Picture transmitted in the 1920s by electric telegraph

1939–1945 **John Logie Baird** *used television for facsimile transmission of maps and written material in WWII.*

1958 *Slow-Scan Television (SSTV).*

1971 *First prototype of laser fax.*

1980 *Public international fax services.*

1984 *Japanese introduce high-quality facsimile.*

V. Recording and Reproducing of Sound and Light (1796–1990)

(Photography, Audio and Video recording and playback, analog and digital methods, cinematography)

1796 *The Music Box was invented in Geneva by the watchmaker* **A. Favre**.

1839 **Louis Jacques Daguerre** *(France) announced his process of making photographs (a silver image on a copper plate), known as the daguerreotype.* **William Talbot** *(England) invented photographic paper for making negatives*

Edmund Becquerel *discovered the electrochemical effects of light: the photovoltaic effect; he observed that shining light on an electrode in an electrolytic cell increased the generation of the electric current between the electrodes. His discovery, however, remained a curiosity of pure science for the next 65 years.*

1840–1856 **Joseph Max Petzval** *(Hungary, 1840) and* **Phillipp Ludwig von Seidel** *(Germany, 1856) lay the foundation for the design and construction of aberration-free objective lenses with large aperture and wide fields. It had great impact on the design of modern cameras and telescopes.*

1855–1857 **Leon Scott de Martinville** *(France) developed a device that produces a graphical image of sound - the Phonautograph. It enables people to "see" sound. The system used a mouthpiece horn and a membrane fixed to a stylus that recorded sound waves on a rotating cylinder with smoked blackened paper. It could not play the 'record' back.*

1856 *The first aerial photos of Paris taken by the French photographer* **Cespard Felix Touranchon** *(1820–1910), from a balloon.*

1873 **Louis Joseph May** *(England) and* **Willoughby Smith** *(1828–1891, England) discovered electrical photoconductivity of selenium, thus enabling the transformation of visual images into electrical signals. They found that the electrical conductively of a bar of the element selenium changes* *when it is exposed to light such that the ensuing current is proportional to the amount of light hitting the bar. May then used selenium to send a signal through the Atlantic telegraph cable.*

1875 **John Kerr** *(Scotland) discovered the electro-optical Kerr effect (1875) through which intensive electric fields cause certain isotropic amorphous substances to become doubly reflecting (anisotropic). In 1876 he discovered the magneto-optic Kerr effect.*

1877–1878 **Werner Siemens** *(1877) and* **Oliver Joseph Lodge** *(1878, England) independently patented the loudspeaker; music had yet to be converted into electrical signals that would enable a speaker to work.*

1877 **Thomas A. Edison** *and* **John Kruesi** *(USA) made the first recording of a human voice – the "talking machine", alias the 'tinfoil cylinder phonograph'. Edison then (1878) recorded sound onto discs and cylinders.*

In the same year, **Charles Cros** *(1842–1888, France) developed a gramophone record from which, through a photogalvanoplastic process, the sound could be reproduced.*

W.G. Adams *and* **R.E. Day** *observed the photovoltaic effect in solid selenium. Made the first selenium cell.*

1880 **Constantin Selencq de Ardres** *(France) announced his method of telegraphic transmission of images via his Telectroscope.*

1880–1904 **Charles Summer Tainter** *(England) constructed the first system that utilized a selenium cell to convert sound into light signals (1880).*

The first practical photocell was devised by **Julius Elster** *and* **Hans Geitel** *in Germany (1900–1904). They patented a system for "recording and reproduction of sound", using a magnetic induction sensing device (1886). This was a precursor of* **Berliner***'s Gramophone.*

1881 **Clement Ader** *(France) built an ultra-sensitive microphone and with it discovered the stereo effect. He used 12 of these microphones to transmit sounds of the Paris Opera, via lines laid through the Paris sewers, to the Exhibition Hall at the "Palais de l'Industrie". Up to 48 listeners* *could hear the opera using two receivers each, one for each ear.*

1883 **Charles Fritts** *(US) described the first solar cell made from selenium wafers.*

1886 **Louise Aimé Augustine Le Prince** *(1842–1890) developed in Leeds, England, the first camera and projector system suitable for cinematography. Mysteriously disappeared in Sep. 1890 on the train from Dijon to Paris, and was never found.*

1887–1901 **Emile Berliner** *(Germany and USA) introduced the modern gramophone with the groovy audio discs, the needle, and the mechanical loudspeaker horn (1887). He later invented (1897) the shellac disc that could be mass-produced.*

1887–1905 **Heinrich Hertz** *discovered the photoelectric effect. Hertz found that electric sparks pass more readily between electrodes illuminated by ultraviolet light given off by another spark (i.e. light altered the lowest voltage capable of causing a spark to jump between two metal electrodes).*

Wilhelm Hallwachs *(1859–1922, Germany) discovered that a combination of copper and cuprous oxide is photosensitive.*

Augusto Righi *(1850–1920, Italy) related (1897) the proportionality of the current of liberated particles to the intensity of the incident light.*

Phillipp Lenard *(1902) found experimentally that the maximum velocity of the released electrons is independent of the intensity of the incident light but gave a wrong interpretation within the framework of classical physics.*

Albert Einstein *(1905) opened the floodgates of modern quantum physics, arguing from Planck's blackbody radiation law that the maximum energy of the released electrons is proportional to the frequency of the incident light.*

1887 **Charles Cros** *(France) innovated the essential ideas behind a disc-based audio recording and playback system. He also developed the idea behind the duplication system in the manufacture of records today. He was, however unable to construct his own machines because of the cost.*

1888 **Oberlin Smith** *(1840–1926, USA) was first to suggest the use of permanent magnetic impressions for sound recording. According to his plan, cotton or silk in which steel dust (or short clippings of fine wire) were suspended, will serve in the role of particle to be magnetized in accordance with the undulatory current delivered from a microphone. Thus, he reckoned, a magnetic pattern will be established which is a replica of the microphone current. Smith never built such an instrument.*

1895 **Auguste** *and* **Louis Lumiéré** *(France) invented a camera and projector system later patented and known as the Cinematographe. The first public showing to an audience of invited specialists was on March 22, at 44 Rue des Rennes, Paris.*

1898–1901 **Valdemar Poulsen** *(Denmark) developed the first practical magnetic sound recorder and reproducer, the telegraphon. It recorded (on a magnetized steel piano-wire), the varying magnetic fields induced by the sound, and played it back in a reverse process.*

William Bu Bois Duddell *(England) and* **E. Ruhmer** *(Germany) independently suggested the photographic recording of sound. Their progress, though incomplete, inspirited further development by* **Lee de Forest** *(1920), whose work, in turn, served as a basis for the 'talkies' of the cinematography industry.*

1902 **Enrico Caruso** *recorded his voice on a* **Berliner** *gramophone disc.*

1907 **Henry Joseph Round** *(1881–1966, England) discovered the phenomenon of electroluminescence (emission of light from a semiconductor diode: after the application of a potential of ten volts between two points on a crystal of Carborundum, the crystal gave out a yellowish bright light).*

1913 *Beethoven's Fifth symphony was recorded in its entirety on a gramophone disc; The performance was by the Berlin Philharmonic Orchestra under the baton of* **Arthur Nickish**. *It is the first full recording of a symphonic piece ever.*

1920 **Lee de Forest** *(USA) pioneered a process of optical sound recording: First, sound waves were recorded on magnetic tape which in turn produced an electrical waveform with corresponding amplitude variation. The electrical impulses were then transformed into mechanical vibration of a mirror. A source of light shone a beam of light on the vibrating mirror which then fell through a slit on an exposed moving sound track film. The pattern of light falling on the film was a transcript of the original sound. To play back, light from another lamp was controlled into a beam that shone through the sound track and then struck a photoelectric cell, converting the modulated beam into electric impulses. These reconstructed impulses were then amplified and fed into a loudspeaker behind the screen. The sound track could be synchronized with an accompanying motion picture.*

In a few years this method served as a basis for the 'talkies' of the cinematography industry. For the next 59 years, optical sound recording was used only in cinema. Then numerous inventors endeavored to apply it to music records.

1924 **C.W. Rice** and **E.W. Kellog** *(US) introduced the modern loudspeaker.*

1927 **Oleg Vladimirovich Losev** *(1903–1942, Russia). Discovered what we now know as LED (Light Emitting Diode) and foresaw its use in telecommunications. His discovery languished for half a century before being recognized in the late 20th century and early 21th century. Losev published his results in Phil. Mag,* **6**, *1024–1044 (1928) and issued a Soviet Patent #12191 in 1929.*

In 1962, four research groups in the US simultaneously reported a functioning LED semiconductor laser based on gallium arsenide crystals, thus opening the field of solid-state optoelectronics.

1927 **J.A. ONeill** *(USA) replaced Poulsen's wire (1898) with a magnetically coated ribbon.* **John Logie Baird** *(England) first recorded TV pictures on shellac discs.*

Fritz Pfleumer *(1897–1945, Germany) built the Magnetophone, the first magnetic tape–recorder, using paper tape coated with steel-powder. The all-electric record-player* *with electronic amplifier and built-in loudspeaker went on sale.*

W.L. Carlson and **G.W. Carpenter** *(USA) invented the "a-c biasing" method, the next great milestone in magnetic recording.*

End of the silent film; The enormous success of the movie The Jazz Singer (starring Al Jolson), introduced the era of talking motion pictures with the projection of synchronized sound.

1928 **Warner Brothers** *in Hollywood adopt the Movietone: the direct photographic recording and reproduction of sound on film:*

Sound is picked up by a microphone, amplified, and the current from the power amplifier is passed through a gas discharge tube, which emits an amount of light proportional to the current. The light is focused into a narrow beam perpendicular to a rolling film, marking upon it a track of constant width but variable density. In the reproducer, a tungsten filament emits a narrow beam of light at right angles to the sound track, and after passing through the track the light is focused upon the cathode of a photocell. The current through the cell is proportional to the pressure of the original sound wave. The photo-current is then passed through a receiver and sets up across it a potential difference, whose fluctuations have the waveform of the original sound. The potential difference is amplified and used to drive a loudspeaker.

1929 *Ten million gramophone records are sold in Germany alone.*

1930–1935 *With the advent of the electric amplifier and a-c biasing it became possible to make quite satisfactory magnetic records and play them back at the desired volume. At that time however, the motion-picture industry was the main client, using magnetic records as a sound-recording medium for talking pictures. Magnetic sound recording was also used in dictating machines and recorders of phone conversations. Thus the magnetic recorder went under the guise names Dailygraph, Telegraphone and Textophone.*

1931 ***Alan D. Blumlein*** *(1903–1942, England) Produced the first stereophonic record and patented the stereo sound system. Bell Telephone Laboratories (USA) also produced stereo recordings for experimental use.*

RCA Victor USA introduced coarse groove discs that ran at $33\frac{1}{3}$ *rpm, but these failed to replace popular 78 rpm discs. The first such record: Beethoven's Fifth was conduced by* **Leopold Stokowski**. *For the first time a complete orchestral piece was issued on a standard long-playing record.*

1935 *Development of the Tweeter and Woofer in loudspeaker technology to reduce distortion.*

1936 **Eduard Schüller** *(1904–1976, Germany) produced a commercial version of the first true magnetic tape recorder.*

1937 *German Telefunken Company issued the first electromagnetic light-weight tone-arm with a sapphire stylus.*

Alec Reeves *(England) reinvented the principle of pulse code modulation (PCM) of digitally encoded signals. It later revolutionize the transmission, recording and processing of voice, fax, data and video signals. PCM protects signals against noise and interference. By using a defined digital format it enables differing types of signals to be assembled into a common multiplex, i.e. an integrated services digital network (ISDN).*

C.N. Nickman *(Bell Telephone, USA) demonstrated a magnetic tape recorder of excellent quality, with a tape speed of 40 cm/sec and a recording medium called Vicalloy. It marks the true beginning of the modern period in magnetic recording.*

1938 **Chester Carlson** *(1906–1968, USA) invented Xerography, the first method of photocopying.*

George Harold Brown *developed the vestigial sideband filter for use in television transmission, doubling the horizontal resolution of television picture at any given bandwidth.*

1940 ***Hans-Joachim von Braümuhl*** *(1900–1980, Germany) and* **Walter Weber** *(1907–1944, Germany) introduced the method of high-frequency premagnetization and ac bias of tapes for better sounding tape recording.*

1941 **Russell Ohl** *(USA) invented the silicon solar cell.*

1947 **Peter C. Goldmark** *(1906–1977, Hungary and USA) developed the first long-playing high-frequency record ($33\frac{1}{3}$ rpm) with 23 min per side capacity [compared to 5^m of the 78 rpm records]. In the new LP vinylite record sound quality was considerably improved. The old records soon disappeared from the market.*

Dennis Gabor *(1900–1979, Hungary and England) pioneered holography.*

Edwin H. Land *(USA) launched the first Polaroid Camera on the US market. It weighted 2.7 kg and took 8 prints per film pack, developing each picture in 60 seconds.*

Invention of the transistor at Bell Laboratories, USA by **William B. Shockley**, **John Bardeen** *and* **W. Brattain**.

1948 **Claude Shannon** *(USA) founded information theory.*

1958 *LP stereo records and record players*[10] *reached the market.*

1958–1964 *First video recorders developed by Ampex (1958) for use by television stations. It is a device to read pictures and sound signals from a television camera onto a magnetic tape. First domestic video recorder was produced by Sony (1964).*

1963 *First compact audio cassette (Phillips).*

Ray Dolby *(US) developed his noise-suppressing technique.*

---

[10] LP *stereo sound reproduction*: The *stylus*, mounted in a *cartridge*, follows the molded contours of the record. These contours are analog representations of the sound waveform. The *pickup* converts the movement of the stylus (imparted to it by the groove) into corresponding electrical signals in the form of alternating voltage.

In a *magnetic pick up* (most commonly used) the electrical output is induced by the relative motion of a magnetic field and a coil (or 2 coils in the case of a stereophonic pick up) located in the field. The movable system in the pickup must be mounted so that the stylus can follow every rapid change of direction of the groove, virtually without resistance. For the same reason the mass of the moving parts must be as small as possible to reduce groove wear. The stylus must have a rounded tip suited to the cross-sectional dimensions of the groove, so as to assure that the tip is maintained in contact with the sloping sidewalls of the groove and clear of the bottom. The wavy pattern of the groove determines both *the frequency and the amplitude* of the sound vibration.

1965 **James T. Russell** *invented the digital compact disc. It has since then become an essential component of audio, video and computer systems.*

1969 **Klass Compaan** (Holland) *conceived the idea of a digitally encoded disc.*

James Fergason (USA) *invented liquid crystal display (LCD). It completely redefined many industries, such as computer displays, medical and industrial devices, and a vast array of consumer electronics.*

LCDs were originally based on a concept which used a large amount of power, provided a limited lifetime, and provided a poor visual contrast. Fergason overcame these obstacles in 1969 with his discovery of the twisted nematic field effect, which forms the basis of modern LCDs.

Digital watches were among the first consumer items to use LCDs. Other items included calculators and computer displays. LCDs are used in over five billion items a year.

1969–1975 *VCR (video cassette recorder) was produced by Phillips. The type of domestic video recorder most used now is the VHS (video home system) format, launched by JVC (Japan) in 1975.*

1969–1990 *Charged-coupled device (CCD) was first demonstrated at Bell Labs (1969). It is a solid-state chip which transforms light into electricity and decomposes an optical image into pixels, whose gray-scale values are then digitized. The CCD contains a bank of light detectors, each registering variation of intensity as small changes in voltage.*

It has the form of a small capacitor, composed of metal oxide and semiconductor layers, capable of both photodetection and memory storage. When the subtle changes in voltage (created by the photoconductor electrons) are applied to the metal layer (called the 'gate'), electron-hole pairs created in the semiconductor (by absorption of photons) are separated by an electric field, and the electrons become trapped in the region under the gate. This trapped charge represents a small piece of the image known as a pixel (picture element).

The complete image can be recreated by reading out a sequence of pixels from an array of CCD's. These arrays are used to capture images in video and digital cameras included in telescopy and microscopy.

1972 *First domestic videodisc demonstrated by Phillips.*

1975 *First laser printer introduced by IBM.*

1979 *Videodisc read by laser in Holland.*

1981 *Hologram technology improved; now in video games*

1982–1984 *The Camera-Recorder combination (camcorder): first coupling of the camera and tape (Japan).*

1982 *Kodak camera used film on a disc cassette.*

1983 *First audio compact discs (CD's) marketed in the UK and the US. It is a plastic disc, 12 cm in diameter, that can hold on a single side over an hour's worth of digitally encoded sound recording, stored as a succession of pits and plateaux in tracks. The disc is coated with a reflective material (usually aluminum) which either scatters or reflects back into the photoelectronic detector a laser beam used to 'read' the encoded sound when the disc is rotated at a high constant linear speed. The advantage of the compact discs over the LP records is its freedom from surface blemishes, so that it can approach perfection in sound reproduction.*

The CD player became available from Sony and Phillips, who collaborated in both research and production. The optical sound reading in compact discs differed from the de Forest process (1920) in that sound was no longer in analog form, but instead it was coded in a digital (binary) form; the micro-hollows on the disc were processed at a rate of 4 million per second. Consequently it produces much better clarity of sound than the microgroove.

The invention of the compact disc, launched in 1982, was the direct result of research on the videodisc, also invented by Phillips and marketed in 1980.

1990 *28% of U.S. homes own CD players; sales total 9.2 million players and 288 million compact discs in the US alone.*

1990–91 *The lossy compression standard JPEG (Joint Photographic Experts Group) is set, and commercial applications based upon it begin to appear.*

1990's *Digital cameras, digital scanners, storage of photos and movies on PC's, and email transmission of imagery become widespread among the general public in developed countries.*

VI. Sonar, Radar and Satellite Radionavigation Systems (1887–1971)

Communication using electromagnetic radiation (except for light) began early in the 20th century, and most early practical systems used very long wavelengths which traveled great distances. Eventually, electronics were developed, including the vacuum tube (valve), which allowed controlled frequencies and modulation schemes. This led to the use of higher carrier *frequencies, many channels, and commercial and industrial radio. During the 1930's and 1940's, various experimenters discovered that higher frequencies could bring other advantages to communications. Some of these experimenters worked for government agencies and the military, some were at universities and some were private individuals. Quantitatively, the high frequency bands divide as follows: (λ = wavelength):*

Television and FM radio	*50–600 MHz*	λ = *6 m–50 cm*
Cellular Phones	*800–900 MHz*	λ = *37.5 cm–33.3 cm*
Microwaves (Radar)	*1000–300,000 MHz*	λ = *30 cm–1 mm*
Cordless Phones	*multi-GHz: 2.4 GHz, etc.*	

Thus every communication service uses a part of the spectrum that is suitable for its needs. Microwaves, for example are easier to control than longer wavelengths because small antennas could direct the waves very well (i.e. the diffraction limit is less constraining). One advantage of such control is that the energy could be easily confined to a tight beam (expressed as narrow beam width). This beam could be focused on another antenna dozens of miles away, making it very difficult for someone to "eavesdrop" on the conversation. Another characteristic is that because of their high frequency, greater amounts of information could be put on them (expressed as increased modulation bandwidth). Both of these advantages (narrow beamwidth and modulation bandwidth) make microwaves very useful for radar.

All of these qualities led to the use of microwaves by the telephone companies. They placed towers every 30 to 60 miles, each with antennas, receivers and transmitters. These would relay voice conversations across the country. The ability to modulate with a wide bandwidth permitted a large number of conversations on just one signal, and the reduction in beamwidth reduced interference and hampered casual eavesdropping.

Amateur radio interests in microwaves have mostly been for the challenge of working with such esoteric frequencies that require specialized techniques in design, fabrication and testing. Furthermore, in order to reach beyond LOS (line-of-sight) amateurs have spent countless hours carefully measuring propagation phenomena. Amateurs have carried on conversations using 10 GHz well over 1,000 miles, and have bounced signals at that frequency off the moon.

Radar (Radio Detection and Ranging) *is an instrument for object detection based on the principle of electromagnetic wave reflection: locating an object at a distance and studying its echo.*

It helps airplanes, ships and ground stations detect other objects before they can be visually identified. Radar can be used for military and civilian purposes.

It works by using a radio transmitter and a receiver. The transmitter sends out a high frequency radio wave. If the radio wave hits a metallic or other conducting object (such as a ship), some of it will reflect back to the receiver. By measuring the time it takes for the wave to return and the power of the scattered radio waves which are picked up by the receiver, the position and size of the plane or ship can be estimated and projected onto a radar screen. If a lot of the signal is bounced back then the object is close; if only a little bit of the signal is bounced back then the radar is far from the object, and/or the object is small. The distance is simply derivable from the round trip signal time, since the radio waves in air travel essentially at the speed of light in vacuum. This process is similar to a bat's use of echo-location to locate insects in the dark. The only difference is that echo-location uses sound waves as opposed to the radio waves sent out by the radar.

Radar can "see" through great distances despite fog, rain, snow, clouds or darkness: It can find and accurately locate missiles, aircrafts, ships, submarines, cities, rainstorm and mountains.

Radar also prevents planes from crushing into runways, control towers, and other planes, and is an essential tool for analyzing and predicting the weather.

Microwave radar, as well as LIDAR (Light Detection And Ranging) are routinely used for communications and to measure atmospheric pollution.

Radar was developed for military use in the 1930's by several countries. The invention of radar just prior to WWII changed forever the nature of warfare. Radar eliminated many of the limitations that were imposed on the human ability to see. It virtually eliminated the possibility of sneak attack by enemy aircraft. For those and many other reasons, radar has grown in prominence, so that today it is an absolute necessity.

During WWII radar was used widely on planes and ships. Battles were won by the side that was first to spot enemy airplanes, ships, and submarines. To give the Allies an edge, British and American scientists developed radar technology to "see" for hundred of kilometers, even at night. The research that went into improving radar helped set the stage for post-war invention of the transistor. During the war, radar was tremendously useful for:

- *Precision bombing of targets under conditions of poor visibility.*
- *Enhancing defense of convoys against German submarines in the Atlantic battles of 1943. Subs were detected by airborne radar and subsequently destroyed from the air.*
- *Communication between aerial defense headquarters and planes equipped with airborne radar receivers.*

Had it not been for radar, England might have been invaded by the Germans and Allied bombers would not have been able to attack Germany successfully. Indeed, it was radar that turned the tide of the war. Winston Churchill was quoted as saying: "The atomic bomb ended the war, but radar won it".

Sonar is an acronym for SOund Navigation And Ranging. It is a detection and ranging system based on the reflection of underwater sound waves.

A typical sonar system emits ultrasonic pulses by using a submerged radiating device. It listens with a sensitive microphone for reflected pulses from potential obstacles or submarines.

Modern submarines rely on sonar for detecting the presence of enemy vessels. The most advanced system, called a towed array, uses long cable to which microphones are attached, and the submarine drags this trailing cable far behind as it travels through the sea. Airplanes are used to deploy a different sonar. This system uses a device called a sonobuoy, consisting of a microphone mounted in a floating buoy. It is designed so that when a sound such as that of a submarine engine is picked up, the detector can transmit signals to patrolling antisubmarine planes.

Some animals use sonar to locate their prey and also for navigation. Thus, bats are equipped with a biotransmitter in the frequency range 15–40 kHz [wavelength 20 mm–3 mm], emitting ultrasonic squeaks that bounce back from obstacles. Dolphins also use sonar. Human sonars generally use sound frequency in the range 5–25 kHz.

In peacetime sonar can be used to determine water depths, or to locate fish schools. The signals are generated by a transmitter within the ship. A transducer changes the electric signals to sound waves and sends them through the water. These waves strike a target and are reflected back (or scattered) to the ship, where they are transformed again into electric signals which activate the indicator that calculates the target's range, direction and size. Scanning sonar employs a rotating beam for rapid search.

The Global Positioning System (GPS) is a space-based radio-navigation system, consisting of 24 satellites and ground support. GPS provides users with accurate information about their position and velocity, as well as the time, anywhere in the world and in all weather conditions.

GPS Satellites fly in circular orbits at an altitude of 10,900 miles (17,500 km) and with a period of 12 hours. The orbits are tilted to the earth's equator by 55 degrees to ensure coverage of polar regions. Powered by solar cells, the satellites continuously orient themselves to point their solar panels toward the sun and their antennae toward the earth. Each satellite contains four atomic clocks.

There are often more than 24 operational satellites, as new ones are launched to replace older satellites. The satellite orbits repeat almost the same ground track once each day (as the earth turn beneath them). The orbit altitude is such that the satellites repeat the same track configuration over any point approximately each 24 hours (4 minutes earlier each day). There are six orbital planes [with nominally four space vehicles (SV) in each, equally spaced (60 degrees apart) and inclined at about 55° w.r.t. the equatorial plane. This constellation provides the user with between 5–8 SVs visible from any point on earth.]

Precise positioning is possible using GPS receivers at reference locations, providing corrections and relative positioning data for remote receivers. Surveying, geodetic control, and plate tectonic studies are examples of applications of GPS.

Time and frequency dissemination, based on the precise clocks on board the SVs and controlled by the monitor station, is another use for GPS. Astronomical observatories, telecommunications facilities and laboratory standards can be set to precise time signals or controlled to accurate frequencies by special purpose GPS receivers.

GPS determines location by computing the difference between the time that a signal is sent and the time it is received. GPS satellites carry atomic clocks that provide extremely accurate time. The time information is placed in the coded broadcast by the satellite so that a receiver can continuously determine the time the signal was broadcast. The signal contains data that a receiver uses to compute the locations of the satellites and to make other adjustments needed for accurate positioning. The receiver uses the time difference between the time of signal reception and the broadcast time to compute the distance, or range, from the receiver to the satellite. The receiver must account for propagation delays, or decreases in the signal's speed caused by the ionosphere and the troposphere. With information about the ranges to three satellites and the location of the satellites when the respective signals were sent, the receiver can compute its own three-dimensional position.

An atomic clock synchronized with GPS is required in order to compute ranges from these three signals. However, by taking a measurement from a fourth satellite, the receiver avoids the need for an atomic clock. Thus, GPS is available in two basic forms: the standard positioning service (SPS) and the precise positioning service (PPS). The atomic-clock time measured on board the GPS satellites is so precise, that the software must correct for STR and GTR effects (!), which amount to about 38 μs per day.

Timeline

_1886_ **_Heinrich Hertz_** *(Germany) conducted experiments on reflection of electromagnetic waves in the microwave region (around 500 MHz with a wavelength of about 60 cm, and around 2 GHz, corresponding to a wavelength of 15 cm). Yet he did not see its potential application.*

_1895_ **Guglielmo Marconi** *(Italy) experimented with transmission of microwave radio signals.*

_1900_ **N. Tesla** *(USA) was first to describe the possibility of locating a moving object using continuous radio-wave echoes. However, technology at that time was insufficient to implement this detection system.*

_1904_ **Christian Huelsmeyer** *(Germany) patented a radio detector based on Tesla's principle. This engineer proposed the use of radio echoes in a detecting device to avoid collisions in marine navigation.*

_1915_ *First sonar invented during WWI by British, American and French scientists. It was used to locate submarines and icebergs via passive listening devices. The first submarine to be sunk after being detected by hydrophone was the German U-boat UC-3 in the Atlantic (1916). Active sonar, using a pinging device, was developed by* **Paul Langevin** *(France; 1918). It was installed in submarines and ships to detect underwater objects by echo sent and received by a transducer. The beam spread out until it detected an object and then reflected (or scattered) back an image echo to the ship. The reflected waves were then converted into electric pulses that formed an image of the object on a screen.*

_1921_ **Albert Wallace Hill** *(US) The Magnetron diode – an electron tube that produced microwaves. It underwent two basic modification (1927, 1941) to become an efficient generator of microwave radiation power.*

_1922_ **G. Marconi** *(Italy),* **A.H. Taylor** *(US), and* **L.C. Young** *(US) suggested detection of a moving object using pulsed waves to facilitate the determination of the times of emission and reception of echoes.*

1924 **Edward Victor Appleton** *(England) used radio echoes to determine the height of the ionosphere (an ionized layer of the upper atmosphere that reflects longer radio waves). It was the first successful radio range-finding experiment.*

1925–1935 *Great efforts by scientists around the world to find a practical method of detecting and locating objects by radio echoes. Most of this research was motivated and sponsored by the military.*

1927 *Split-anode magnetron was introduced in Japan.*

1931 **W. Butement** *and* **P. Pollard** *(England) built the first experimental radar system to measure the range of ships from shore.*

1934–1937 *British scientists, led by* **Robert Alexander Watson-Watt** *developed a working radar system and succeeded in detecting radio waves reflected from a flying aircraft (1935). At the time, Watson-watt suggested linking together radio-detection stations to a fighter-control network. His work was vital in developing radar for use in the Second World War in Britain. German scientists, working independently, had already tested their radar system successfully in 1934.*

The equipment used by **Watson-Watt** *in 1935 consisted of a high-voltage radio pulse generator (thermionic valves) and a cathode-ray tube display. However, since the initial beam could not be strongly focused*[11]*, the echo thrown back from the surface of the water or the earth sometimes overshadowed the much weaker echo from the target. Satisfactory focalization was achieved only in 1936 when Watson-Watt employed a magnetron with an electron gun installed in the cathode emission tube and an electronic amplifier for the echoes. In 1937 he was able to erect a chain of 20 radar station along the British east coast.*

1938–1940 *Telefunken in Germany built a 560 MHz radar system for the Luftwaffe. Over 8 military research centers and 200 German institutes were working on improving the radar, under the somewhat vague direction of Göring. In 1940,*

[11] The size and shape of the antenna depends on its function: e.g. a vertical antenna is used to find the height of aircraft, etc.

however, Hitler banned all electronic research, however fundamental it was, under the pretext that electronics was a 'Jewish science'. His total lack of scientific education and antisemitic zeal cost him the war.

1938–1939 **Russel Harrison Varian** *and* **Sigurd Fergus Varian** *(US) invented the klystron tube for the generation and amplification of microwaves. It basically converts the dc kinetic energy of cathode-emitted electrons into UHF electromagnetic energy. The klystron played a crucial role as part of radar systems during WWII. The invention spawned a whole new microwave industry, and came to play a key role in particle accelerators, used for high-energy fundamental physics research and to collimate x-ray synchrotron radiation for applied research.*

1940 **John T. Randall** *and* **Howard A.H. Boot** *(England) invented the multi-cavity magnetron. This tube is capable of generating high-frequency (microwave) radio pulses at high power levels in the wavelength band around 1 cm. In fact, the magnetron allowed a 20-fold increase in radiated power over the existing electron-valve system. Ground experiments in 1940 demonstrated that detection ranges of order 10 km were possible.*

The first radar stations used aerials over 100 m in height to produce a directional beam of radio waves. But if aerial were much smaller and could be steered, they would be much more useful. However, to make smaller aerials meant using radio waves of shorter wavelengths. The cavity magnetron was created to generate such waves. It converted the dc kinetic energy of electrons, accelerated by a dc potential difference, into UHF electromagnetic energy.

In the magnetron, the tank circuit of the valve oscillator was replaced by a cavity or several cavities. It generated pulses for the radar, each pulse lasting for a few microseconds with wavelengths of 1–10 cm and corresponding frequencies of 3,000–30,000 MHz, and typical peak power of 150 kW at a pulse rate of 1000 sec^{-1} *were attained during WWII. [The magnetron has found applications in television transmission, satellite communication, industrial heating, home cooking (microwave oven), medical imaging, radiation therapy and asteroid tracking.]*

The British had an excellent ground-based-radar system using the magnetron transmitters, but these were too heavy and big to be fitted into an aircraft. A compromise was sought in mounting only the radar receiver *in an aircraft, tuned to the same frequency as a ground radar to hopefully pick up reflected radar echoes. By 1940, when the Germans switched to night bombing, the British equipped their night fighters with* klystron radar *as part of their* airborne radar*, which helped win the Battle of Britain.*

The Germans had developed a ground-based radar which was rather bulky but essentially worked well for their defense *purposes. However, they payed little attention to the klystron know-how (found in their military archives after the war), depending instead on their V-2 rockets to defeat England.*

1940–1945 *Establishment (1940) of the* MIT Radiation Laboratory *(US) which became the center for most radar developments during WWII. In this Lab, United States and British radar scientists cooperated closely. American scientists gave the British the* klystron *generator and the* duplex switch*, and the British gave the United States the multicavity magnetron. The Allied scientists scored a big achievement by developing receivers, high-power transmitters, and microwave antennae. This made it possible to develop narrow-band, highly accurate radars with small antennae for aircraft, ships and mobile ground stations. The new equipment enabled detection of submarines in the Atlantic (1943), development of long-range navigation systems, early warning radar, and radar for night bombing of remote targets.*

Radar became so effective in aiming anti-aircraft guns that each side tried to jam the other's radar. Allied bombers carried radios to send signals that confused or blanked out enemy radar. They also dropped tons of aluminum foil strips that reflected false echoes to enemy radar screens. In the Pacific Ocean, radar gave the US Navy superiority over Japanese forces in night naval battles. In the Atlantic, airborne radar enabled the Allies to inflict crippling losses on the German submarine fleet.

The research that went into improving radar helped set the stage for the post-war invention of the transistor: radar systems used a semiconductor crystal rectifier diode. These crystals often could not adjust to the quickness and intensity of a rapidly fluctuating radar signal, and they would frequently burn. A number of institutions including Purdue University, Bell Labs, MIT and the University of Chicago joined forces to build better crystals.

1942 **Seymour Benzer** *at Purdue University (US) discovered that Germanium crystals made the best detectors. Germanium was used to make the first working transistor 5 years later. Scientists also learned new technique on how best to grow and dope the crystals. Within the decade, this superb understanding of crystal growth would pay dividends in unexpected areas, not the least of which were the insights necessary to allow the solid state researchers at Bell Labs to grow the germanium semiconductors that were at the heart of the first transistors.*

1943–1945 *Radar operators continually saw on their radar screens images of precipitation (like rain and snow). Scientists then realized, for the first time, that radar would be sensitive enough to detect precipitation. After the war radar become an essential tool for analyzing and predicting the weather.*

1945–1989 *Advances in sonar technology (used for tracking and identification of nuclear submarines) during the Cold War.*

1953 *Invention of the ammonia maser (Microwave Amplification by Stimulated Emission of Radiation). It emits ultra-short wave beams at the frequency 23,900 MHz and wavelength 1.3 cm.*

1960 *The US Navy, using the moon as a reflector, sent a radio messages from Pearl Harbor to Washington DC in a pioneering communication experiment.*

1968 *Asteroid 1566 Icarus tracked by radar.*

1970 *Pulsed ultrasonic Doppler used for blood-flow sensing.*

1971 *Doppler radar is used by meteorologists to study storm systems.*

VII. Television technology – or "A picture is better than ten thousand words" (1817–1990)

The possibility of vision at a distance had occupied scholars' minds long before the idea of sound broadcasting. Webster's dictionary defines television as "the process of transmitting images by converting light to electrical signals and then back again". Television is constructed of two elements – video and audio. Video comes from the Latin word "I see" and audio is derived from the Latin word "I hear". Historically, no single person or invention is credited with the development of television.

The modern day television set can be traced back to the discovery of selenium in 1817 by **Jons Berzelius** *(Sweden). Television is based on photoelectric technology. Television's initial developments were linked to pioneering attempts to improve the transmission of still images down a telegraph wire. In the mid 1800's, sending still images by telegraph wire was an electrochemical process. The concepts of synchronized scanning and the use of photoelectric technology evolved over a fifty year period.*

The idea of sending still images via the telegraph traces its roots to 1839. At that time **Edmond Becquerel**, *a French physicist interested in the study of light, found that when two pieces of metal were immersed in an electrolyte, an electrical charge developed when one of the pieces was illuminated. Although Becquerel had discovered the electrochemical effects of light he did not offer any practical suggestion for its use.*

In 1842, **Alexander Bain** *proposed a facsimile telegraph transmission system based on Becquerel's discovery. Bain proposed that metallic letters of the alphabet could be illuminated in solution, and the resulting modulated voltages scanned and transmitted electrically. The electrified metal letters could be scanned by a pendulum device and reproduced at the other end of the telegraph wire by a synchronized pendulum contacting a piece of chemical*

1907–1911 *Early attempts to combine mechanical scanning with cathode-ray tube by* **Boris Lwowitsch Rosing** *(1869–1933, Russia) and* **Alan Archibald Campbell-Swinton** *(1863–1930, England). Rosing (1907) used a rotating mirror drum for scanning and Braun's tube to reproduce television images. Swinton (1908–1911) utilized cathode ray tube as image converter for transmission, designing in effect, the first electronic camera tube.*

1922–1927 **Philo Taylor Farnsworth** *(1906–1971, USA) first invented the all-electronic television (1922). Produced (1927) the first successful television transmission by wholly electronic means.*

1923–1935 **John Logie Baird** *(1888–1946, England) developed a working electromechanical television system based on Nipkow disc. First to harness television for military uses.*

1925–1929 **Herbert Eugene Ives** *(1882–1952, US),* **Charles Francis Jenkins** *(1867–1934, US) and* **Dénes von Mihàly** *(1894–1953, Hungary) improve on the transmission of television images.*

1926–1928 **Kálmán Tihanyi** *(1897–1947, Hungary and USA). Patented a fully electronic television on the basis of continuous electron emission with accumulation and storage of released secondary electrons during the entire scanning cycle.*

1926 **Kenjito Takayanagi** *(Japan) operated a working electronic television system, using a CRT to transmit an image of Japanese writing.*

1927 *Early attempts by AT&T (US) to combine the television and the telephone into a system of videophone with no specific additional equipment.*

1928–1935 **Vladimir Kosma Zworykin** *(1889–1982, Russia and USA). Constructed a complete all-electronic television system composed of the iconoscope (an electronic tube that converts light rays into electric signals and acts as a television camera suitable for broadcasting, 1923) and the kinescope (the picture tube used in television receivers).*

By 1933 the system reached a resolution of 240 lines per inch. His television provided the final impetus for the development of modern television as an entertainment and educational medium. He also developed a color television system (1928).

1929–1931 **René Bartholemy** *(1889–1954, France) developed a mechanical television scanner (using 30-line scanning) with a disc receiver. He gave two demonstrations (1931).*

1934–1936 **Isaac Schoenberg** *(England) developed a camera tube similar to the iconoscope (1934).*

Advent of public television in Germany (1935) and England. The BBC employed a regular 405-line monochrome TV service using Schoenberg's Emitron camera tube, which converted light from a scene in a studio into a television signal (1936). The Nazis in Germany built a TV station in order to broadcast the Berlin Olympic games. But in both countries, mass-market manufacture of television receivers was halted when WWII broke out.

A videophone service is set up in Germany with overhead cables between Berlin-Witzleben, Leipzig, Nuremberg and Hamburg. Transmission took place using the Nipkow mechanical scanning system; the image to be transmitted was imaged by 90 objective lenses in 1/25-th of a second and consisted of 180 lines at reception.

1938 *Sideband filter used in television broadcasting is developed for doubling resolution of pictures.*

1940 **Peter Carl Goldmark** *(Hungary and USA) demonstrated a color television system of his invention. CBS broadcasted the world's first color TV signals.*

One million television sets in the United States alone.

1956 *The Ampex Company (US) demonstrated the first viable video recording*[12] *of television.*

---

[12] *Video recording*: In the early days, film was the only medium available for recording television programmes. Owing to the specific needs of American television networks, however (broadcasting at different times on the Atlantic and Pacific coasts) researchers were led to investigating more flexible systems. Thoughts turned to magnetic tape,which was already being used for sound, but the greater quantity of information carried by the television signal demanded new studies. During the 1950s, a number of American companies began investigating the problem. In April 1956, the Ampex company demonstrated the first viable product.

1956 **Robert Adler** *(USA) invented the television remote control.*

1958 *Silicon controlled rectifier (SCR) introduced by General Electric (US). This semiconductor device was about to revolutionize dimming applications for theater and television lighting around the world. Previous to this time, dimming systems were large, generally inefficient and mechanically very complex.*

The SCR allowed the design of compact remote controlled dimming systems – with no moving parts in the dimmer. The typical modem SCR dimmer employs two PNPN semiconductor devices commonly known as silicon control rectifiers, or thyristors, connected in inverse parallel and in series with the lamp load. A signal applied to the control gates of these devices is utilized to control their conduction period. The dimmer thereby controls the effective power dissipated in the lamp load, and thus the intensity of the lamps. The dimmer is completely inert and requires no maintenance.

1961–1967 *Development of color television in Europe.*

In 1961, **Henri de France** *put forward the SECAM system (Sequentiel Couleur à Memoire) in which the two chrominance components are transmitted in sequence, line after*

It recorded in black and white. Its rival RCA followed suit in 1957, with equipment designed for color. The mechanical principle adopted was the same and was to remain in use for a long time. The system had four heads on a disc rotating obliquely across the width of the tape, thus tracing an oblique track pattern. The tape was 50.8 mm (2 inches) wide. With the development of editing equipment, the initial "delayed broadcasting" function gradually gave way to "production" functions. The first all-electronic editing equipment avoiding the need for splicing tape was introduced in the late 1960s. Slow-motion and variable-speed playback techniques were impossible with the "four head" system. The situation changed with the advent of helical-scan video recorders (Toshiba, 1959) which at last provided editing facilities analogous to those of film. Helical scanning is now used in all video recorders: each track contains one entire field (or a major part of it), and the tape can be "read" at different speeds, even when it is stationary. The magnetic tape is 25.4 mm (1 inch) wide. In 1986 the first digital video cassette recorder meeting the international digital television standards was presented by the Sony Corporation.

line, using frequency modulation. In the receiver, the information carried in each line is memorized until the next line has arrived, and then the two are processed together to give the complete color information for each line.

In 1963, **Walter Bruch** *(Germany) proposed a variant of the American NTSC system, known as PAL (Phase Alternation by Line). It differs from the NTSC system by the transmission of one of the chrominance components in opposite phase on successive lines, thus compensating for phase errors automatically. Both solutions found application in the color television services launched in 1967 in England, Germany and France, successively.*

1964–1971 *Picture-phone* system, combining television and telephone, was exhibited by AT&T but failed commercially.

1964 **Marshall McLuhan** *(Canada) foresaw the impact of television on man's social behavior; Television aims at our most immediate perception. Pictures are to see, almost to feel; they present the whole world before us. TV offers us entertainment games, sports and news. It offers something of everything for everybody.*

Our senses are assailed every day by the attraction of the visual message – "The medium is the message" as McLuhan succinctly put it. Its all-pervasiveness and instantaneity are tuned to our way of thinking, and vice versa. We expect from it effortless pleasure and hot news. But the stupefaction takes its toll and we thirst for more. Images pour over us in a never-ending torrent.

Television has already modified our social behavior. It fosters, for example, our taste for things visual through the impact of the picture and its colors. It encourages in us a yearning for the big spectacle and the forthright declaration. The effect can be seen in the way we react to one another and in the world of advertising.

But television cannot yet be said to have enriched our civilization, mainly because it has not yet become interactive.

Television cannot, on its own, serve as an instrument of culture. It must be appreciated that it is not well suited for detailed analysis or in-depth investigation. The way it operates and its hi-tech infrastructure are such that it cannot do justice to the words of the poet.

In the flood of images from the silver screen the less good accompanies the best, just as in the cinema or literature. The factor which distinguishes television from the cinema and books, however, is that the full quality range, down to the very worst, is offered to us round the clock, in our own homes. Unless we take particular care to preserve our sense of values, we let it all soak in. We have not yet become "diet conscious", as regards our intake of television fare. Without this self-control our perception becomes blurred and the lasting impression we have ceases to be governed by a strict process of deliberate reflection.

However, by the end of the 20th century, TV seems destined to merge with the internet (if not be replaced by it altogether). The return path *in cable television networks, and VOD (Video On Demand) services – as well as commercial movie-download online databases and the World Wide Web itself – suggest that TV, or its replacement, will be as interactive as the internet.*

1964 *Color television sets become popular in the United States.*

1967 *There are in the US 3,895 AM stations, 1,336 FM stations and 770 television stations.*

BBC and ITV began regular color TV transmissions.

1970–1980 *High-definition television (HDTV) developed by Sony in Japan.*

1972 *Digital[13] TV operating in US laboratories.*

[13] *Digital television*: The values of the brightness or color signals (*luminance* and *chrominance* respectively) of picture elements along a television line can be represented by a series of numbers. If these are expressed in base 2, each value can be transformed into a sequence of digital (binary) electrical pulses.
The conversion from the "analog" world to the "digital" world comprises two stages:
- *sampling*: in which the pixel values are measured at regular intervals
- *quantization*: in which each measurement is converted into a binary number.

The last operation is carried out by an Analogue to Digital Converter (ADC).
The series of "1" and "0"s obtained after quantization can be modified (i.e. coded) to counteract more effectively the disturbances the signal will undergo during transmission.
Digital television technology is an extension of computer and image processing technology. Advantages are high fidelity, easy storage and great scope for image

1977 *Renewed efforts to establish a low-cost videophony by digital transmission of images via telephone lines using optical fibers and image compression methods: the image is broken into pixels (=* **Pic***ture* **El***ements), as for television, each pixel being defined by one numerical value of luminance and two for chrominance (color). In order not to exceed the transmission capacity of optical fibers at that time, the image resolution was limited to 34 megabits per second.*

1978 *Rapid expansion of Cable Television systems.*

1979 *300 million television sets in the US.*

1982 *Japanese introduced wristwatch-size television with 3 cm screen.*

1983 *600 million television sets in the world.*

1985 *Sony produces a television set with a screen size of 24 m × 48 m.*

1991 *BBC World System Television is launched via satellite.*

1993–1995 *America's first high-power Direct Broadcast Satellite (DBS) is launched. It is followed by another in the same year and a third in 1995.*

1976–1990 *Advent of CNN television worldwide news network.*
1976 – Ted Turner delivered nationwide TV in the U.S. via satellite.
1980 – 24-hour news channel.
1990 – Domination of worldwide news coverage.

1990

- *1446 Television stations broadcasting in the United States.*
- *There are more television sets in the world than there are telephones.*
- *In Belgium, the average time spent watching television by children from 10 to 13 years was 210 minutes per day.*

---

processing.
Each picture element is isolated and can be called up independently.

VIII. Fiber Optics (1870–1988)

1870 **John Tyndall** *(UK) demonstrates that light can travel along a curved transparent waveguide by total internal reflection.*

1876 **Alexander Graham Bell** *(USA) invented the photophone, an instrument for transmitting sound by its modulation of a beam of light.*

1880 **William Wheeler** *(USA). Engineer in Concord, MA. Patents a system of internally reflective pipes to guide light through a building.*

1887 **Charles Vernon Boys** *(1855–1944, England) describes the concept of guiding light through glass fibers.*

1893 **Edward Drummond Libbey** *(USA) introduced fiberglass, a material made of very thin strands of glass. The fibers may be many times finer than human hair, and may look and feel like silk. The flexible glass fibers are stronger than steel, and will not burn, stretch, rot or fade.*

1917 **Albert Einstein** developed the theory of stimulated emission of radiation, a process by which an incoming photon beam can be *amplified* by stimulated emission of photons of the same frequency.

1931–1939 *Experiments conducted by the Owens Illinois Glass Company and the Corning Glass Works, led to development of practical methods of making fiberglass commercially.*

1927 **John Logie Baird** *(UK) experimented with light propagation along strands of flexible glass. A pioneer in fiber optics.*

1955 **Narinder S. Kapany** *(UK). The father of optical fiber communication. First to develop the theory of light propagation by total internal reflection in transparent glass fibers which contain a core encased in a cladding of lower refractive index. Tests begin to demonstrate communication via fiber optics.*

1957 **Basil Isaac Hirschowitz** *(South Africa) first used a new kind of an endoscope – the fiberscope: a flexible tube, made of an optical fiber, to explore the inside of a human body.*

1957–1960 *Creation of the LASER (Light Amplification by Stimulated Emission of Radiation) by* **Gordon Gould**, **Charles Townes**, **Arthur Schawlow** *and* **Theodore H. Maiman** *(USA).*

The characteristic peculiar to the laser effect is that the photons are emitted exactly in phase with the stimulating beam. This coherence *means that the laser beam is as directed (non divergent) as allowed by the theoretical diffraction limit, and can deliver considerable power to a point of microscopic dimensions. (The name* laser *was coined by* **Gould**.*)*

1966 **Charles Kao** *and* **G. Hockham** *of Standard Telecom Labs (UK) presented design principles of fiber-optic cables instead of copper wires to carry telephone conversations, using laser light and lossless hair-thin glass fibers. This concept was still a 'leap of faith' since at that time losses in the glass exceeded 10 db/km. At transmission a modulator transforms the electric signals into light signals, and at reception these are transformed by photo-diodes back into electrical signals.*

Optical fibers carry signals with much less energy loss than twisted copper-wire pairs or coaxial copper cables, and with much higher bandwidth. *This means that fibers can carry more channels of information over longer distances and with fewer repeaters required. Optical fibers are much lighter and thinner than copper wires with the same bandwidth. This means that much less space is required, per unit bandwidth, in underground cable ducts. They are immune to electromagnetic interference from radio signals, car ignition system, lightning etc. They can be routed safely through an explosive or flammable atmosphere.*

1970 *Corning Glass researchers* **Robert Maurer**, **Donald Keck** *and* **Peter Schultz** *designed and produced the first optical fiber with optical losses low enough for wide use in telecommunications (1 db/km). Their fiber-optic wire was capable of carrying 65,000 times more information than* *conventional copper wire. This technology would revolutionize communications in coming decades, replacing copper wires with cheaper and less bulky glass fibers. Normal electrical wiring limits the amount of telephone, television and computer traffic that can be carried by wire. But, in theory, a single laser beam should be able to carry all radio, television and telephone conversation of the entire world.*

1975–1977 *Transmission of* video signals *by optical fibers in UK and USA. Bell Telephone System installed the first working laser cable system beneath the streets of downtown Chicago (1977). It was the first to carry phone calls, computer data and video signals on pulses of light.*

1985–1986 *AT&T's Labs achieve the equivalent of sending 300,000 simultaneous telephone conversations (or 200 high-resolution TV channels) at once over a single optical fiber; optical fiber attenuation is reduced to a loss of 0.154 db/km, close to the 0.13 db/km silica theoretical limit.*

1988, Dec 14 *World's first trans-Atlantic optical fiber system. It extends 5800 km and has a capacity to carry 40,000 telephone conversations simultaneously, doubling trans-Atlantic cable capacity in a stroke.*

Because the information-carrying capacity of a signal increases with frequency, the use of laser light offers many advantages, and therefore fiber-optic laser systems are currently being used in military communications networks, cable television, and other applications.

One advantage of optical fiber systems is the long distances that can be maintained before signal repeaters are needed to regenerate signals. These are currently separated by about 100 km, compared to about 1.5 km for electrical systems. Newly developed optical fiber amplifiers can extend this distance even farther.

Optical fiber networks also offer a more private transfer of data, as it is quite difficult to 'tap' into a fiber, even compared to a 'private' phone line.

IX. Computer-based Communication Systems (1922–1998)

A. First Generation *(1822–1959):* Principle of programmable computer; mechanical, electromechanical and vacuum-tube based calculating machines; punched card and paper input/output, stored program, machine and assembler languages, high-level language.

1943 *The earliest programmable digital Electronic Computer runs for the first time in Britain. It contained 2400 vacuum-tubes ("electron valves") for logic, and was called the Colossus. It was built by* **Thomas Flowers** *to crack the German coding 'Enigma' machines and used at Bletchly Park during WWII. It translated 5000 characters a second and used punched tape for input.*

1944 ***H.A. Aiken*** *and* ***Grace Hopper*** *created the Harvard MARK 1 computer.*

1946 *ENIAC (Electronic Numerical Integrator and Computer): the first totally electronic, electron-valve driven, digital computers. Development started in 1943 and finished in 1946 at the Ballistic Research Laboratory by* ***John W. Mauchly*** *and* ***John P. Eckert****. ENIAC weighed 30 tons and contained 18,000 electronic valves, consuming around 25 kW of electric power. It could do around 100,000 calculations a second and was used for calculating ballistic trajectories and testing theories behind the hydrogen bomb.*

1947 ***Norbert Wiener*** *(US) published Cybernetics.*

William B. Shockley*,* ***John Bardeen*** *and* ***Walter H. Brattain*** *(US) invent the transistor at Bell Laboratories. This invention greatly impacted the history of computers.*

1949–1952 *EDVAC (Electronic Discrete Variable Computer), the first computer to use magnetic tape. Proposed by* **John von Neumann** *and completed (1952) at the Institute of Advanced Study, Princeton, USA.*

1950 **Yoshiro Nakamata** *invented the Floppy Disc at the Imperial University in Tokyo. The sales license for the disc was granted to IBM, who introduced it only in 1971.*

1951 *UNIVAC 1, the first commercially available electronic computer designed to handle both numeric and textual information. Designed by* **John P. Eckert** *and* **John W. Mauchly**.

1953 *There are about 100 computers in the world.*

1957 *The development of FORTRAN (***For***mula* **Tan***slation) completed by* **John Backus** *and his team at IBM – the first scientific computer programming language.*

1958 **Jack St.Clair Kilby** *(at Texas Instruments) and* **Robert N. Noyce** *invented the integrated circuit ('microchip').*

B. Second Generation *(1959–1965):* ADVENT OF TRANSISTORS, MAGNETIC TAPE, PRINTED CIRCUIT, HIGH-LEVEL LANGUAGES, RESULTING IN MUCH SMALLER, MORE POWERFUL AND 'USER FRIENDLY' COMPUTERS.

1960 *About 6000 computers are in operation in the US.*

1963 *The first PDP-8 minicomputer built.*

1965 *Fuzzy Logic designed by* **Lofti Zadeh** *at the University of Berkeley. BASIC (***B***eginners* **A***ll-Purpose* **S***ymbolic* **I***nstruction* **C***ode) developed at Dartmouth college, USA by* **T.E. Kurtz** *and* **J. Kemeny**. *Not implemented on minicomputers until 1975.*

James T. Russell *(USA) created the compact disc (CD) – the optical digital recording and playback process which has become an essential component of computer systems.*

C. Third Generation *(1965–1971):* INTEGRATED CIRCUITS, MONITORS AND KEYBOARDS, MOUSE AND WINDOWS, HYPERTEXT, OPERATING SYSTEMS, FAMILIES OF COMPUTERS (IBM 360 SERIES ETC).

1965–1968 **Douglas Engelbart** *of SRI (US), invented the computer mouse and Windows; pioneered work in the creation and design of modern interactive computer environments. His NLS (o***N***-***L***ine-***S***ystem) introduced two-dimensional computerized text editing using the mouse (to position a pointer into the text on the screen), a keyboard, a word processing program and a hypertext system. The mouse was conceived in 1965 but did not become popular until 1983 with the Apple computers, and was adopted by IBM only by 1987.*

The first supercomputer, the Control Data CD6660, was developed (1965).

ARPA Sponsored a study on 'cooperative networks of time-sharing computers' (1965). It then asked for bids to build a prototype link of 4 host computers in California and Utah (1968–1971).

Theodor Holm Nelson *coined the term 'hypertext' to describe nonsequential writing-text that branches and allows choices to the reader; best used at an interactive screen.*

The first supercomputers were first used for cryptography and nuclear physics, but have come into use for highly complex physical and mathematical computations in fields including oil and mineral prospecting, analysis of subatomic an subnuclear physics, studying the earth's ozone layer and simulating processes including weather systems and nuclear reactions.

1967 *First construction of a computer with parallel processor is proposed.*

1969

- *Bubble-memory system for computers was invented; it retains information even after computer is turned off.*
- *Work began on the ARPAnet, grandfather to the Internet. Computers began to talk to each other.*

1969–1971 *Advent the microprocessor – a computer on a chip:* ***Marcian Hoff*** *(US) proposed (at Intel) that a single chip general-purpose computer known as Central Processor Unit (CPU), would be programmed to perform most of the calculator function.*

Further refinement in architecture and logic design were made by ***Stanley Mazor*** *and* ***Frederico Faggin****. The first microprocessor, the 4004 was released in 1970. It contained the equivalent of about 2300 transistors and was capable of about 60,000 operations per second, running at a clock rate of 108 kHz. This single chip had as much computing power as the first computer, ENIAC (1946) which filled a room! The microprocessor is one of the most important developments of the second half of the* 20th *century. It is now found virtually in every automobile, household appliance, and computer.*

1970–1971 *Development of the UNIX operating system started by* ***Ken Thomson*** *and* ***Dennis Ritchie****. It was created at Bell Laboratories (US) for multi-user computers. It can run on a variety of different computers and thus rapidly became a worldwide standard, especially in universities and research institutions. It also speeded the development of Unix to Unix computer communication via phone and high-speed data lines.*

Intel 1103 – the first available dynamic RAM chip. (RAM = Random Access Memory.)

1970–1999 *From ARPANET to INTERNET – story of the 'Information superhighway'.*

*A military network called ARPAnet (***A***dvanced* ***R***esearch* ***P***roject* ***A***gency) was formed. It became open to non-military users in 1979–1980 when Academia was allowed to connect. Its name was then changed to INTERNET (1979) and many universities and large businesses went "on-line". The number of linked host computers increased exponentially during the years: 1969 (4); 1971 (23); 1981 (213); 1984 (1000), 1987 (10,000); 1988 (100,000); 1990 (300,000); 1992 (1,000,000); 1996 (10,000,000). Significant milestones in the development of the INTERNET are:*

- *In the first International Conference on Computer Communication at Washington DC (1972) the ARPANET was in the public eye for the first time.*
- *The engineer* **Ray Tomlinson** *(US) sent the first electronic mail between two computers at Cambridge, MA (1972). His software was rapidly incorporated into ARPANET's file-transfer protocol ('ftp') to facilitate communication via e-mail. He also created the @ symbol in the e-mail address to separate the user's name from that of his or her machine or domain.*
- *ARPANET became international with the first international link to University College, London (1973).*
- *BITNET (=* '**B***ecause* **I***t's* **T***ime'* **Net***work) launched (1981) as a cooperative network of the City University of New York (CUNY). It used e-mail. One of the first wide-area networks.*
- *The term INTERNET used for the first time in 1982. It has grown to be an international network of networks.*
- **W***orld* **W***ide* **W***eb (WWW) was launched in 1992.*
- *INTERNET encompassed 13,000 regions and national networks; there are some 10 terabytes of publicly available data (1993).*
- *NETSCAPE browser 1.0 (1994).*
- *NETSCAPE 'NAVIGATOR' 2.0 (1996).*
- *150 countries are connected to the INTERNET (1996).*

1971 *The Floppy disc is reintroduced by* **Alan Shugart** *at IBM for storing data used by computers. It is a 20 cm disc coated with iron oxide.*

D. Fourth Generation *(1972–):* MAGNETIC DISC, MICROCOMPUTERS, LSI (= LARGE SCALE INTEGRATION) BASED ON MICROPROCESSORS, TYPICALLY 500 OR MORE COMPONENTS ON A CHIP; VLSI (VERY LARGE SCALE INTEGRATION), TYPICALLY 10,000 COMPONENTS ON A CHIP.

Modern chips may now contain millions of components. This has led to very small, yet incredibly powerful computers. The fourth generation is generally viewed as running right up to the present time, since although computing power has increased, the basic technology has remained virtually the same. By the late 1990's people began to suspect that this technology has built-in limitations, and that further miniaturization could only proceed so much. Thus, 64 megabit RAM chips have circuit so small that it can be measured in atoms. Circuits this small pose many technical problems – notably the heat created, but they are also very susceptible to influence by temperature and radiation.

1972 *C programming language developed.*

First electronic pocket calculator (Texas Instruments).

Alan Kay *conceived the laptop computer (introduced by 1981).*

1973 *the microcomputer was born in France.*

1974 *The first computer with parallel architecture.*

1975–1981 *First personal (consumer) computer (PC), the Altair 8800 is introduced in kit form in the US (1975). It has 256 bytes of memory. About 100 computers connected to the ARPANET. IBM launches the 5100 portable computers.*

Apple computer, Inc. founded (1976). Designed by **Stephen Wozniak** *and* **Steve Jobs**.

Apple II, the first personal computer available in assembled form, was introduced (1976). In 1978, Apple brought out the first disc hard drive used in personal computers. In 1981, IBM announced their 5150 PC.

Its operating system soon became an industry-standard **D***isc* **O***perating* **S***ystem (DOS) with 64 kB of RAM and 40 kB of ROM.*

1973 **Robert Metcalfe** *at Xerox (US) created the Ethernet – a local area network for connecting computers within a building, using hardware running from machine to machine. The patent described it as "multinet data communication system with collision detection". It differs from the Internet* *which connect remotely located computers by telephone line, software protocol and some hardware.*

1976–1992 *CRAY 1, the first commercially developed supercomputer, built by* **Seymour Cray** *(US). It contained 200,000 integrated circuits and was freon-cooled. It could perform 150 million floating-point operations per second. The circuits of CRAY 2 (1992) featured gallium arsenide instead of silicon chips, and were submerged in a cooling bath of liquid fluorocarbon to prevent heat from the gallium arsenide from melting the machine. It could do about 250 million floating-point operations per second.*

1979 *Advent of word processors.*

1980 *More than 1 million computers are in use in the US.*

1981 *Microsoft's MS-DOS operating system.*

1983–1985 *The CD-ROM (Compact Disc Read Only Memory) is developed by Phillips (1983) and marketed by Sony (1985). It is an extension of audio CD technology for use in computers.*

Apple's Lisa brings the old mouse (1965) and pull-down menu to the PC (a mouse is a devise that moves the cursor on the screen as a result of moving the mouse on a hard surface; pressing the button (or one of the buttons) on the mouse sends a command to the computer, depending on where the cursor is located).

IBM's PC-XT is the first PC with a hard-disc drive built into it – a magnetic memory device then capable of storing 10 MB of information. By 2001 PC disc drives would hold many tens of gigabytes.

1984 *Apple Macintosh released (16 MB (megabytes) of RAM).*

1984 *Optical discs for storage of computer data were introduced.*

1985–1998 *Microsoft Windows launched (1985); it was followed by Windows 3.0 (1990), Windows 95 (1995), Windows 98 (1998), and later versions.*

1987 *A bandgap-engineered Heterojunction Bipolar Transistor made of epitaxial SiGe (Silicon-Germanium) alloys was demonstrated for the first time. Because the band structure and transport properties of electrons and holes can be* *engineered in these alloys, cutoff frequencies can be made larger than in devices based on doped Silicon alone; by 2004, unity-gain cutoff frequencies well above 200 GHz have been achieved for SiGe devices. SiGe HBT's can be used in electronic circuits for RF applications, as well as in high-speed optical networks.*

1987–1988 *Massive Connection Machine: a supercomputer which instead of integration of circuits operates up to 64,000 fairly ordinary microprocessors (using parallel architecture) at the same time. It can perform about 2 billion operations per second.*

1988–1990 *First optical chip, the optical computer developed by AT&T; it uses light (photons) instead of electricity (electrons) to carry data and can potentially reach calculation speed orders of magnitude faster than existing electronic computers.*

1989 *World Wide Web, invented by* **Tim Berners-Lee** *(at CERN, the European particle-physics laboratory). He saw the need for a global information exchange that would allow physicists to collaborate on research and exchange text, visual and other files interactively. The Web was a result of the integration of hypertext (coded in 'HTML': Hypertext Mark-up Language) and the internet.*

1992 *First U.S. website launched at SLAC (Stanford Linear Accelerator Center).*

1993–1998 *Pentium chip released (1993). The 166 MHz version contains the equivalent of 3.3 million transistors. The 1995 version of Pentium Pro achieved a clock speed of 200 MHz and contained 5.5 million transistors. In 1998, Intel produced the 333 MHz Pentium II processor.*

1996 *Texas Instruments announced they can now pack 125 million transistors into a single silicon chip the size of a thumbnail.*

E. Fifth Generation: CD-ROM, OPTICAL DISCS, PHOTONIC COMPUTERS.

1985 *CD-ROM can put 275,000 pages of text on a single CD record.*

1997 *Seiko Epson Japan invented a hand-cranked PC. When the battery starts to fade, a red-alert message alerts the user to crank the main spring.*

1998 *Sales of Personal Computers (percentage of all US homes): 1995 (27%), 1997 (43%), 1998 (50%); Driven by strong sales of lower-priced PC (ca 2100 dollar per unit), half of all US households have a PC. The PC is rapidly becoming a standard household appliance.*

1998 *Internet details:*

- *The total world wired population is 150 million people.*
- *The number of computer hosts in the global network is 36 million.*
- *Internet provides electronic mail, file transfer, a dynamic document distribution and presentation system (WWW). It is the greatest information and intellectual resource in the world, has a growing presence in entertainment, commerce, chats and other social interactions, and is the most visible manifestation of the information society.*

2004 *The Intel corporation released the LGA775 platform for the Pentium-4 chip. Its clock speed is 3.6 GHz, it packs 1 MB (MegaByte) of L2 cache, and the width of its circuit strips is 90 μm (900 Angstrom – one-sixth the wavelength of yellow light!). Chips based on bandgap-engineered devices – e.g. utilizing SiGe (Silicon-Germanium alloys) – could enter later generation CPU chips, because they will enable clock speeds of hundreds of GHz.*

X. Artificial Intelligence (AI) (1917–1997)

(*inter* = between; *legere* = to choose; Intelligence = to choose between)

1917–1921 **Karel Capek** *(1890–1938, Czechoslovakia) coined the term 'Robot'* (robota *= forced labor in Czech) to describe the mechanical people in his science fiction story, R.U.R. His intelligent machines, intended as servants for their human creators, end up taking over the world and destroying all mankind. "Robot" is defined as an automatic, autonomous device that performs functions normally ascribed to humans, or a machine in the form of humans.*

1928 **John von Neumann** *(USA) discovered the minimax theorem central to Game Theory.*

1937 **Alan Turing** *(England) introduced the abstract Turing Machine.*

1941 **Konrad Zuse** *(Germany) completed the world's first fully programmable* digital computer.

1942 **Isaac Asimov** *(1920–1992, US), Science Fiction author, introduced his fictional 'three laws of robotics'.*

1943 **Warren McCulloch** *and* **Walter Pits** *discuss neural-network architectures for intelligence.*

1946 ***John P. Eckert*** *(US)* and ***John W. Mauchly*** *(US) developed ENIAC, the world's first fully electronic, programmable digital computer.*

1947 **Norbert Wiener** *(US) published Cybernetics, a seminal book on information and control theory.*

1947 *The field of Artificial Intelligence (AI) began as the first computers were developed. It is often defined as a multidisciplinary field encompassing computer science, neuroscience, philosophy, psychology, robotics, and linguistics. It attempts to reproduce with machines the methods and results of human reasoning, speech and other brain activities.*

1949 **Maurice Wilkins** *(US) built EDSAC, the world's first stored-program computer.*

1950 **Alan Turing** *described a criterion for determining whether a machine is intelligent (the Turing Test).*

Claude E. Shannon *proposed a computer chess program.*

1956 **Stanislaw Ulam** *(US) wrote MANIAC 1, the first computer program to beat a human being in a chess game.*

The first use of the term Artificial Intelligence was given by **John McCarthy** *and* **Claude Shannon**, *at a computer conference at Dartmouth College, US.*

1956 **D. Devol** *and* **J.F. Engelberger** *(US): the first industrial robot 'Unimate'. It was later installed at a General Motors plant to work with heated die-casting machines.*

1957 **Allen Newell**, **Herbert Simon** *and* **J.C. Shaw** *(US) developed their Logic Theorist program, one of the first AI programs that allow a computer to reason abstractly (it could, for example, prove theorems in the 'Principia Mathematica' of Russell and Whitehead.*

1959 **Arthur Samuel** *of IBM wrote a checkers-playing program that performs as well as some of the best players at that time.*

1962 **Engelberger**'*s 'Unimation' company began marketing industrial robots.*

1963 **Marvin Minsky** *(US) published "Steps Toward Artificial Intelligence".*

1964 **Marshall McLuhan** *(Canada) in his book 'Understanding Media', foresaw electronic media as creating a 'global village' in which "the medium is the message".*

1966 *The SRI (**S**tandard **R**esearch **I**nstitute) robot represents an attempt to combine learning programs, pattern recognition programs, problem-solving programs and programs that represent information about the outside world.*

1969 *First man landed on the moon.*

1971 *The first microprocessor was introduced in the US.*

1974 *The first computer-controlled industrial robot was developed.*

1977 *Voyagers 1 and 2 launched.*

1980 *Philosopher* **John R. Searle** *(USA) criticized computationalism and the pretensions of the strong Artificial Intelligence program. He maintains that computers cannot think because a system's behaving as if it had mental states is insufficient to establish that it does in fact have these states. Moreover, he claims that there is no fact intrinsic to the physics of computers that makes their operations syntactic or symbolic: rather, the ascription of syntax or symbolic operations to a computer program is a matter of human interpretation.*

1982 *Second-generation robots were built, with the ability to precisely effect movements with 5 or 6 degrees of freedom. They are used for industrial welding and spray painting.*

Defense robots used by Israel in Lebanon.

1984 *Waseda University in Tokyo completed Wabot-2, a 100 kg robot that reads sheet music through its camera eyes and plays organ with its ten fingers and two feet.*

Marvin Minsky *published The Society of Mind, in which he presents a theory of the mind in which intelligence is seen to be the result of proper organization of a very large number of simple mechanisms, each of which is by itself unintelligent.*

1986 *Third-generation robots with limited intelligence and some vision and tactile sensing.*

Dallas police used a robot to break into an apartment. The fugitive ran out in fright and surrendered.

1986 *'Optical character recognition' (OCR)* technology is growing.

New *medical imaging systems* are creating a revolution in medicine.

The university of Pennsylvania developed a robotic *ping-pong player* that wins against human beings.

1988 Population of *industrial robots* has increased from a few hundred (1970) to several thousand, most of them in Japan.

1992– Military strategies of the leading industrial nations increasingly rely on flying *smart weapons* which incorporate electronic copilots, pattern-recognition techniques, and advanced technologies for tracking, identification, homing and destruction.

A multi-hundred-billion-dollar *computer and information-processing industry* is emerging, together with a generation of ubiquitous machine intelligence that works intimately with its human creators.

Reliable person identification, using pattern-recognition techniques applied to visual and speech patterns, replace locks and keys in many instances.

AI technology is of greater strategic importance than man-power, geography, and natural resources.

1997, May 11 The IBM computer *DEEP BLUE* wins $3\frac{1}{2} : 2\frac{1}{2}$ against the world chess champion Garry Kasparov; first machine of human-competitive intelligence created by man. A computer is the world chess champion. The machine will be able to handle (ca 2000 CE) 1 billion nodes per second and will have a chess master rating of 3,400, about 500 points higher than the world's best human player.

XI. Geo Satellite in the service of telecommunication (1945–1996)

1945 **Arthur C. Clarke** *proposed the idea of communication satellite that would appear to hover motionless over a single point on earth; called synchronous satellite* (or satellite in *a geosynchronous orbit*). *They became the principal means of intercontinental communication starting in 1965. Newtonian physics predicts a 24 hour period for a mass 35,880 km above the equator. Clarke also recognized that a net of three such satellites, suitably placed, could provide radio relays between any points on earth (except regions very close to the poles).*

1957, Oct 4 *Soviet satellite Sputnik 1 ("fellow traveler") was launched: a little steel ball weighing 83.6 kg, containing a radio transmitter and batteries, transmitting a steady series of beeps, but doing nothing more.*

1958, Jan 31 *First US satellite Explorer 1 was launched: a two meter long body, 15 cm in diameter, weighing 14 kg. It carried a Geiger-Mueller counter to detect cosmic rays. It detected the Van Allen radiation belt girdling the earth.*

1960, Apr *The weather satellite Tiros 1 (military, in civilian guise) discovers the beginnings of a tropical storm in the South Pacific. It sent back nearly 23,000 photographs of earth's cloud cover.*

On Aug 12, NASA launched ECHO 1, a spherical balloon with a metalized skin. Once in orbit the balloon was inflated until it reached its intended diameter of 30 m, and it was then used as a reflector to bounce radio signals across the oceans – simply a radio mirror in the sky.

1962, July 10 *US "Telstar 1" was launched, the first commercially developed true communications satellite. It was a multifaceted sphere just under 1 m in diameter and weighing about 77 kg. It was able to receive messages from the ground, amplify them and then retransmit them immediately. This made it possible to send high-quality data, for instance television pictures, from place to place. On July 23 Telstar 1* *carried a TV program between USA and Europe that was watched by 200 million viewers. It ceased to function in March 1963.*

1964, Aug 19 *US launched "Syncom 3" into a geostationary orbit; the first successful demonstration of the principle of a geostationary communication satellite; It was used with great effect to broadcast the coverage of the 1964 Olympic games from Japan. It was a cylindric satellite with a built-in rocket motor. Upon reaching the altitude of 35880 km it fired its motor to circularize the orbit at that height and become geostationary.*

1964, Aug 20 *The formation, by 11 countries, of INTELSAT (International Telecommunications Satellite Organization) with headquarters in Washington DC (119 members in 1990). Tokyo Olympic Games are replayed to North America.*

1965, April 6 *INTELSAT 1 (or Early Bird) launched and placed in geostationary orbit over the Atlantic Ocean. It went into service on June 28 – the world's first commercial communication satellite. Early Bird was a cylinder, 72 cm wide and 52 cm high, with solar cells around its circumference. It was spun around the axis to provide stability in space. It could relay 24 telephone lines or one television channel and remained in service for $3\frac{1}{2}$ years.*

1967–1991 *INTELSAT developed bigger and bigger satellites in response to continuing increase in demand for services: INTELSAT 2 (1967) weighed 86 kg. INTELSAT 3 (1968) weighed 150 kg, could carry 1200 telephone circuits or four Television channels (or a combination of both) and its antenna could point continually toward the earth.*

INTELSAT 4 (1971) weighed 720 kg with an overall height (including extended antennas) reaching 5 m. It was able to carry 6000 telephone circuits and color television channels.

INTELSAT 5 (1980) weighed 1000 kg and was able to transmit 12,000 telephone channels plus color television. It was kept stable about all 3 axes by gas jets, and had a pair of large solar wings to generate electricity. This system allowed all the solar panels to be exposed to sunlight all the time and produced more power, which in turn increased the capabilities of the satellite.

INTELSAT 6 (1989–1992), weighed 2500 kg (11.7 m high), and could relay 24,000 telephone lines plus 3 television channels. It was launched into orbit by a commercial Titan rocket.

1969 *First live television broadcast from the moon, by the US Apollo 11 moon-landing mission, is transmitted to some 600 million viewers worldwide by satellite feed. The worldwide linkup cost $ 55 billion and involved some 40,000 TV technicians and personnel in 49 nations. Sophisticated transmitting and receiving equipment was part of the compact VHF communication system on board the Apollo command and lunar modules.*

1972–1984 *Landsat 1 (formerly named ERTS 1), dubbed 'eye-in-the-sky', was launched on 23 July, 1972, by a Delta rocket as part of a project to observe the earth, monitoring such problems as crop disease, flooding, icebergs and pollution. It was followed by Landsat 2 (1975), Landsat 3 (1978), Landsat 4 (1982), Landsat 5 (1984) and Landsat 6 (1992).*

Landsat 1 weighed 891 kg. It was placed at an altitude of 900 km in a polar orbit, from which it could observe the same area every 18 days. Landsat 4 and 5 weighed 2000 kg each. Since 1985, Landsat data have been marketed, supplying users with photographs or electronic images suitable for display and image processing on desktop computers.

1977 *Eutelsat links the telecommunication networks of the European countries via communication satellites.*

1987 *Soviet radar satellite (weighing 20 tons) was launched. It has applications in map-making, oceanography, crop predictions, ice monitoring and prospecting for minerals.*

1993 *NASA's* **A***dvanced* **C***ommunications* **T***echnology* **S***atellite (ACTS) is launched from the Shuttle Discovery during Mission STS-51 on Sept 12. ACTS is a testbed for future communications satellite technology, such as multi-beam antennae and advanced signal handling.*

1995 *World Wide Web has become big business as NETSCAPE Communications goes public. About 6.6 million computers are connected to the INTERNET.*

1996 *13 million host computers on the INTERNET.*
The INTERNET includes by now 18 satellites in orbit.

XII. Is Anyone out There? – Spacebound Odyssey in Search of Extraterrestrial Intelligence (1959–1989)

(a) Introduction

Until very recently, man's ancient dream to free himself of the chains of gravity and fly away from earth in search of other worlds, materialized only in myth, literature, and science fiction.

However, the rapid development of aeronautical science, nuclear physics, solid state physics, computer technology and fuel chemistry in the wake of WWII, to which the cold war added a strong element of competition – pushed the science of rocketry to the verge of realization.

In 1944, within two years of the historic first self-sustaining nuclear chain reaction (Dec. 2, 1942), **Stanislaw Ulam** *and* **Frederick de Hoffman** *mused at Los Alamos that the power of the atomic explosion might somehow be controlled to launch space vehicles. Ulam and de Hoffman were following the thoughts of the great master of science fiction,* **Jules Verne**, *who, in his 1865 novel, De la Terre la Lune, wrote that the Baltimore Gun Club fired a manned projectile to the moon from a huge cannon emplaced near Cape Canaveral, now Cape Kennedy, Florida.*

But ideas rarely come to fruition unless a practical need beckons; near the close of WWII, German V-2 rockets had proved they could carry some 750 kg of amatol explosives 360 km from Nazi-held territory in Europe to London. What if the V-2's had carried atomic bombs? The thought was unsettling. Soon designs for Intercontinental Ballistic Missiles (ICBMs) began to take shape on drawing boards around the United States. Some of these big rocket design were nuclear at both ends – they had nuclear warheads and nuclear engines. Secret reports issued in July 1946 by North American Aviation, Inc., and Douglas Aircraft Company are landmarks in the history of nuclear rockets. The reports underlined the great promise of the "heat transfer" nuclear rocket, noting its high exhaust velocity[14]*, its attainment of very high temperatures, and its high rate of heat transfer.*

The military rocket work was naturally classified as secret by the U.S.

---

[14] Chemical rocket engines, jet engines, automobile engines – in fact, most of mankind's engines – extract heat from a fuel and turn it into macroscopic motion through the expansion of hot gases. The nuclear rocket also creates hot high-pressure gas and turns it into reaction thrust. The higher the propellant *velocity*, the more thrust we get from each kilogram of gas that roars out the nozzle each second. We want to have a high exhaust velocity for good rocket performance, because we can thereby accomplish a space mission with less propellant.
The nuclear rocket produces about *twice* the exhaust velocity of the best chemical rocket, for the following reason: the exhaust velocity V of any rocket is proportional to $\sqrt{\frac{T}{M}}$, where T is the temperature of the hot gases just before they enter the nozzle throat, and M is the average molecular weight of the exhaust gases. Now, chemical rockets already operate at temperatures close to 3000 °K; nuclear rocket reactors operate at the same general temperature level. However, chemical rocket exhaust velocities are limited by the high molecular weight of the combustion products [water ($M = 18$), methane ($M = 16$), ammonia ($M = 17$)] whereas a nuclear rocket, where combustion is not required, can make use of a propellant with low molecular weight, such as molecular hydrogen ($M = 2$). [Fissile-material nucleus release kinetic energy without chemical stimulation, and the propellant is not an engine fuel but a separate substance that is heated by the fissioning nuclei in a nuclear reactor.] With $M = 2$ (instead of 18 as in hydrogen-oxygen chemical engines), the nuclear rocket exhaust velocity will be more than double that of the best chemical rocket for the same temperature. Using the equations: thrust $= \frac{dm}{dt}V$, power $= \frac{1}{2}\frac{dm}{dt}V^2$, we see that if the thrust is held fixed and exhaust velocity V is doubled, propellant flow $\frac{dm}{dt}$ will be halved, but the power required will double. The price of increasing the exhaust velocity is the need for increased power production by the engine. From this relationship arises another important advantage of the nuclear rocket: The great reservoir of energy contained in its nuclear fuel can be turned into high exhaust velocity.
The three basic facts about nuclear rockets are:

- They convert fission-generated heat into the kinetic energy of rocket propellant.
- Chemical combustion is not needed, and so they can use lower molecular weight propellant to attain high exhaust velocities.
- Their reactor fuel has a more potential energy packed into it.

Air Force. However, at the Applied Physics Laboratory of the John Hopkins University in Maryland a group of engineers, who were unaware of the ICBM work and did not have access to the secret reports, innocently proceeded to duplicate all the important findings of North American and Douglas. Their unclassified report was published in January 1947. It was obvious that any competent with a slide rule and a few scraps of paper could discover the essentials of the nuclear rocket without much help.

In 1948 and 1949, two British space enthusiasts, **A.V. Cleaver** *and* **L.R. Shepherd**, *again duplicated most of the secret nuclear rocket fundamentals in a classic series of papers published in the Journal of the British Interplanetary Society. Not long before the English report appeared, the American-educated Chinese scientist* **H.S. Tsien** *had reported his studies on the application of nuclear energy to rockets and other "thermal jets" at a M.I.T. seminar. The basic principles of the nuclear rocket could no longer be concealed. [It is interesting that Shepherd went on to become a key man in Britain's atomic energy program, and Tsien later returned to China where he was a principal figure in the development of the Beijing government's atomic bomb.]*

A nuclear rocket engine is considerably more than a heater of hydrogen. It is true that the engine is built around the reactor core, the wellspring of energy, but in addition hydrogen propellant must be transported from tank to reactor; heat must be partially converted to thrust. The engine, in fact, has five major segments:

- *The nuclear reactor heat source.*
- *The pump that pulls liquid hydrogen from its tanks and forces it through the reactor.*
- *The nozzle, which transforms heat to thrust.*
- *The structure that physically holds the pieces together.*
- *The controls that force all engine components to march in step at the command of the spacecraft pilot.*

In a nuclear rocket engine intended for space travel inside the solar system, the reactor must generate at least 5000 megawatts of thermal power (more than the output of 50,000 V.W. cars or 160,000 home-heating furnaces), and still not encumber the rocket with too much inertia.

Four requirements control reactor design:

- *The need to attain a critical mass (i.e., the smallest mass of fissionable material that will sustain a chain reaction).*

Photo 2: Elsewhere in the Universe (courtesy of Steve Satushek)

- *The need to remove all generated heat.*
- *The need to raise the power level at will, or control the reactor.*
- *The need to maintain structural integrity at high temperatures and under the forces exerted by the high thrust.*

Attainment of a critical mass is a matter of "neutron economics", because it is the cloud of invisible neutrons coursing through the reactor that stimulates nuclear fission and thus power production. Each fissioned atom of uranium-235 produces $2\frac{1}{2}$ *new neutrons, on the average. If the rate at which fission occur (and consequently the power level) is to remain constant, exactly one of these* $2\frac{1}{2}$ *neutrons (on average) has to go on and cause another fission. Reactor "criticality" occurs at just this point. This balance sheet leaves* $1\frac{1}{2}$ *neutron per fission, each of which may either escape the reactor altogether or be absorbed in nonfission nuclear reactions. To prevent too* many *neutrons from escaping, a material is placed around the reactor to reflect some errant neutrons back into the core.*

Excessive neutron absorption can be avoided by using core and reflector materials that have little appetite for neutrons; fortunately, graphite *has just such a low neutron absorption cross section. Naturally, enough uranium-235 atoms must be dispersed throughout the graphite core so that the neutrons can find and fission them. Core design requires a balancing of all these considerations, plus one more.*

Almost all the energy released by uranium is first incorporated in the kinetic energy of two large fission fragments that fly off in opposite directions as the nucleus splits. In a few millionths of a second, the kinetic energy of the heavy fragments is transferred to the nearby atoms and molecules, setting them in motion. A heat pulse *flows outwards to the boundary of the* nuclear fuel *(uranium carbide dispersed throughout the graphite), where it heats the hydrogen propellant. If this heat is not removed, core temperatures will quickly rise beyond the sublimation point of the fuel. The reactor core, therefore, has to be designed in such a way that all this fission-generated heat is transferred to the hydrogen gas that is to be driven through the reactor by the pump. This means that the reactor must be perforated with holes that carry the hydrogen (serving here as a reactor coolant) through and past the hot fuel to the nozzle. Coolant holes are coated with niobium carbide to prevent chemical corrosion of graphite by the hot hydrogen.*

To raise or lower the reactor power, the neutron economy must be upset, or altered. Control drums *help perform this task. These are cylinders covered on one side with neutron "poison", i.e. materials whose nucleus readily absorb neutrons at the typical speeds they acquire in the reactor.* Boron *is one such element. When all the drums' absorbing faces are turned inward, neutron that would otherwise be reflected back into the core to cause new fissions are absorbed by the poison instead. To start the reactor, motors slowly rotate the control drums, moving the poison away from the core regions, thus giving the neutron economy a boost. Unless the drums are returned to the exact point where criticality occurs, reactor power will rise exponentially. Because neutron generations are only milliseconds apart, neutron "population explosion" (and thus reactor power changes) can come about very quickly.*

A small portion of the hot hydrogen stream is "bled" off, diluted with a little cold hydrogen and directed through the turbine that powers the pump. This pump must raise the pressure of liquid hydrogen by about 94 atmospheres while delivering about 8 tons of it per minute to the reactor.

The hydrogen leaving the hot end of the nuclear rocket reactor is laden with the thermal energy that first must be converted into gas kinetic energy and then into rocket kinetic energy. This is the job of the nozzle. A constriction called the nozzle "throat" starts the process. First the throat speeds up the hydrogen velocity until it is traveling at the speed of sound (Mach 1). Beyond the throat, the nozzle opens up into a carefully contoured divergent section. Here, the hydrogen expands and cools rapidly as heat energy is converted to gas velocity (kinetic energy). The hydrogen, now traveling at supersonic velocities, pushes against the divergent sides of the nozzle, thrusting the rocket in the opposite direction through airless space. The expansion of hot gas in the nozzle is analogous to gas expansion against a piston or a turbine blade. To achieve a high exhaust velocity, the exit area of the divergent section must be as large as possible in comparison to the throat area. This ratio (100: 1) is limited only by the length and weight of the nozzle. Powerful forces act on the nozzle because it has to carry the entire thrust load up to the rocket body proper. What material can withstand such stresses in the presence of super-hot hydrogen rushing past it at supersonic speeds?

The solution is a high temperature alloy such as stainless steel, covered by a solid phalanx of cooling tubes that keep the nozzle temperatures well below the melting point of the alloy. The coolant that is pumped through these tubes is the supercooled liquid hydrogen which is returned to reactor's core. Without this cooling the nozzle would not survive more than a few seconds.

Besides the system unity imposed by the structure and controls, the engine parts have to fit together thermodynamically. To breath "life" into a nuclear rocket there must be a starter, like that in an automobile, intrinsic in the system. The engine must "catch", become self-sustaining, and generate useful power. For this to happen, two energy sources must be found: One to start the engine and another to power the pump that keeps propellant flowing through the engine.

In a nuclear rocket, the heavy reactor replaces the empty combustion chamber of the chemical rocket. Heavy though the nuclear engine may be, it is still but a small appendage on a much larger structure consisting mainly of the huge propellant tank.

The doubled exhaust velocity of the nuclear rocket in comparison with the chemical rocket, means that the nuclear rocket uses only half as much propellant each second of operation. Thus, the nuclear rocket's great economy in propellant consumption make it superior for missions in which most of the spacecraft weight is allotted to propellant. This is subject to the condition that the payload (anything carried that is not necessarily for the flight of the vehicle) is not a great deal smaller than the weight of the nuclear engine itself; manned expeditions to Mars, or the ferrying of supplies to the moon, are right down the nuclear rocket's alley. Generally speaking, the more ambitious the mission, the better the nuclear rocket looks.

On Oct. 04, 1957, the Soviet Union launched the first artificial satellite, SPUTNIK (Russian: fellow traveler). It was a metal sphere with diameter of ca 58 cm, circling the earth with a velocity ca 29,000 km/h, once every 95 minutes[15]*. Its remains fell to earth on Jan. 04, 1958. In 1959, a Russian*

---

[15] Let a mass m be fired off a planet of mass M and radius R. To achieve orbit, the initial speed should exceed the *first cosmic velocity*, V_1:

$$\frac{mV_1^2}{R} = mg = G\frac{mM}{R^2}, \therefore V_1 = \sqrt{\frac{GM}{R}} = \sqrt{gR}$$

For earth $V_1 = 7.9$ km/sec.
Second cosmic velocity (=*escape velocity*):

$$\frac{1}{2}mv_2^2 = \frac{GmM}{R} \therefore V_2 = \sqrt{\frac{2GM}{R}} = \sqrt{2gR}$$

For earth $V_2 = 11.3$ km/sec. Above, G is the universal gravitational constant and g is the surface gravity.
For a mass m orbiting the said planet at radial distance $r > R$, Newton's 2nd law, $\frac{GMm}{r^2} = \frac{mV^2}{r}$, yields $V = \sqrt{\frac{GM}{r}}$. The period of revolution is $T = \frac{2\pi r}{V} = 2\pi\sqrt{\frac{r^3}{GM}}$, namely $T^2 \propto r^3$ (Kepler's third law). Substituting $GM = gR^2$, we find $\left(\frac{r}{R}\right)^3 = \frac{gT^2}{4\pi^2 R}$. If T is chosen to be the planet's own period of rotation about its axis, we obtain for the earth $r = 6.618\ R$, or altitude $= r - R = 5.618\ R = 35,890\ km$. This is the suitable altitude for a *synchronous communication satellite*.

space probe televised the first pictures of the dark side of the moon hidden from earth. On April 12, 1961, the Russians launched Yuri Gagarin, the first man ever in space, on the spacecraft Vostok I. He circled the earth in 1^h48^m.

The United States responded to the Soviet challenge in space by establishing the National Aeronautics and Space Administration (NASA) in 1958 to conduct and coordinate the U.S. nonmilitary research into problems of flight within and beyond the earth's atmosphere. It has more than 10,000 scientists, engineers and technicians. Its installations include the John F. Kennedy Space Center (Cape Canaveral, FA), the Lyndon B. Johnson Space Center (near Houston, TX), the Goddard Space Flight Center (Greenbelt, MD), the George C. Marshall Space Flight Center (Huntsville, AL), Flight Research Center (Edwards AFB, CA), Jet Propulsion Laboratory (Pasadena, CA) and other centers.

The results of this gigantic effort culminated in the first lunar landing of Apollo 11 on July 20, 1969. On that day (10:56 UT pm), after 4 days of travel at an average speed of 4000 km/h, **Neil A. Armstrong** *set foot on the rocky plain of the moon's Sea of Tranquility.*

The commitment to land the first man on the moon, successfully fulfilled in 1969, again boosted scientists onto the crest of a wave of popularity. Caught up in the excitement of the space race, no fewer than 20 U.S. Federal agencies were supporting research and development in the 1960's.

This multiplicity of funding sources produced spectacular results. It yielded exciting new knowledge about the nature of the planets and of our place in the universe. Fundamental physics, astronomy, electronics and computer technology have benefited, as well as chemistry, material science and the life sciences. Not only did the U.S. decisively win the race to the moon, but it has acquired new and better electronic and medical services that owe their existence, at least in part, to the needs of the space exploration.

On March 03, 1972, Pioneer 10 was launched from Cape Canaveral, FL. The 260 kg craft made its way safely through the asteroid belt, a region between the orbits of Mars and Jupiter littered with rocky debris. It flew within 130,000 km of Jupiter's cloudtops on Dec. 02, 1973, returning the first close-up images of the sun's largest planet. It proceeded to traverse the orbits of Saturn, Uranus, Neptune and Pluto. On June 13, 1983, Pioneer became the first spacecraft to depart the realm of the known planets.

Note that the net energy of the orbiting mass (kinetic plus potential) is

$$\frac{1}{2}mV^2 - \frac{GmM}{r} = \frac{1}{2}m\frac{GM}{r} - \frac{GmM}{r} = -\frac{1}{2}\frac{GmM}{r} < 0.$$

It becomes *more* negative, the closer the mass is to the planet.

Radio messages from Pioneer tell us that it moves in cold, dark and empty space. 'Solar wind' particles, moving with speeds of 1.5 million km/h, blow at its tail while cosmic rays race inward past it. It may be able to detect gravitational waves or locate the putative Planet X perturbing Uranus and Neptune. As the years go by, its radio will go dead and its guidance sensors will lose sight of the sun. It will then cruise on, mankind's first emissary to the universe. About 10,506 years from now it will pass within 3.8 LY of Barnard's star. In 862,063 years Pioneer will approach the vicinity of Altair, a star nine times brighter than the sun. In case Pioneer is intercepted by intelligent beings, it carries a plaque with images of a man and a woman, a diagram of the solar system and other symbols that might enable 'others' to locate the origin of the little craft. Five billion years from now, Pioneer should be wandering about the outer rim of the Milky Way galaxy. The sun is expected to burn out and die in 5 billion years, and with it the earth.

At the turn of the 21th century, the engineering problems encountered in the design of interstellar space vehicles seem to present a number of apparently insurmountable obstacles. It is nevertheless quite interesting to examine the feasibility of interstellar travel from the point of view of the theory of relativity.

The interstellar distances involved are of the order of a few light years ($\approx 10^{16}$ m) to the nearest star, of order 10^4 light-years ($\simeq 10^{20}$ m) to the center of our own galaxy, and of order 10^6 light-years ($\approx 10^{22}$ m) to the nearest neighboring galaxies. In all cases the limited lifetime of the crew requires either space vehicles capable of attaining speeds close to the speed of light, or a multi-generational ('space ark') ship.

Supposing one can solve the engineering problem of constructing a vehicle that can accelerate (as measured by the crew) at the rate $a = g \approx 10$ m/sec^2 for the entire duration of the trip, then the velocity reached by the vehicle after a time t has elapsed in an inertial frame stationary w.r.t. the earth is, according to a standard relativistic calculation,

$$v = \frac{gt}{\sqrt{1 + (gt/c)^2}}$$

and as seen from earth it will have traveled the distance

$$d = \frac{c^2}{g}\left(\sqrt{1 + \left(\frac{gt}{c}\right)^2} - 1\right).$$

The crew benefits, however, from the time dilation effect. One finds for the time t' elapsed, as measured by the crew during earth time t, the expression

$$t' = \frac{c}{g} ln \left[\frac{gt}{c} + \sqrt{1 + \left(\frac{gt}{c} \right)^2} \right] .$$

Assuming the crew wants to land at a distance D from earth, the ideal way of traveling would consist of accelerating at rate g up to $d = \frac{1}{2}D$ and then decelerating at that rate over the other half of the total distance, so that the total traveling time will be $T = 2t$ in earth time, and $T' = 2t'$ in crew time. By applying the formulae given above to this case one finds:

Distance traveled $D = 2d$	*Traveling time as measured on earth,* $T = 2t$	*Traveling time as elapsed for crew,* $T' = 2t'$
2×10^{16} *m*	*3.6 yr*	*2.7 yr*
2×10^{20} *m*	2.2×10^4 *yr*	*20 yr*
2×10^{22} *m*	2.2×10^6 *yr*	*29 yr*

Clearly, because of the relativistic time dilation effect, one cannot rule out absolutely the feasibility of interstellar or even intergalactic travel within a single current human lifetime, even without cryogenics-assisted suspended animation for the crew.

Besides nuclear-powered impulse rockets of the kind described here, other, even more revolutionary designs have been proposed and studied for interplanetary and interstellar space missions. Among these:

- *ion rocket (tested in a 1998 NASA experiment)*
- *antimatter engines (power and/or thrust via annihilation of stored antimatter with matter and subsequent particle production)*
- *photon rocket (shining backward – propagating light beams for thrust)*
- *ramjet engine based on nuclear fusion fueled by scooped interstellar hydrogen*

(B) Timeline

Scientists have been sending regular radio waves out into space since 1920. All of our radio, TV, satellite, and radar signals are currently spreading out, slowly sweeping through the ca. 10^{11} *(100 billion) stars of the Milky Way Galaxy.*

1959 *The birth of SETI (acronym for the* **S***earch for* **E***xtra***T***errestrial* **I***ntelligence) – the science of searching the skies for signals from alien civilizations using radio and optical telescopes. Incepted with a paper in Nature by* **Giuseppe Cocconi** *and* **Philip Morrison** *in which detection by radio waves was suggested as the earliest method of communication.*

1960–1990 **Frank Drake** *(radio astronomer, US) embarks on Project Ozma: a search for extraterrestrial intelligence in the form of radio signals from other civilizations. With an antenna diameter of 25 m he listened, at the edge of the "spectral water hole", at frequency 1420 MHz (the natural frequency of the 21 cm – wavelength hyperfine – splitting line of atomic hydrogen). In 1975, Drake and* **Carl Sagan** *(1934–1996, US astronomer) listened by means of an antenna with diameter 300 m in the frequency range 1420–2380 MHz. An ongoing SETI program has been privately funded since the 1980s. So far, no alien signals have been detected. Considering that there are millions of frequency bands to sift through and the relatively short period of listening, it may be that we just have not yet looked at the right place at the right time.*

1972–1973 *The first space probes Pioneer 10 (launched March 02, 1972) and Pioneer 11 (launched 05 April 1973) to leave the solar system. Each carries a plaque that will allow whoever finds them to trace them back to earth, using pulsars as astronomical signposts in time and space. Each plaque, 155 mm × 229 mm in size, is of gold-anodized aluminum plate, into which is etched a diagram of the solar system, a pulsar map and a picture of man and a woman. The plaques are mounted in a position on the spacecraft's antenna mount which is expected to protect them from erosion by interstellar dust for at least 100 million years. By 1974 the Pioneers coasted Jupiter and by 1994 they were 9 billion km from the sun – far beyond the Solar system.*

1977–1989 *Voyager 1* (Sept. 05) and *Voyager 2* (Aug 20) sent from USA to explore the edge of the solar system.

Voyager 1 arrived at *Jupiter* (spring 1979), *Saturn* (Nov 1980) and its largest satellite *Titan*.

Voyager 2 flew by *Jupiter* (July 1979), *Saturn* (Aug 1981), past *Uranus* (1986), *Neptune* (Aug 1989) and Neptune's largest moon, *Triton*. On Nov 05 2003, Voyager 1 reached 90 AU and is the first human-built craft to explore and report back on the *interstellar medium*, with Voyager 2 close on its heel.

1990's–early 2000 Over 120 *exoplanets* (planet around other nearby stars) discovered indirectly. Indirect methods include:

- *Radial Doppler shifts* (periodical shifts in a star's spectrum caused by the wobble of the star about the COM of its planetary system).
- *Precision astrometry* (minute movements of a star perpendicular to its line-of-sight from earth, again due to planet-caused wobble).
- *Photometric detection* of the effects of exoplanet transits upon the parent star's spectrum.
- *Single-star gravitational lensing.*
- *Space-based interferometry.*

Numerous exoplanet-hunting space missions are in the works, scheduled to be launched during the first two decades of the 21th century. Some of them will try to look *directly* at exoplanets, by either dimming their star's disc or looking in the infrared.

Most of the exoplanets thus far found are gas giants, and all are either too cold or too hot to be likely harbors of life. It is hoped that some of the planned missions will be able to identify *earth-like* exoplanets, and even analyze their chemistry (and possible biochemistry!) via their atmospheric spectroscopy.

A Timeline of Discoveries in Particle Physics

Bubble tracks are left in a bubble chamber by tiny electrically charged subatomic particles as they travel through the chamber's depressurized cryogenic fluid. The ions left along the track nucleate bubbles because the chamber's rapid de-pressurization – synchronized with the arrival of the primary (projectile) particle beam – renders the fluid superheated.

The bubble chamber functions as both a target for the projectile particles (furnished by the liquid's nuclei-protons in the case of liquid hydrogen) and as a detector for all charged particle tracks, whether projectiles or reaction products.

Bubble chamber images have the topology of branched trees: tracks (prongs) successively branch off at interaction vertices.

An electrically neutral particle leaves no bubble tracks, but its path can be reconstructed between the vertex which created it (if any) and the vertex at which it ceased to exist. An ambient magnetic field bends the charged tracks; the curvature is used to calculate the particle's momentum and charge. The charged particle's energy affects its rate of energy and momentum loss along the track, and can thus be deduced from the bubble track's appearance. From a track's momentum $\boldsymbol{p}$ *and energy* E*, the corresponding particle's rest mass can be deduced via the special-relativistic formula:*

$$m_0^2c^4 = E^2 - \boldsymbol{p}^2c^2.$$

Particle physicists thus have to decode cascades (vertex-track-vertex trees) – captured in bubble chamber images and other types of detectors arrayed around mammoth particle accelerators – in order to deduce basic information about the observed particles, such as electric charge, particle spin, mass, lepton number, baryon number, parity and other quantum numbers that turn out to be useful in describing the elementary particle side of nature.

Before the invention of the bubble chamber, particle tracks were traced and interpreted via other means, such as photographic emulsions and cloud chambers. The latter is based on an inverse principle to that of the bubble chamber: the charged particle is allowed to enter a super-saturated vapor chamber, and its ionized track furnishes nucleation centers for liquid drops (instead of bubbles), which can then be photographed and the resulting visible tracks interpreted.

Here is a short history of elementary particle observations, which began with the discovery of the electron in 1897:

1896 *X rays and other forms of radioactivity were observed.*

1897 *The electron was discovered. Electrons (e^-) were first called cathode rays by their discoverer.*

1899 *Alpha particles were discovered, and later shown to be helium nuclei consisting of two neutrons and two protons.*

1911 *Nuclear model of the atom, with heavy nucleus in the center and light electrons orbiting around it, was proposed, and became accepted. C.T.R. Wilson (Cambridge, England) observes for the first time the tracks of ionizing particles (alpha and beta particles) in a cloud chamber, a device he had invented already in 1896. The cloud chamber played an important role in early nuclear and particle physics.*

1911 *Electron charge measured in an oil drop experiment indicates that all electrons carry the same electric charge.*

1932 *The neutron directly observed for the first time.*

1932 *The positron (e^+), predicted by* **Paul A.M. Dirac** *in 1928, was discovered by* **Carl D. Anderson** *in a mountain-top cloud chamber tracks.*

1934 *Radioactive nuclei were produced in the laboratory.*

1937 *The muon (μ^+), a charged lepton[16] like the electron (only about 200 times heavier and unstable), was observed.*

---

[16] *Lepton*, *meson* and *baryon* are Greek-derived neologisms meaning, respectively: light (small), intermediate, and heavy (large). The muon was originally classified as a meson, as were the "pions" (charged and neutral π mesons), since they are all intermediate in mass between the electron and the *nucleons* (proton and neutron). Nowadays, the classification is based on quantum numbers and interaction types, not mass. The electro; muon; tau lepton; plus their neutrinos, and the antiparticles of all these, comprise the 12 *leptons*; *mesons* are particles made of a valence quark and antiquark; and a baryon has 3 valence quarks. An anti-meson is still a meson, while the anti-baryon is a distinct category. A *hadron* is a meson, baryon or anti-baryon.

1947 *Two charged π mesons, with positive and negative charge, were discovered.*

1950 *The neutral π meson was discovered.*

1952 *Invention of the bubble chamber charged-particle track detector by* **D. Glaser**.

1953 *The lambda baryon and K meson were discovered.*

1956 *The existence of the electron neutrino (ν_e) and its antiparticle ($\bar{\nu}_e$), predicted by weak-interaction theory in 1930, was confirmed by experiments in which the missing neutral track form the vertex of a weak nuclear decay (such as beta decay,*

$$n \text{ (neutron)} \to p \text{ (proton)} + e^- + \bar{\nu}_e$$

creates a vertex at which an inverse process occurs (e.g. $\bar{\nu}_e + p \to n + e^+$).

1950s–60s *Many baryons and mesons were discovered, and their properties recur in regular patterns that look as if baryons and mesons were made of smaller building blocks.*

1961 *The muon neutrino was discovered and shown to be a different particle from the electron neutrino..*

By the mid-1960's, physicists realized that their previous understanding, whereby all matter is composed of the fundamental protons, neutrons, and electron, was insufficient to explain the myriad new particles being discovered. Quark theory solved these problems. Over the last forty years, the theory that is now called the Standard Model of particle physics has gradually emerged and gained increasing acceptance with new evidence from new particle accelerators.

1964 **Murray Gell-Mann** *and* **George Zweig** *tentatively put forth the idea of quarks. They suggested that baryons and antibarions are respectively composed of 3 quarks and 3 antiquarks of types called up, down, or strange (u, d, s) with spin 1/2 and electric charges (in units of proton's charge) 2/3, -1/3, -1/3, respectively; mesons consist of quark-antiquark pairs. Since fractional charges had never been observed, the introduction of quarks was treated more as a mathematical explanation of patterns of particle masses,* *decays and reactions than as a postulate of actual physical objects. Later theoretical and experimental developments allow us to now regard quarks as real physical objects, even though they cannot be isolated.*

1968 *Particle physicist* **G. Veneziano** *(Italy, Israel) used the Euler beta function to create a class of models of the strong nuclear interactions called Dual Resonance Models (DRM).*

These models are at first heuristic, but exhibit a duality of high-energy scattering amplitudes which is in accord with experiments. The duality property is between unstable excited hadrons resonantly produced during collisions on the one hand, and virtual particles exchanged during scattering, on the other. Work by L. Susskind, Y. Nambu and others in the late 1960's showed that DRM's can be explained if hadrons are viewed as relativistic strings. But the high-energy interactions of hadrons with leptons, studied experimentally at accelerators at SLAC, Brookhaven, CERN and other labs, failed to agree with string theories, and they were abandoned in the 1970's in favor of local non-abelian (Yang-Mills type) gauge quantum field theories of the strong, weak and electromagnetic interactions.

Meanwhile, however, **Michael Green**, **John Schwartz** *and other researchers noted that some string theories predict a spin-2 massless particle which couples to the energy-momentum 4-tensor – the graviton! These workers thus suggested that string theories might be related to grand unification and quantum gravity. These ideas were not then embraced by the particle physics community. But starting in 1984, another reversal occurred. Many classes of string theories with several curled-up (Kaluza-Klein) spatial dimensions were found to be mathematically consistent, and even free of the short-distance divergences that plague all quantum field theories.*

From that time on, various versions of String Theory have been viewed as leading contenders for an ultimate "Theory of Everything", although mathematical difficulties and lack of data at high enough energies has so far (2008) prevented this very active field of theoretical research from making any direct contact with experimental physics.

1968–1970s *Deep inelastic electron-proton scattering experiments revealed more of the quark structure inside protons and other hadrons, and that quarks' strong nuclear interactions decrease with decreasing distance (asymptotic freedom).*

1974 *A fourth flavor of quark (beyond u, d and s) – named charm – was detected in a newly discovered meson, confirming a theoretical prediction. The Glashow-Salam-Weinberg theory, unifying Quantum Electrodynamics (QED) with the weak nuclear forces in the framework of a spontaneously broken, Non-Abelian Gauge Theory, was shown to be mathematically consistent ('renormalizable').*

Another, unbroken, non-abelian gauge theory – called Quantum Chromodynamics (QCD) – was shown to account for accelerator experiment result such as 'asymptotic freedom', leading to increasing acceptance of QCD as the theory of the strong nuclear force. QCD predicts that each quark flavor comes in three 'colors', but that only colorless combinations can exist as separate, free particles.

The Glashow-Salam-Weinberg theory and QCD theories together compose the Standard Model.

1975 *The tau lepton was discovered, marking a third generation of leptons after the electron generation (e^-, ν_e and their antiparticles) and the muon generation (μ_-, ν_μ and antiparticles). The tau neutrino would only be observed in 2000.*

1979 *A fifth flavor of quark, named bottom, was found in the newly discovered Upsilon meson. This pattern leads particle physicists to believe they will find a sixth and final flavor of quark. This predicted last flavor of quark is called top and would only be detected experimentally in 1995.*

1982 *The massive gauge bosons that carry the weak nuclear force, called the W^+, W^- and Z^0, were discovered and their masses measured, confirming key predictions of the Glashow-Salam-Weinberg model.*

1989 *The lifetime of the Z^0 weak nuclear gauge boson was measured, and agrees precisely with the existence of only three kinds of neutrinos.*

1995 *The top quark was finally directly observed and measured, confirming the predictions of theorists that there are six flavors of quarks, as described in the current version of the Standard Model.*

Future *The search goes on for the Higgs boson (the only particle predicted by the Standard Model that hasn't been seen yet), for supersymmetric particles predicted by some extensions of the Standard Model, for proton decay and for magnetic monopoles predicted by Grand Unified Theories, and for new kinds of exotic particles.*

Timeline History of Quantum Theory

At the start of the twentieth century, scientists believed that they understood the most fundamental principles of nature. Atoms were solid building blocks of matter; people trusted Newtonian laws of motion; most of the problems of physics seemed to be solved.

However, starting with Einstein's theory of relativity which replaced Newtonian mechanics, scientists gradually realized that their knowledge was far from complete. Of particular interest was the growing field of quantum mechanics, which completely altered the fundamental precepts of physics.

1900 **Max Planck** *suggested that radiation as emitted and absorbed by atoms is quantized (it comes in discrete amounts of energy proportional to the frequency of the radiation).*

1905 **Albert Einstein**, *one of the few scientists to take Planck's ideas seriously, proposed a quantum of light (later named the photon) which behaves like a particle carrying energy, momentum and angular momentum. Einstein's other theories explained the equivalence of mass and energy, the particle-wave duality of photons, the equivalence principle, and special relativity.*

1909–1911 **Johannes Geiger** and **Ernest Marsden**, *under the supervision of* **Ernest Rutherford**, *scattered alpha particles off a gold foil and observe large angles of scattering, suggesting that atoms have a small, dense, positively charged nucleus.*

1912 **Albert Einstein** *explained gravitation as the curvature of space-time.*

1913 **Niels Bohr** *succeeded in constructing a provisional theory of atomic structure based on quantum ideas.*

1919 **Ernest Rutherford** *found the first evidence for a proton.*

1921 **James Chadwick** and **E.S. Bieler** *concluded that some strong forces holds the nucleus together.*

1923 **Arthur Compton** *discovered the quantum (particle) nature of X rays, thus confirming photons as particles.*

1924 **Louis de Broglie** *proposed that matter has wave properties.*

1925(Jan) **Wolfgang Pauli** *formulated the exclusion principle for electrons.*

1925(April) **Walther Bothe** and **Johannes Geiger** *demonstrated that energy and mass are conserved in atomic processes.*

1926 **Erwin Schrödinger** *developed wave mechanics, which describes the behavior of quantum systems in terms of a matter-wave equation which reduces to Newtonian dynamics in the short-wavelength limit.*

Max Born *gave a probability interpretation of quantum mechanics.*

G.N. Lewis *proposed the name "photon" for a light quantum.*

1927 *Certain materials had been observed to emit electrons (nuclear beta decay). Since both the atom and the nucleus have discrete energy levels, the observed continuous spectrum of emitted electrons implies that another, invisible particle is emitted as well.*

1927 **Werner Heisenberg** *formulated the uncertainty principle, which may be roughly formulated as follows: the more you know about a particle's location, the less you know about its momentum (and vice versa).*

1928 **Paul A.M. Dirac** *combined quantum mechanics and special relativity to describe the electron, and his new theory predicts the positively-charged positron,* e^+ *(the anti-electron).*

1930 *Quantum mechanics and special relativity are well established. There are thought to be just three fundamental particles: protons, electrons, and photons.*

Max Born, *after learning of the* **Dirac** *equation, said, "Physics as we know it will be over in six months."*

1930 ***Wolfgang Pauli*** *suggested the neutrino to explain the continuous electron spectrum for beta decay.*

Carl D. Anderson *discovers the positron (e^+) in cosmic-ray tracks left in a magnetic cloud chamber atop a mountain (Pike's Peak, Colorado).*

1931 ***James Chadwick*** *discovered the* neutron. *The mechanisms of nuclear binding and decay become primary problems of modern physics.*

1933–1934 ***Enrico Fermi*** *put forth a theory of beta decay that introduces the weak (nuclear) interaction. This is the first theory to explicitly incorporate neutrinos and what would later be called* flavor-changing charged currents, *which play the role that* electric *currents do in classical and quantum electrodynamics.*

1933–1934 ***Hideki Yukawa*** *combined relativity and quantum theory to describe nuclear interactions by an exchange of new particles (mesons called "pions") between protons and neutrons. From the size of the nucleus and the uncertainty principle, Yukawa concluded that the mass of the conjectured particles (mesons) is about 200 electron masses. This was the beginning of the meson theory of nuclear forces.*

1937 *A particle with mass about 200 electron masses was duly discovered in cosmic rays. While at first physicists thought it was Yukawa's pion, it was later discovered to be a muon, the second-generation charged lepton.*

1938 ***E.C.G. Stuckelberg*** *observed that protons and neutrons do not decay into any combination of electrons, neutrinos, muons, or their antiparticles. The stability of the proton cannot be explained in terms of energy or charge conservation; he proposes that heavy particles (what are now called* baryons) *are endowed with an independently conserved quantum number.*

1941 ***C. Moller*** *and* ***Abraham Pais*** *introduced the term "nucleon" as a generic term for protons and neutrons.*

1946–1947 *Physicists realize that the cosmic ray particle thought to be Yukawa's meson is instead a "muon," the first particle of the second generation of matter particles to be found. This discovery was completely unexpected –* ***I.I. Rabi*** *comments* *"who ordered that?" The term "lepton" was introduced to describe objects that do not interact via the strong nuclear force but do interact via the weak interactions (electrons and muons are both leptons, as well as their then-hypothetical associated neutrinos and the antiparticles of all these. Later (1975) the tau particles ($\tau^{\pm}$) and the corresponding neutrino ν_τ and antineutrino, $\bar{\nu}_\tau$, would complete the list of known leptons).*

1947 *A meson that does interact strongly was found in cosmic rays, and is determined to be the pion.*

1948 **Willis Lamb** *used molecular beam and RF technology to measure the small split (Lamb Shift) between the $2S_{1/2}$ and $2P_{1/2}$ energy hydrogen levels (transition frequency: 1058 MHz). This splitting results from QED "radiative correction" effects involving virtual electrons, positrons and photons in the atomic vacuum. Lamb's measurement spurred physicists to reformulate QED (Quantum Electrodynamics) in order to circumvent its inherent short-distance divergencies and extract unambiguous predictions for such radiative corrections. Their final version of QED not only agreed with Lamb's measured shift, but also led to other, extremely high precision predictions (such as for the electron-positron gyromagnetic ratio) that were found to be in agreement with increasingly accurate experiments ever since.*

1948 *Physicists* **R. Feynman**, **Julian Schwinger** *and* **S. Tomonaga** *developed procedures to calculate electromagnetic properties of electrons, positrons, and photons. Introduction of* **Feynman** *diagrams.*

1948 *The Berkeley synchro-cyclotron produced the first artificial pions.*

1949 **Enrico Fermi** *and* **C.N. Yang** *suggested that a pion is a composite structure of a nucleon and an anti-nucleon. This idea of composite "elementary" particles was quite radical.*

1949 *Discovery of K^+ (charged strange meson) via its weak nuclear decay.*

1950 *The neutral pion was discovered.*

1951 *Two new types of particles are discovered in cosmic rays. They were discovered by looking at V-like tracks and reconstructing the electrically-neutral object that must have decayed to produce the two charged objects that left the tracks. The particles were named the Λ (lambda) and the K^0 (neutral K meson).*

1951 **Erwin Wilhelm Mueller** (Germany) developed the *field ion microscope.*

1952 *Discovery of particle called Δ: there were four similar particles with different electric charges (Δ^{++}, Δ^{+}, Δ^{0}, and Δ^{-}.)*

1952 **Donald Glaser** invented the *bubble chamber.*

The Brookhaven Cosmotron, a 1.3 GeV accelerator, started operation.

1953 *The beginning of a "particle explosion" – a proliferation of "elementary" particles.*

1953–1957 *Scattering of electrons off nuclei revealed a charge density distribution inside protons, and even neutrons. Description of this electromagnetic structure of protons and neutrons suggested some kind of internal structure to these objects, though they were still regarded as fundamental particles.*

1954 **C.N. Yang** and **Robert Mills** developed a new class of nonlinear field theories called *non-abelian gauge theories* by combining ideas from differential geometry, electrodynamics and general relativity. Although not realized at the time, this type of theory now forms the basis of the Standard Model, underlying both the weak and strong nuclear forces.

1956 **C.N. Yang** (USA) and **T-D Lee** (USA) discovered that parity is not conserved for week interactions.

1957 **C.G. Wu** (China and U.S.A.) verified the **Yang-Lee** theory of parity violation in weak nuclear decay.

She did this by detecting a spatial asymmetry in beta particle (electron) emission during the radioactive decay of a Cobalt-60 nucleus, and establishing that this asymmetry depends upon the nuclear spin in the precise manner predicted by the theory.

1957 ***Julian Schwinger*** *proposed the unification of weak and electromagnetic interactions.*

1957 ***John Bardeen*** *(USA),* ***L.N. Cooper*** *(USA) and* ***J.R. Schrieffer*** *(USA) explained the phenomenon of superconductivity, first observed by* ***Kamerling Onnes*** *(1911), using a quantum mechanical theory. Accordingly, the superconducting current is carried by electron pairs ("Cooper pairs") weakly bound together through lattice vibration quanta (phonons). At low enough temperatures, these pairs' kinetic energy cannot be dissipated through scattering (the usual mechanism for electrical resistance in conductors).*

1957–1959 ***Julian Schwinger***, ***Sidney Bludman***, and ***Sheldon Glashow***, *in separate papers, suggested that all weak interactions are mediated by charged heavy bosons (later called* W^+ *and* W^-*) and augmented by the neutral* Z*. Actually it was* ***Hideki Yukawa*** *who first discussed bosons exchange twenty years earlier, but he proposed the pion as the mediator of the strong nuclear force.*

1961 *As the number of known particles keep increasing, a mathematical classification scheme to organize the hadrons helps physicists recognize patterns of particle types and properties.*

1961 ***Robert Hofstadter*** *(USA) discovered the inner structure of protons and neutrons.*

1967 ***Steven Weinberg*** *(USA),* ***Abdus Salam*** *(England) and* ***Sheldon Glashow*** *(USA) developed a theory of unification of the weak force and the electromagnetic force.*

1975 *Gravitational physicist* ***Stephen W. Hawking*** *(Cambridge, England) investigates a quantized field in the background of a Schwartzschield spacetime metric. His theoretical calculations predict that, due to the uncertainty principle and quantum causality violations near the event horizon of a black hole, black holes are in fact not black; they should be sources of thermal radiation, with effective temperature proportional to the inverse of the black hole mass. Since the mass gradually decreases due to this Hawking radiation, the temperature increases as the black hole shrinks and evaporates (negative specific heat!), finally resulting in* *an explosion when the black hole mass is reduced to order 10 micrograms (the Planck mass). Actual black hole candidates thus far observed by astronomers (at galactic centers and remnants of heavy-star supernova explosions) glow for other reasons – infalling matter from companion stars and ambient dust and gas – and their predicted Hawking radiation is many orders of magnitude weaker than the EM radiation due to this infall; thus Hawking radiation has not yet been observed, and is not expected to until we either find a much smaller* primordial *black hole, or develop technology to manufacture one ourselves.*

Hawking's result does not depend upon the details of the (as yet unknown) theory of Quantum Gravity, and thus is a very robust prediction which should constrain any such theory, including string theories.

Hawking's calculation confirmed the earlier (1973–74) ideas of **J. Bekenstein** *(Israel), who used a purely classical argument – the increase of total black hole horizon area when two black holes merge – to suggest an analogy with the* second law of thermodynamics*, with the event-horizon area playing the role of entropy. Hawking's field-theoretical model indeed revealed that the entropy of the thermal radiation is proportional to the event horizon area – suggesting that* bits of information are stored on the horizon *at a density of order one bit per square Planck length (one bit per about 10^{-66} cm^2!). Black hole entropy remains an active field of (thus far only theoretical) research in string theories.*

1980 **Klaus Von Klitzing** *(Germany) discovered the quantum Hall effect: A plate kept in a transverse magnetic field near absolute zero shows changes in transverse (Hall) impedance in discrete steps instead of continuously. It is one of the few examples of quantum behavior that is directly observed.*

1984 String theory *was accepted by the mainstream of the theoretical particle physics community as a candidate for a theory unifying quantum mechanics the Standard Model and gravity.*

1986 *A team of physicists from the US National Bureau of Standards observed individual quantum jumps in individual atoms.*

1986–1987 **Alex K. Müller** and **George J. Bendorz** *developed materials that become superconductive at substantially higher temperature then liquid Helium. The new materials are ceramics that display superconductivity between 90° and 120° K, above the boiling point of liquid nitrogen.*

1995 *At the Stanford Linear Accelerator Center (SLAC), a high power compressed-pulse laser beam (multi-TeraWatt peak power) is made to collide almost head-on with the 50 GeV (Giga-electron-Volts) electron beam. The peak transverse electric fields experienced by electrons in their rest frame are of order the Schwinger critical field* $(1.3\times10^{18}$ *Volt/meter). Such high fields result in coherent production of multiple back-scattered gamma-ray photons and electron-positron pairs, and probe non-perturbative nonlinear quantum electrodynamics.*

Photo 3

Astronomer by Candlelight (1658 CE)

by

Geritt Dou (1613–1675, Holland)

Working by candlelight an astronomer measures the distance between two points on a celestial globe.

The preeminent candle and the hourglass are traditional symbols of the brevity of life, suggesting that the astronomer symbolizes the vanity of human ambition seeking to comprehend the infinite.

Dou was Rembrandt's first student. Always admired for his nightscenes and his depicting of figures in niches, the artist combined the two genres in this meticulously executed painting.

Relativistic Astrophysics and Cosmology

History and Structure of the Universe

As far as we can tell, the expansion of the universe started ca 14 billions of years ago from a very hot and dense state of uniform elementary-particle plasma, too hot for even nuclei to exist. From that initial state, it mushroomed and evolved into the universe we know today. Cosmologists call that process of expansion the Big Bang because in some phases, especially in the beginning, the process was rather like an explosion.

Much of understanding of the Big Bang is based on extrapolating between knowledge of particle physics today, and projections from the mathematical model of an expanding universe in general relativity. The Einstein field equations give us a mathematical model for calculating how fast the universal expansion would be accelerating or decelerating at a given age (epoch), given the energy density and equation of state of matter and radiation at that time. We base our estimates about the matter and radiation density of the early universe on ancient light and radio waves reaching us from the past and collected in telescopes, and what we have learned about elementary particle physics, through theory and experiment.

The history of the universe divides roughly into three regimes which reflect the status of our current understanding:

- *Standard cosmology*
- *Particle cosmology*
- *Quantum cosmology*

The standard cosmology is the most reliably elucidated epoch, spanning the epochs from about $\frac{1}{100}$ of a second after the Big Bang through to the present day. The composition of the universe during this stage evolved from a soup of neutrons, protons, electrons, positrons, photons and neutrinos, through nucleosynthesis (formation of light nuclei) and positron annihilation, through the recombination era (ca 300,000 yr after the Big Bang) when the universe became transparent and the formation of stars and galaxies. The standard model for the evolution of the universe during this epoch has successfully faced many stringent observational tests, including the $2.7^\circ K$ blackbody cosmic *microwave background radiation* and the abundances of hydrogen, helium, deuterium and lithium.

Particle cosmology builds a picture of the universe prior to this, but at temperature regimes which still lie within known physics. For example, high energy particle accelerators at CERN and Fermilab allow us to test physical models for processes which would occur only 10^{-11} *seconds after the Big Bang. This area of cosmology is more speculative, as it involves at least some extrapolation from the Standard Model of particle physics, and often faces intractable calculational difficulties. Many cosmologists argue that reasonable extrapolations can be made to times as early as a grand unification phase transition (temperatures of order* 10^{27} *or* 10^{25} $°K$, *a time ca* 10^{-23} *sec after the Big Bang). This stage in the evolution of the universe includes the baryogenesis epoch, during which a symmetry developed between the abundances of matter and antimatter.*

Quantum cosmology considers questions about the origin of the universe itself, including the spacetime manifold in which it is embedded. It endeavors to describe quantum processes at the earliest times that we can conceive of in a classical space-time, that is, the Planck epoch at ca 10^{-43} *sec after the Big Bang. Given that we do not as yet have a theory of quantum gravity, this area of cosmology is extremely speculative.*

The four key observational successes of the standard Hot Big Bang model are the following:

- *Expansion of the universe*
- *Origin of the cosmic background radiation*
- *Nucleosynthesis of the light elements*
- *Formation of galaxies and large-scale structure*

The Big Bang model makes accurate and scientifically testable predictions in each of these areas, and the remarkable agreement with the observational data gives us considerable confidence in the model.

Expansion of the Universe

The universe began about 14 billion years ago in a violent explosion, at temperatures in excess of 10^{30} $°K$ *and correspondingly high mass-energy densities. The fact that galaxies are receding from us in all directions is a consequence of this initial explosion, and was first discovered observationally by*

Hubble. *There is now excellent evidence for Hubble's law, which states that the recessional velocity v of a galaxy is proportional to its distance d from us, that is, $v = Hd$ where H is Hubble's constant*[17]*. Projecting galaxy trajectories backwards in time means that they converge to a high density state – the initial fireball.*

The cosmological principle states that the universe appears the same in every direction from every point in space. It amounts to asserting that our position in the universe – with respect to the largest scales – is in no sense preferred. There is considerable observational evidence for this assertion, including the measured distributions of galaxies and faint radio sources. The best evidence comes from the near-perfect isotropy of the relic cosmic microwave background radiation (microwave photons detected by us now, were redshifted from near-visible photon emitted by hydrogen atoms during the recombination era, 3×10^5 yr after the Big Bang). This means that any observer anywhere in the universe will enjoy much the same view as we do, including the observation that galaxies are moving away from them, in accordance with Hubble's law.

The fact that the universe is expanding – about every point in space – can be a difficult concept to grasp. The analogy of an expanding balloon may be helpful: Imagine residing in a curved flatland on the surface of a balloon. As the balloon was inflated, the geodesic distance between any two points grew; the two-dimensional universe grew but there was no preferred center. For many decades it was unclear whether the universe is topologically closed (a 3D version of the flatland balloon surface), or open (albeit curved). Data collected by observatories in the 1990's, including the Hubble Space Telescope, indicates that the universe we inhabit is open, i.e. infinite in spatial extent and in (future) temporal extent, too.

Origin of the cosmic background radiation

About 300,000 years after the Big Bang, the temperature of the Universe had dropped sufficiently for electrons and protons to combine into hydrogen atoms ("recombination era"). From this time onwards, the matter in the universe came to gravitationally dominate over radiation and the hydrogen-filled

[17] This simple law requires nonlinear corrections due to gravitational, relativistic and time-delay effects; these corrections are computable, within a given cosmological model, from GTR.

universe became largely transparent to electromagnetic radiation (decoupling of matter and radiation). The last photon emitted in that era have propagated freely ever since, while constantly losing energy and increasing in wavelength because its wavetrains are stretched by the expansion of the universe. At the recombination decoupling, the radiation temperature was about 3000 degrees Kelvin, whereas today it has fallen to only 2.7 °K.

Observers detecting this radiation today are able to see the universe at a very early stage on what is known as the 'surface of last scattering'. Photons in the cosmic microwave background radiation (CMBR) have been traveling towards us for over 13 billion years.

Timeline – History

1576 **Thomas Digges** *modified the Copernican system by removing its outer edge and replacing the edge with a star-filled unbounded space.*

1610 **Johannes Kepler** *uses the darkness of the night sky to argue for a finite universe.*

1720 **Edmund Halley** *also formulated an early form of* **Olbers**' *paradox.*

1744 **Jean Philippe de Cheseaux** *also formulated an early form of* **Olber**'s *paradox.*

1862 **Heinrich Olbers** *enunciated* Olbers' paradox: *If stars are distributed throughout the universe they must have been shining for a finite time interval, or else our night sky on earth, here and now, would be uniformly ablaze with the brightness of the sun's disc.*

By the beginning of the 20th century, it was generally accepted that our galaxy was disc-shaped and isolated. But what about spiral nebulae like M31 (Andromeda) – were they inside or outside the Milky Way? **Immanuel Kant** *had speculated that they were 'island' universes.*

1912 **Vesto Slipher** *measured spectra from spiral nebulae, showing that many were Doppler-shifted. That is to say: the wavelengths of the observed line spectrum from many of them stretched (scaled up) relative to laboratory-observed spectra of the corresponding atoms, or relative to the sun's spectrum. And these wavelength stretchings were consistent with Doppler* red shifts *due to motions of the nebulae* away *from the solar system, at various speeds (different nebulae receding at different speeds). Such electromagnetic Doppler shifts are well known from Maxwell's theory, and are routinely encountered in the laboratory, satellite telemetry, traffic enforcement, meteorology, etc. The optical Doppler effect is analogous to the way the pitch of a train's whistle or a car's sound is modulated by its velocity relative to the listener.*

1915–1917 **Albert Einstein** *created the General Theory of Relativity. It became the cornerstone of massive stars' astrophysics and of all future studies of the large scale structure of the universe.*

1917 **Willem de Sitter** *derived from Einstein's GTR equations an isotropic static cosmology with a cosmological constant as well as an empty expanding cosmology with a cosmological constant.*

1918 *A key advance in cosmology came with the development of means to measure the distance to these nebulae.* **Harlow Shapley** *used Cepheid variables, bright stars which pulsate at regular intervals ranging from a few days to about a month. The period of their variation is correlated with their absolute luminosity, which he calibrated in the nearby Large Magellanic Cloud. Comparison of nebulae's apparent and absolute luminosities then gave him their distances from earth.*

1922 **Vesto Slipher** *summarized his findings on the spiral nebulae's systematic redshifts.*

1922 **Alexander Friedmann** *found a solution to the Einstein field equations which suggests the general expansion of space.*

1923–1929 **Edwin Hubble** *was able to resolve Cepheid variable stars in M31 (The Andromeda nebula or galaxy) with the 100″ telescope at Mt. Wilson. He developed a new distance calibration method using the brightest stars in more distant galaxies (as the nebulae now recognized as external to the Milky Way galaxy were now increasingly referred to). He correlated these measurements with Slipher's nebulae to discover a proportionality between velocity v and distance d, that is, Hubbles law $v = Hd$. The constant of proportionality H is called Hubble's constant (it was significantly overestimated by Hubble himself). He than concluded (1929) that the universe is expanding.*

1927 **Georges-Henri Lemaître** *discussed the creation event of an expanding universe governed by the Einstein field equations.*

1928 **Harold Robertson** *briefly mentioned that Vesto Slipher's redshift measurements combined with brightness measurements of the same galaxies indicate a redshift-distance relation.*

1933 **Edward Milne** *named and formalized the cosmological principle.*

1934 **Georges-Henri Lemaître** *interpreted the cosmological constant as due to a "vacuum" energy with an unusual, perfect fluid equation of state.*

1938 **Paul A.M. Dirac** *presented a cosmological theory where the gravitational constant slowly decreases so that the age of the universe divided by the time light takes to traverse the atomic nucleus is of order the ratio of the electric and gravitational forces between a proton and electron.*

1948 **Ralph Alpher**, **Hans Bethe**, and **George Gamow** *examined element synthesis (nucleosynthesis) in a rapidly expanding and cooling universe and suggested that the elements were produced by rapid neutron capture.*

1948 **Herman Bondi**, **Thomas Gold**, and **Fred Hoyle** *proposed steady state cosmologies based on the perfect cosmological principle, i.e. one in which the average isotropy and uniformity of the universe to observers in any galaxy is augmented by a time invariance (static universe).*

1960 **Robert V. Pound** *(USA) and* **Glen A. Rebka** *(USA) made a laboratory measurement of the change in frequency of gamma-ray photons as they fall in a gravitation field [gravitational 'Red Shift' of light], reconfirming the first of Einstein GTR predictions. Thus began the renaissance of interest in GTR.*

1960 **Rudolph L.B. Minkowski** *(USA). Using the 200 inch Palomar telescope, obtained spectra for a cluster of galaxies, receding at nearly half the speed of light. This would put their distance at some 7 billion light years. At this point the range to which galaxies could be observed is about, or somewhat less than, half the estimated "radius" (reckoned via light time-of-flight distance coordinate) of the observable universe.*

1960 **Thomas Mathews** *(USA) and* **Allan Sandage** *(USA), at Palomar, examined radio source 3C48 and discovered a 'quasar'. This discovery thrusts GTR to the forefront of astronomy. This quasar recedes with* $\frac{1}{3}$ *the speed of light and its brightness is 100 times that of our galaxy.*

The source of this power was suggested as gravity, which is the strongest force on a cosmic scale. The source has to be very compact; Since the source was varying coherently over a period of one hour, it could not be much longer then the distance traveled by light in one hour, in order for one side of the source to 'know' what the other side is doing, and thus behave in unison.

A strong gravitational field requires a very dense mass confined to a space with a diameter of the orbit of Jupiter.

1961 *Reflection (echo) of radar pulses from Venus (at time of 'inferior conjunction' (closest to earth)), enabled a precise determination of the average distance of the sun from earth: 149,500,000 km. This ended a 2250 year pursuit after the scale of the solar system from* **Aristarchos** *and* **Hipparchos** *through the parallax-seekers* **Richer-Cassini** *(1671),* **J.F. Encke** *(1835) and the international team of 1931.*

1963–1965 **R. Kerr** *discovered a family of exact solutions to Einstein's vacuum field equations; these solutions describe uncharged, rotating black holes, and reduce to the Schwartzschield solution as a special case.*

The charged generalization was found as a solution of the Einstein-Maxwell field equations by **E.T. Newman** *et al. (1965). Only later was the connection of theses mathematical solutions to black holes recognized.*

The Kerr-Newman geometry described by these solutions provides a unique and complete description of the external gravitational and EM fields of a stationary, rotating black hole.

1963 **Maarten Schmidt** *(USA) discovered that absorption lines in the spectra of object 3C273 were 'red shifted' by an extraordinary amount. This 'red shift' corresponds to a recession velocity of 47400 km/sec and constitutes the first recognition of quasars. It opened a new field of relativistic astrophysics.*

1965 **Arno Penzias** (USA) and **Robert Wilson** discovered the *cosmic microwave background radiation (CMBR)*, a radio-wave remnant of the 'Big-Bang' previously suggested by **George Gamow**.

1965 **Martin Rees** and **Dennis Sciama** analyzed quasar source count data and discover that the quasar density increases with redshift.

1965 **Edward Harrison** resolved Olber's paradox by noting the finite lifetime of stars.

1966 **Stephen W. Hawking** and **George Ellis** showed that any plausible general relativistic cosmology is singular.

1966 **Jim Peebles** showed that the Hot Big Bang predicts the correct helium abundance in the universe.

1967 **Andrei Sakharov** derived, from first principles, the requirements for cosmological generation of a baryon-antibarion asymmetry in the universe. Such an symmetry, is necessary to explain the universe in the present epoch, in which antimatter is very rare.

1967 **John Bahcall**, **Wal Sargent**, and **Maarten Schmidt** measured the fine-structure splitting of spectral lines in object 3C191 and thereby showed that the fine-structure constant does not vary significantly with time.

1967 **Jocelyn Bell** (England) and **Anthony Hewish** (England) discovered the first *pulsar* (CP1919) in the middle of the Crab Nebula. It is recognized as a *neutron star* rotating with period $t = 1.3373011$ sec and emitting beams of *synchrotron radiation* (coherent regular pulses, in this case in the radio wavelength region of the EM spectrum). Coherence and pulse-rate indicate a source small compared to normal stars. The theorized emission mechanism involves a scenario of synchrotron radiation emitted by charged particles near the star's magnetic poles: neutrons near the star's surface decay into protons and electrons and these are driven by powerful radial electric fields on the star's surface, along the curved intense magnetic field lines, with relativistic velocities.

Thus strong directional beams in the direction tangential to the charged particles' motion are radiated. As the star *rotates, the narrow beams sweeps around the galaxy like a searchlight. If the earth happens to lie in the path of the beams, the pulsar can be detected every time it passes the earth. The star loses angular momentum during radiation. Hence its rotation and pulsation rates slow down [CP1919 increases its period by 42×10^{-9} sec/year]. Thus, pulsars are cosmic lighthouses with rotating beacons of radio waves (and in some cases of visible light, X-rays and gamma rays).*

The Crab Nebula pulsar has a period of 0.033 sec. It was formed in the 1054 CE supernova explosion.

1968 **Brandon Carter** *speculated that perhaps the fundamental constants of nature must lie within a restricted range to allow the emergence of life – first use of the weak anthropic principle.*

1968 **Joseph Weber** *(USA) made pioneering efforts to detect gravitational waves in the laboratory by means of a bar antenna. These experiments stimulated a worldwide search for this weak and elusive radiation. Improvement in sensitivity of detectors by several orders of magnitude will likely be needed before this goal is achieved.*

1970 **Stephen W. Hawking** *(England) and* **Roger Penrose** *(England) showed that the equations of GTR in their classical form (without allowing for quantum effects) absolutely require that there was a singularity at the birth of the universe. Hence, there is no way around the singularity problems within the framework of GTR. If singularities are to be avoid in the real universe, the only hope is to improve General Relativity theory by including the effects of quantum theory and developing a quantum theory of gravity.*

The singularity theorem assumes, in addition to the validity of GTR, that the following 4 conditions are met:

(1) The mass density and pressure of early cosmological matter never became negative.

(2) There are no closed time-like or light-like curves in the universe (i.e. it is impossible to visit one's own past, such as in Gödel's universe solution of GTR).

(3) The universe is either closed or there is enough matter in the universe to refocus light via 'light bending'.

(4) A reasonable mathematical condition is satisfied.

1972–1979 *Radio-wave deflection experiments (1972), lunar laser-ranging results (1976), radio wave time-delay results that came in through the missions of Mariner 6, 7, 9 (1975–1978) and finally the Viking mission (1975) – all sided decisively with Einstein's GTR and against the rival Brans-Dicke theory.*

1973–1974 *USA astronauts made observations with a large telescope mounted on the Skylab space station.*

1973 **Edward P. Tyron** *(USA) suggested that the entire universe may have been created from absolutely NOTHING as a result of the probabilistic laws of quantum mechanics: an allowable fluctuation in a quantum vacuum could result in the creation of energy.*

The great medieval Hebrew philosopher-poet **Shlomo Ibn-Gabirol (Avicebron)** *(1021–1058, Spain) expounded the vision of the creation of the universe ex-nihilo in his metaphysics poem The Royal Crown, in the following words (1050 CE):*

> *"Calling unto the void and it was cleft,*
> *And unto existence and it was urged,*
> *And to the universe and it was spread out."*

1974–1978 **Russell A. Hulse** *(USA) and* **Joseph A. Taylor** *discovered the first binary pulsar PSR 1913+16 at the Arecibo Radio Telescope, Puerto Rico. The system consists of a neutron star pulsar, in orbit about a companion dark star, of nearly equal mass and probably also a neutron star. Its discovery opened up a new area for experimental relativity.*

After four years of observation it was concluded that the source consisted of two neutron stars, each of mass near the Chandrasekhar limit (1.42 solar masses). They move in elliptical orbits about their common center of mass. Only one of them beams its pulses toward earth, while the other's radiation is beamed elsewhere, if at all. The intrinsic pulse period is 0.05902 99952 71 sec, increasing at the rate of 0.273×10^{-9} *sec/year. The orbital period was 27906.98163 sec (sept 01, 1974) and the estimated distance of the system from earth is 16,000 LY. The radio frequency of the observations was 430 megahertz.*

Since the individual masses of the pulsars can not be measured independently , observations could not be used as a direct test of GTR. However, assuming the validity of GTR, all observed quantities could be explained and reconciled with known results.

On account of the precision of the pulsar's period, relativistic effects can be measured with great accuracy. Thus, the periastron advance of the orbit, 4.2263°/year, is some 36,000 times larger than the perihelion advance of Mercury! Other effects are:

(1) Ordinary Doppler-shift of pulsar's period (affects 5th decimal place).

(2) Special-relativistic time dilation: pulsar's clock run slow (as seen by us) on account of its velocity. Because the orbital speed varies during the motion (being maximal at periastron and minimal at apoastron), the amount of slowing down will be variable but repeat itself each new orbital passage. (affects 8th decimal).

(3) Gravitational red shift (equivalence principle). The pulsar moves in the gravitational field of its companion, while we observe it at great distance. Consequently the pulsars period is red-shifted (lengthened). However, this lengthening varies with the distance between the pulsar and its companion as the system changes from periastron to apoastron, and also repeats itself each orbital passage. The combined effect of (2) and (3) results in a periodic up and down variation of the pulsars period, by at most 58×10^{-9} sec.

(4) GTR predicts a decrease of the orbital period at a rate of 75×10^{-6} sec/year. Using data through Aug 1983, Taylor and Colleagues reported in 1984 an observed value of $(76 \pm 2) \times 10^{-6}$ sec /year.

The predicted decrease could be a result of continual energy loss due to gravitational wave emission, which in turn results from the orbital acceleration. This loss would manifest itself in the speed-up of the two bodies and decrease in their orbital separation. The combined effect of these two changes will cause the time required for a complete orbital period to decrease.

In 1978, Taylor announced what he claims to be the discovery of gravitational radiation, on the strength of the fit of the observed orbital decay compared with the theoretical prediction of GTR. Although the agreement seemed impressive, alternative mechanisms have not been ruled out, candidates including tidal friction, presence of other stars, or the possible invalidity of the linear approximation in the process of emission *of gravitational waves. Thus, one could envisage the possibility that the agreement is purely coincidental, resulting from a conjunction of nonlinear gravitational wave 'back reaction' combined with tidal effects involving the observed pulsar and its companion.*

1974 **Robert Wagoner**, **William Fowler**, and **Fred Hoyle** *showed that the Hot Bing Bang predicts the correct deuterium and lithium abundances.*

1975–1977 **Vera Rubin** *(USA) and* **Kent Ford** *(USA) established the anisotropy of the universe over a scale of 400 million LY. It is detected by measurements of different recession velocities for distant galaxies in different directions. The amplitude of this deviation from large scale uniformity amount to at most 10 percent of the* **Hubble** *recession velocity. Over larger scales, the astronomical evidence suggests isotropy. Rubin and Ford discovered that the net velocity of the Milky Way galaxy relative to the cosmological reference frame (as determined by the cosmic microwave background radiation) is about 600 km/sec [the velocity of the earth relative to the background radiation is 39 km/sec; the earth's motion around the sun is at 30 km/sec; the solar motion around the galactic center is 250 km/sec, and the motion of the Milky was galaxy toward the Andromeda Galaxy is 100 km/sec. A vector addition of these velocities will yield the above-quoted velocity of the Milky Way relative to the local CMBR frame].*

1976 **A.I. Shlyakhter** *used samarium ratios from the primordial natural fission reactor in Gabon to show that some laws of physics have remained unchanged for over two million years.*

1977 **Gary Steigman**, **David Shramm**, and **James B. Gunn** *examined the relation between the primordial helium abundance and number of neutrino types, and deduced that at most five lepton families can exist.*

1980 ***Alan H. Guth*** *(USA) proposed a new cosmological model of the birth of the universe, called inflationary universe. Accordingly, the universe expanded exponentially rapidly for a very short epoch during the early part of the "particle cosmology" stage of expansion following the Bing Bang.*[18]

1985 ***Mark Morris*** *(USA) discovered at the center the Milky-Way galaxy a number of strong string-shaped radio sources, candidates for low-energy 'cosmic strings'. These are long thin remnants of the original energy of the Bing-Bang. These 'cosmic strings' could supply part of the 'missing mass' needed to form the observed nearly-flat universe.*

1986 *A team of Astronomers discovered that our galaxy cluster (the local group), and other components of the local super-cluster of galaxies, move toward the 'Great Attractor' - a point in the direction of the Southern Cross.*

1986 *Deep redshift galaxy surveys demonstrated the existence of huge bodies, filaments and sheets on scales from 25 Mpc to over 100 Mpc. Subsequent galaxy surveys are providing detailed information about the distribution of large-scale structures. Radio galaxy and quasar surveys indicate that homogeneity (or uniformity) is approached only on scales of several hundred Mpc (that is, nearly a billion light years).*

---

[18] Inflationary cosmology was motivated by several previously unexplained observations:

- The high degree of isotropy of the CMBR, in apparent violation of causality ("horizon problem").
- Why has the universe's spatial curvature, ever since the late "particle cosmology" era, been so small? ("flatness problem").
- How can the observed upper bound on the abundance of *magnetic monopoles*, predicted by many Grand Unification extensions of the Standard Model of Particle Physics, be so low? ("monopole problem").

The exponentially rapid expansion predicted by inflationary models – perhaps by a factor of ca. 10^{50} – explains all three puzzles. GTR predicts that inflation would occur if the expanding vacuum is initially trapped in a metastable "false vacuum" quantum state; the subsequent release of vacuum energy when the vacuum "rolls" to its lowest-energy state created a temporarily large cosmological constant, which caused the inflationary expansion. The "standard cosmology" stages of Big Bang cosmology are largely unaffected by inflation.

1987 **Francisco Paresce** and **Christopher Burrows** *discovered a disc of protoplanets (gas, dust and debris around the star Beta Pictoris, 53 light years away) that will eventually crash into each other and coalesce to form planets.*

1987 **Roger C. Lynds** and **Vahe Petrosian** *discovered in Abell 370 an image of a far-distant unseen galaxy; the image is believed to have been formed by gravitational lensing, an effect predicted by GTR.*

1987 *Neutrinos from a supernova explosion 1987A in the Large Magelanic Cloud reached earth (150,000 LY away).*

1992 *In April 1992, the COBE satellite team announced the discovery of anisotropies in the cosmic microwave background radiation at the level of one part in 100,000. These are thought to be a snapshot, at* $t = 300,000$ *years after the Big Bang, of the primordial fluctuations that led to galaxy formation. This map of the sky is also the best evidence for the high degree of isotropy (or spherical symmetry) of the universe.*

1992, Feb *The Hubble Space Telescope (HST) revealed a black hole in Galaxy M-87, in Virgo, at distance* $R = 52$ *million light years away.*

1995–1996 *The Hubble Space Telescope (HST) was able to resolve Cepheid variable stars in the Virgo cluster, ensuring a much better calibration of cosmological distance measures. This has allowed more accurate estimates to be made of Hubble's constant* H*, and thus, of the age of the universe. Early galaxies and quasars have also been observed by the HST, raising serious doubts about current structure formation models.*

Modern Microscopy

A microscope (Greek: micron = small, scopos = aim) is an instrument for viewing and magnifying very small, close objects, too small to be seen by the unaided eye. The first to be invented was the *optical microscope*, containing one or more lenses that produce an enlarged image of an object placed in the focal plane of the lens(es).

The principle of the *simple microscope* (uses only one lens for magnification) was known already to the Romans in the form of water–filled glass bowls (1st century CE).

Microscopy is the technical field of using microscopes to view samples or objects. There are three well-known branches of microscopy: optical, electron and scanning probe microscopy.

Optical and electron microscopy involve the diffraction, reflection, or refraction of electromagnetic radiation incident upon the subject of study, and the subsequent collection of this scattered radiation in order to build up an image. This process may be carried out by wide field irradiation of the sample (for example standard light microscopy and transmission electron microscopy) or by scanning of a fine beam over the sample (for example confocal microscopy and scanning electron microscopy). Scanning probe microscopy involves the interaction of a scanning probe with the surface or object of interest. The development of microscopy revolutionized biology and remains an essential tool in that science, along with many others.

Optical microscopy techniques include:

- Bright field optical microscopy
- Dark field optical microscopy
- Phase–contrast optical microscopy
- Differential interference contrast microscopy
- Fluorescence microscopy

- *Confocal laser scanning microscopy*
- *Bright field optical microscopy*
- *Deconvolution microscopy*
- *X–ray microscopy*

For light microscopy, the wavelength of light limits the resolution to around 0.2 micrometers (2000 Å). In order to gain higher resolution, the use of an electron beam width for smaller wavelength is used in electron microscopes.

- *Transmission electron microscopy* (TEM) is principally quite similar to the compound light microscope, by sending an electron beam through a very thin slice of the specimen. The resolution (2005) is around 0.05 nanometer (0.5 Å).
- *Scanning electron microscopy* (SEM) visualizes details on surfaces of cells and particles and gives a 3D view.
- *The atomic de Broglie microscope* uses neutral Helium atoms as probe particles, could provide a resolution at nanometer scale and be absolutely non-destructive.

Scanning probe microscopy is a sub-diffraction technique. It includes:

- The atomic force microscope (AFM)
- The scanning tunneling microscope (STM)
- The photonic force microscope (PFM)

All such methods imply a solid probe tip in the vicinity (near field) of an object, which is supposed to be almost flat.

Modern Telescopy

By 2004, old mothballed Mount Wilson Observatory – which was used to revolutionize our understanding of the cosmos in the first half of the 20th century – was refurbished with some 21st century technology. On that year, the Georgia State University (USA) began to operate CHARA (Center for High Angular Resolution Astronomy) array. It uses optical interferometry to combine signals from six one-meter telescopes for a combined angular resolution (in IR) of 5×10^{-4} arc-sec. The largest single terrestrial optical telescopes are presently the twin Keck telescopes (Mauna Kea, HI; ten meters diameter each; 1993, 1996). Each of their mirrors comprises 36 hexagonal segments, constantly realigned – via computerized control – to an accuracy of four nanometers. Both Keck telescopes are equipped with adaptive optics to cancel atmospheric turbulence-caused blurring. When operating together as the Keck-Interferometer, the twin telescopes can achieve an angular resolution of 5×10^{-3} arc-sec at a wavelength of 2.2 micron.

Timeline – History

1267 **Roger Bacon** *described experiments with hand-held magnifying glasses.*

1608–1609 **Hans Janssen** *and his son* **Zacharias Janssen** *(1588–1630), Dutch lens grinders and spectacle makers, invented the first compound microscope. In its simplest form (as used by* **Robert Hooke***) it would have a single glass lens of short focal length for the objective, and another single glass lens for the eyepiece or ocular.* **Galileo Galilei** *developed (1609) a compound microscope with a convex and a concave lens.* **Christiaan Huygens** *developed a simple 2-lens ocular system (late 1600's) that was achromatically corrected.*

1610 **Johannes Kepler** *invented the modern compound microscope.*

1674 **Anton von Leeuwenhoek** *was first to bring the simple microscope to the attention of biologists. His microscopes consisted of a single, small, convex lens mounted on a plate with a mechanism to hold the biological specimen. With a magnification of about 270, he was able to see highly detailed images. Thus, he was first to describe cells and bacteria.*

18*th* century *– Several technical innovations make microscopes better and easier to handle, which leads to microscopy becoming more and more popular among scientists. An important discovery is that lenses combining two types of glass could reduce the chromatic effect, with its disturbing halos resulting from differences in refraction of light.*

1826 **Dames Smith** *(ca 1800–1870, England) constructed a microscope with much reduced chromatic and spherical aberrations.*

1830 **Joseph Jackson Lister** *reduced the problem with spherical aberration by showing that several weak lenses used together at certain distances gave good magnification without blurring the image.*

1878 ***Ernst Abbe*** *formulates a mathematical theory correlating resolution to the wavelength of light. Abbes formula (sine condition) make calculations of maximum resolution in microscopes possible. The company of* ***Carl Zeiss*** *exploited this discovery and became the dominant compound microscope manufacturer of its era.*

Modern microscopes of this kind are usually more complex, with multiple lens components in both objective and eyepiece assemblies. These multi-component lenses are designed to reduce aberrations, particularly chromatic aberration and spherical aberration. In modern microscopes the mirror is replaced by a lamp unit providing stable, controllable illumination.

1903 ***Richard Adolf Zsigmondi*** *(1865–1929, Austria) invented the ultramicroscope, for seeing small particles in a colloidal solution. Improved by* ***Joseph Barnard*** *(1870–1949, England) in 1912.*

1932 ***Frits Zernike*** *(1888–1966, Holland) invented the phase–contrast microscope that allows the study of colorless and transparent biological materials.*

1933 ***Ernst A.F. Ruska*** *(1906–1988, Germany) built the first electron–microscope, using* ***de Broglie*** *(1924) electron waves: electrons are emitted from heated metal and accelerated through a vacuum. They are then focused by powerful magnets onto the specimen. The magnified image appears on a screen. Ruska obtained a magnification of 12,500. Modern electron microscopes can reach a magnification of 1 million.*

1936 ***Erwin Wilhelm Mueller*** *(1911–1977, Germany and USA). Invented the field–emission*[19] *microscope (FEM). It is a type of electron microscope in which a high negative voltage is applied to a metal tip emitter, placed in an evacuated vessel some distance from a detector: a glass screen*

[19] The emission of electrons from cold metals by electric fields. In order to build up sufficiently large electric fields, the metal is usually shaped to a sharp needle point. Field emission is an example of the "tunnel effect" in quantum mechanics, with an electron in the metal being in a potential barrier. In field emission the probability of tunneling, which can be calculated using the "semiclassical approximation" is related to the "work function" of the metal.

with a fluorescent coating. The tip produces electrons by "field emission". The emitted electrons form an enlarged pattern on the fluorescent screen, related to the individual exposed planes of atoms. Since the resolution of the instrument is limited by the vibrations of the metal atoms, it is helpful to cool the tip in liquid helium.

Although the individual atoms forming the point are not displayed, individual absorbed atoms of other substances can be, and their activity is observable.

1948 **Paul Kirkpatrick** *(1894–1992, USA) and* **Albert Baez** *developed X-ray reflection microscope.*

1955 **Erwin Wilhelm Mueller** *developed the field–ion microscope (FIM). The first instrument that can picture individual atoms. It is, again, a type of electron microscope that is similar in principle to the "field–emission microscope", except that a high positive voltage is applied to the metal tip, which is surrounded by a low pressure gas (usually helium) rather than a vacuum.*

The image is formed in this case by field ionization: ionization at the surface of an unheated solid as a result of a strong electric field creating positive ions by electron transfer from surrounding atoms or molecules. The image is formed by ions striking the fluorescent screen. Individual atoms on the surface of the tip can be resolved and, in certain cases, absorbed atoms may be detected.

1955 **George Nomarski** *published the theoretical basis of Differential interference contrast microscopy.*

1969 **Manfred von Ardenne** *(1907–1997, Germany) built the first Scanning electron microscope (SEM).*

In contradistinction to the transmission electron microscope (Ruska) the beam of primary electrons scans the specimen, and those electrons that are reflected, together with any secondary electrons emitted, are collected. This current is used to modulate a separate electron beam in a TV monitor, which scans the screen at the same frequency, consequently building up a picture of the specimen.

1981 **Gerd Binnig** and **Heinrich Rohrer** *invented the scanning tunneling microscope that gives 3-dimensional images of objects down to the atomic level. The surface of the specimen is scanned by measuring a current between a very small tip and the specimen. Individual atoms can thus be detected.*

1986 **Gerd Binnig** *and colleagues invented the atomic force microscope: a small probe, consisting of a tiny chip of diamond, is held on a spring–cantilever in contact with the surface of the sample. The probe is moved slowly across the surface and the tracking force between the tip and the surface is monitored. The probe is raised and lowered so as to keep this force constant, and a profile of the surface is produced. Scanning the probe over the sample gives a computer generated contour map of the surface. The instrument is similar to the "scanning tunneling microscope", but uses mechanical forces rather than electrical signals. It can resolve individual molecules and, unlike the scanning tunneling microscope can be used with nonconducting samples, such as biological specimens.*

1987 **Arthur Rich** *and* **James van Hoch** *develop the positron microscope, using positrons emitted from a radioactive source.*

From Cell to Biosphere — Hierarchy of the Living World

* *
*

"Many worlds might have been botched and bungled, throughout an eternity, ere this system was struck out; much labor lost: Many fruitless trials made: And a slow, but continued improvement carried on during infinite ages in the art of world-making."

David Hume, 1779

* *
*

"If it could be demonstrated that any complex organ existed which could not possibly have been formed by numerous, successive, slight modifications – my theory would absolutely break down."

Charles Robert Darwin

* *
*

"No theory of evolution can be formed to account for the similarity of molecules, for evolution necessarily implies continuous change, and the molecule is incapable of growth or decay, of generation or destruction.
None of processes of Nature, since the time when Nature began, have produced the slightest difference in the properties of any molecule.
... They continue this day as they were created – perfect in number and measure and weight; and from the ineffaceable character impressed on them we may learn that those aspirations after accuracy in measurement, and justice in action, which we reckon among our noblest attributes as men, are ours because they are essential constituents of the image of Him who in the beginning created, not only the heaven and the earth, but the materials of which heaven and earth consist."

James Clerk Maxwell, 1873

* *
*

"The eternal mystery of the world is its comprehensibility."

Albert Einstein, 1915

* *
*

"As a result of a thousand million years of evolution, the Universe is becoming conscious of itself, able to understand something of its past history and its possible future. This cosmic self-awareness is being realized in one tiny fragment of the universe – in a few of us human beings... "

Julian Huxley

* *
*

"The probability of life originating at random is so utterly miniscule as to make it absurd."

Francis Crick

* *
*

The match between our intelligence and the intelligibility of the world is no accident. Nor can it properly be attributed to natural selection, which places a premium on survival and reproduction and has no stake in truth or conscious thought. Indeed, meat-puppet robots are just fine as the output of a Darwinian evolutionary process.

William A. Dembski

Contents

(*) *These two rubrics are relevant at both cellular and organism levels*

PREFACE

The following monograph is a bold attempt to present a compact and lucid outline of the vast field of biology and biological history in just 160 pages.

This is, no doubt, a pretentious and somewhat perilous undertaking through which important details are sometimes sacrificed for the sake of over-simplified generalizations.

Yet, this unavoidable pitfall, notwithstanding such an effort carries its own hidden benefits:

First, the attentive reader never looses temporal and spatial perspective of his field of view, and is seldom consumed with overwhelming details. He may always complete the missing information regarding a specific mechanism or particular substance (molecule, protein etc.), using a proper textbook, article or a computer database.

Second, I explicitly narrate the accomplishments of some 300 scientists, who during the past 2300 years since Aristo, have created and developed the medical and biological sciences to its present state.

It is hoped that with a constructive feedback flowing from our readers and reviewers, we would be able to improve this outline in the future.

Introduction

BIOLOGY is the branch of science dealing with the study of life. It encompasses a broad spectrum of fields that together address phenomena related to living organisms over wide range of scales. It is thus concerned with the characteristics, classification, and function of organisms, how species come into existence, and the interactions they have with each other and with the environment.

At the molecular scale, life is studied in the disciplines of MOLECULAR BIOLOGY, BIOCHEMISTRY *and* MOLECULAR GENETICS.

At the next level, that of the cell, it is studied in cell biology. At the multicellular scale, it is examined in PHYSIOLOGY, ANATOMY, *and* HISTOLOGY.

DEVELOPMENTAL BIOLOGY *studies life at the level of individual organism's development (ontogeny).*

Moving up the scale towards more then one organism, ETHOLOGY *considers the behavior of groups of organisms.* POPULATION GENETICS *operates at the level of an entire population, and systematics considers multi-species scale of lineages. Independent populations and their habitats are examined in* ECOLOGY *and evolutionary biology. This last discipline is concerned with the origin and descent of species, as well as their change over time.*

EVOLUTIONARY BIOLOGY *is mainly based on paleontology, which searches for evidence in the fossil record.*

The two major traditional taxonomically-oriented disciplines are BOTANY *and* ZOOLOGY. *Botany is the scientific study of plants. Botany encompasses a wide range of scientific disciplines that study the growth, reproduction, metabolism, development, diseases, and evolution of plant life. Zoology involves the study of animals, including the study of their physiology within the fields of anatomy and embryology.*

Biology is subjected to the same physical laws operating in branches of science (such as the laws of chemical thermodynamics and conservation of mass). Yet, because of the inherent complexity of most biological systems a straightforward application of the physics and mathematical physics is not always feasible. Nevertheless, the biological sciences are characterized and unified by several major underlying principles and concepts: UNIVERSALITY, EVOLUTION, DIVERSITY, CONTINUITY, GENETICS, HOMEOSTASIS, *and* INTERACTIONS.

Some striking examples of biological universality include life's carbon-based biochemistry and its ability to pass on characteristics via genetic material, using a DNA and RNA based genetic code with only minor variations across the range of living things.

Another universal principle is that all organisms (that is, all forms of life on Earth except for viruses) are made of cells. Similarly, all organisms share common developmental processes. For example, in most animals, the basic stages of early embryonic development share similar morphological characteristics and include similar genes.

The central organizing concept in biology is that all life has a common origin ancestor and has changed and developed through the process of evolution. This is thought to have led to the observed similarity of processes.

Charles Darwin *established evolution as a viable scenario by articulating its driving force, natural selection. Genetic drift was embraced as an additional mechanism of evolutionary development in the modern synthesis of the theory.*

The evolutionary history of a species — which describes the characteristics of the various species from which it descended — together with its genealogical relationship to every other species is called its phylogeny. Widely varied approaches to biology generate information about phylogeny. These include the comparisons of DNA sequences conducted within molecular biology or genomics, and comparisons of fossils or other records of ancient organisms in paleontology.

Despite its underlying unity, life exhibits an astonishingly wide diversity in morphology, behavior, and life histories. In order to grapple with this diversity, biologists attempt to classify all living things. Scientific classification seeks to reflect the evolutionary trees (phylogenetic trees) of the organism being classified. Classification is the province of the disciplines of systematics and taxonomy. Taxonomy places organisms in groups called taxa, while systematics seeks to define their relationships with each other. This classification technique has evolved to reflect advances in cladistics and genetics, shifting the focus from physical similarities and shared characteristics to phylogenetics.

Classification systems generally begin with the three-domain system: Archaea, Bacteria, Eukaryote. These domains reflect whether the cells have nuclei or not, as well as differences in the cell exteriors. Further, each kingdom is broken down continuously until each species is separately classified.

The hierarchy is: 1) KINGDOM, *2)* PHYLUM, *3)* CLASS, *4)* ORDER, *5)* FAMILY, *6)* GENUS, *7)* SPECIES. *The scientific name of an organism is obtained from its Genus and Species. For example, humans would be listed as Homo sapiens.* Homo would be the Genus and Sapiens is the species. Whenever writing the scientific name of an organism it is proper to capitalize the first letter in the genus and put all of the species in lowercase; in addition the entire term would be put in italics. The term used for classification is called Taxonomy.

There is also a series of intracellular parasites that are progressively "less alive" in terms of metabolic activity: Virus, Viroid, Prions.

Up into the 19th century, it was commonly believed that life forms could appear spontaneously under certain conditions. This misconception was challenged by **William Harvey's** dictum that "all life [is] from [an] egg" (from the Latin "Omne vivum ex ovo"), a foundational concept of modern biology. It simply means that there is an unbroken continuity of life from its initial origin to the present time.

A group of organisms is said to share a common descent if they share a common ancestor. All organisms on earth are thought to have descended from a common ancestor or an ancestral gene pool. This last universal common ancestor of all of today's organisms is believed to have appeared about 3.5 billion years ago. Biologists generally regard the universality of the genetic code as definitive evidence in favor of the theory of universal common descent (UCD) for all BACTERIA, ARCHAEA, and EUKARYOTES.

Homeostasis is the ability of an open thermodynamical system to regulate its internal environment to maintain a stable condition by means of multiple dynamic equilibrium adjustments controlled by interrelated regulation mechanisms. All living organisms, whether unicellular or multicellular, exhibit homeostasis. Homeostasis manifests itself at the cellular level through the maintenance of a stable internal acidity (pH); at the organismic level, warm-blooded animals maintain a constant internal body temperature; and at the level of the ecosystem, as when atmospheric carbon dioxide levels rise and plants are theoretically able to compensate by removing more of the gas from the atmosphere. Tissues and organs can also maintain homeostasis.

Every living thing interacts with other organisms and its environment. One reason that biological systems can be difficult to study is that so many different interactions with other organisms and the environment are possible, even on the smallest of scales. A microscopic bacterium responding to a local sugar gradient is responding to its environment as much as a lion is responding to its environment when it searches for food in the African savanna. For any given species, behaviors can be co-operative, aggressive, parasitic or symbiotic. Matters become more complex when two or more different species interact in an ECOSYSTEM. Studies of this type are the province of ECOLOGY.

Whether we study a single one-celled organism or the world of life as a whole, we can identify a hierarchy of biological organization:

- *Atoms join to form molecules of varying size, including very large macromolecules such as proteins and DNA.*
- *Atoms and molecules form organelles, such as cell nuclei and mitochondria (the site of energy transformations).*
- *Many organelles work together to perform the various functions of the cell.*
- *Cells associate to form tissues, such as bone marrow, skin, etc.*
- *Tissues form organs, such as bones, heart, liver, etc, that in turn comprise organ systems.*
- *The skeletal system and other structural systems work together to make up the functioning organism.*
- *A population of different species that inhabit a particular area make up a community, which together with the nonliving environment form an ecosystem.*
- *Those parts of earth's atmosphere, bodies of water and crust that support life, together with all its ecosystems, constitutes the biosphere.*

That aspects of the natural sciences whose methodologies aim at structures by studying their parts is known as reductionism. However, the whole is more then the sum of its parts. Each level of structural scale & description reveals emergent properties — characteristics not found at lower levels.

Table 6.8 correlates the characteristic spatial size of each organizational level with the corresponding subfield of Biology which aims to study phenomena at this level. It spans 15 orders of magnitude ($10^{-9}m$ – $10^{6}m$).

Table 6.8: LIFE'S HIERARCHY

	LEVEL OF ORGANIZATION *(entities or system)*	TYPICAL LINEAR SIZE	SUBFIELD
Molecular Biology	*Molecule (protein, enzyme, sugar, fatty acid, nucleic acid, solutes, etc.)*	2–10nm	*Biochemistry, Biophysics, Molecular genetics, Molecular evolution*
	Virus	20–400nm	*Virology*
	Chromosome	300–700nm *(unfolded)*	
	Bacterium	200–8000nm	*Bacteriology, Microbiology*
	Living cell	1–50μm	*Origins of Life, cell-Biology (cytology), Mycology, Immunology*
Organizational Biology	*Multicellular systems*		*Histology, Physiology, Anatomy, Nuclear Medicine*
	Individual organism or organ	1mm–30m	*Botany, Zoology, Paleontology, Anatomy, Medicine, Systematics*
	Ecosystems and Populations	1m–1000kn	*Ecology and animal behavior (Ethology), Biogeography, Population genetics, Evolutionary Biology, Developmental Biology*

Life at the Molecular Level (Molecular Biology[20])

Molecular biology is the study of biology at the molecular level. The field overlaps with other areas of biology and chemistry, particularly genetics and biochemistry. Molecular biology chiefly concerns itself with understanding the interactions between the various systems of a cell, including the interrelationship of DNA, RNA and protein synthesis and, learning how these interactions are regulated.

Biochemistry is the study of the chemical substances and thermodynamic processes occurring in living organisms.

Genetics is the study of the physico-chemical properties, variations, mutations, combinations, heredity and expression of genetic materials & the information & instructions they encode. Gene expression involves the extent and manner in which individual genetic differences affect individual organism (via protein synthesis). Often this can be inferred by the absence of a normal component (e.g. one gene). Genetics includes the study of "mutants" – organisms which lack (or posses an extra) one or more functional or structural components with respect to the so-called "wild type" or normal phenotype. Genetic interactions such as epistasis can often confound simple interpretations of such "knock-out" studies.

Molecular biology is the study of molecular underpinnings of the process of replication, transcription and translation of the genetic material. The central dogma of molecular biology holds that genetic material is transcribed into RNA and then translated into protein, with each gene coding for a single & unique protein. This "dogma", despite being an oversimplified picture of molecular biology, still provides a good starting point for understanding the field. This picture, however, is undergoing revision in light of emerging novel roles for RNA.

Much of the work in molecular biology is quantitative, and recently much work has been done at the interface of molecular biology and computer science

[20] The term was first coined by **Warren Weaver** (1938).

in bioinformatics and computational biology. As of the early 2000s, the study of gene structure and function, molecular genetics, has been amongst the most prominent subfields of molecular biology.

Increasingly many other fields of biology focus on molecules, either directly studying their interactions in their own right such as in cell biology and developmental biology, or indirectly — as when the techniques of molecular biology are used to infer historical attributes of populations or species, as in evolutionary biology, population genetics and phylogenetics. There is also a long tradition of studying biomolecules "from the ground up" in biophysics.

Since the late 1950s and early 1960s, molecular biologists have learned to characterize, isolate, and manipulate the molecular components of cells and organisms. These components include DNA, the repository of stored genetic information; RNA, a close relative of DNA whose functions range from serving as a temporary working copy of DNA to actual structural and enzymatic functions as well as a functional and structural part of the translational apparatus; and proteins, the major structural and enzymatic type of molecule in cells.

The successes of MOLECULAR BIOLOGY *derive from the exploration of that unknown world via the new technologies & methodologies developed by chemists, physicists & mathematicians during the* 20th *century; these studies revealed the structure and function of biotic macromolecules (as well as higher-level structures).*

The salient technologies and methodologies are:

Chromatography	*1901*	*(**M.S. Tsvet**)*
X-ray crystallography	*1912*	*(**L. Bragg**)*
Ultracentrifugation	*1925*	*(**T. Svedberg**)*
Electron microscopy	*1932*	*(**Knoll and Ruska**)*
Electrophoresis	*1933*	*(**Arne Tiselius**)*
Radionuclide imaging	*1938*	*(**Glenn Seaborg**)*
Ultrafast Laser spectroscopy	*1960*	*(**T. Maiman**); 1987 (**Steven Chu**)*
Single-photon emission tomography	*1964*	
CT Scanning	*1972*	*(**Godfrey Hounsfield**)*
NMR Spectroscopy	*1973*	*(**Paul Lauterbur**)*
Atomic force microscopy	*1986*	*(**Gerd Binnig** et al.)*
Optical trapping nanometry		
Scanning tunneling electron microscopy	*1989*	
Optical trapping interferometry	*1993*	
Cryo-electron microscopy	*2000*	
Two-photon microscopy		
FRET (Fluorescent Resonance Energy Transfer) microscopy		
Fluorescence microscopy using chromophores		
Microarray chips & microfluidics		
Statistical methods (to reconstruct genome sequences from oligonucleotide fragments)		

I. Biophysics and Biochemistry

Biophysics

Biophysics (also biological physics) is an interdisciplinary science that applies the theories and methods of physical sciences, especially those of physics, to questions of biology.

Biophysics research today comprises a number of specific lines of biological studies, which don't share a unique identifying factor, or admit of clear-cut and concise definitions. This is the result of biophysics' relatively recent appearance as a scientific discipline. The studies included under the umbrella of biophysics range from sequence analysis through fluid mechanics (e.g. to study blood flow) to neural networks.

In the recent past, biophysics included creating mechanical limbs and nanomachines to regulate biological functions. Nowadays, these are more commonly referred to as belonging to the fields of bioengineering and nanotechnology respectively. We may expect these definitions to further refine themselves. Traditional studies in biology are conducted using statistical ensemble experiments, typically using femto- to micro-molar concentrations of macromolecules.

Because the molecules that comprise living cells are so small, techniques such as PCR amplification, gel blotting, fluorescence labeling and in vivo staining are used so that experimental results are observable with an unaided eye or, at most, optical magnification. Using these techniques, biologists attempt to elucidate the complex systems of interactions that give rise to the processes that make life possible.

By drawing knowledge and experimental techniques from a wide variety of disciplines, biophysicists are able to indirectly observe or model the structures and interactions of individual molecules or complexes of molecules.

In addition to things like solving a protein structure or measuring the kinetics of single molecule interactions, biophysics is also understood to encompass research areas that apply models and experimental techniques derived from physics (e.g. electromagnetism and fluid dynamics) to larger systems such as tissues or organs (hence the inclusion of basic neuroscience as well as more applied techniques such as fMRI).

The interdisciplinary nature of biophysics is manifested in the wide range of biological fields in which these researchers are active:

Of these, the fields of cellular biophysics, molecular motors, protein unfolding, *biopolymer mechanics* and *receptor-ligand interactions have been very active in the past. Indeed, the advent of biophysical techniques for the manipulation of single biological molecules has made possible a large number of significant breakthroughs in biology.*

BIOCHEMISTRY

During the 21st *Century "big 4" of X-ray, radionuclide imaging, ultrasound and MRI continue to dominate, in their many variants, but many other interesting developments in other techniques are occurring, especially when we consider "imaging" to include microscopic as well as macroscopic biological structures (thermal imaging, electrical impedance tomography, scanned probe techniques, etc.) In addition, the emphasis in the future will increasingly be on obtaining functional and metabolic information simultaneously with structural (image) information. This can already be done to some extent with radioactive tracers (e.g. PET) and magnetic resonance spectroscopy.*

Biochemistry then, is the science dealing with the chemical constitutions of living systems and the dynamics of living chemical processes (metabolism). Although the word "biochemistry" was coined by **F. Hoppe-Seyler** *(1877), the full-fledged and institutionalized discipline emerged only in 1903. It was formed from:*

- *chemist's animal and vegetable chemistry*
- *biologists, cytologists and physicians physiological, zoological or biological chemistry.*

It rapidly also formed a base for immunologists, nutritionists and chemists working on fermentation.

Biochemistry interfaces with *biology and chemistry* and is concerned with *the chemical processes that take place within living cells. Modern biochemistry developed out of and largely came to replace what in the nineteenth and early twentieth centuries was called physiological chemistry, which dealt more with extracellular chemistry, such as the chemistry of digestion and of body fluids. Biochemistry as such is largely, though not exclusively, a twentieth-century discipline.*

Molecular biology, on the other hand, has come to mean the study of the function and the three-dimensional structure of such biologically important macromolecules as proteins and nucleic acids. Molecular biology is as much an interface of *biology with physics* as of *biology with chemistry*. In many respects biochemistry and molecular biology represent the realization of the dream of early twentieth-century mechanistic biologists, who were convinced that the most fundamental biological processes could ultimately be understood in terms of the laws of physics and chemistry.

Researchers in biochemistry use specific techniques native to biochemistry, but increasingly combine these with techniques and ideas from *genetics*, *molecular biology* and *biophysics*. There has never been a hard-line between these disciplines in terms of content and technique, but members of each discipline have in the past been very territorial; today the terms molecular biology and biochemistry are nearly interchangeable.

Biochemical research is involved in three main fronts:

- Components of cells: structure and function
- Energetics and metabolism
- Working of the genetic code

Living organisms, and even the individual cells of which they are composed, are enormously complex and diverse. Nevertheless, certain unifying features are common to all things that live. All make use of the same types of biomolecules, and all use energy. As a result, organisms can be studied via the methods of chemistry and physics. The belief in "vital forces" (forces thought to exist only in living organisms) held by 19th-century biologists has long since given way to awareness of an underlying unity throughout the natural world.

Disciplines that appear to be unrelated to biochemistry can provide answers to important biochemical questions. An example is the discovery, made by physicist in the early 20th century, that x-rays can be diffracted by crystals. The resultant experimental method of x-ray crystallography led to the elucidation of the three-dimensional structures of molecules as complex as proteins and nucleic acids. Biochemistry is a field that draws on many disciplines, and its multidisciplinary nature allows it to use results from many sciences to answer questions about the *molecular nature of life processes*. Enormously important applications of this kind of knowledge are made in medically related fields; an understanding of health and disease at the molecular level leads to more effective treatment of illnesses of all sorts.

The activities within a cell are analogous to the transportation system of a city. The cars, buses, and taxis correspond to the molecules involved in *reactions (or series of reactions) within a cell. The routes traveled by these vehicles are likewise comparable to the reaction & transport pathways that occur in the life of the cell. Note particularly that many vehicles travel more that one route — for instance, cars and taxis can go anywhere — whereas other, more specialized modes of transportation such as subways and streetcars are confined to single paths.*

Similarly, some molecules play multiple roles, whereas others take part only in specific series of reactions. And in terms of spatial transport, each cell's cytoplasm is criss-crossed by a complex 3-D network of myriad microtubules — nearly invisible "subway tunnels" along which molecules are pushed and pulled (between the membrane & interior of the cell) by molecular nano-motors. Also, all the routs (both spatial transport paths and chemical pathways) operate simultaneously.

The fundamental similarity of cells of all types makes it interesting and illuminating to speculate on the origins of life. Even the structures of comparatively small biomolecules consist of several parts. Large biomolecules such as proteins and nucleic acids have complex structures, and living cells are enormously more complex.

Even so, both molecules and cells must have arisen ultimately from very simple molecules such as water, methane, carbon dioxide, ammonia, nitrogen, and hydrogen[21]*. In turn, these simple molecules must have arisen from atoms. The way in which the universe itself, and the atoms of which it is composed, came to be, belongs to the discipline of cosmology.*

Since, life-forms alive today are believed to have descended from the same common ancestor, they certainly have similar biochemistries, even in matters which would appear to be essentially arbitrary, such as the genetic code or handedness of various biomolecules. It is unknown whether alternate biochemistries are possible or practical.

Biochemistry is the study of the structure and function of cellular components, such as proteins, carbohydrates, lipids, nucleic acids, and other biomolecules. Chemical biology aims to answer many questions arising from biochemistry by using tools developed within synthetic chemistry.

Although there are a vast number of different biomolecules, they tend to be composed of the same repeating subunits (called monomers), in different orders. Each class of biomolecules has a different set of subunits. Recently, biochemistry has focused more specifically on the chemistry of enzyme-catalyzed reactions, and on the properties of proteins.

---

[21] The biochemistry of all life-forms on earth is carbon and water based.

The biochemistry of cell metabolism and the endocrine system has been extensively described. Other areas of biochemistry include the genetic code (DNA, RNA), protein synthesis, cell membrane transport, and signal transduction.

Originally, it was generally believed that life was not subject to the laws of science the way non-life was. It was thought that only living beings could produce the molecules of life (from other, previously existing biomolecules). Then, in 1828, **Friedrich Wöhler** *published a paper about the synthesis of urea, proving that organic compounds can be created artificially. The dawn of biochemistry may have been the discovery of the first enzyme, diastase (today called amylase), in 1833 by* **Anselme Payen**.

Eduard Buchner *contributed the first demonstration of a complex biochemical process outside of a cell in 1896: alcoholic fermentation in cell extracts of yeast. Although the term biochemistry seems to have been first used in 1882, it is generally accepted that the formal coinage of biochemistry occurred in 1903 by* **Carl Neuberg**.

Since then, biochemistry has advanced, especially since the mid-20th century, with the development of new techniques such as chromatography, X-ray diffraction, NMR spectroscopy, radioisotopic labeling, electron microscopy and molecular dynamics simulations. These techniques allowed the discovery and detailed analysis of many molecules and metabolic pathways of the cell, such as glycolysis and the Krebs cycle (citric acid cycle).

Today, the findings of biochemistry are used in many areas, from genetics to molecular biology and from agriculture to medicine.

In the 1940s, following up on **Griffith***'s experiment,* **Avery, MacLeod** *and* **McCarty** *definitively identified deoxyribonucleic acid (DNA) as the "transforming principle" responsible for transmitting genetic information. In 1953,* **Francis Crick** *and* **James D. Watson** *published their famous paper on the structure of DNA, based on the research of* **Rosalind Franklin** *and* **Maurice Wilkins**. *These developments ignited the era of molecular biology and transformed the understanding of evolution by enabling a description of it as a molecular process: the mutation of segments of DNA.*

During this era of molecular biology, it also became clear that a major mechanism for variation within a population, once again, is mutations of DNA. In the mid-1970s, **Motoo Kimura** *formulated the neutral theory of molecular evolution, firmly establishing the importance of genetic drift as a major mechanism of evolution. The theory sparked the "neutralist-selectionist" debate, partially solved by the development of Tomoko Ohta's nearly neutral theory of evolution.*

II. Molecular Genetics — from Gene to Genome (1859–2008)

Genetics is the study of genes on all levels: from the level of molecules up to the level of populations.

It is the science of genes, heredity, and the variation of organisms. In modern research, genetics provides important tools in the investigation of the function of a particular gene, or the analysis of genetic interactions. Within organisms, genetic information generally is carried in chromosomes, where it is represented in the chemical structure (such as base sequences) of particular stretches of the DNA molecule.

Genes encode the information necessary for synthesizing proteins, which in turn play a large role in influencing (though, in many instances, not completely determining) the final phenotype of the organism.

Heredity and Genetics

In about 1902, the chromosome was identified as being the site of the genes, and its central position in heredity and development were finally realized. Linkage of genes and the crossing over of chromosomes during cell division were explored, particularly in **Thomas Hunt Morgan***'s fly lab in Columbia University. Early in the twentieth century, a unification of the idea of evolution by natural selection with Mendelian genetics to produce the modern synthesis occurred. These ideas continued to be developed in the discipline of population genetics and in the second half of the century began to be applied in the new discipline of the genetics of behavior, sociobiology, and, especially in human's, evolutionary psychology.*

By the end of the 19th century all of the major pathways of drug metabolism had been discovered. In the early decades of the twentieth century, the role of minor components of foods in human nutrition, the vitamins, began to be isolated and synthesized. Then in the 1920s and 1930s, the metabolic pathways of life, such as the citric acid cycle, glycogenesis and glycolysis finally began to be worked out by biochemists.This work continued to be very actively pursued for the rest of the century and into the next. During 1939–1941 **Fritz Lipmann** *showed that ATP is the universal carrier of energy in the cell, and then in the mid-1950's the power generators of the cell, the mitochondria, also began to be understood.*

Oswald Avery *conclusively showed in 1943 that DNA was the genetic material of the chromosome, not its protein. By 1953* **James D. Watson** *and* **Francis Crick** *had shown that the structure of DNA was a double helix, and its probable connection to replication. The nature of the genetic code was* *unraveled experimentally starting with the work of* **Nirenberg Khorana** *and others in the late 1950's. This discovery and others — gave rise to the vigorous science that we know today as molecular biology.*

The largest, most costly single biological study ever undertaken, the Human genome project, began in 1988 under the leadership of James D. Watson, and a first draft of the human DNA sequence announced in 2000. By 2003, 99% of the genome had been sequenced to an accuracy of one part in ten thousand. The HapMap project to determine patterns of differences in the human genome began in 2002 and by 2005 completed its first phase work.

The advent of whole-genome sequencing and surveys of their variation in different populations (races), together with new statistical methods, permitted researches by 2006 to systematically identify candidate loci for recent natural selection during evolution in humans. Some of these genes were also shown to be ancestry-informative markers which came to be used in genealogical studies and to understand ancient human migrations.

The study of organisms, their reproduction, and the functions of their organs had increasingly become the study of molecules. Reductionism was triumphant. Even the methods of scientific classification of organisms, especially cladistics, began in the last quarter of the century to use RNA and DNA sequences as characters. By the mid 1980's even the overall division of the tree of life into three domains (as opposed to the classical two), the Archaea, the Bacteria, and the Eukaria, became generally accepted in the scientific community.

While cloning in plants was known for millennia it was only in 1951 that the first animal, the tadpole, was cloned by nuclear transfer. Within a few years, several other animals, including dogs, cats, horses and cattle were cloned by similar methods.

In 1965 it was shown that normal cells in culture divide only a fixed number of times. Then aged and died. About the same time, stem cells were shown to be exceptions to this rule and began to be studied in earnest. Toward the end of the century, stem cells came to be recognized as crucial for the understanding of developmental biology and raised hopes for new medical applications.

In 1983 the unity of much of the morphogenesis of organisms from fertilized egg to adult began to be unraveled, first in fruit fly, then in other insects and animals, including man.

Genetic Engineering (GE)

Genetic engineering, genetic modification (GM) and gene splicing are terms the process of manipulating genes, usually outside the organism's normal reproductive process.

It involves the isolation and reintroduction of DNA into cells or model organisms, usually to express a protein. The aim is to introduce new characteristics or attributes physiologically or physically, such as making a crop resistant to herbicide, introducing a novel trait or producing a new protein or enzyme.

Since a protein is specified by a segment of DNA called a gene, future versions of that protein can be modified by changing the gene's underlying DNA. One way to do this is to isolate the piece of DNA containing the gene, precisely cut the gene out, and then reintroduce (splice) the gene into a different DNA segment. **Daniel Nathans** *and* **Hamilton Smith** *received the 1978 Nobel Prize in physiology or medicine for their isolation of restriction endonucleases, which are able to cut DNA at specific sites. Together with ligase, which can join fragments of DNA together, restriction enzymes formed the initial basis of recombinant DNA technology.*

The first Genetically Engineered drug was human insulin, approved by the USA's FDA in 1982. Another early application of GE was to create human growth hormone as replacement for a drug that was previously extracted from human cadavers. In 1986 the FDA approved the first genetically engineered vaccine for humans, for hepatitis B. Since these early uses of the technology in medicine, the use of GE has expanded to supply many drugs and vaccines.

One of the best known applications of genetic engineering is the creation of genetically modified organisms (GMOs).

Although there has been a tremendous revolution in the biological sciences in the past twenty years, there is still a great deal that remains to be discovered. The completion of the sequencing of the human genome, as well as the genomes of most agriculturally and scientifically important plants and animals, has increased the possibilities of genetic research immeasurably. Expedient and inexpensive access to comprehensive genetic data has become a reality with billions of sequenced nucleotides already online and annotated.

Human Genome Project (HGP)

In biology the genome[22] *of an organism is its whole hereditary information and is encoded in the DNA (or, for some viruses RNA). This includes both the genes and non-coding sequences of the DNA.*

More precisely, the genome of an organism is a complete DNA sequence of one set of chromosomes; for example, one of the two sets that a diploid individual carries in every somatic cell. The term genome can be applied specifically to mean the complete set of nuclear DNA (i.e., the "nuclear genome") but can also be applied to organelles that contain their own DNA, as with the mitochondrial genome or the chloroplast genome.

When people say that the genome of a sexually reproducing species has been "sequenced," typically they are referring to a determination of the sequences of one set of autosomes and one of each type of sex chromosome, which together represent both of the possible sexes.

Even in species that exist in only one sex, what is described as "a genome sequence" may be a composite from the chromosomes of various individuals. In general use, the phrase "genetic makeup" is sometimes used conversationally to mean the genome of a particular individual or organism. The study of the global properties of genomes of related organisms is usually referred to as genomics, which distinguishes it from genetics which generally studies the properties of single genes or groups of genes.

Most biological entities more complex than a virus sometimes or always carry additional genetic material besides that which resides in their chromosomes. In some contexts, such as sequencing the genome of a pathogenic microbe, "genome" is meant to include this auxiliary material, which is carried in plasmids. In such circumstances then, "genome" describes all of the genes and non-coding DNA that have the potential to be present.

In vertebrates such as sheep and other various animals however, "genome" carries the typical connotation of only chromosomal DNA. So although human mitochondria contain genes, these genes are not considered part of the genome. In fact, mitochondria are sometimes said to have their own genome, often referred to as the "mitochondrial genome".

Comparison of different genome sizes is shown in Table 6.9

---

[22] The term coined in 1920 by **Hans Winkler** (Germany) as a portmanteau of the words *GEN*e and chromos*OME*.

Table 6.9: Comparison of different genome sizes

Organism	**Genome size (base pairs)**	**Note**
Virus, Phage -X174	*5386*	*First sequenced DNA-genome*
Virus, Phage λ	5×10^4	
Archaeum, *Nanoarchaeum equitans*	5×10^5	*Smallest non-viral genome Dec, 2005*
Bacterium, *Buchnera aphidicola*	6×10^5	
Bacterium, *Wigglesworthia glossinidia*	7×10^5	
Bacterium, *Escherichia coli*	4×10^6	
Amoeba, *Amoeba dubia*	6.7×10^{11}	*Largest known genome, Dec 2005*
Plant, *Arabidopsis thaliana*	1.2×10^8	*First plant genome sequenced, Dec 2000*
Plant, *Fritillaria assyrica*	1.3×10^{11}	
Plant, *Populus trichocarpa*	4.8×10^8	*First tree genome, Sept 2006*
Yeast, *Saccharomyces cerevisiae*	2×10^7	
Nematode, *Caenorhabditis elegans*	9.8×10^7	*First multicellular animal genome, December 1998*
Insect, *Drosophila melanogaster* aka *Fruit Fly*	1.3×10^8	
Mammal, *Homo sapiens*	3×10^9	

Note: The DNA from a single human cell has a length of 1.8 m (but at a width of 2.4 nanometers).

Genomes are more than the sum of an organism's genes and have traits that may be measured and studied without reference to the details of any particular genes and their products. Researchers compare traits such as chromosome number (karyotype), genome size, gene order, codon usage bias, and GC-content to determine what mechanisms could have produced the great variety of genomes that exist today.

Duplications play a major role in shaping the genome. Duplications may range from extension of short tandem repeats, to duplication of a cluster of genes, and all the way to duplications of entire chromosomes or even entire genomes. Such duplications are probably fundamental to the creation of genetic novelty.

The information generated by the human genome project is expected to be the source book for biomedical science in the 21st century and will be of immense benefit to the field of medicine. It will help us to understand and eventually treat many of the more than 4000 genetic diseases that afflict mankind, as well as the many multifactorial diseases in which genetic predisposition plays an important role.

III. Structural Biology — the Molecular Architecture of Life (1853–2008)

Structural biology is a branch of molecular biology concerned with the study of the architecture and shape of biological macromolecules – proteins and nucleic acids in particular – and what causes them to have the structures they have. This subject is of great interest to biologists, because macromolecules carry out most of the functions of a cell, and because typically it only is by coiling into a specific three-dimensional shape that they are able to perform their functions. This shape, which is called the "tertiary structure" of a molecule, depends in a complicated way on the molecule's basic composition, or "primary structure."

Biomolecules are too small to see in detail even with the most advanced light microscopes. The methods that structural biologists use to determine their structures generally involve measurements on vast numbers of identical molecules at the same time. These methods include crystallography, NMR, ultra fast laser spectroscopy, electron microscopy, electron cryomicroscopy (cryo-EM), and circular dichroism. Most often researchers use them to study the static "native states" of macromolecules. But variations on these methods are also used to watch nascent or denatured molecules assume or re-assume their native states.

Proteins are amino acid chains, made up from 20 different amino acids that fold into unique 3-dimensional protein structures. The shape into which it folds is determined by its sequence of amino acids (aa). Below about 40 aa the term peptide is frequently used. A certain amount of aa is necessary to perform a particular biochemical function, and about 40–50 aa appear to be the lower limit for a functional domain size. Protein sizes range from this lower limit to several thousand aa in multi-functional or structural proteins. However, the current estimated for the average protein length is about 300 aa.

A third approach that structural biologists take to understanding structure is bioinformatics to look for patterns among the diverse sequences that give rise to particular shapes. Researchers often can deduce aspects of the structure of membrane proteins based on the membrane topology predicted by hydrophobicity analysis.

In the past few years it has become possible for highly accurate physical molecular models to complement the study of biological structures. Rapid prototyping technologies such as those used by 3D Molecular Design, or the creation of molecular models in glass, are examples of recent advances in this field.

Some of the 20 standard proteinogenic amino acids are called essential amino acids because the human body cannot synthesize them from other compounds through chemical reactions, but therefore must be obtained from food. Histidine and arginine are generally considered essential only in children, because the metabolic pathways that synthesize these amino acids are not fully developed in children.

Structural biology aims to explain the activity of biological important molecules in terms of their atomic structure – and to use this knowledge to design new therapies or vaccines.

When scientists understand the precise structure of some components of the surface of cancer cells, for instance, they may be able to tailor new drugs to fit these components as accurately as a key in a lock. With a really tight fit, very low doses of drugs could kill the cancer cells effectively without harming the normal cells nearby.

Similarly, researchers are trying to understand the three-dimensional structure of some natural chemicals that attach themselves to the genetic material DNA. These chemicals selectively control the activity of specific genes at specific times and in specific places. Deciphering their structure could lead to laboratory-made chemicals that do exactly the same thing.

Structural biologists use information gathered from chemistry, physics, genetics, cell biology, and mathematics, plus supercomputers and other highly specialized equipment. In recent years these scientists have relied increasingly on the use of computer graphics to visualize how various subunits of molecules fit together or move.

They have made rapid strides in the analysis of proteins (the extraordinarily varied molecules that do most of the work of the body); nucleic acids (DNA and a related molecule, RNA); carbohydrates (sugars and starches); lipids (fats); and the complex combinations of these substances.

Structural biology burst upon public attention for the first time in 1953, when ***Francis H.C. Crick*** *and* ***James D. Watson*** *announced that they had deciphered the structure of DNA – work for which they later won the Nobel Prize. Their model, the famous double helix, paved the way for many advances in genetics, including the development of recombinant DNA technology, which allows scientists to cut and splice together pieces of DNA from different sources.*

According to this model, DNA is a twisting ladder made up of different sequences of four components called nucleotides. Each rung of the ladder consists of a pair of nucleotides (which can pair only in certain ways). The specific sequence of these pairs contains all the information necessary for the development and survival of an organism such as a human being. The model explained for the first time how genetic information is transferred from the parent archive of DNA to daughter strands of nucleic acid: the nucleotide pairs separate in the middle of the rung, as if a zipper had opened, making two complementary strands, and each of these strands participates in forming a new, complete double helix.

The structures of many other biologically important molecules have been revealed since then, particularly the structures of those marvelously versatile substances called proteins.

Nonscientists tend to think of proteins as just something we eat. But in fact we depend on tens of thousands of different proteins in our bodies to keep us alive. Each protein is a complex biochemical machine with its own specialty. The reason we eat proteins from plants or animals in order to make proteins of our own.

Some of the proteins we depend on are enzymes that increase the speed of chemical reactions up to a million times, without themselves undergoing any change. In this way, enzymes control the pathways and the timing of billions of chemical operations. They regulate our growth from a single cell to a mature organism. They make our cells differentiate into eyes, blood, or brain cells. They deftly break down or build up other proteins.

There are other proteins that sit on the surface membranes of our cells, where they receive messages from distant cells, or control the flow of molecules into or out of the cell. The surfaces of cancer cells are dotted with proteins that play a large role in cells' runaway growth. The coats of viruses are made up largely of proteins.

Some of the neurotransmitters that carry urgent massages from one nerve cell to another are also proteins. So are many of the hormones that regulate our growth, sex drive, and reactions to stress.

The thousands of antibodies that recognize and fight foreign substances such as viruses and bacteria are proteins. The chemicals that interact with DNA that turn specific genes on or off are proteins. One protein, hemoglobin has the key job of transporting oxygen throughout the body via the bloodstream.

Different kinds of proteins act as structural materials, making up our skin, hair, nails, muscles, tendons, and bones. They also maintain the inner structure of the millions of cells in our bodies.

Some proteins are relatively small and compact; others are bulky conglomerates. Some are globular; other are long and narrow. But all of these different types of proteins are polymers, substances composed of many smaller units linked in chains. The subunits of proteins are 20 different amino acids, which can be linked together in any order. Thus, for any protein made up of 100 amino acids there can be a huge number of possible structures. But each protein has a specific order, and one protein can be distinguished from another by the order of its amino acids.

Difficult as it was to "solve" the structure of DNA, protein structures are far more difficult to decipher – not only because proteins have more subunits, but because they have irregular shapes. DNA structures tend to be more regular, although variations can still occur. By contrast, each type of protein molecule has a different shape, which determines its function. This shape defines what other molecules will bind to the protein and what chemical activity will take place.

All proteins are constructed in accordance with instructions coded by the DNA inside our cells.

Nevertheless, we have only begun to understand how proteins function. The toughest problem has been to decipher their 3-D structure, which holds the key to their normal or abnormal activity.

Analyzing the structures of proteins and nucleic acids down to the precise arrangement of their atoms, and trying to understand the basis of each protein's activity, are major goals of structural biology. For a while these goals seemed very distant, but now the field is in a period of great excitement because of the new possibilities opened up by several recent developments:

- *Recombinant DNA technology is allowing scientists to obtain large quantities of specific proteins or nucleic acids for study. In the past, research was severely limited by the shortage of experimental material.*
- *Researchers can now determine the sequence of subunits of these substances – for example, the sequence of amino acids in specific proteins, or the sequence of nucleotides on specific stretches of DNA – without an extraordinary expenditure of time and effort.*
- *Major improvements in X-ray crystallography, nuclear magnetic resonance (NMR) spectroscopy, and other techniques are enabling researchers to decipher the 3-D structures of large molecules more precisely and rapidly.*
- *Faster and less expensive computers are giving scientists new powers to handle enormous quantities of data and – through computer graphics – to visualize the structures and movements of proteins.*

The coming together of these separate developments has made structural biology ready for a major leap forward. So promising is this work that many research laboratories in universities and drug companies are rushing into the field.

DNA and protein molecules are called "macro", or "large", in comparison with molecules of water, fats, nucleotides,or amino acids, which consist of fewer that 50 atoms. A smaller protein molecule has about 1,000 atoms, while a more complex protein such as hemoglobin may have roughly 10,000 atoms. Some very large proteins contain as many as 100,000 atoms. Towering above them, an average molecule of human DNA has as many as 4.4 billion atoms.

But in fact, even these molecules are still exceedingly small – so small that their images appear weak and fuzzy under the most powerful electron microscope. How can one determine the structure of anything so tiny?

The main sources of information have been the laws of chemistry that bear on the strength and geometry of the chemical bonds between atoms, and data from instruments developed by physicists. Some of the most useful data have come from X-ray crystallography, a technique based on the fact that the atoms in a crystal are arranged in a definite pattern that is repeated regularly in three dimensions.

As early as 1912, **Max von Laue**, *realized that the wavelengths of X-rays – 3,000 to 4,000 times shorter than those of visible light – were just about the same size as the spaces between atoms in a crystal. Therefore, he reasoned, X-rays that were passed through a crystal should be affected by the regularly spaced layers of atoms in just the same way as light waves are affected by the slats of a metal grid. Von Laue's idea proved correct: crystals scattered X-rays in specific patterns. But he mistakenly thought that some peculiarities in these patterns were the result of peculiarities in the X-rays.*

At about the same time, **William Lawrence Bragg**, *looked at the same patterns and concluded instead that their peculiarities were vital clues to the arrangement of atoms in the crystal. He realized that the orderly array of atoms scattered the X-rays in a orderly way, and that this caused a repeating series of overlapping circles of waves. When intercepted by a photographic plate, the peaks and troughs of these waves reinforced each other at some points and canceled each other out at other points, producing a pattern of spots of varying intensity – an interference pattern that could be translated into information about the structure that produced it.*

Bragg rapidly worked out equations with which scientists could get an image of the arrangement of atoms in a crystal from the X-rays diffracted by a particular axis of the crystal. If one keeps realigning the crystal so that a different axis of the crystal is parallel to the X-ray beam each time, one can make a set of photographs which, with the aid of Bragg's equations, allow one to deduce the 3-D arrangements of the molecules in the crystal.

Bragg soon tried out his technique on crystals of sodium chloride (table salt) and potassium chloride and succeeded in working out their atomic structures.

He thus invented X-ray crystallography and "solved" the first structures of crystals.

In order to do X-ray diffraction studies of a biological substance, one must get the molecules neatly aligned to form straight planes that can reflect the X-ray beams. This generally means that the substance must take the form of a crystal. But in some cases the fibers of substance will naturally be so neatly aligned that they behave as though they were well-ordered crystals. Fortunately, this is what happened in some of the first experiments with X-ray diffraction of DNA.

DNA is an extremely long, thin, and fragile thread that scientists have had great difficulty in extracting from cell nuclei in pieces that are large enough to study. While working with a highly viscous solution of DNA extracted from calf thymus cells in 1950, **Maurice Wilkins** *noticed that every time he touched this solution with the tip of a glass rod and then drew the rod away, he "had spun a very thin fibre of DNA, almost invisible, like a filament of spider web." The fibers seemed highly uniform, so Wilkins took them to a graduate student for X-ray crystallography. These first X-ray diffraction pictures of DNA offered many clues to its structure.*

Very little was then known about DNA's vital statistics – its diameter, length, density, chemical bonds, or the angle at which it twisted. About the only information one could rely on came from X-ray diffraction patterns. These showed the distance between the nucleotides in DNA and pointed to a large repeated structure of some kind. But the X-ray patterns did not reveal how many strands of nucleotides the DNA molecule contained, nor whether the backbone of the structure was on the inside or outside of the strands.

On the basis of these X-ray diffraction pictures, **James Watson** *and* **Francis Crick** *built models of what the DNA molecule might look like and then testing these models. Did every precisely scaled piece of the model fit what was known from the X-ray data? Did it obey standard rules about chemical bonds?*

Their first models failed the tests. But eventually – after learning more about DNA from some new X-ray pictures taken by **Rosalind Franklin** *at King's College – they devised a double helix whose basic structure repeated itself every 34 hundred-millionths of a centimeter (34 angstroms), exactly 10 times the distance between one nucleotide and the next. They put two backbones on the outside and had the nucleotides meet in the center.*

While shifting around cardboard cutouts of nucleotide bases (key component of nucleotides), Watson realized that one of these bases, adenine (A), could form two hydrogen bonds with another one, thymine (T). Moreover, the bond lengths were correct for the model. Another base, guanine (G), could make similar hydrogen bonds with the fourth base, cytosine (C). When Watson compared the two cardboard pairs (A with T and G with C), they turned out to form nearly identical shapes that could fit snugly inside the backbones. This conformed with – and explained – a finding made 3 years earlier by **Erwin Chargaff** *of Columbia University that number of A bases in DNA was equal to that of T bases, while the number of G bases was equal to that of C bases.*

From the way the bases attached to the backbones, which consisted of chains of sugars and phosphates, Crick then saw that the two chains must run in opposite directions. The bases could appear in any order on one strand of DNA, but this order determined the sequence of the complementary bases on the other strand. He also saw a built-in means of replication. Prior to the cell's duplication, the hydrogen bonds (connecting the base pairs) are broken, and the two chains unwind and separate. Each chain then acts as a template for the formation on to itself of a new companion chain, so that eventually we shall have two *pairs of chains, where we only had one before. Moreover, the sequence of the pairs of bases will have been duplicated exactly.*

The precise sequence of the bases is the code which carries the genetical information. This meant that the four bases were a kind of alphabet – a small set of "letters" with which an infinite number of instructions could be written.

The instructions contained in DNA sequences tell the cell how to manufacture the thousands of enzymes and other proteins on which life depends. But these instructions are not transmitted directly. The DNA remains safely in the nucleus, somewhat like the printing block in a printing press. Meanwhile a copy of one strand of DNA is made in the nucleus, leaves it, and directs the production of proteins in other parts of the cell. This "working copy" of the DNA is a nucleic acid called RNA.

The language in which the DNA's and RNA's instructions are transmitted – the genetic code – was deciphered by **Marshall W. Nirenberg** *and* **H.G. Khorana**. *The code is based on triplets of nucleotides, or "codons", which are read in sequence. Each codon specifies either one of 20 amino acids or a signal to start or stop constructing an amino acid chain. Each gene consists of a series of codons that contain the instructions for building a specific protein, which influences a specific trait, or that make the RNA used to carry out the DNA's instructions.*

For years scientists had thought that nucleotides occurred in regular, repeating sets. Watson and Crick's model made it clear that nucleotides occur in infinitely varied sequences. There are 3 billion pairs of nucleotides in the DNA of a typical human cell, and each person's genetic material has a unique nucleotide sequence, duplicated only in identical twins.

The sequence of nucleotides in a piece of DNA can now be determined through increasingly efficient techniques, some of which are being automated. Scientists know that these sequences contain the instructions for making proteins. They can "read" these instructions. They can also know how to determine the sequence of amino acids in a given protein. But to date nobody understands precisely how the sequence of amino acids leads to the remarkably complex and irregular three-dimensional structures of proteins.

IV. Biotechnology and Bioinformatics (1950–2008)

Biotechnology[23] *can be defined as the manipulation of organisms to do practical things and to provide useful products.*

Early cultures also understood the importance of using natural processes to breakdown waste products into inert forms. From very early nomadic tribes to pre-urban civilizations it was common knowledge that given enough time organic waste products would be absorbed and eventually integrated into the soil. It was not until the advent of modern microbiology and chemistry that this process was fully understood and attributed to bacteria.

The most practical use of biotechnology, which is still present today, is the cultivations of plants to produce food suitable to humans. Agriculture has been theorized to have become the dominant way of producing food since the Neolithic Revolution. The processes and methods of agriculture have been refined by other mechanical and biological sciences since its inception. Through early biotechnology farmers were able to select the best suited and high-yield crops to produce enough food to support a growing population.

Other uses of biotechnology were required as crops and fields became increasingly large and difficult to maintain. Specific organisms and organism byproducts were used to fertilize, restore nitrogen, and control pests. Throughout the use of agriculture farmers have inadvertently altered the genetics of their crops through introducing them to new environments, breeding them with other plants, and by using artificial selection. In modern times some plants are genetically modified to produce specific nutritional values or to be economical.

---

[23] The word "biotechnology" was first used by the agricultural engineer **Karl Ereky** (1919).

The process of Ethanol fermentation lead to one of the first forms of biotechnology. Cultures such as those in Mesopotamia, Egypt, and Iran developed the process of brewing which consisted of combining malted grains with specifics yeasts to produce alcoholic beverages. In this process the carbohydrates in the grains were broken down into alcohols such as Ethanol. Later other cultures produced the process of Lactic acid fermentation which allowed the fermentation and preservation of other forms of food. Fermentation was also used in this time period to produce leavened bread. Although the process of fermentation was not fully understood until **Louis Pasteur's** *work in 1857, it is still the first use of biotechnology to convert a food source into another form.*

Combinations of plants and other organisms were used as medications in many early civilizations. Since as early as 200 BC people began to use disabled or minute amounts of infectious agents to immunize themselves against infections. These and similar processes have been refined in modern medicine and have lead to many developments such as antibiotics, vaccines, and other methods of fighting sickness.

One can distinguish between:

- *Red biotechnology is applied to medical processes. Some examples are the designing of organisms to produce antibiotics, and the engineering of genetic cures to cure diseases through genomic manipulation.*

- *White biotechnology, also known as grey biotechnology, is biotechnology applied to industrial processes. An example is the designing of an organism to produce a useful chemical. White biotechnology tends to consume less in resources than traditional processes when used to produce industrial goods.*

- *Green biotechnology is biotechnology applied to agricultural processes. An example is the designing of transgenic plants to grow under specific environmental conditions or in the presence (or absence) of certain agricultural chemicals. One hope is that green biotechnology might produce more environmentally friendly solutions than traditional industrial agriculture. An example of this is the engineering of a plant to express a pesticide, thereby eliminating the need for external application of pesticides. Whether or not green biotechnology products such as this are ultimately more environmentally friendly is a topic of considerable debate.*

- *Blue biotechnology is used to describe the marine and aquatic applications of biotechnology.*

Traditional pharmaceutical drugs are small molecules that treat the symptoms of a disease or illness - one molecule directed at a single target. Biopharmaceuticals are large biological molecules known as proteins and these target the underlying mechanisms and pathways of a malady; it is a relatively young industry. They can deal with targets in humans that are not accessible with traditional medicines. A patient typically is dosed with a small molecule via *a tablet while a large molecule is typically injected.*

Small molecules are manufactured by chemistry but large molecules are created by living cells: for example, - bacteria cells, yeast cell, animal cells.

Modern biotechnology is often associated with the use of genetically altered microorganisms such as E. coli *or yeast for the production of substances like insulin or antibiotics. Genetically altered mammalian cells, such as Chinese Hamster Ovary (CHO) cells, are also widely used to manufacture pharmaceuticals. Another promising new biotechnology application is the development of plant-made pharmaceuticals.*

Biotechnology is also commonly associated with landmark breakthroughs in new medical therapies to treat diabetes, hepatitis B, hepatitis C, cancers, arthritis, haemophilia, bone fractures, multiple sclerosis. Cardiovascular as well as molecular diagnostic devices than can be used to define the patient population. Herceptin, is the first drug approved for use with a matching diagnostic test and is used to treat breast cancer in women.

A more recent field in biotechnology is that of genetic engineering. Genetic modification has opened up many new fields of biotechnology and allowed the modification of plants, animals, and even humans on a molecular level.

Informatics[24] *is defined as the study of the structure, behavior, and interactions of natural and artificial computational systems. It encompasses the study of systems that* represent, process *and* communicate *information, including all computational, cognitive and social aspects. The central notion is the transformation of information – whether by computation or communication, whether by organisms or artifacts. In this sense, informatics can be considered as encompassing computer science, cognitive science, artificial intelligence, information science and related fields, and as extending the scope of computer science to encompass computation in natural, as well as engineered, computational systems.*

---

[24] In 1957 the **Karl Steinbuch** (1917–2008) published a paper called "Informatik: Automatisch Informationsverarbeitung" (i.e. "Informatics: automatic information processing").
The term was coined as a combination of "information" and "automation", to describe the science of automatic information processing. The morphology — *informat*-ion + *-ics* — uses "the accepted form for names of sciences, as conics, linguistics, optics, or matters of practice, as economics, politics, tactics", and so, linguistically, the meaning extends easily to encompass both the science of information and the practice of information processing.

Informatics includes the science of information the practice of information processing.

Informatics studies the structure, behavior, and interactions of natural and artificial systems that store, process and communicate information. It also develops its own conceptual and theoretical foundations. Since computers, individuals and organizations all process information, informatics has computational, cognitive and social aspects.

Used as a compound, in conjunction with the name of a discipline, as in medical informatics, bioinformatics, *etc., it denotes the specialization of informatics to the management and processing of data, information and knowledge in the named discipline.*

Informatics is broader in scope than: information theory *— the study of a particular mathematical concept of information;* information science *— a field primarily concerned with the collection, classification, manipulation, storage, retrieval and dissemination of information in human society; artificial intelligence — the study and engineering of intelligent behavior, learning, and adaptation, in machines; or* computer science *— the study of the storage, processing, and communication of information using engineered computing devices.*

Bioinformatics and computational biology involve the use of techniques including applied mathematics, informatics, statistics, computer science, artificial intelligence, chemistry, and biochemistry to solve biological problems usually on the molecular level. Research in computational biology often overlaps with systems biology. Major research efforts in the field include sequence alignment, gene finding, genome assembly, protein structure alignment, protein structure prediction, prediction of gene expression and protein-protein interactions, and the modeling of evolution.

The terms bioinformatics and computational biology are often used interchangeably. However bioinformatics more properly refers to the creation and advancement of algorithms, computational and statistical techniques, and theory to solve formal and practical problems inspired from the management and analysis of biological data.

Computational biology, on the other hand, refers to hypothesis-driven investigation of a specific biological problem using computers, carried out with experimental or simulated data, with the primary goal of discovery and the advancement of biological knowledge. Computational biology also includes lesser known but equally important subdisciplines such as computational biochemistry and computational biophysics.

A common thread in projects in bioinformatics and computational biology is the use of mathematical tools to extract useful information from data produced by high-throughput biological techniques such as genome sequencing.

Life at the Cellular Level (Cell Biology)

Cell biology *(also called cellular biology or cytology, from Greek kytos, "container") is an academic discipline that studies cells. This includes their physiological properties, their structure, the organelles they contain, interactions with their environment, their life cycle, division and death. This is done both on a microscopic and molecular level. Cell biology research extends to both the great diversity of single-celled organisms like bacteria and the many specialized cells in multicellular organisms like humans.*

Knowing the composition of cells and how cells work is fundamental to all of the biological sciences. Appreciating the similarities and differences between cell types is particularly important to the fields of cell and molecular biology. These fundamental similarities and differences provide a unifying theme, allowing the principles learned from studying one cell type to be extrapolated and generalized to other cell types. Research in cell biology is closely related to genetics, biochemistry, molecular biology and developmental biology.

Every cell typically contains hundreds of different kinds of macromolecules that function together to generate the behavior of the cell. Each type of protein is usually sent to a particular part of the cell. An important part of cell biology is the investigation of molecular mechanisms by which proteins are moved to different places inside cells or secreted from cells.

I. Circulatory fluid systems — Body's Internal Transport System

There are two main fluid systems in the body: blood and lymph. The blood and lymph systems are intertwined throughout the body and they are responsible for transporting the agents of the immune system.

The Blood System

The 5 liters of blood of a 70 kg (54 lb) person constitute about 7% of the body's total weight. The blood flows from the heart into arteries, then to capillaries, and returns to the heart through veins.

Blood is composed of 52–62% liquid plasma and 38–48% cells. The plasma is mostly water (91.5%) and acts as a solvent for transporting other materials (7% protein [consisting of albumins (54%), globulins (38%), fibrinogen (7%), and assorted other proteins (1%)] and 1.5% other stuff). Blood is slightly alkaline ($pH = 7.40 \pm .05$) and somewhat heavier than water ($density = 1.057 \pm .009$).

All blood cells are manufactured by stem cells, which live mainly in the bone marrow, via a process called hematopoiesis. The stem cells produce hemocytoblasts that mature into three types of blood cells: erythrocytes (red blood cells or RBCs), leukocytes (white blood cells or WBCs), and thrombocytes (platelets).

The leukocytes are further subdivided into granulocytes (containing large granules in the cytoplasm) and agranulocytes (without granules). The granulocytes consist of neutrophils (55–70%), eosinophils (1–3%), and basophils (0.5–1.0%). The agranulocytes are lymphocytes (consisting of B cells and T cells) and monocytes. Lymphocytes circulate in the blood and lymph systems, and make their home in the lymphoid organs.

There are 5000–10,000 WBCs per mm^3 and they live 5–9 days. About 2,400,000 RBCs are produced each second and each lives for about 120 days (they are trapped by the spleen. Once there, that organ scavenges usable proteins from their carcasses). A healthy male has about 5 million RBCs per mm^3, whereas females have a bit fewer than 5 million.

Normal Adult Blood Cell Counts are:

Red Blood Cells	$5.0 * 10^6/mm^3$	
Platelets	$2.5 * 10^5/mm^3$	
Leukocytes	$7.3 * 10^3/mm^3$	
Neutrophil		*50–70%*
Lymphocyte		*20–40%*
Monocyte		*1–6%*
Eosinophil		*1–3%*
Basophil		$< 1\%$

The proteins on RBCs are responsible for the usual ABO blood grouping, among other things. The grouping is characterized by the presence or absence of A and/or B antigens on the surface of the RBCs. Blood type AB means both antigens are present and type O means both antigens are absent. Type A blood has A antigens and type B blood has B antigens.

Some of the blood, but not red blood cells (RBCs), is pushed through the capillaries into the interstitial fluid.

History

As the most extensive visible fluid in the body, blood assumes major medical and symbolic significance. In many societies its shedding, both natural (as in menstruation) and deliberate, is unclean, impious and unlucky except in particular solemn and ritual circumstances. Blood is attributed responsibility not only for thought and sensation but also for life itself.

Greek medical writers confusingly considered it both one of the four humors and the fluid in which the humor blood predominated. It was the source of nutrient for many parts of the body, but any excess, plethora, either of it as humor or of one of its constituent humors was highly dangerous and might require phlebotomy (bloodletting).

The discovery of the circulation of the blood by **Harvey** *(1578–1657) established its primacy and many 18th-century authors attributed to it alone all the properties formerly associated with the other humors. From the end of the 18th century, physiological investigations concentrated more upon its constituent parts and assigned properties to them, and, although modern reliance on blood tests and transfusions has emphasized its role in diagnosis and therapy, it is now viewed primarily as a carrier and transmitter of other, more important, chemical substances around the body, e.g. hormones.*

It was traditionally observed that blood (one of the humors) was not a homogeneous substance, settling into a red clot and a colorless fluid ('plasma') separated by a thin white region non-clotting inside the living body often attributed to vitalism (e.g. by **John Hunter** *(1728–1793)).*

Microscopists like **Jan Swammerdam** *(1637–1680) and* **M. Malpighi** *(1628–1694) described red blood 'particles', the larger 'corpuscles' (leukocytes) being studied by Hunter's pupil* **William Hewson** *(1739–1774). The much-smaller platelets were described (1842) by* **Alfred Donné** *(1801–1878), their role in clot formation elucidated (1882) by* **Giulio Bizzozero** *(1846–1901).*

From the 1830s, the cell theory provided a sharper framework, **J.H. Bennett** *(1812–1875) and* **Rudolf Virchow** *(1821–1902) describing a pathological increase in leukocytes (leukemia), and* **Thomas Addison** *(1793–1860) observed pathological decrease in red blood cells (anemia).* **Paul Ehrlich's** *(1854–1915) staining techniques showed several different kinds of leukocytes, important in inflammation and defense mechanisms. The erythrocytes' function in respiration was uncovered from the 1850s, and the complicated cellular and chemical events in blood coagulation was studied by many, including* **G. Hayem** *(1841–1933) and* **W.H. Howell** *(1860–1945).*

The Lymph System

Lymph is an alkaline (pH > 7.0) fluid that is usually clear, transparent, and colorless. It flows in the lymphatic vessels and bathes tissues and organs in its protective covering. There are no RBCs in lymph and it has a lower protein content than blood. Like blood, it is slightly heavier than water (density $= 1.019 \pm .003$).

The lymph flows from the interstitial fluid through lymphatic vessels up to either the thoracic duct or right lymph duct, which terminate in the subclavian veins, where lymph is mixed into the blood. (The right lymph duct drains the right sides of the thorax, neck, and head, whereas the thoracic duct drains the rest of the body.)

Lymph carries lipids and lipid-soluble vitamins absorbed from the gastrointestinal (GI) tract. Since there is no active pump in the lymph system, there is no back-pressure produced. The lymphatic vessels, like veins, have one-way valves that prevent back-flow. Additionally, along these vessels there are small bean-shaped lymph nodes that serve as filters of the lymphatic fluid. It is in the lymph nodes where antigen is usually presented to the immune system.

The human lymphoid system has the following:

- *primary organs: bone marrow (in the hollow center of bones) and the thymus gland (located behind the breastbone above the heart), and*
- *secondary organs at or near possible portals of entry for pathogens: adenoids, tonsils, spleen (located at the upper left of the abdomen), lymph nodes*

(along the lymphatic vessels with concentrations in the neck, armpits, abdomen, and groin), Peyer's patches (within the intestines), and the appendix.

History

Ancient writers (e.g. ***Herophilos****, fl 290 BCE) probably noted the lacteals (abdominal lymphatics) and* ***Gaspare Aselli*** *(1581–1625) definitely described them as ending in the liver.* ***Jean Pecquet*** *(1622–1674) noted the thoracic duct in 1647, its connection to the lacteals established (1652) by* ***Olaf Rudbeck*** *(1630–1702).* ***Francis Glisson*** *(c1597–1677) suggested they carried fluid lubricating the body cavities back to the blood vessels.*

Lymph glands' (nodes) were described by ***M.A. Severino*** *(1580–1656) and* ***Johann Peyer*** *(1653–1712), and associated with the lymphatic system by* ***Marcello Malpighi*** *(1628–1694). Claims for priority in establishing the absorbent functions of smaller lymph vessels (through injection experiments) were contested by* ***William Hunter*** *(1718–1783) and* ***Alexander Monro secundus*** *(1733–1817). The role of the lymphatic system in the body's defense mechanisms has been elucidated by modern immunology.*

II. The Immune System — Body's Internal Defense

Introduction

The immune system, our internal defense system, protects the body against disease-causing organisms and certain toxins. Disease-causing organisms, or pathogens, include certain viruses, bacteria, fungi, and protozoa. Pathogens enter the body with air, food, and water; during copulation; and through wounds in the skin. The immune system recognizes pathogens and toxins and responds to eliminate them. Derived from the Latin for "safe", the word immune *refers to the early observation that when a person recovered from smallpox and other serious infections, they were safe from contracting the same illnesses again. Immunology, the study of internal defense systems of humans and other animals, is one of the most rapidly changing, challenging, and exciting fields of biomedical research today.*

For more than 50 years, immunologists based their work on the hypothesis that internal defense depends on the animal's ability to distinguish between self *and* nonself*. Such recognition is possible because each individual is biochemically unique. Cells have surface proteins different from those on the cells* of other species or even other members of the same species. An animal's immune system recognizes its own cells and can identify those of other organisms as foreign. Thus when a pathogen invades an animal, its distinctive macromolecules stimulate the animal's defensive responses. A single bacterium may have from 10 to more than 1000 distinct macromolecules on its surface.

Immunologists are aware of several limitations of the self-nonself hypothesis. For example, the immune system does not typically respond to foreign molecules that are harmless. In 1994, the U.S. National Institutes of Health, proposed the danger model, which hypothesizes that the immune system does more than distinguish between self and nonself. It responds to danger signals from injured tissues, such as proteins released when cell membranes are damaged.

Most immunologists now agree that internal defense relies on a combination of factors, including the ability to identify foreign molecules and to respond to chemical clues from injured tissues. The immune system is a collection of many types of cells and of tissues scattered throughout the body. Immune responses require communication among cells, or cell signaling. Cells of the immune system communicate directly by means of their surface molecules and indirectly by releasing messenger molecules. Understanding the complex signaling systems of the immune system is a major focus of research.

Sometimes pathogens overcome the body's internal defenses, resulting in disease. Some diseases, as well as certain genetic mutations, prevent or compromise immune function. HIV, the retrovirus that causes AIDS, infects T cells, an important component of the immune system. The immune system may overfunction, as in allergic reactions, or it may respond in ways that are clinically important, such as in Rh incompatibility or the destruction of the cells of organ transplants. In about 5% of adults in highly developed countries, certain immune responses are directed against self tissues, resulting in autoimmune disease.

Among the greatest accomplishments of immunologists are the development of vaccines that prevent disease, and techniques for successful tissue and organ transplantations. Sophisticated research tools, such as gene transfer, have enabled immunologists to expand their knowledge of the cells and molecules that interact to generate immune responses, and to develop new approaches to the prevention and treatment of disease. Much has been learned, and many challenges lie ahead.

An *immune response* is the process of recognizing foreign or dangerous macromolecules and responding to eliminate them. Two main types of immune responses protect the body: nonspecific and specific. *Nonspecific immune responses*, or *innate immunity*, provide general protection against pathogens, parasites, some toxins and drugs, and cancer cells.

Nonspecific immune responses prevent most pathogens from entering the body and rapidly destroy those that do penetrate the outer defenses. For example, the cuticle or skin provides a physical barrier to pathogens that come in contact with an animal's body. Phagocytosis, another nonspecific defense, destroys bacteria that invade the body.

Some of the molecules important in nonspecific immune responses recognize and attack certain pathogen- associated molecular patterns, which are shared by whole groups of viruses, bacteria, or fungi.

Specific immune responses, also referred to as adaptive or acquired immunity, are highly specific. Any molecule that cells of the immune system specifically recognize as foreign is called an antigen. Proteins are the most powerful antigens, but some polysaccharides and lipids can be antigenic.

Antibodies are highly specific proteins that recognize and bind to specific antigens. Specific immune responses are directed toward particular antigens and typically include the production of antibodies. In complex animals, an important characteristic of specific immune responses is immunological memory, the capacity to respond more effectively the second time foreign molecules invade the body.

All invertebrates species that researchers have studied demonstrate the ability to distinguish between their own cells and those of other species. For example, sponge cells have specific glycoproteins on their surfaces that enable them to recognize their own species. When cells of two different species are mixed together, they reassort according to species. When two different species of sponges are forced to grow in contact with each other, tissue is destroyed along the region of contact. Cnidarians (such as corals), annelids (such as earthworms), arthropods (such as insects), and echinoderms (such as sea stars) reject tissue grafted from other animal, even from the same species.

Invertebrates have very efficient nonspecific immune mechanisms. For example, many invertebrates (cnidarians, annelids, and mollusks) are covered by mucus that traps and kills pathogens. Tough external skeletons, such as shells or cuticles, shield the body of many invertebrates. Most invertebrate coelomates (animals with a coelom) have amoeba-like phagocytes that engulf and destroy bacteria and other foreign matter. In mollusks, substances in the hemolymph (blood) enhance phagocytosis.

Researches have identified antimicrobial peptides in all eukaryotes (including plants) that have been studied, suggesting an early common origin of these molecules. More than 800 of these peptides that inactivate or kill pathogens have been described! When researches inject an antigen into an insect, as many as 15 antimicrobial peptides are produced within a few hours. Antimicrobial peptides are very effective because of their small size (a dozen or fewer amino acids), which facilitates their rapid production and diffusion.

What stimulates an animal to produce antimicrobial peptides and other immune defenses? Animal cells have receptors that recognize certain types of pathogen molecules, and then signal the cell to produce antimicrobial peptides.

One important family of these signaling receptors, the Toll group, is a focus of current research. Immunologists first identified the Toll group in the fruit fly Drosophila. Toll receptors recognize some common molecular features of classes of pathogens called pathogen-associated molecular patterns, or PAMPs. Examples of PAMPs include the double-stranded RNA of certain viruses and peptidoglycan in Gram-positive bacteria.

Certain cnidarians, arthropods, some echinoderms and simple chordates (such as tunicates) appear to remember antigens for a short period. As mentioned earlier, immunological memory enables the body to respond more effectively when it encounters the same pathogens again. Although certain invertebrates demonstrate some specificity and memory, their immune responses are primarily nonspecific.

A specialized lymphatic system, including lymphocytes, (white blood cells specialized to carry out immune responses), evolved in the vertebrates. The lymphatic system performs the sophisticated specific immune responses of vertebrates.

The Immune System: Cellular Aspects

A major component of the immune system is the class of white blood cells called lymphocytes. Like all blood cells, they arise from common precursor cells (stem cells) in the bone marrow. Unlike other blood cells, however, they can leave the blood vessels and circulate in the lymphatic system. Lymphoid tissues such as lymph nodes, (as spleen, and, above all, the thymus gland) play important roles in the working of the immune system.

Two kinds of lymphocytes can be distinguished: T cells and B cells. T cells develop primarily in the thymus gland and B cells primarily in the bone marrow, accounting for their names. Much of the cellular aspect of immunity is the province of the T cells, whereas much of the molecular aspect depends on the activities of the B cells.

T cells can have a number of functions. As T cells differentiate, each becomes specialized for one of the possible functions. The first of these possibilities, that of killer T cells, involves surface receptors that recognize and bind to antigens, the foreign substances that trigger the immune response. The antigens are present to the T cell by other white blood cells called macrophages. The macrophages ingest and process antigens, and then present them to T cells.

T cells that bind to a given antigen, and only to that antigen, grow when these conditions are fulfilled. Note the specificity of which the immune system is capable. Many substances, including ones that do not exist in nature, can be antigens. The remarkable adaptability of the immune system in dealing with so many possible challenges is another of its main features. The process by which only those cells that respond to a given antigen grow in preference to other T cells is called clonal selection. The immune system can thus be versatile in its responses to the challenges it meets.

As their name implies, killer T cells destroy antigen-infected cells. They do so by binding to them and by releasing a protein that perforates the plasma membrane of the infected cell. This aspect of the immune system is particularly effective in preventing the spread of viral infection by killing virus-infected host cells. In a situation such as this, the antigen can be considered to be all or part of the coat protein of the virus. When the infection subsides, some memory cells remain, conferring immunity against later attacks from the same source.

The Immune System: Molecular Aspects

Antibodies are Y-shaped molecules, consisting of two identical heavy chains and two identical light chains, held together by disulfide bonds. They are glycoproteins, with oligosaccharides linked to their heavy chains. Each light chain and each heavy chain has a constant region and a variable region. The variable region (also called the V domain) is found at the prongs of the Y and is the part of the antibody that binds to the antigen. The binding sites for the antibody on the antigen are called epitopes.

Most antigens have several such binding sites, so that the immune system will have several possible avenues of attack of naturally occurring antigens. Each antibody can bind to two antigens, and each antigen usually has several binding sites for antibody, giving rise to a precipitate that is the basis of experimental methods for immunological research. The constant region (the C domain) is located at the hinge and the stem of the Y; it is this part of the antibody that is recognized by phagocytes and by the complement system (the portion of the immune system that destroys antibody-bound antigen).

How does the body produce so many highly diverse antibodies to respond to essentially any possible antigen? The number of possible antibodies is virtually unlimited, as is the number of words in the English language. In a language, the letters of the alphabet can be arranged in countless ways to give a variety of words, and the same possibility for enormous numbers of rearrangements exists with the gene segments that code for portions of antibody chains.

Each B cell (and each progeny plasma cell) produces only one kind of antibody. In principle each such cell should be a source of a supply of homogeneous antibody by cloning. This is not possible in practice because lymphocytes do not grow continuously in culture. In the late 1970s **Georges Köhler** *and* **César Milstein** *developed a method to circumvent this problem, a feat for which they received the Nobel Prize in physiology or medicine in 1984.*

The technique requires fusing lymphocytes that make the desired antibody with mouse myeloma cells. The resulting hybridoma (hybrid myeloma), like all cancer cells, can be cloned in culture and produces the desired antibody. Since the clones are the progeny of a single cell, they produce homogeneous monoclonal antibodies. In this way it is possible to produce antibodies to almost any antigen in quantity. Monoclonal antibodies can be used to assay for biological substances that can act as antigens. A striking example of their usefulness is in testing blood for the presence of HIV; this procedure has become routine to protect the public blood supply.

We have been considering active immunity, immunity that develops following exposure to antigens. Active immunity can be naturally or artificially induced. If someone with chickenpox sneezes near you and you contract the disease, you develop active immunity naturally.

Active immunity can also be artificially induced by immunization, that is, by exposure to a vaccine. When an effective vaccine is introduced into the body, the immune system actively develops clones of cells, produces antibodies, and develops memory cells.

The first vaccine was prepared in 1796 by British physician **Edward Jenner** *against vaccinia, the cowpox virus. The term vaccination was thus derived. Jenner's vaccine provided humans with immunity against the deadly disease smallpox. Jenner had no knowledge of microorganisms or of immunology, and it remained for French chemist* **Louis Pasteur** *to begin to develop scientific methods for preparing vaccines 100 years later. Pasteur showed that inoculations with preparations of attenuated (weakened) pathogens could be used to develop immunity against the virulent (infectious) form of the pathogen.*

However, not until 20th century advances in immunology – for example, Burnet's clonal selection theory in 1957 and the discovery of T and B cells in 1965 – did scientists gain a modern understanding of vaccines. Effective vaccination stimulates the body to launch an immune response against the antigens contained in the vaccine. Memory cells develop, and future encounters with the same pathogen are dealt with rapidly.

Microbiologists prepare effective vaccines in a number of ways. They can attenuate a pathogen so it loses its ability to cause disease. When pathogens are cultured for long periods in non-human cells, mutations adapt the pathogen to the nonhuman host so that they no longer cause disease in humans. This is how the **Sabin** *polio vaccine and the measles vaccine are produced.*

Whooping cough and typhoid fever vaccines are made from killed pathogens that still have the necessary antigens to stimulate an immune response. Tetanus and botulism vaccines are made from toxins secreted by the respective pathogens. The toxin is altered so that it can no longer destroy tissues, but its antigenic determinants are still intact.

Most vaccines consist of the entire pathogen, attenuated or killed, or of a protein from the pathogen. Researches are investigating several approaches that would reduce potential side effects. For example, they are developing DNA vaccines (or RNA vaccines), made from a part of the pathogen's genetic material. The DNA of the pathogen is altered so that it transfers genes that specify antigens. When injected into a patient, the altered DNA is taken up by cells and makes its way to the nucleus. The encoded antigens are manufactured and stimulate both cell-mediated and antibody-mediated immunity. Several DNA vaccines, including vaccines to prevent and treat HIV infections, are in clinical trials.

Specific Immune Responses

An antibody molecule, also called immunoglobulin (Ig), has two main functions: It combines with antigen, and it activates processes that destroy the antigen that binds to it. For example, an antibody may stimulate phagocytosis. Note that an antibody does not destroy an antigen directly; rather, it labels the antigen for destruction.

The basic structure of the immunoglobulin molecule was clarified by **Rodney Porter**, *of the University of Oxford in England, and* **Gerald Edelman**, *of Rockefeller University in New York, during the 1960s. Porter used the plant enzyme papain, a protease, to split Ig into fragments. Based on his findings, Porter developed a working model of the structure of the Ig molecule and was the first to suggest that it is Y-shaped. These researchers then constructed an accurate model of the antibody molecule. Porter and Edelman won the 1972 Nobel Prize in Medicine for their contributions.*

Two fragments of the antibody molecule bind antigen and are referred to as Fab fragments (Fab stands for antigen-binding fragment). The fragment that interacts with cells of the immune system is the Fc fragment (c indicates that this fragment crystallizes during cold storage). Many cells of the immune system have Fc receptors.

A typical antibody is a Y-shaped molecule in which the two arms of the Y (the Fab portions) bind with antigen. This shape enables the antibody to combine with two antigen molecules and allows the formation of antigen-antibody complexes. While the arms of the Y bind to antigen, the tail of the Y, the Fc portion, interacts with cells of the immune system, such as phagocytes, or binds with molecules of the complement system.

The antibody molecule consists of four polypeptide chains: two identical long chains called heavy chains, and two identical short chains called light chains. Each chain has a constant region and a variable region. In the constant (C) region, of the heavy chains, the amino acid sequence is constant within a particular immunoglobulin class. One can think of the C region as the handle portion of a door key. Like the pattern of bumps and notches at the part of a key that slides into a lock, the variable (V) region has a unique amino acid sequence. The variable region of the immunoglobulin extends outward from the B cell, whereas the constant region anchors the molecule to the B cell.

At its variable regions, the antibody folds three-dimensionally, assuming a shape that enables it to combine with a specific antigen. When they meet, antigen and antibody fit together somewhat like a lock and key. They must fit in just the right way for the antibody to be effective. A given antibody can bind with different strengths, or affinities, to different antigens. In the course of an immune response, higher-affinity antibodies are generated.

In an antigen, specific sequences of amino acids make up an antigenic determinants, or epitope. These sequences give part of the antigen molecule a specific shape that is recognized by an antibody or T cell receptor. Usually, an antigen has many different antigenic determinants on its surface; some have hundreds.

How can the immune system recognize every possible antigen? In 1956, the clonal selection hypothesis was developed which proposes that before a lymphocyte ever encounters an antigen, the lymphocyte has specific receptors for that antigen on its surface. When an antigen bins to a matching receptor in a lymphocyte, it activates the lymphocyte, which then gives rise to a clone of cells with identical receptors. A major problem with this hypothesis was its suggestion that our cells must contain millions of separate antibody genes. Since each human cell has a large amount of DNA, it is not enough to provide a different gene to code for each of the millions of possible specific antibody molecules.

In 1976, immunologists demonstrated that three separate families of genes code for immunoglobulins and that each gene family contains a large number of DNA segments that code for V regions. Recombination of these DNA segments during the differentiation of B cells is responsible for antibody diversity.

Note that rearrangement of these DNA segment produces an enormous number of potential combinations! Millions of different type of B (and T) cells are produced. By chance, one of those cells may produce just the right antibody to destroy the pathogen that invades the body.

Before 1975, the only method for obtaining antibodies for medicine and research was immunizing animals and collecting their blood. Then, immunologists developed monoclonal antibodies – identical antibodies produced by cells cloned from a single cell.

There are three memory systems in the human body:

- *Central nervous system memory, located in the brain.*
- *Immunological memory, located in nonlymphatic tissues including the lung, liver, kidney, and intestine.*
- *Genetic memory.*

Immunological memory manifest itself in "memory B cells" that continue to live and produce small amounts of antibody long after the body has overcome an infection. If the same pathogen enters the body again, the antibody immediately targets it for destruction. At the same time, specific memory cells are stimulated to divide, producing new clones of plasma cells that produce the same antibody.

III. The Nervous System — Body's Communication Network

I. Introduction

Neurobiology is the study of cells of the nervous system and the organization of these cells into functional circuits that process information and mediate behavior. It is a subdiscipline of both biology and neuroscience. Neurobiology differs from neuroscience, a much broader field that is concerned with any scientific study of the nervous system. Neurobiology should also not be confused with other subdisciplines of neuroscience such as computational neuroscience, cognitive neuroscience, behavioral neuroscience, biological psychiatry, neurology, and neuropsychology despite the overlap with these subdisciplines.

Neuroscience is an interdisciplinary program of study that provides an interconnectedness of the sciences and requires the synthesis and integration of different focal areas of investigation such as, neuroanatomy, neuroimaging, neuropsychopharmacology, neurophysiology, molecular neurobiology, neuroendocrinology, and how brain structures function to regulate behavior.

The integrative nature of neuroscience requires the tools provided by experience and training in general biology, genetics, physiology, molecular biology, chemistry (general, organic and biochemistry), physics, psychology (behavior, memory, cognition, sensation & perception) and research design and analysis.

More information has been discovered about the brain in the last ten years of investigative study than has ever been known before, in part, due to the fact that the 1990s were proclaimed as the "Decade of the Brain." Neuroscience is one of the most exciting and rapidly expanding fields of study that is only limited by the boundaries of the advancements in technology.

Below, are listed some of the most notable discoveries and achievements in Neuroscience in the 20th century.

Neuroscience subfields: Neurobiology, Cognitive Neuroscience, Computational Neuroscience, Neural Engineering, Neuroanatomy, Neurochemistry, Neuroimaging, Neurolinguistics, Neurology, Neuropharmacology, Neurophysiology, Neuropsychology, Psychopharmacology, Systems Neuroscience.

II. Neural signaling

An organism's ability to survive and to maintain homeostasis[25] depends largely on how effectively it detects and responds to stimuli — changes in the environment. Stimuli within the body include internal signals such as hunger or lowered blood pressure. Stimuli from the outside world include changes in temperature, or light, an odor, or movement that may indicate the presence of a predator or of prey.

In all animals except the sponges, responses to stimuli depend on cell signaling by networks of nerve cells, or neurons. These cells are specialized for transmitting propagating and receiving electrochemical impulses, electrical signals and chemical messages. In the human brain alone, there are over 10^{11} (a hundred billion) neurons.

---

[25] The capacity of an organism to maintain internal stability of equilibrium. Coined (1926) by **W.B. Cannon** (1871–1945). Much modern physiological work on homeostasis concerns itself with *bio-feedback* mechanisms which are controlled by the autonomic nervous system.

In most animals, neurons and supporting cells are organized as a nervous system that, like a computer takes in information, integrates it, and responds. Just how animals respond to stimuli depends on how their neurons are organized and connected to one another. A single neuron in the vertebrate brain may be functionally connected to thousands of other neurons. The endocrine system works with the nervous system to regulate many behaviors and physiological processes. The endocrine system generally provides relatively slow long-lasting regulation, whereas the nervous system typically permits more rapid, but brief, responses.

Neurobiology is one of the most exciting areas of biological research. Many investigations are studying neurotransmitters, the chemical messengers used by neurons to signal other neurons, and the receptor that bind with the neurotransmitters.

Another active area of research is the role of glial cells in the nervous system. These cells support and protect the neurons and have many regulatory functions. The glial cells provide glucose for neurons and also help regulate the composition of the extracellular fluid in the brain and spinal cord. Neurobiologists recently demonstrated that astrocytes induce and stabilize synapses (connections between neurons) in the brain. Although astrocytes can generate weak electrical signals, they communicate with one another and with neurons mainly with chemical signals.

A nerve consist of many parallel, independent signal paths, each of which is a nerve cell or fiber. Each cell is capable of transmitting signals in only one direction; separate cells carry signals to or from the brain. Each cell has an input end, a long conducting portion or axon, and an output end. It is these ends that give the cell its unidirectional character. The input end may be a transducer (stretch receptor, temperature receptor, etc.) or a junction (synapse) with another cell. A threshold mechanism is built into the input end; when an input signal exceeding a certain level is received, the nerve fires and an impulse of fixed size and duration travels down the axon. There may be several inputs at the synapse which may either aid or inhibit each other, depending on the nature of the synapse.

The axon is a long tail on the nerve cell that transmits the impulse without change of shape. It may be more than a meter in length, extending in the human, for example, from the brain to low in spinal cord or from the spinal cord to a finger or toe. Bundles of axons constitute what we usually think of as a nerve. The output end of the axon branches out in fine nerve endings, which appear to be separated by a gap from the next nerve or muscle cell which they drive.

The long, cylindrical axon has properties similar to those of an electric cable. Its diameter may range from less than a micron to 500 μm for the giant *axon of a squid; in humans, the upper limit is about 20 μm. Pulses travel along it with speeds ranging from 0.6 ms^{-1} to 100 ms^{-1}, depending, among other things, on the diameter of the axon. The axon core may be surrounded by either a membrane (for an* unmyelinated *fiber) or a much thicker sheath of fatty material (*myelin*), wound on like electrical tape. A myelinated fiber has its sheath interrupted at intervals and replaced by a short segment of membrane similar to that on an unmyelinated fiber. These interruptions are called the* nodes of Ranvier.

The axon may be removed from the rest of its cells and still will conduct nerve impulses. Its conduction properties depend on the membrane; the interior fluid (axoplasm) has been squeezed out of squid giant axons and replaced by an electrolyte solution without altering appreciably the propagation of the impulses. The axoplasm does contain chemicals essential to the long-term metabolic requirements of the cell.

Most animal cells have a difference in electrical charge across the plasma membrane — a more negative electrical charge inside the cell compared with the electrical charge of the extracellular fluid outside. The plasma membrane is said to be electrically polarized, *meaning that one side, or pole, has a different charge from the other side. When electrical charges are separated in this way, a potential energy difference exists across the membrane. This difference in electrical charge across the plasma membrane gives rise to an electrical voltage gradient.*

The voltage measured across the plasma membrane is called the membrane potential. *If the charges are permitted to come together, they have the ability to do work. Thus the cell can be thought of as a biological battery. In excitable cells, such as neurons and muscle cells, the membrane can transmit signals to other cells.*

The membrane potential in a resting (not excited) neuron or muscle cell is its resting potential. *The resting potential is generally expressed in units called* millivolts *(mV). (A millivolt equals one thousandth of a volt.) Voltage is the force that causes charged particles to flow between two points. Like other cells that can produce electrical signals, the neuron has a resting potential of about 70 mV. By convention this is expressed as -70 mV because the cytosol close to the plasma membrane is negatively charged relative to the extracellular fluid.*

Biologists cam measure the potential across the membrane by placing one electrode inside the cell and a second electrode outside the cell, and connecting through a very sensitive voltmeter or oscilloscope. If one places both electrodes on the outside surface of the neuron, no potential difference between them is registered. (All points on the same side of the membrane have the same charge.) However, once one of the electrodes penetrates the cell, the voltage changes from zero to approximately $-70mV$.

Two main factors determine the magnitude of the membrane potential: (1) differences in the concentrations of specific ions inside the cell compared with the extracellular fluid, and (2) selective permeability of the plasma membrane to these ions. The distribution of ions inside neurons and in the extracellular fluid surrounding them is like that of most other cells in the body. The potassium ion (K^+) *concentration is about 10 times greater inside than outside the cell. In contrast, the sodium ion* (Na^+) *concentration is about 10 times greater outside than inside. This asymmetric distribution of ions across the plasma membrane at rest is brought about by the action of selective ion channels and ion pumps. In vertebrate neurons (and skeletal muscle fibers), the resting membrane potential depends mainly on the diffusion of ions down their concentration gradients.*

Ions cross the plasma membrane by diffusion through ion channels that are formed by membrane proteins. Net movement of ions occurs from an area of higher concentration of that ion to one of lower concentration. Typically, these channels allow only specific types of ions to pass.

Neurons have three types of ion channels: passive ion channels, voltage-activated channels, and chemically activated ion channels. Passive ion channels permit the passage of specific ions such as Na^+, K^+, Cl^-, *and* Ca^{2+}. *Unlike voltage-activated and chemically activated ion channels, passive ion channels are not controlled by gates.*

Potassium channels are the most common type of passive ion channel in the plasma membrane, and cells are more permeable to potassium than to other ions. In fact, in the resting neuron, the plasma membrane is up to 100 times more permeable to K^+ *than to* Na^+. *Sodium ion pumped out of the neuron cannot easily pass back into the cell, but* K^+ *pumped into the neuron easily diffuse out.*

Potassium ions leak out through passive ion channels following their concentration gradient. As these positively charged ions diffuse out of the neuron, they increase the positive charge in the extracellular fluid outside the cell relative to the charge inside the cell. The resulting change in the electrical gradient across the membrane influences the flow of ions. This electrical gradient forces some of the positively charged potassium ion back into the cell.

At the input end of a nerve cell, the response to chemicals from a synapse is often an increase in membrane permeability to sodium ions, which causes an increase in the interior potential. In other cases the interior potential becomes more negative and firing is inhibited. If the potential becomes high enough (that is, more positive or less negative), the regenerative action of the membrane takes over and the cell initiates an impulse.

If the input end of the cell acts as a transducer, the interior potential rises as the cell is stimulated. If the input is from another nerve, the signal may cause the nerve to fire, or it may cause the potential to increase by a subthreshold amount so that two or more stimuli must be received simultaneously to cause firing, or it may decrease the potential and inhibit stimulation by another nerve at the synapse. Comparison of the axoplasm with the interstitial fluid surrounding each axon shows an excess of potassium and a deficit of sodium and chloride ions within the axon.

A typical axon might have a radius of 5 μm = 5000 nm If the axon is not myelinated, the thickness of the cell membrane might be 5–10 nm; a myelin layer might be 2000 nm thick, with nodes of Ranvier spaced every 1–2 mm.

The inside of the cell has an electrical potential about 70 mV less than outside the cell.

As the pulse passes by, the potential at a fixed point on the axon rises in a millisecond or less to about +40 mV. Then it falls back to about –90 mV and finally recovers slowly to the resting value of –70 mV. The membrane is said to depolarize and then repolarize. The regenerative action that produces these sudden changes of membrane potential is caused by changing permeability of the membrane to sodium and potassium ions.

At the end of a nerve cell the signal passes to another nerve cell or to a muscle cell across a synapse or junction. There are gaps of 10–20 nm between presynaptic and postsynaptic nerve cells and gaps of 50–100 nm of the neuromuscular junction. In some cases such as the heart, the transmission may be electrical; yet in many cases the signal is carried by chemicals.

At most vertebrate neuromuscular junctions, the nerve impulse is followed by an electrical impulse which propagates throughout the muscle fiber and initiates contraction. There is good experimental evidence that acetylcholine is released by the nerve endings when the nerve fires. It increases the permeability of the nearby muscle membrane to sodium, which then leaks in and depolarizes the membrane.

III. Mathematical Models

*Neuron physiology describes the electrical properties of nerve cell membrane. Models of nerves are based on the Nernst equation (**W.H. Nernst**, 1864–1941) that determines a cell membrane's potential from the ion concentrations near it. A neuron consists of dendrites that receive signals, a cell body that synthesized incoming signals and generates new ones, an axon that* *transmits new signals away from the cell body, and a synapse that transmits the signals to other cells. The structure and function of synapses play important roles in this study. Neurotransmitters (chemical molecules) released at synapses in response to changes in membrane voltage communicate these changes to the neuron's environment.*

Attempts to describe a nerve cell's electrical behavior have been based on electrical circuit analogies and their mathematical models.

The Hodgkin-Huxley model of nerve membrane (1952), was based on observations of neural signals propagating in the nerve axon of a giant squid. While it does not accurately describe all aspects of membrane behavior, it has been instrumental in suggesting and promoting understanding of a variety of important experiments.

An electrical potential is established across a cell's membrane by having different concentrations of electrically charged chemical species inside and outside of a semipermeable membrane separates two regions of space that have *concentrations of ions*, e.g. C_1 and C_2, inside and outside, respectively, then Nernst equation states that the resulting potential E is given by $E = (\frac{RT}{q})\log(\frac{C_1}{C_2})$, when R is a gas constant, T is the absolute temperature. This shows how to compute the membrane potential once the ion concentrations inside and outside are known.

The ions most important to the cell membrane potential are sodium (Na^+) and potassium (K^+). Each ionic species has associated with it a membrane potential that is maintained by a pump in the equilibrium, the Nernst equation shows that

$$E_{Na} = (\frac{kT}{q})\log(\frac{C_0^{Na}}{C_i^{Na}}) = 55mV,$$
$$E_K = (\frac{kT}{q})\log(\frac{C_0^{K}}{C_i^{K}}) = -75mV.$$

These are referred to as the sodium and potassium resting potentials.

The main parts of interest in the neuron are the *dendrites*, which receive signals from impinging axons, the *cell body* that can generate electrical activity, usually in the from of a voltage pulse called an *action potential*, the *axon*, which carries an action potential from the cell body to the synapse, and the *synapse*, which causes chemical signals to be released outside the cell in response to the arrival of an action potential.

Chemical signals are molecules called neurotransmitters. These diffuse across the synaptic gap to a dendrite and cause an electrical potential to be created across the dendrite's membrane. Eventually, this either causes the cell body to fire (i.e., creates an action potential) or inhibits it from *firing. Neuron membranes at rest are impermeable to* (Na^+). *Therefore, the observed potential is near* E_K. *When excited,* (Na^+) *channels open rapidly and the membrane potential approaches* E_{Na}. *The* (K^+) *channels opens more slowly, but eventually returns the membrane potentials to near* E_K.

The mechanisms that control these channels remain unknown, although many models of membrane potentials and currents have been derived and used effectively to suggest experiments and to interpret data.

Action potentials arriving at a synapse, called presynaptic potentials, cause vesicles containing chemical neurotransmitters to migrate to the synapse membrane and release their contents into the synaptic gap.

The neurotransmitters diffuse across the synaptic gap, though some are lost from the gap. The molecules that arrive at the postsynaptic membrane interact with it to modify its membrane potential. If $c(t)$ *denotes the concentration of neurotransmitter in the gap at time* t, *this increases in response to the arrival of action potential and decreases because of chemical binding with the postsynaptic membrane and diffusion out of the gap. These chemical kinetics are modeled by the equation*

$$\mathrm{dc/dt} = -k_{dif}\,\mathrm{c} - k_{post}\,\mathrm{c} + S$$

Basically the axon is a long cylindrical tube which extends from each neuron and electrical signals propagate along its membrane, about 50–70 Ångströms thick. The electrical pulses arise because the membrane is preferentially permeable to various chemical ions with the permeabilities affected by the currents and potentials present. The key elements in the system are potassium (K^+) *ions and sodium* (Na^+) *ions. In the rest state there is a transmembrane potential difference of about* -70 *millivolts (mV) due to the higher concentration of* (K^+) *ions within the axon as compared with the surrounding medium.*

The deviation in the potential across the membrane, measured from the rest state, is a primary observable in experiments. The membrane permeability properties change when subjected to a stimulating electrical current I : *they also depend on the potential. Such a current can be generated, for example, by a local depolarization relative to the rest state.*

Let us take the positive direction for the membrane current, denoted by I, *to be outwards from the axon. The current* $I(t)$ *is made up of the current due to the individual ions which pass through the membrane and the contribution from the time variation in the transmembrane potential, that is the membrane capacitance contribution. Thus we have*

$$I(t) = C\frac{dV}{dt} + I_i, \tag{1}$$

where C is the capacitance and I_i is the current contribution from the ion movement across the membrane. Based on experimental observation **Hodgkin** and **Huxley** *(1952) took*

$$\begin{aligned} I_i &= I_{Na} + I_K + I_L, \\ &= g_{Na} m^3 h (V - V_{Na}) + g_K n^4 (V - V_K) + g_L (V - V_L), \end{aligned} \qquad (2)$$

where V is the potential and I_{Na}, I_K and I_L are respectively the sodium, potassium and "leakage" currents: I_L is the contribution from all the other ions which contribute to the current. The g's are constant conductances with, for example, $g_{Na} m^3 h$ the sodium conductance, and V_{Na}, V_K and V_K are constant equilibrium potentials. The m, n and h are variables, which are determined by the differential equations

$$\begin{aligned} \frac{dm}{dt} &= \alpha_m(V)(1-m) - \beta_m(V)m, \\ \frac{dn}{dt} &= \alpha_n(V)(1-n) - \beta_n(V)n, \\ \frac{dh}{dt} &= \alpha_h(V)(1-h) - \beta_h(V)h, \end{aligned} \qquad (3)$$

where the α and β are given functions of V (again empirically determined by fitting the results to the data). α_n and α_m are qualitatively like $(1+\tanh V)/2$ while $\alpha_h(V)$ is qualitatively like $(1-\tanh V)/2$, which is a "turn-off" switch if V is moderately large.

If an applied current $I_a(T)$ is imposed the governing equation, using (1), becomes

$$C\frac{dV}{dt} = -g_{Na} m^3 h (V - V_{Na}) - g_K n^4 (V - V_K) - g_L (V - V_L) + I_a \qquad (4)$$

The system (4) with (3) constitute the 4-variable model which was solved numerically by **Hodgkin** and **Huxley** *(1952).*

If $I_a = 0$, the rest state of the model (3) and (4) is linearly stable but is excitable. That is, if the perturbation from the steady state is sufficiently large there is a large excursion of the variables in their phase space before returning to the steady state. If $I_a \neq 0$ there is a range of values where regular repetitive firing occurs; that is – the mechanism displays limit cycle characteristics. Both types of phenomena have been observed experimentally.

*Because of the complexity of the equation system, various simpler mathematical models, which capture the key features of the full system, have been proposed, the best known and particularly useful one is the FitzHugh-Nagumo model (***FitzHugh** *1961,* **Nagumo** et al. *1962).*

The time scales for m, n *and* h *in* (3) *are not all of the same order. The time scale for* m *is much faster than the others, so it is reasonable to assume it is sufficiently fast such that it relaxes immediately to its value determined by setting* $dm/dt = 0$ *in* (3). *If we also set* $h = h_0$*, a constant, the system still retains many of the features experimentally observed. The resulting 2-variable model in* V *and* n *can then be qualitatively approximated by the dimensionless system*

$$\frac{du}{dt} = f(u) - \omega + I_a, \qquad \frac{dv}{dt} = bu - \gamma v, \tag{5}$$

$$f(u) = v(a-u)(u-1), \tag{6}$$

where $0 < a < 1$ *and* b *and* γ *are positive constants. Here* u *is like the membrane potential* V*, and* v *plays the role of all three variables* m, n *and* h *in* (3).

It can be shown that this system of equations governs the existence of traveling pulses which only propagate if a certain threshold perturbation is exceeded. By a pulse here, we mean a wave which represent an excursion from a steady state and back to it – like a solitary wave on water.

If in (5) *we assume* $I_a = 0$ *(no applied current) and allow spatial diffusion in the transmembrane potential*

$$\frac{du}{dt} = f(u) - v + D\frac{\partial^2 u}{\partial x^2}, \qquad \frac{dv}{dt} = bu - \gamma v, \tag{7}$$

$$f(u) = u(a-u)(u-1), \tag{8}$$

Here, u *is directly related to the membrane potential* V *and* v *plays the role of several variables associated with terms in the contribution to the membrane current from sodium, potassium and other ions. The "diffusion" coefficient* D *is associated with the axial current in the axon. The parameters* $0 < a < 1$*,* b *and* γ *are all positive.*

IV. The Endocrine System — Body's Messengers and Regulators.

Introduction

A caterpillar becomes a butterfly. A crustacean changes color to blend with its background. A young girl develops into a woman. An adult copes with chronic stress. These physiological processes and many other adjustments of metabolism, fluid balance, growth and development, and reproduction are regulated by the endocrine system. *This system works closely with nervous system to maintain homeostasis, the steady state of the body.*

The endocrine system is a diverse collection of cells, tissues, and organs, including specialized endocrine glands *that produce and secrete* hormones, *chemical messengers responsible for the regulation of many body processes. Hormones excite, or stimulate, changes in specific tissues.*

Endocrine glands have no ducts; they secrete hormones directly into the interstitial fluid or blood. Hormones are typically transported by the blood and produce a characteristic response only after they reach target cells and bind with specific receptors. Target cells, the cells influenced by a particular hormone, may be another endocrine gland or in an entirely different type of organ, such as a bone or the kidney. Target cells may be located far from the endocrine gland. For example, the vertebrate thyroid gland secretes hormones that stimulate metabolism in tissues throughout the body. Several types of hormones may be involved in regulating the metabolic activities of a particular type of cell. In fact, many hormones produce a synergistic effect in which the presence of one hormone enhances the effects of another.

Endocrinologists extract the suspected compound from the endocrine tissue of one animal and inject it into an experimental animal from which the tissue producing the compound has been removed. Deficiency symptoms should be relieved by replacing the suspected hormone. Researchers then isolate the active compound and determine its chemical structure. Finally, the compound is synthesized in the laboratory and injected into experimental animals. If its effects are those predicted, the researchers have data to support their hypothesis.

Using such procedures, endocrinologists have identified about 10 discrete endocrine glands. More recently investigators have identified specialized cells in the digestive tract, heart. kidneys, and many other parts of the body that also release hormones. In addition, some neurons release hormones. As a result of these discoveries, the scope of endocrinology has been broadened to include the production and actions of chemical messengers produced by a wide variety of organs, tissues, and cells.

One active focus of research is the study of the mechanisms of hormone action, which includes characterizing receptors and identifying the molecules involved in signal transduction. Some endocrinologists now use a reverse strategy for discovering new hormones and signaling pathways within the cell. They focus on "orphan" nuclear receptors, those for which ligands (the molecules that bind with them) are not yet known. Some of these "orphan" receptors are receptors for hormones that have not yet been identified. Using this strategy, researchers have identified intracellular signaling pathways for steroids, fatty acids, and several other compounds.

Cell signal one another with neurotransmitters, hormones, and local regulators. Some chemical compounds function as all three of these types of signals. Thus a neuron, endocrine gland, or some other cell type all may secrete the same chemical message. However, the same message can have different meanings for various target cells.

Pheromones are chemicals messengers that animals produce for communication with other animals of the same species. Because pheromones are generally produced by exocrine glands and do not regulate metabolic activities within the animal that produces them, most biologists do not classify them as hormones.

Endocrine glands differ from exocrine glands (such as sweat glands and gastric glands) that release their secretions into ducts. Endocrine glands have no ducts, and secrete their hormones into the surrounding interstitial fluid or blood. Typically, hormones diffuse into capillaries and are transported by the blood to target cells. biologists have discovered that in addition to the discrete classical endocrine glands, specialized cells in many tissues and organs (such as kidneys and heart) also release hormones or hormone-like substances.

The complexity of animal physiology challenges simplistic definitions. As new chemical signals and their modes of action have been discovered, the traditional definition of a hormone as a substance secreted by an endocrine gland and transported by the blood has become inadequate.

Certain neurons, known as neuroendocrine cells, are an important link between the nervous and endocrine systems. Neuroendocrine cells produce neurohormones that are transported down axons and released into the interstitial fluid. they typically diffuse into capillaries and are transported by the blood. invertebrate endocrine systems are largely neuroendocrine. In vertebrates, the hypothalamus produces several neurohormones that link the nervous system with the pituitary gland, an endocrine gland that secretes several hormones.

A local regulator is a signaling molecule that diffuses through the interstitial fluid and acts on nearby cells. Certain chemical compounds that are indisputably hormones because they are typically transported by the blood, under some conditions acts as local regulators. In autocrine regulation, a hormone, or other regulator, acts on the very cells that produce it. For example, the female hormone estrogen, which functions as a classical hormone, may also exert an autocrine effect that stimulates additional estrogen secretion. Estrogen can also act on nearby cells, a type of local regulation known as paracrine regulation.

Local regulators include local chemical mediators such as histamine, growth factors, and prostaglandins.

Histamine is stored in mast cells and is released in response to allergic reactions, injury, or infection. Histamine causes blood vessels to dilate and capillaries to become more permeable. More than 50 growth factors, typically peptides, stimulate division and normal development in specific types of cells. Growth factors have autocrine or paracrine effects. Nitric oxide (NO), another local regulator, is produced by many types of cells, including those lining blood vessels. It relaxes nearby smooth muscle fibers, dilating the blood vessel.

Prostaglandins are modified fatty acids released continuously by the cells of most tissues. Biologists have grouped them into nine different classes. Although present in very small quantities, these local regulators affect a wide range of body processes. Prostaglandins are paracrine regulators that act on cells in their immediate vicinity. They modify cyclic adenosine monophosphate (cAMP) levels and interact with other hormones to regulate various metabolic activities.

The major prostaglandin target is smooth muscle. Some prostaglandins stimulate smooth muscle to contract, whereas others cause relaxation. Thus some reduce blood pressure, whereas others raise it. Prostaglandins synthesized in the temperature-regulating center of the hypothalamus cause fever. In fact, nonsteroidal anti-inflammatory drugs (NSAIDs) such as aspirin and ibuprofen reduce fever and decrease inflammation by inhibiting prostaglandin synthesis.

Because prostaglandins are involved in the regulation of so many metabolic processes, they have great potential for a variety of clinical uses. Physicians use them to induce labor in pregnant women, to induce abortion, and to promote the healing of ulcers in the stomach and duodenum. Prostaglandins may someday be used to treat a wide variety of illnesses, including asthma, arthritis, kidney disease, certain cardiovascular disorders, and some forms of cancer.

Although hormones are chemically diverse, they generally belong to one of four different chemical groups: (1) fatty acid derivatives, (2) steroids, (3) amino acid derivatives, or (4) peptides or proteins.

The endocrine system provides an electrochemical connection from the hypothalamus of the brain to all the organs that control the body metabolism, growth and development, and reproduction.

There are two types of hormones secreted in the endocrine system: steroidal and non-steroidal, or protein based hormones. Signal transduction of some hormones with steroid structure involves nuclear hormone receptor proteins that are a class of ligand activated proteins that, when bound to specific sequences of DNA serve as on-off switches for transcription within the cell nucleus.

These switches control the development and differentiation of skin, bone and behavioral centers in the brain, as well as the continual regulation of reproductive tissues. They also bind to receptor sites, and activate second messenger systems for more rapid responses. Nonsteroidal hormones bind to receptor sites on the external surface of the cell membrane and use a second messenger method of altering internal cell functions, by altering the pathways already existing in the cells, by activating or deactivating enzymes which modify existing proteins.

The endocrine system regulates its hormones through negative feedback. Increases in hormone activity decrease the production of that hormone. The immune system and other factors contribute as control factors also, altogether maintaining constant levels of hormones.

There are three different kinds of hormones based on their chemical composition:

Amines

Amines, such as norepinephrine, epinephrine, and dopamine, are derived from single amino acids, in this case tyrosine. Thyroid hormones such as 3,5,3'-triiodothyronine (T3) and 3,5,3',5'-tetraiodothyronine (thyroxine, T4) make up a subset of this class because they derive from the combination of two iodinated tyrosine amino acid residues.

Peptide and Protein

Peptide hormones and protein hormones consist of three (in the case of thyrotropin-releasing hormone) to more than 200 (in the case of follicle-stimulating hormone) amino acid residues and can have molecular weights as large as 30,000. All hormones secreted by the pituitary gland are peptide hormones, as are leptin from adipocytes, ghrelin from the stomach, and insulin from the pancreas.

Steroid

Steroid hormones are derived from cholesterol and are subdivided into those with an intact steroid nucleus (gonadal and adrenal steroids) and those with a broken steroid nucleus (vitamin D). Steroid hormones include estrogen and progesterone from the ovary, testosterone from the testes, and cortisol and aldosteron from the adrenal gland.

Table 6.10 lists the human endocrine glands and their hormones.

Table 6.10: ENDOCRINE GLANDS AND THEIR HORMONES

A = Amino acid derivatives
S = Steroids
P = Polypeptides

GLAND	HORMONE	TARGET TISSUE	PRINCIPAL ACTIONS
Hypothalamus	*Releasing and inhibiting hormones*	*Anterior lobe of pituitary*	*Regulate secretion of hormones by the anterior pituitary*
Posterior pituitary	*Oxytocin (P)*	*Uterus*	*Stimulates contraction*
		Mammary glands	*Stimulates ejection of milk into ducts*
	Antidiuretic hormone (ADH)	*Kidneys (collecting ducts)*	*Stimulates reabsorption of water*
	Growth hormone (GH)(P)	*General*	*Stimulates growth of skeleton and muscle*
	Prolactin	*Mammary glands*	*Stimulates milk production*
	Thyroid-stimulating hormone (TSH)(P)	*Thyroid gland*	*Stimulates secretion of thyroid hormones*
	Adrenocorticotropic hormone (ACTH)(P)	*Adrenal cortex*	*Stimulates secretion of adrenal cortical hormones*

Table 6.10: (Cont.)

GLAND	HORMONE	TARGET TISSUE	PRINCIPAL ACTIONS
	Gonadotropic hormones (follicle-stimulating hormone[FSH](P); Luteinizing hormone[LH](P))	*Gonads*	*Stimulates gonad function and growth*
Thyroid gland	*Thyroxine* (T_4) *and triiodothyronine* (T_3)	*General*	*Stimulates metabolic rate; regulate energy metabolism*
	Calcitonin	*Bone*	*Lowers blood-calcium level*
Parathyroid glands	*Parathyroid hormone*	*Bone, kidneys, digestive tract*	*Regulates blood-calcium level*
Pancreas	*Insulin(P)*	*General*	*Lowers blood glucose concentration*
	Glucagon(P)	*Liver, adipose tissue*	*Raises blood glucose concentration*
Adrenal Medulla	*Epinephrine and Norepinephrine(A)*	*Muscle; blood vessels; liver; adipose tissue*	*Help body cope with stress; increase metabolic rate; raise blood glucose level;increases heart rate and blood pressure*
Adrenal cortex	*Mineralocorticoids (S)*	*Kidney tubules*	*Maintain sodium and potassium balance*

Table 6.10: (Cont.)

GLAND	HORMONE	TARGET TISSUE	PRINCIPAL ACTIONS
	Glucocorticoids (S)	*General*	*Help body cope with long-term stress; raise blood-glucose level*
Pineal gland	*Melatonin*	*Hypothalamus*	*Important in biological rhythms*
Ovary	*Estrogens(S)*	*General; uterus*	*Develop and maintain sex characteristics in female; stimulates growth of uterine lining*
	Progesterone	*Uterus; breast*	*Stimulates development of uterine lining*
Testis	*Testosterone*	*General; reproductive structures*	*Develops and maintains sex characteristics in males; promotes spermatogenesis*

Life at the Organizational Level (Developmental and Evolutional Biology)

The complicated reconciliation of Darwinian natural selection with Mendelian genetics involved contributions from naturalists, experimentalists and population geneticists. Naturalists, intimately acquainted with adaptation, geographic variation and speciation, helped promote the idea of species as populations rather than ideal types. Experimentalists provided the important distinction between genotype and phenotype, showed that mutations could be small and inherited in a Mendelian fashion, and demonstrated, that selection could work on continuous variations to modify the characters of a population. Population genetics constructed mathematical models demonstrating how mutations, selection, migration and other factors could affect the frequencies of genes in populations.

The genetic analysis of natural populations, pioneered in Russia by **Sergei Chetverikov** *(1880–1959), was developed in the West by his student,* **Theodosius Dobzhansky** *(1900–1975). With Dobzhansky,* **Julian Huxley** *(1887–1975),* **Ernst Mayr** *(1904–2005), and* **George Gaylord Simpson** *(1902–1984), the modern, "synthetic" theory of evolution took shape. This synthetic or "neo-Darwinian" theory of evolution sees natural selection, working on small, Mendelian variations produced by mutations and recombination, as the major agent in evolutionary change.*

Thus, in the early 20th century, the rediscovery of Mendel's work led to the rapid development of genetics by **Thomas Hunt Morgan** *and his students, and by the 1930s the combination of population genetics and natural selection led to the "neo-Darwinian synthesis" and the rise of the discipline of evolutionary biology. New biological disciplines developed rapidly, especially after* **Watson** *and* **Crick** *discovered the structure of DNA in 1953.*

Following the cracking of the genetic code and the establishment of the Central Dogma, biology was largely split between organismal biology – consisting of ecology, ethology, systematics, paleontology, evolutionary biology, developmental biology, and other disciplines that deal with whole organisms and groups of organisms – and the constellation of disciplines related to molecular biology – including cell biology, biophysics, biochemistry, neuroscience, immunology, and many other overlapping subjects.

Developmental biology studies the process by which organisms grow and develop. Originating in embryology, today developmental biology studies the genetic control of cell growth, differentiation and "morphogenesis," which is the process that gives rise to tissues, organs and anatomy. Model organisms for developmental biology include the round worm Caenorhabditis elegans, the fruit fly Drosophila melanogaster, the zebrafish Brachydanio rerio, the mouse Mus musculus, and the weed Arabidopsis thaliana.

Molecular data regarding the mechanisms underlying development started to accrue quickly during the 1980's and '90's. As scientists began to compare the developmental mechanisms in different organisms, they realized that these mechanisms are conserved through deep evolutionary time. By combining the disciplines of phylogenetics, paleontology and comparative developmental biology, scientists try to infer the way in which early organisms developed, thus spawning the new discipline of "evo-devo."

I. Botany and Zoology

Botany is the scientific study of plantlife. As a branch of biology, it is also sometimes referred to as plant science(s) or plant biology. Botany covers a wide range of scientific disciplines that study the structure, growth, reproduction, metabolism, development, diseases, ecology and evolution of plants.

As with other life forms in biology, plant life can be studied from different perspectives, from the molecular, genetic and biochemical level through organelles, cells, tissues, organs, individuals, plant populations, and communities of plants. At each of these levels a botanist might be concerned with the classification (taxonomy), structure (anatomy and morphology), or function (physiology) of plant life.

Historically, botany covers all organisms that were not considered to be animals. Some of these "plant-like" organisms include fungi (studied in mycology), bacteria and viruses (studied in microbiology), and algae. Most algae, fungi, and microbes are no longer considered to be in the plant kingdom. However, attention is still given to them by botanists, and bacteria, fungi, and algae are usually covered in introductory botany courses.

The study of plants has importance for a number of reasons. Plants are a fundamental part of life on Earth. They generate the oxygen, food, fibres, fuel and medicine that allow higher life forms to exist. Plants also absorb carbon dioxide, a significant greenhouse gas, through photosynthesis. A good understanding of plants is crucial to the future of human societies as it allows us to:

- *Feed the world*
- *Understand fundamental life processes*
- *Utilize medicine and materials*
- *Understand environmental changes*

Virtually all of the food we eat comes from plants, either directly from staple foods and other fruit and vegetables, or indirectly through livestock, which rely on plants for their nutrition. In other words, plants are at the base of nearly all food chains, or what ecologists call the first trophic level.

Understanding how plants produce the food we eat is therefore important to be able to feed the world and provide food security for future generations, for example through plant breeding. Not all plants are beneficial to humans, some weeds are a considerable problem in agriculture and botany provides some of the basic science in order to understand how to minimize their impact.

However, other weeds are pioneer plants which start an abused environment back on the road to rehabilitation, underlining that the term 'weed' is a very relative concept and that, broadly defined, a weed is simply a plant which is too successful. Ethnobotany is the study of this and/or other relationships between plants and people.

Plants are convenient organisms in which fundamental life processes (like cell division and protein synthesis for example) can be studied, without the ethical dilemmas of studying animals or humans. The genetic laws of inheritance were discovered in this way by **Gregor Mendel***, who was studying the way pea shape is inherited. What Mendel learnt from studying plants has had far reaching benefits outside of botany.*

Additionally, **Barbara McClintock** *discovered 'jumping genes' by studying maize. These are a few examples that demonstrate how botanical research has an ongoing relevance to the understanding of fundamental biological processes.*

Many of our medicinal and recreational drugs, like cannabis, caffeine, and nicotine come directly from the plant kingdom. Aspirin, which originally came from the bark of willow trees, is just one example. There may be many novel cures for diseases provided by plants, waiting to be discovered. Popular stimulants like coffee, chocolate, tobacco, and tea also come from plants. Most alcoholic beverages come from fermenting plants such as barley malt and grapes.

Plants also provide us with many natural materials, such as cotton, wood, paper, linen, vegetable oils, some types of rope, and rubber. The production of silk would not be possible without the cultivation of the mulberry plant. Sugarcane and other plants have recently been put to use as sources of biofuels, which are important alternatives to fossil fuels.

Plants can also help us understand changes in on our environment in many ways:

- *Understanding habitat destruction and species extinction is dependent on an accurate and complete catalog of plant systematics and taxonomy.*

The active search for knowledge by means of observation and experiment found its natural home in the universities. Owing to the connection of medicine with these seats of learning, it was natural that the study of the structure and functions of the human body and of the animals nearest to man should take root there; the spirit of inquiry which now for the first time became general, showed itself in the anatomical schools of the Italian universities of the 16th century, and spread fifty years later to Oxford.

The discovery of the microscope in the 1500's contributed a lot to the study of biology. In 1555, **Conrad Gesner** *wrote the first of a series of several books called Historia Animalum, which became the new standard for the next two hundred years. During this time there were also number of books about specialized topics, such as birds, fish, and others. Gesner, and other writers like* **Thomas Browne** *(1605–1682) and* **Ulisse Aldrovandus** *(1522–1602) began to subject biology and zoology to scientific scrutiny.*

In the 17th century the investigators of nature by means of observation and experiment, banded themselves into academies or societies for mutual support and intercourse. The first founded of surviving European academies, the Academia Naturae Curiosorum (1651)[27] *especially confined itself to the description and illustration of the structure of plants and animals; eleven years later (1662) the Royal Society of London was incorporated by royal charter, having existed without a name or fixed organization for seventeen years previously (from 1645).*

A little later the Academy of Sciences of Paris was established by Louis XIV. The influence of these great academies of the 17th century on the progress of zoology had the effect of bringing together of the museum-men and the physicians or anatomists, which was just what needed for further development.

As the amount of knowledge grew quickly, it was necessary to develop a classification system. **Carl Linnaeus** *(1707–1778) created a classification system that used two Latin names – the species, and the genus. This is the system that is still used today. But it still took many years for scientists to understand how systems of biology and zoology worked.*

Whilst the race of collectors and systematizers culminated in the latter part of the 18th century in Linnaeus, a new type of student made its appearance in such men as **John Hunter** *and other anatomists, who, not satisfied with the superficial observations of the popular "zoologists", set themselves to work to examine anatomically the whole animal kingdom, and to classify its members by aid of the results of such profound study.*

[27] The Academia Secretorum Naturae was founded at Naples in 1560, but was suppressed by the ecclesiastical authorities.

Under the influence of the touchstone of strict inquiry set by the Royal Society, the marvels of witchcraft, sympathetic powders, and other relics of medieval superstition disappeared like a mist before the sun, whilst accurate observations and demonstrations of a host of new wonders accumulated. Among these which were numerous contributions to the anatomy of animals, and none perhaps more noteworthy than the microscope observations, of ***Leeuwenhoek****, (1683).*

It was not until the 19th century that the microscope, thus early applied by ***Leeuwenhoek****,* ***Malpighi****,* ***Hooke****, and* ***Swammerdam*** *to the study of animal structure, was perfected as an instrument, and accomplished for zoology its final and most important service.*

The perfecting of the microscope led to a full comprehension of the great doctrine of cell structure and the establishment of the facts —
(1) that all organisms are either single cells of living material (microscopic animalcules, etc.) or are built up of an immense number of such units;
(2) that all organisms begin their individual existence as a single unit or corpuscle of living substance, which multiplies by binary fission, the products growing in size and multiplying similarly by binary fission; and
(3) that the life of a multicellular organism is the sum of the activities of the corpuscular units of which it consists, and that the processes of life must be studied in and their explanation obtained from an understanding of the chemical and physical changes which go on in each individual corpuscle or unit of living material or protoplasm.

Paleontology is a newer science than zoology. Some ancient Greek writers believed that fossils were from prehistoric creatures, but ***Aristotle*** *thought they were merely formed by a mud slide. For hundreds of years, nobody understood fossils, and thought they were freaks of nature. Paleontology did not become a science until the early 1800's when* ***George Cuvier*** *(1769–1832) founded comparative anatomy and brought about a synthesis of anatomy and physiology. This science studied the development, functions, and structure of internal organs, and allowed scientists to identify and understand fossilized remains.*

Meanwhile the astronomical theories of development of the solar system from a gaseous condition to its present form, put forward by ***Kant*** *and by* ***Laplace****, had impressed men's minds with the conception of a general movement of spontaneous progress or development in all nature.*

The science of geology came into existence, and the whole panorama of successive stages of the earths history, each with its distinct population of strange animals and plants (unlike those of the present day and simpler in proportion as they recede into the past) was revealed by ***Cuvier****,* ***Agassiz****, and others.*

The history of the crust of the earth was explained by **Lyell** *as due to a process of slow development, with no cataclysmic agencies and no mysterious forces differing from those operating at the present day.*

Thus he carried on the narrative of orderly development from the point at which it was left by Kant and Laplace – explaining by reference to the ascertained laws of physics and chemistry the configuration of the earth, its mountains and seas, its igneous and its stratified rocks, just as the astronomers had explained by those same laws the evolution of the Sun and planets from diffused gaseous matter of high temperature. The suggestion that living things must also be included in this great development was obvious.

The delay in the establishment of the doctrine of organic evolution was due, not to the ignorant and unobservant, but to the leaders of zoological and botanical science.

Knowing the almost endless complexity of organic structures, realizing that man himself with all the mystery of his life and consciousness must be included in any explanation of the origin of living things, they preferred to regard living things as something apart from the rest of nature, specially cared for, specially created by a Divine Being.

Thus it was that the so-called "Natur-philosophen" of the last decade of the 18th century, and their successors in the first quarter of the 19th, found few adherents among the working zoologists and botanists.

Lamarck, **Treviranus**, **Erasmus Darwin**, **Goethe**, *and* **Saint-Hilaire** *preached to deaf ears, for they advanced the theory that living beings had developed by a slow process of transmutation in successive generations from simpler ancestors, and in the beginning from simplest formless matter, without being able to demonstrate any existing mechanical causes by which such development must necessarily be brought about.*

They were met by the criticism that possibly such a development had taken place; but, as no one could show as a simple fact of observation that it had *taken place, nor as a result of legitimate inference that it* must *have taken place, it was quite as likely that the past and present species of animals and plants had been separately created or individually brought into existence by unknown and inscrutable causes. It was held that scientific man would refuse to occupy himself with such fancies facts.*

In 1859, **Charles Darwin** *(1809–1882) placed the whole theory of organic evolution on a new footing. Indeed, he gave a new direction to morphology and physiology, by uniting them in a common biological theory: the theory of organic evolution. The result was a reconstruction of the classification of animals upon a genealogical basis, fresh investigation of the development of animals, and early attempts to determine their genetic relationships.*

Moreover, he discovered a mechanical cause actually existing and demonstrable by which organic evolution must be brought about.

After publication of the Origin of the Species, Darwin became interested in the animal and plant mechanism that confer advantages to individual members of a species.

Darwin's evolutionary theory resolved many zoological problems, not least because he established definite connections between extinct organisms (fossils) and those of today.

Darwin upset the theological vision of the "economy of nature"[28]*, stimulating the development of disciplines like biogeography, ecology and ethology.*

Nowadays, zoology is particularly important for genetic investigations (genetic code, mutation, recombination) and for population studies.

[28] **Linnaeus'** (1707–1778), *Oeconomia Naturae* (1749) suggested that each creature has its allotted place in nature, having been assigned its peculiar food and geographic range. Competition with other creatures was thereby avoided, ensuring harmony and plenty. Different creatures were linked together in elaborate food chains: the excess of one species sustaining another. Thus, predation and high reproductive rates amongst prey species were functional to maintain a just proportion between all species (no competitive struggle for existence).

II. Dietetics, Hygiene, Metabolism and Nutrition

Dietetics[29] *is the systematic control of food and drink in order to conserve health and combat disease.*

Metabolism is the chemical process within an organism, whereby new substances are synthesized (anabolism) or broken down (catabolism) for purposes such as regulational (homeostasis), growth, tissue repair and energy supply.

Nutrition is the body of science that seeks to explain metabolic and physiologic responses to diet. To this end it studies the relationship between diet, health and disease. It is thus concerned with the process of eating, digesting and using food and the determination of an optimal diet for purposes of health, body building and other purposes.

Hygiene encompasses theories and activities for preserving individual, communal and public health discussed under prevention of disease.

With advances in molecular biology, biochemistry, and genetics, nutrition is additionally developing into the study of integrative metabolism, which seeks to connect diet and health through the lens of biochemical processes.

The human body comprises chemical compounds such as water, amino acids (proteins), fatty acids (lipids), nucleic acids (DNA/RNA), and carbohydrates (e.g. sugars). These compounds in turn consist of elements such as carbon, hydrogen, oxygen, nitrogen, and phosphorus, and may or may not contain minerals such as calcium, iron, and zinc. Minerals ubiquitously occur in the form of salts and electrolytes. All of these chemical compounds and elements occur in various forms and combinations (e.g. hormones/vitamins, phospholipids, hydroxyapatite), both in the human body and in organisms (e.g. plants, animals) that humans eat.

The human body necessarily comprises the elements that it eats and absorbs into the bloodstream. The digestive system, except in the unborn fetus, participates in the first step which makes the different chemical compounds and elements in food available for the trillions of cells of the body. In the digestive process of an average adult, about seven liters of liquid, known as

---

[29] Already **Hippocrates** (450–370 BCE), **Asclepiades** (13–140 BCE) and their followers elaborated a philosophy of living based on moderation, instructing on diet, exercise, massage, sex, dress, bathing etc. Ancient philosophers (e.g. **Plato** (427–347, BCE), the Pythagoreans, Stoics) were also concerned with rules of conduct and living. Early Christians eventually integrated Greek Hygienic precepts with Old and New Testament Instructions.

digestive juices, exit the internal body and enter the lumen of the digestive tract. The digestive juices help break chemical bonds between ingested compounds as well as modulate the conformation and/or energetic state of the compounds/elements. However, many compounds/elements are absorbed into the bloodstream unchanged, though the digestive process helps to release them from the matrix of the foods where they occur. Any unabsorbed matter is excreted in the feces. But only a minimal amount of digestive juice is eliminated by this process; the intestines reabsorb most of it; otherwise the body would rapidly dehydrate.

The body requires amino acids to produce new body protein (protein retention) and to replace damaged proteins (maintenance) that are lost in the urine. In animals, amino acid requirements are classified in terms of essential (an animal cannot produce them) and non-essential (the animal can produce them from other nitrogen containing compounds) amino acids. Consuming a diet that contains adequate amounts of essential (but also non-essential) amino acids is particularly important for growing animals, which have a particularly high requirement.

Mineral and/or vitamin deficiency or excess may yield symptoms of diminishing health such as goiter, scurvy, osteoporosis, weak immune system, disorders of cell metabolism, certain forms of cancer, symptoms of premature aging, and poor psychological health (including eating disorders), among many others.

As of 2005, twelve vitamins and about the same number of minerals are recognized as "essential nutrients", meaning that they must be consumed and absorbed - or, in the case of vitamin D, alternatively - to prevent deficiency symptoms and death. Certain vitamin-like substances found in foods, such as carnitine, have also been found essential to survival and health, but these are not strictly "essential" to eat because the body can produce them from other compounds. Moreover, thousands of different phytochemicals have recently been discovered in food (particularly in fresh vegetables), which have many known and yet to be explored properties including antioxidant activity (see below). Other essential nutrients include essential amino acid, choline and the essential fatty acids.

In addition to sufficient intake, an appropriate balance of essential fatty acids – omega-3 and omega-6 fatty acids – has been discovered to be crucial for maintaining health. Both of these unique "omega" long-chain polyunsaturated fatty acids are substrates for a class of eicosanoids known as prostaglandins which function as hormones. The omega-3 eicosapentaenoic acid (EPA) (which can be made in the body from the omega-3 essential fatty acid alpha-linolenic acid (LNA), or taken in through marine food sources), serves as building block for series 3 prostaglandins. The omega-6 dihomo-gamma-linolenic acid (DGLA) serves as building block for series 1 prostaglandins, whereas arachidonic acid (AA) serves as building block for series 2 prostaglandins.

Both DGLA and AA are made from the omega-6 linoleic acid (LA) in the body, or can be taken in directly through food. An appropriately balanced intake of omega-3 and omega-6 partly determines the relative production of different prostaglandins, which partly explains the importance of omega-3/omega-6 balance for cardiovascular health. In industrialized societies, people generally consume large amounts of processed vegetable oils that have reduced amounts of essential fatty acids along with an excessive amount of omega-6 relative to omega-3.

Because different types and amounts of food eaten/absorbed affect insulin, glucagon and other hormones to varying degrees, not only the amount of omega-3 versus omega-6 eaten but also the general composition of the diet therefore determine health implications in relation to essential fatty acids, inflammation and mitosis.

Obesity can unfavorably alter hormonal and metabolic status via resistance to the hormone leptin, and a vicious cycle may occur in which insulin/leptin resistance and obesity aggravate one another. The vicious cycle is putatively fueled by continuously high insulin/leptin stimulation and fat storage, as a result of high intake of strongly insulin/leptin stimulating foods and energy. Both insulin and leptin normally function as satiety signals to the hypothalamus in the brain; however, insulin/leptin resistance may reduce this signal and therefore allow continued overfeeding despite large body fat stores. In addition, reduced leptin signaling to the brain may reduce leptin's normal effect to maintain an appropriately high metabolic rate.

Antioxidants are another recent discovery. As cellular metabolism /energy production requires oxygen, potentially damaging (e.g. mutation causing) compounds known as radical oxygen species or free radicals may form. For normal cellular maintenance, growth, and division, these free radicals must be sufficiently neutralized by antioxidant compounds, some produced by the body with adequate precursors (glutathione, Vitamin C in most animals) and those that the body cannot produce may only be obtained through the diet through direct sources (Vitamin C in humans, Vitamin A, Vitamin K) or produced by the body from other compounds (Beta-carotene converted to Vitamin A by the body, Vitamin D synthesized from cholesterol by sunlight).

Different antioxidants are now known to function in a cooperative network, e.g. vitamin C can reactivate free radical-containing glutathione or Vitamin

E by accepting the free radical itself, and so on. Some antioxidants are more effective than others at neutralizing the free radicals. Some cannot neutralize certain free radicals. Some cannot be present in certain areas of free radical development (Vitamin A is fat-soluble and protects fat areas, Vitamin C is water soluble and protects those areas). When interacting with a free radical, some antioxidants produce a different free radical compound that is less dangerous or more dangerous than the previous compound.

Since the Industrial Revolution some two hundred years ago, the food processing industry has invented many technologies that both help keep foods fresh longer and alter the fresh state of food as they appear in nature. Cooling is the primary technology that can help maintain freshness, whereas many more technologies have been invented to allow foods to last longer without becoming spoiled. These latter technologies include pasteurization, autoclavation, drying, salting, and separation of various components, and all appear to alter the original nutritional contents of food.

Pasteurization and autoclavation (heating techniques) have no doubt improved the safety of many common foods, preventing epidemics of bacterial infection. But some of the (new) food processing technologies undoubtedly have downfalls as well.

Modern separation techniques such as milling, centrifugation, and pressing have enabled up-concentration of particular components of food, yielding flour, oils, juices and so on, and even separate fatty acids, amino acids, vitamins, and minerals. Inevitably, such large scale up-concentration changes the nutritional content of food, saving certain nutrients while removing others. Heating techniques may also reduce food's content of many heat-labile nutrients such as certain vitamins and phytochemicals, and possibly other yet to be discovered substances.

Because of reduced nutritional value, processed foods are often "enriched" or "fortified" with some of the most critical nutrients (usually certain vitamins) that were lost during processing. Nonetheless, processed foods tend to have an inferior nutritional profile than do whole, fresh foods, regarding content of both sugar and high GI starches, potassium/sodium, vitamins, fibre, and of intact, unoxidized (essential) fatty acids. In addition, processed foods often contain potentially harmful substances such as oxidized fats and trans fatty acids.

A dramatic example of the effect of food processing on a population's health is the history of epidemics of beri-beri in people subsisting on polished rice. Removing the outer layer of rice by polishing it removes with it the essential vitamin thiamine, causing beri-beri. Another example is the development of scurvy among infants in the late 1800s in the United States. It turned out that the vast majority of sufferers were being fed milk that had been heat-treated (as suggested by Pasteur) to control bacterial disease. Pasteurization was effective against bacteria, but it destroyed the vitamin C.

As mentioned, lifestyle- and obesity-related diseases are becoming increasingly prevalent all around the world. There is little doubt that the increasingly widespread application of some modern food processing technologies has contributed to this development. The food processing industry is a major part of modern economy, and as such it is influential in political decisions (e.g. nutritional recommendations, agricultural subsidizing). In any known profit-driven economy, health considerations are hardly a priority; effective production of cheap foods with a long shelf-life is more the trend. In general, whole, fresh foods have a relatively short shelf-life and are less profitable to produce and sell than are more processed foods.

Thus the consumer is left with the choice between more expensive but nutritionally superior whole, fresh foods, and cheap, usually nutritionally inferior processed foods. Because processed foods are often cheaper, more convenient (in both purchasing, storage, and preparation), and more available, the consumption of nutritionally inferior foods has been increasing throughout the world along with many nutrition-related health complications.

History of Dietetics, Hygiene, Metabolism and Nutrition

Humans are believed to have evolved as omnivorous hunter-gatherers over the past 250,000 years. Early diets were primarily vegetarians with infrequent game meats and fish where available.

Agriculture developed about 10,000 years ago in multiple locations throughout the world, providing grains such as wheat, rice, and maize, with staples such as bread and pasta. Farming also provided milk and dairy products, and sharply increased the availability of meats and the diversity of vegetables. The importance of food purity was recognized when bulk storage led to infestation and contamination risks. Cooking developed as an often ritualistic activity to strict recipes and procedures, and also contributed to demands for food purity and consistency (e.g. the laws of ceremonial purity and dietary restrictions given in the Old Testament book of **Leviticus** *chapters 11–16, which were established ca 1300 BCE and written down in the days of King Hezekiah, ca 710 BCE).*

The first recorded nutritional experiment is found in the Bible's Book of Daniel. Daniel and his friends were captured by the king of Babylon during an invasion of Israel. Selected as court servants, they were to share in the king's fine foods and wine. But they objected, preferring vegetables and water in accordance with their Jewish dietary restrictions. The king's chief steward reluctantly agreed to a trial. Daniel and his friends received their diet for 10 days and were then compared to the king's men. Appearing healthier, they were allowed to continue with their diet.

The idea that changes that we now call chemical, occur in the human body is very old. According to **Aristotle** *(384–322 BCE) and* **Galen** *(129–200), food entering the alimentary tract undergoes processes resembling fermentation and through it turn into blood. This theory was generally accepted for more than 20 centuries. On its basis, numerous diets designed to make easier food transformations into blood and provide it with higher nutritive properties were developed for healthy and sick subjects. According to this classical tradition alimentation was one of the exterior forces which acted upon the human body.*

Considered from the therapeutic point of view the proper regulation of these forces was subsumed under the general Greek term diaita meaning "way of life" or regimen. Foodstuffs were classified in terms of their qualities of hotness, coldness, moistness and dryness in accordance with the dominant humoral theory of physiological action. As a consequence of their particular qualities foods might be purgative, constipating, strengthening or weakening. Their role in the production of the humors was also a major consideration.

With their increasing search for chemical mechanisms, the iatrochemists of the 16th *and* 17th *centuries tried to explain body functions by purely chemical reactions such as the neutralization of acids and bases. Nevertheless, ideas of chemicophysiological mechanisms remain vague and disorganized even while* **Vesalius** *(1514–1564) was describing in a scientific manner the details of anatomy, and the physicomechanical details of physiology were developing toward the discovery of the circulation of the blood by* **William Harvey** *(1578–1657).*

In 1747, Dr. **James Lind** *(1716–1794), a physician in the British navy, discovered that lime juice saved sailors (who had been at sea for years) from scurvy*[30]*, a deadly and painful bleeding disorder. The discovery was ignored*

---

30 Ascorbic acid, or Vitamin C, was discovered after scientists had searched centuries for a cure for the disease known as scurvy. The name Ascorbic acid comes from word "anti-scurvy" acid, because it was known to dramatically cure this disease. This disease was caused by a serious deficiency of Vitamin C, and it caused its victim's small blood vessels to rupture, bones to weaken, and joints to swell, among other symptoms. These symptoms were due to the fact that without a source of Vitamin C one developed severe problems concerning the body's connective tissues, which is found in bones, skin, muscles, teeth, blood vessels, and cartilage. This disease would eventually lead to death if it went untreated,

for 40 years. The essential Vitamin C within the lime juice would not be recognized by scientists until the 1930's.

During the 18th *century, classical dietary theories began to be superseded by ideas derived from the new chemistry. Qualitative grades tended to be replaced by such considerations as "acidity or alkalinity". In 1770,* **Antoine Lavoisier** *(1743–1794), the "Father of Chemistry" discovered the details of metabolism, demonstrated that oxidation of food is the source of body heat. In 1790,* **George Fordyce** *(1736–1802) recognized, that calcium was necessary for fowl survival.*

Early 19th*-century scientists distinguished "animals" and plants as complementary, the former only breaking down material synthesized by plants. From the middle of the* 19th *century, the chemistry of food and biochemistry started to develop successfully.*

In the early 1800s, the elements carbon, nitrogen, hydrogen and oxygen were recognized as the primary components of food, and methods to measure their proportions were developed.

In 1816, **Francois Magendie** *(1783–1855) discovers that dogs fed only carbohydrates and fat lost their body protein and died in a few weeks, but dogs fed also protein survived, identifying protein as an essential dietary component.*

In 1840, the German chemist **Justus Liebig** *(1803–1873) discovered the chemical makeup of carbohydrates (sugars), fats (fatty acids) and proteins*

---

and was not uncommon, especially during the winter months of the year. The disease often plagues armies, explorers, and crusaders, since these men's diets normally consisted of biscuits and salted meat that could easily be stored and kept unspoiled on a ship.

Lind published his findings as *Treatise on the Scurvy* in 1753, and as a result, in 1795 daily doses of lime juice were prescribed to all the sailors in the British navy and Scurvy quickly vanished. However, the British were the only people who accepted the idea that Scurvy was the result of a dietary deficiency, and Great Britain was the only place where there was a decline in the cases of Scurvy. In America, during the Civil war, many men on both sides of the war died from this disease due to the lack of a source of Vitamin C in their diet.

In 1907 **Axel Holst** and **Theodore Frolich**, proved that Scurvy's symptoms could be produced in a guinea pig when denied certain foods, and that these symptoms would vanish when the food was restored. However, even on into the early 20th century, explorers were still dying of Scurvy. One example was Robert Scott's expedition to the South Pole, when he and his crew were affected by the lack of Vitamin C containing foods in their diets, not the harsh conditions and temperatures.

(amino acids). In 1842 he applied a "black-box" model to animal metabolism, analyzing intake of food (e.g. protein, sugar) and excretion (urea, carbon dioxide) and assuming a simple one-way path from one to the other.

In the 1860's, **Claude Bernard** *(1813–1878) discovered that body fat can be synthesized from carbohydrate and protein, showing that energy in blood glucose can be stored as glycogen.*

His discovery of glycogen synthesis and breakdown by the liver showed animal metabolism to be rather complicated. This induced further research of metabolism parameters by **Carl von Voit** *(1831–1908),* **Max von Pettenkofer** *(1818–1901),* **T.L.W. von Bischoff** *(1807–1882),* **Max Rubner** *(1854–1932) and others.* **Edward Frankland** *(1825–1899),* **F.K.A. Stohmann** *(1832–1887) and others established accurate calorific values for many foods and the discovery of "cell-free" fermentation by* **Eduard Buchner** *(1860–1917) implicated intracellular metabolic processes, regulated by enzymes.*

Controlled dietary studies by **T.B. Osborne** *(1859–1929),* **L.B. Mendel** *(1872–1935) and* **E.V. McCollum** *(1879–1967) showed that some amino acids (the building blocks of proteins) could not be synthesized by animals and were essential components of the diet, as were the vitamins.*

Modern biochemistry and molecular biology have elucidated many molecular mechanisms, including anabolic and catabolic pathways and the genetic control of enzyme synthesis ('one gene – one enzyme hypothesis'). The study of inherited metabolic disorders was pioneered by **A.E. Garrod***'s (1857–1936) in 1909.*

Thus, from the second half of the 19th *century dietetic theory took new forms based on an understanding of the role of food substances in body metabolism.*

The following timeline summarizes the progress made during 1800–1950:

- *1816* **Francois Magendie** *(1783–1855, France). Physiologist. Showed for the first time that nitrogenous foods were needed for life.*
- *1819–1820* **Henri Braconnot** *(1781–1854, France). Naturalist. Obtained glucose (1819) from sawdust, linen and bark of trees. Studied and isolated the first amino acid (1820). [Earlier (1812),* **Gottlieb Sigismund Constantin Kirchhoff** *(1764–1833, Russia), chemist, produced glucose by heating starch with a small amount of sulfuric acid as catalyst.]*

- *1824* **Joseph Louis Gay-Lussac** *(1778–1850, France). Wrote the chemical equation for glucose fermentation.*
- *1827* **William Prout** *(1785–1850, England). Chemist. Classified food components into fats, carbohydrates and proteins.*
- *1835* **Anselm Payen** *(1795–1871, France). Chemist. Isolated and studied the first plant enzyme diastase from grain. It catalyzes the breakdown of starch.*
- *1836* **Theodor Schwann** *(1810–1882, Germany). Discovered and isolated pepsin, a digestive enzyme in the human stomach.*
- *1843* **Carl Schmidt** *(1822–1894, Germany). Physiological chemist. Coined the name carbohydrates. Found that blood contains small quantities of glucose.*
- *1840* **Justus von Liebig** *(1803–1873, Germany). Published "Die Organische Chemie und ihre Anwendung auf Agrikultur und Physiologie", explaining his theory of the exchange of carbon and nitrogen in plants and animals — the first crude model of the carbon and nitrogen cycles in the biosphere.*
- *1843–1856* **Claude Bernard** *(1813–1878, France). Physiologist. Discovered (1843) that the liver serves as a source of blood sugar by converting glycogen to glucose. Coined the word glycogen. Isolated glycogen from the liver (1856). [This was done independently in the same year by* **Viktor Hensen** *(1835–1924, Germany).]*
- *1856* **Louis Pasteur** *(1822–1895, France). Discovered that fermentation is caused by microorganisms (yeast).*
- *1868–1902* **Carl von Voit** *(1831–1908, Germany). Physiologist. Showed (1868) that energy conversion in the body takes place through intermediary substances which are formed from the original food before final union with oxygen occurred. Discovered (1891) that various ingested sugars are converted to glycogen in the body for storage until needed.*
- *1870* **Friedrich Miescher** *(1844–1895, Switzerland) first identified nucleic acid in the nuclei of puss cells.*
- *1871* **Ernst Hoppe-Seyler** *(1825–1895, Germany). Physiologist and chemist. Discovered invertase, an enzyme that speeds up conversion of sucrose into glucose and fructose.*
- *1884–1894* **Max Rubner** *(1854–1932, Germany). Physiologist. Discovered that the body gets energy from carbohydrates, fats and proteins after stripping away nitrogen for other uses.* **Voit** *and* **Rubner**, *independently measured caloric energy expenditure in different species of animals, applying principles of physics in nutrition.*

- *1889* **Oskar Minkowski** *(1858–1931, Germany), physiologist, and* **Joseph von Mering** *(1849–1908, Germany), physician, discovered that the pancreas supplies a hormone (insulin) essential to glucose metabolism.*

- *1890–1907* **Emil Hermann Fischer** *(1852–1919, Germany). Chemist. First to synthesize a simple protein molecule (1907) and demonstrate the peptide-bond. Pioneered in understanding nitrogen metabolism. Proposed the theory of lock and key to explain stereospecific interaction of enzyme with substrate.*

- *1896* **Eugen Baumann** *(1846–1896, Germany) discovered that the thyroid gland is rich in iodine.*

- *1897* **Eduard Buchner** *(1860–1917, Germany). Chemist. Demonstrated that alcoholic fermentation of sugars is due to action of enzymes contained in yeast. That meant that enzymes extracted from yeast are effective in converting sugar into alcohol. Awarded the 1907 Nobel Prize for chemistry.*

- *1897* **Christiaan Eijkman** *(1858–1931), a Dutch physician in Java cured natives of beriberi by feeding them brown rice. Over two decades later, nutritionists learned that the outer rice bran contains vitamin B1, also known as thiamine. Together with biochemist* **Frederick G. Hopkins** *(1861–1947) he was awarded the 1929 NP for Physiology or Medicine.*

- *1897–1901* **Jokichi Takamine** *(1854–1922, Japan), physiologist and* **John Jacob Abel** *(1857–1938, U.S.A.), physiological chemist, independently isolated adrenaline.*

- *1905* **Arthur Harden** *(1865–1940, England). Chemist. First to detect and identify inorganic phosphate in metabolic intermediates.*

- *1906* **F.G. Hopkins** *(1861–1947, England) recognized "accessory food factors" other than calories, protein and minerals, as organic materials essential to health but which the body cannot synthesize.*

- *1912* **Casimin Funk** *(1884–1967, Poland) coined the term vitamin, a vital factor in the diet, from the word "vital" and "amine", because the unknown substances preventing scurvy, beriberi, and pellagra, were thought then to be derived from ammonia.*

- *1913* **Elmer V. McCollum** *(1879–1967) discovered the first vitamins, fat soluble vitamin A, and water soluble vitamin C (as the then-unknown substance preventing scurvy). In 1922 he discovered vitamin D in cod liver oil, which prevents rickets.*

- *1913* **Archibald Vivian Hill** *(1886–1977, England). Physiologist. Discovered that muscle cells use oxygen after contraction is finished in anaerobic glycolysis.*
- *1918* **Otto Fritz Meyerhof** *(1884–1951, Germany and U.S.A.). Biochemist. Showed that muscular activity involves anaerobic conversion of glucose glycogen (glycolysis). During muscle rest, lactic acid combines with oxygen to restore glycogen level. First to note that metabolic pathways of all organisms are essentially similar.*
- *1921* **Frederick Grant Banting** *(1891–1941, Canada). Physician. Isolated the hormone insulin.*
- *1922* **H.M. Evans** *(1882–1971, USA) and* **K.S. Bishop** *discovered vitamin E in green leafy vegetables.*
- *1926* **James Batcheller Sumner** *(1887–1955, U.S.A.). Biochemist. Proved that the enzyme urease was indeed a protein. It quickly became apparent that all enzymes are proteins.*
- *1927* **Adolf O.R. Windaus** *(1876–1959, Germany) synthesized vitamin D (NP 1928).*
- *1928–1935* **Albert Szent-Györgi** *(1893–1986, USA). Physiologist and biochemist. Isolated ascorbic acid (1928) and proved that it is vitamin C (1932). Discovered fundamental processes in cellular energy metabolism. Concurrently elucidated much of the citric acid cycle. (NP 1937).*
- *1929* **Albert Lipmann** *(1899–1986, Germany and USA), biochemist, and* **K. Lohmann** *isolated adenosine triphosphate (ATP) from muscle tissue.*
- *1931–1939* **William C. Rose** *(1887–1985, USA) identified essential amino acids, necessary proteins which the body cannot synthesize.*
- *1931* **Linus Pauling** *(1901–1994, U.S.A.). Discovered the hydrogen bond.*
- *1932–1937* **Hans Adolf Krebs** *(1900–1981, England). Biochemist. Discovered the most important metabolic pathway and energy producer in living organism (Krebs' cycle), the second stage in the aerobic glucose metabolism.*
- *1933* **Walter N. Haworth** *(1883–1950, England) and* **Tadeus Reichstein** *(1897–1996, Poland and Switzerland) synthesized vitamin C, becoming the first vitamin to be artificially made. Haworth won the NP in 1937 jointly with* **Paul Karrer** *(1889–1971, Switzerland). Reichsten won his NP in 1950.*

- *1935* **E.J. Underwood** *(1905–1980, England) and* **H.R. Marston** *(1900–1965, England) independently discover the necessity of cobalt.*
- *1936* **Gerty** and **Carl Cori** *(1896–1957, 1896–1984, U.S.A.). Biochemists. Discovered the Cori cycle of carbohydrate metabolism: the complementary process of glycolysis in the muscle and gluconeogenesis in the liver and the process of phosphorolysis; showed that glycogen in the body is broken down by the use of phosphoric acid.*
- *1938* **Erhard Fernholz** *discovered the chemical structure of vitamin E.*
- *1938* **Paul Karrer** *(1889–1971) synthesized vitamin E.*
- *1946* **Linus Pauling** *suggested that enzymes work by lowering the energy-barrier of a reaction.*
- *1947* **Alexander Robertus Todd** *(1907–1997, England). Biochemist. Synthesized ADP and ATP.*
- *1947–1951* **Albert Lipmann** *discovered the coenzyme A, a key substance in human body metabolism which is involved in the control of energy in cells. Proposed that the high-energy phosphate-bond and its transformation to and from ATP was a 'common currency' of energy transfer in biology (1947). Discovered acetylcoenzyme A, an essential part of body chemistry that is especially important in breaking down carbohydrates, fats and proteins to obtain energy for cells (1951).*

III. Evolutionary Biology

(a) Introduction

Evolutionary biology is a sub-field of *biology* concerned with the origin and descent of species, as well as their change, multiplication, and diversity over time. One who studies evolutionary biology is known as an *evolutionary biologist*, or less formally, an *evolutionist*.

Evolutionary biology is an *interdisciplinary* field because it includes scientists from a wide range of both field and lab oriented disciplines. For example, it generally includes scientists who may have a specialist training in particular organisms such as *mammalogy*, *ornithology*, or *herpetology*, but use those organisms as *case studies* to answer general questions in evolution.

It also generally includes *paleontologists* and *geologists* who use *fossils* to answer questions about the tempo and mode of evolution, as well as theoreticians in areas such as *population genetics*.In the 1990s *developmental biology* made a re-entry into evolutionary biology from its initial exclusion from the modern synthesis through the study of *evolutionary developmental biology*.

Its findings feed strongly into new disciplines that study mankind's *socio-cultural evolution* and *evolutionary behavior*. Evolutionary biology's frameworks of ideas and conceptual tools are now finding application in the study of a range of subjects from *computing* to *nanotechnology*.

Artificial life is a sub-field of *Bioinformatics* that attempts to model, or even recreate, the evolution of organisms as described by evolutionary biology. Usually this is done through mathematics and computer models.

Historically, when Mendel's work was "rediscovered" in 1900, it led to a conflict between Mendelians and biometricians, who insisted that the great majority of traits important to evolution must show continuous variation that was not explainable by Mendelian analysis.

Eventually, the two models were reconciled and merged, primarily through the work of the biologist and statistician **R.A. Fisher**. This combined approach, applying a rigorous statistical model to Mendel's theories of inheritance via genes, became known in the 1930s and 1940s as the modern synthesis of Darwin's theory.

Evolutionary biology as an *academic discipline* in its own right emerged as a result of the *modern evolutionary synthesis* in the 1930s and 1940s. It was not until the 1970s and 1980s, however, that a significant number of universities had departments that specifically included the term evolutionary biology in their titles. In the United States, as a result of the rapid growth of *molecular and cell biology*, many universities have split (or aggregated) *their biology departments into molecular and cell biology-style departments and ecology and evolutionary biology-style departments (which often have subsumed older departments in paleontology, zoology and the like).*

Microbiology has recently developed into an evolutionary discipline. It was originally ignored due to the paucity of morphological traits and the lack of a species concept in microbiology. Now, evolutionary researchers are taking advantage of our extensive understanding of microbial physiology, the ease of microbial genomics, and the quick generation time of some microbes to answer evolutionary questions. Similar features have led to progress in viral evolution, particularly for bacteriophage.

Notable contributors to evolutionary biology include:

Pierre Louis Maupertuis	*1698–1759*
Jean-Baptiste Lamarck	*1744–1829*
Charles Darwin	*1809–1882*
Alfred Russel Wallace	*1823–1913*
August Weismann	*1834–1914*
Ernst Haeckel	*1834–1919*
Sewall Wright	*1889–1988*
R.A. Fisher	*1890–1962*
J.B.S. Haldane	*1892–1964*
Theodosius Dobzhansky	*1900–1975*
Ernst Mayr	*1904–2005*
Gustave Malécot	*1911–1998*
James F. Crow	*1916–*
John Maynard Smith	*1920–2004*
Motoo Kimura	*1924–1994*
George C. Williams	*1926–*
Carl Woese	*1928–*
Edward Osborn Wilson	*1929–*
Richard Lewontin	*1929–*
Allan Wilson	*1934–1991*
W.D. Bill Hamilton	*1936–2000*
Danil Janzen	*1939–*
Stephen Jay Gould	*1941–2002*
Robert Trivers	*1943–*
Niles Eldredge	*1943–*
Richard D. Alexander	*1948–*

(B) Molecular Evolution and Population Genetics

Mendel (1866) cross-pollinated purebred pea plants RR with purebred wrinkled pea plants (rr). The first filial generation peas were all round (Rr), but those in the next generation included RR (round), Rr (round), Rr (round) and rr (wrinkled). Thus, the genotype Rr produced the phenotype round peas.

This work of Mendel and later geneticists has shown that the development of the individual organism is controlled by hereditary regulation known as genes. Genes are constructed of the nucleic acid called DNA (deoxyribonucleic acid) and are normally located in the cell nucleus, where they are organized into larger, paired, thread-like units called chromosomes, each of which may contain thousands of genes.

The number of chromosomes is usually constant for each species, but varies between species, ranging from as few as one pair to as many as several hundred pairs. The usual number is between 5 and 30 pairs. Man, for example, has 23 pairs. When a cell divides during normal growth, the chromosomes reproduce themselves exactly to give, in the two new cells, the same number and kinds of chromosomes as in original parent cell. This process of exact chromosome reduplication is called mitosis. In organisms with sexual reproduction, a more specialized kind of cell division called meiosis takes place in the organs where gametes (specialized reproductive cells, such as eggs and sperm) are produced.

In this second process there are two cell divisions. In the first the chromosome pairs are divided in a random fashion so that each of the two new cells receives one number of each original pair, or exactly half the number of chromosomes found in the original cell. Then these two cells reproduce themselves exactly, each of which carries half the number of chromosomes necessary for the final organism.

When two gametes meet in the process of fertilization, a new organism is produced, one that receives half of its chromosomes from each of the two parents. Thus, meiosis and subsequent fertilization provide a means of interchanging genetic material between organisms, whereas mitosis provides a means of exactly duplicating cells within an individual organism. In man, all cell divisions are achieved by mitosis except those in the testes of the male and ovaries of the female where sperm and eggs are produced by meiosis.

The chromosomes are the larger units of heredity, but it is the smaller genes that ultimately determine the nature of the individual organism. Genes may exist in different expressions, called alleles, each of which leads to a different hereditary result in the adult organism. In man, for example, blue eyes are the result of one eye-color allele, whereas brown eyes result from another. Recall that chromosomes normally occur in pairs. It is a fundamental fact of heredity that paired chromosomes have analogous sets of genes that control the development of the same structures in the adult organism.

Cells produced by mitosis therefore have two complete sets of genes, one set contained in each member of the paired chromosomes. Such cells are called diploid cells. In meiosis, on the other hand, only one member of each chromosome pairs is transmitted to the daughter cell, which thus has only half as many chromosomes as the diploid parent cell. Such cells are called haploid cells. Haploid cells contain one complete set of genetic instructions, whereas diploid cells contain two complete sets. Diploid cells, with their double set of chromosomes, may have the same or different alleles for a particular gene in each set. In man, for example, an individual may have the blue-eye allele in both chromosomes of the pair which contains the eye-color gene, or he may have the brown-eye allele in both chromosomes, or, finally, he might have the brown-eye allele in one chromosome and the blue-eye in the other.

If both chromosomes contain the same allele, then all gametes produced by dividing the pairs during meiosis will also have the same eye-color allele. Individuals with the same allele on both chromosomes thus produce only one kind of gamete and are said to be homozygous for that particular gene. When the two chromosomes of a pair contain different alleles, the gametes produced by dividing the pairs will be mixed; half will contain the blue-eye allele and half the brown-eye allele. In this case the individual is said to be heterozygous for that particular gene.

Note that each character of the adult organism is not always controlled by a single gene. Some few characters, such as human eye color, are inherited in this simple way, but most characters are determined by the combined effects of many genes. As the eye-color example shows, predictions about the effects of differing combinations of alleles on the next generation are easily made when only a single gene is responsible for a character, but predictions become increasingly difficult when more genes, each with differing alleles, are involved.

We now return to our evolutionary theme and consider the actual causes if the variations found among individuals of the same species. Such variations are of two kinds: those due to heredity, and those due to environmental influences operating during the lifetime of the individual. In man, differences in eye color are hereditary variations, whereas the differences in muscle size between an athlete and an office worker of comparable physical build would be an example of variation caused by environmental influences.

Only those variations caused by hereditary differences are important in evolution, for they alone can be passed on to the next generation. The children of the athlete will inherit their eye color from him, but they will not have his physique if they choose to become office workers.

Inherited variations, in turn, result from two interrelated processes: mutation and genetic recombination. Mutation is the sudden, spontaneous appearance of a new allele for a particular gene or group of genes. Apparently, mutations are continuously taking place in all organisms, but normally they occur at a very low rate. In the fruit-fly genus Drosophila, which has long been a favorite animal group for genetic study, there is about one gene mutation for every 20 gametes produced. Because each gamete includes about 20,000 genes, the rate of mutation is only about one per 400,000 genes. Mutations may have little or no effect on the adult organism; some, however, are lethal, while others lead to small but advantageous changes. The causes for mutations are obscure, but in most organisms the rate of production can be changed (usually accelerated) by artificial exposure to certain kinds of radiation (gamma, ultraviolet, cosmic), to various chemicals, or to changes in temperature. Apparently, these agents alter the chemical structure of the DNA which makes up the genes, thus producing new alleles.

Mutations are the only source for new alleles but are too rare to be directly responsible for most of the constantly appearing variation found in individuals of the same species. These variations result from simple recombination, during meiosis and fertilization, of the alleles already present in the parent organisms. Because the chromosomes of most kinds of organisms contain tens of thousands of genes, each of which may have several alleles, there are almost limitless possibilities for recombination of alleles to produce individuals with differing genetic patterns. It is these differing patterns that lead to the variations seen in adult organisms of the same species.

The genotype of any organism is total makeup of its genetic material, i.e the particular set of genes it possesses. Two organisms whose genes differ at even one position of their genome are said to have different genotypes. The term "genotype" refers, then, to the full hereditary information of an organism.

The phenotype of an organism, represent its actual physical properties, such as height, weight, hair color, and so on.

Mendelian genetics has shown that because of the dominance of some characters, organism with the same phenotype may have different genotypes. Only by studying their offsprings can we distinguish between them. Likewise, organisms with identical genotype differ in their phenotypes (e.g identical twins).

Biological evolution seems to violate our common sense awareness that in general, disorder tends to increase as time passes. All around us, we see configurations of matter move from order to disorder, from improbable states to more probable ones. Paint weathers; rocks crumble; iron rusts; wood decays; stars radiate away their energy. But here on Earth, life continues to combine elements into specific molecules and monomers into lengthy polymers, making ever-greater complexity and order from simplicity and disorder.

A single DNA molecule, which may be a million times longer than it is wide, represents an exceedingly nonrandom bit of matter. Such molecules store the tremendous amounts of information needed to carry out life's activities, information that can be preserved undamaged through thousands of replications. Hence, a great degree of order is required to keep matter alive. This order not only persists but has actually increased *as life has evolved to ever more complex forms on earth.*

How can we explain this apparent contradiction – the maintenance, and even the increase, in the order and complexity of life in a universe that is inexorably evolving toward increasing disorder? The resolution lies in a consideration of the total system, of which life is just one part.

Life on earth does not form a closed system. Instead, life can maintain its highly improbable configuration only at the expense of its environment; that is, life can become highly organized only by increasing the disorganization of its surroundings. The disorder of the total system increases, *while the disorder of living creatures within it* decreases. *Here "disorder" refers not to pollution but to the way that life acquires the energy it needs.*

Molecular evolution is the process of evolution at the scale of *DNA*, *RNA*, and *proteins*. Molecular evolution emerged as a scientific field in the 1960's as researchers from *molecular biology*, *evolutionary biology* and *population genetics* sought to understand recent discoveries on the structure and function of nucleic acids and protein. Some of the key topics that spurred development of the field have been the evolution of enzyme function, the use of nucleic acid divergence as a *"molecular clock"* to study species divergence, and the origin of non-functional or *junk DNA*.

Recent advances in genomics, including whole-genome sequencing, high-throughput protein characterization, and bioinformatics have led to a dramatic increase in studies on the topic. In the 2000s, some of the active topics have been the role of *gene duplication* in the emergence of novel gene function, the extent of adaptive molecular evolution versus neutral drift, and the identification of molecular changes responsible for various human characteristics especially those pertaining to *infection*, *disease*, and *cognition*.

Mutations are permanent, *transmissible* changes to the *genetic material* (usually *DNA* or *RNA*) of a *cell*. Mutations can be caused by copying errors in the genetic material during *cell division* and by exposure to *radiation*, chemicals, or *viruses*, or can occur deliberately under cellular control during the processes such as *meiosis* or *hypermutation*. Mutations are considered the driving force of *evolution*, where less favorable (or *deleterious*) mutations are removed from the gene pool by *natural selection*, while more favorable (or *beneficial*) ones tend to accumulate. *Neutral mutations* do not affect the organism's chances of survival in its natural environment and can accumulate over time, which might result in what is known as *punctuated equilibrium;* the modern interpretation of classic evolutionary theory.

There are four known processes that affect the survival of a characteristic; or, more specifically, the frequency of an *allele;*

- *Mutation*
- *Genetic drift* describes changes in gene frequency that cannot be ascribed to selective pressures, but are due instead to events that are unrelated to inherited traits. This is especially important in small mating populations, which simply cannot have enough offspring to maintain the same gene distribution as the parental generation.
- *Gene flow:* or gene admixture is the only one of the agents that makes populations closer genetically while building larger gene pools.
- *Selection,* in particular *natural selection* produced by differential mortality and fertility. Differential mortality is the survival rate of individuals before their reproductive age. If they survive, they are then selected further by differential fertility – that is, their total genetic contribution to the next generation. In this way, the alleles that these surviving individuals contribute to the gene pool will increase the frequency of those alleles. *Sexual selection,* the attraction between mates that results from two genes, one for a feature and the other determining a preference for that feature, is also very important.

The production and redistribution of variation is produced mostly by three of the four agents of evolution: mutation, genetic drift, and gene flow. Natural selection, in turn, acts on the variation produced by these agents. One important goal is to understand (using both data from *molecular biology* and theory from *population genetics*) the main force driving molecular evolution.

Currently, three main positions are defended. (1) *Neutralism* and *near-neutralism* (**Kimura**, 1983), where neutral or nearly-neutral mutations, along with *random genetic drift* and purifying selection, explain most of evolution. (2) *Selectionism*, where balancing selection is considered the main force, and finally, (3) mutationism, where mutational input and random genetic drift are thought to be more important.

Population genetics is the study of changes in gene allele frequency distribution within interbreeding populations under the influence of the four evolutionary forces: natural selection, genetic drift, mutation, and migration. It also takes account of population subdivision and population structure in *space. As such, it attempts to explain such phenomena as adaptation and speciation. Population genetics was a vital ingredient in the modern evolutionary synthesis, its primary founders were* **Sewall Wright**, **J.B.S. Haldane** *and* **R.A. Fisher**, *who also laid the foundations for the related discipline of quantitative genetics.*

What was perhaps the most significant discovery concerning such changes was made independently in 1908 by **G. Weinberg**, *a German geneticist, and* **G.H. Hardy**, *a British mathematician, and has come to be known as the Hardy-Weinberg law. Hardy and Weinberg demonstrated by simple algebra that the relative proportion of alleles within a randomly interbreeding population will remain constant unless outside forces work to change it. Intuitively, one would assume that rare alleles would gradually be lost from the population, and that common would tend to become more common. Instead, Hardy and Weinberg showed that there is a natural genetic equilibrium which can preserve even the least common alleles in a randomly interbreeding population. Only through nonrandom processes, such as nutation or selective reproduction, can the proportion of rare alleles be increased or the proportion of common alleles decreased.*

This discovery was particularly significant in re-establishing natural selection as an evolutionary mechanism, because, as you will recall, the basis of natural selection is nonrandom reproduction. Not every individual, but, on the average, more of the better fitted, will survive to produce the next generation. Natural selection is therefore an ideal mechanism for explaining changes in allele frequency and shift away from genetic equilibrium.

Building on the Hardy-Weinberg law, modern population geneticists have developed a body of refined mathematical models, often devised with help of high-speed computers, to stimulate changes in gene frequencies by natural selection in populations that differ in such features as selection pressures, size, original allele ratios, reproductive habits, mutation rates, migration rates, and recombination patterns. In addition to these mathematical formulations, there is now a large body of observational and experimental evidence that confirms the importance of natural selection as a means of changing gene frequencies.

(c) Non–Darwinian Evolution; Genetic Drift (1968–1983)

In order to follow the changes in life through billions of years, one must look more closely at the property of life that are most distinctive: the capacity to reproduce and to evolve. At the molecular level, life's ability to reproduce begins with the replication of DNA, during which two new spirals are created that are exact replicas of the origin molecule.

Sometimes a change in the sequence of nucleotide bases, called a mutation, occurs in the DNA polymer. Such changes arise basically at random, sometimes from the impact on the DNA molecules of high-energy gamma rays or of cosmic-ray particles, or from exposure to various chemical agents called mutagens, or even from rare errors made by cell's own DNA-coping machinery. We do not know which of these processes predominates in causing mutations throughout the history of life on earth. When a mutation arises in a part of the DNA where information for a protein is encoded, it can cause a different, "incorrect" amino acid to be inserted into the protein under construction.

Many mutations are neither helpful nor hurtful. They are simply called "neutral" mutations. If the new protein does its cellular job poorly, the organism may be less fit for survival and reproduction, or it might not survive at all. On other occasions, mutations may actually change a protein in such a way that it does its job better than the original protein did. The lucky organism with this mutation would have some advantage over its fellows – perhaps it can replicate its DNA more quickly, swim more rapidly, sense food more efficiently, have more brightly colored feathers, or smell predators at a greater distance.

Such advantages would give the organism greater reproductive success than its fellows: Because they are healthier, more resistant to cold, prettier, better at escaping from predators, or superior for some other reasons, the more fit organisms will (by definition) produce more surviving offspring than their less fit relatives. As a result, organisms carrying favorable mutations will, over time, come to predominant in a population. "Differential reproduction" – the greater or lesser success that organisms achieve in producing offspring – lies at the heart of the process called natural selection.

Differential reproduction determines whether a given mutation becomes established, or "fixed" in the general population. Thus natural selection, operating through differential reproductive success, causes the characteristics of a species gradually to change when advantageous, or "adaptive", mutations sweep through the population. In this way, differential reproduction allows one species to evolve into a new species.

Sometimes groups within a species become isolated from each other for many generations. When this happens, different mutations, appearing at random, become fixed in the separated populations. Gradually, the populations differ more and more from each other. When populations differ significantly (usually, when they can no longer interbreed to produce fertile offspring) we call them two separate species.

In 1839, **Charles Darwin** *noticed such changes among bird species in the Galápagos Islands. Darwin studied populations of birds resembling, but not identical to, those that he knew well from his native England. He found* *different species and subspecies of finches on the different islands that he visited, and his speculations about how the differences arose led to his theory of evolution, first published in 1859. In this epochal work, Darwin identified what we have called natural selection (differential reproductive success) as the driving force that makes new species on earth.*

During the past 40 years much progress in understanding the process of evolution has resulted from combining Darwin's mechanism of natural selection with the discoveries of geneticists concerning the inheritance of individual variations. Because of the renewed emphasis on natural selection, this modern synthesis is often referred to as Neo-Darwinism. *A fundamental theme of Neo-Darwinism has been the study of inheritance not merely in individual organisms, but in* populations, *which are interbreeding groups of individuals of the same species. It is now recognized that mutation, recombination, and natural selection can lead to evolutionary change only as they act on such groups of individuals; this emphasis has given rise to the science of* population genetics, *a subject that has been responsible for many of the advances of Neo-Darwinism.*

The size of a population has important effects on allele frequencies because random events, or chance, tend to cause changes of relatively greater magnitude in a small populations. If a population consist of only a few individuals, an allele present at a low frequency in the population could be completely lost by chance. Such an event would be unlikely in a large population. For example, consider two populations, one with 10,000 individuals and one with 10 individuals. If an uncommon allele occurs at a frequency of 10%, or 0.1, in both populations, then 1900 individuals in the large population have the allele.[31] *That same frequency, 0.1, in the smaller population means that only about two individuals have the allele.*[32] *From this exercise, it is easy to see that there is a greater likelihood of losing the rare allele from the smaller population than from the larger one. Predators, for example, might happen to kill one or more individuals possessing the uncommon allele in the smaller population purely by chance so these individuals would leave no offspring.*

The production of random evolutionary changes in small breeding populations is known as genetic drift. *Genetic drift result in changes in allele frequencies in a population from one generation to another. One allele may be eliminated from the population purely by chance, regardless of whether that allele is beneficial, harmful, or of no particular advantage or disadvantage. Thus, genetic drift decreases genetic variation* within *a population, although it tends to increase genetic differences* among *different populations. Because*

[31] $2pq + q^2 = 2(0.9)(0.1) + (0.1)^2 = 0.18 + 0.01 = 0.19; \quad 0.19 \times 10,000 = 1900$

[32] $0.19 \times 10 = 1.9$

of fluctuations in the environment, such as depletion in food supply or an outbreak of disease, a population may rapidly and markedly decrease from time to time. The population is said to go through a bottleneck during which genetic drift can occur in the small population of survivors. As the population again increases in size, many allele frequencies may be quite different from those in the population preceding the decline.

Scientists hypothesize that genetic variation in the cheetah was considerably reduced by a bottleneck at the end of the last Ice Age, some 10,000 years ago, Cheetahs nearly became extinct, perhaps from overhunting by humans. The few surviving cheetahs had greatly reduced genetic variability, and as a result, the cheetah population today is so genetically uniform that unrelated cheetahs can accept skin grafts from one another. (Normally, only identical twins accept skin grafts so readily.)

Whereas natural selection describes the tendency of beneficial alleles to become more common over time (and detrimental ones less common), genetic drift refers to the fundamental tendency of any allele to vary randomly in frequency over time due to statistical variation alone, so long as it does not comprise all or none of the distribution.

Genetic drift may be modeled as a stochastic process that arises from the role of random sampling in the production of offspring. The genes of each new generation are not a simple copy of the genes of the successful members of the previous one, but rather a sampling, which includes some statistical error. Drift is the cumulative effect over time of this sampling error on the allele frequencies in the population.

By definition, genetic drift has no preferred direction. A neutral allele may be expected to increase or decrease in any given generation with equal probability. Given sufficiently long time, however, the mathematics of genetic drift predict the allele will either die out or be present in 100% of the population, after which time there is no random variation in the associated gene. In this regard, genetic drift tends to sweep gene variants out of a population over time, such that all members of a species would eventually be homozygous for this gene. Genetic drift is opposed in this regard by genetic mutation which introduces novel variants into the population according to its own random processes.

Like selection, genetic drift acts on populations, altering the frequency of alleles (gene variations) and the predominance of traits. Drift is observed most strongly in small populations and results in changes that need not be adaptive.

Similarly, in a breeding population, if an allele has a frequency of p, probability theory dictates that (if natural selection is not acting) in the following *generation, a fraction p of the population will inherit that particular allele. However, allele frequencies in real populations are not probability distributions; rather, they are a random sample, and are thus subject to the same statistical fluctuations.*

When the alleles of a gene do not differ with regard to fitness, on average the number of carriers in one generation is proportional to the number of carriers in the previous generation. But the average is never tallied, because each generation parents the next one only once. Therefore the frequency of an allele among the offspring often differs from its frequency in the parent generation. In the offspring generation, the allele might therefore have a frequency p', slightly different from p. In this situation, the allele frequencies are said to have drifted. Note that the frequency of the allele in subsequent generations will now be determined by the new frequency p'.

The size of the breeding population (the effective population size*) governs the strength of the drift effect. When the effective population size is small, genetic drift will be stronger.*

Drifting alleles usually have a finite lifetime. As the frequency of an allele drifts up and down over successive generations, eventually it drifts until fixation – that is, it either reaches a frequency of zero, and disappears from the population, or it reaches a frequency of 100% and becomes the only allele in the population. Subsequent to the latter event, the allele frequency can only change by the introduction of a new allele by a new mutation.

The lifetime of an allele is governed by the effective population size. In a very small population, only a few generations might be required for genetic drift to result in fixation. In a large population, it would take many more generations. On average, an allele will be fixed in $4N_e$ *generations, where* N_e *is the effective population size.*

According to the Hardy-Weinberg Principle, *which holds that allele frequencies in a gene pool will not change over time, a population must be sufficiently large to prevent genetic drift from changing allele frequencies over time. This is why the law is unstable in a small population.*

Genetic drift and natural selection *rarely occur in isolation from each other; both forces are always at play in a population. However, the degree to which alleles are affected by drift and selection varies according to circumstance.*

In a large population, where genetic drift occurs very slowly, even weak selection on an allele will push its frequency upwards or downwards (depending on whether the allele is beneficial or harmful). However, if the population is very small, drift will predominate. In this case, weak selective effects may not be seen at all as the small changes in frequency they would produce are overshadowed by drift.

Drift can have profound and often bizarre effects on the evolutionary history of a population. These effects may be at odds with the survival of the population.

In a population bottleneck, where the population suddenly contracts to a small size (believed to have occurred in the history of human evolution), genetic drift can result in sudden and dramatic changes in allele frequency that occur independently of selection. In such instances, many beneficial adaptations may be eliminated even if population later grows large again.

Similarly, migrating populations may see founder's effect, where a few individuals with a rare allele in the originating generation can produce a population that has allele frequencies that seem at odds with natural selection. Founder's effects are sometimes held to be responsible for high frequencies of some genetic diseases. The mathematical foundation of the neutral theory of evolution was first promulgated in 1968 by **Motoo Kimura** *(1924–1994, Japan). It challenged the notion that natural selection was the sole directive force in evolution. Arguing that mutations and random drift account for variations at the level of DNA and amino acids, Kimura advanced a theory of evolutional change. The crux of his theory is in the motion that most genetic diversity is there because it makes no difference, not because it has been picked by natural selection for a purpose. Mutation pumps a continual stream of genetic changes that do not affect anything into the gene pool, and that they are gradually purged again by genetic drift – a random change. So there is constant turnover without adaptive significance.*

According to Kimura, when one compares the genomes of existing species, the vast majority of molecular differences are selectively "neutral." That is, these differences do not influence the fitness of either the species or the individuals who make up the species. As a result, the theory regards these genome features as neither subject to, nor explicable by, natural selection. This view is based in part on the degenerate genetic code, in which sequences of three nucleotides (codons) may differ and yet encode the same amino acid (GCC and GCA both encode alanine, for example). Consequently, many potential single-nucleotide changes are in effect "silent" or "unexpressed"

Such changes are presumed to have little or no biological effect. However, it should be noted that the original theory was based on the consistency in rates of amino acid changes, and hypothesized that the majority of those changes too were neutral.

A second assertion or hypothesis of the neutral theory is that most evolutionary change is the result of genetic drift acting on neutral alleles. A new allele arises typically through the spontaneous mutation of a single nucleotide within the sequence of a gene. In single-celled organisms, such an event immediately contributes a new allele to the population, and this allele is subject to drift.

In sexually reproducing multicellular organisms, the nucleotide substitution must arise within one of the many sex cells that an individual carries. Then only if that sex cell participates in the genesis of an embryo and offspring, does the mutation contribute a new allele to the population. Neutral substitutions create new neutral alleles.

Through drift, these new alleles may become more common within the population. They may subsequently decline and disappear, or in rare cases they may become "fixed" – meaning that their substitution becomes a universal feature of the population or species. When an allele carrying one of these new substitutions becomes fixed, the effect is to add a substitution to the sequence of the previously fixed allele. In this way, neutral substitutions tend to accumulate, and genomes tend to evolve.[33]

The process of change of the frequency of a gene over time in a large random-mating population can be treated as a stochastic process and approximated by a diffusion process. Kimura modeled the dynamic process of gene frequency change over time under different models for mutation and selection by making use of diffusion theory.

Consider a large random-mating population; two alleles exist at a locus with a selectively advantageous allele with frequency p_0 *at generation* 0. *The probability density that the frequency of the favored allele is* x *at generation* t, *denoted by* $\phi(x, t|p_0)$, *satisfies the Kolmogorov forward equation.*

The average change of allele frequency per generation under selection (with selection coefficient $s > 0$*) is approximately* $sx(1-x)$ *according to a diffusion approximation if there is no dominance. The variance of the change of gene frequency due to random drift is* $\frac{1}{2N}x(1-x)$ *and* $\phi(x, t|p_0)$ *can be obtained by solving the following PDE,*

---

[33] As of the early *2000s*, the neutral theory is widely used as a "null model" for so-called *null hypothesis* testing. Researchers typically apply such a test when they already have an estimate of the amount of time that has passed since two species or lineages diverged – for example, from *radiocarbon dating* at *fossil* excavation sites, or from historical records in the case of human families. The test compares the actual number of differences between two sequences and the number that the neutral theory predicts given the independently estimated divergence time. If the actual number of differences is much less than the prediction, the null hypothesis has failed, and researchers may reasonably assume that *selection* has acted on the sequences in question. Thus such tests contribute to the ongoing investigation into the extent to which molecular evolution is neutral.

$$\frac{\partial\phi(x,t|p_0)}{\partial t}=\frac{1}{4N}\frac{\partial^2[x(1-x)\phi(x,t|p_o)]}{\partial x^2}-\frac{\partial[sx(1-x)\phi(x,t|p_o)]}{\partial x}$$

with boundaries $x=0$, and $x=1$ where N is the population size. Kimura obtained the explicit separation of variables solution,

$$\phi(x,t|p_0)=\sum_{k=1}^{\infty}C_k e^{-\frac{\lambda_k^2+c^2}{4N}t+2cx}\sum_{k=0,1}^{1}f_n^k\,T_n^1(1-2x),\qquad(1)$$

where
$r=1-2p_0$, $c=Ns$, $T_n^1=$ the Gegenbauer polynomial, λ_k is the eigenvalue of the oblate spheroidal angular function, f_n^k is the intermediate coefficient of the spheroidal harmonics, and

$$C_k=\frac{(1-r^2)e^{-c(1-r)}\sum'_{n=0,1}f_n^k\,T_n^1(r)}{\sum'_{n=0,1}((n+1)(n+2))/(2n+3)(f_n^k)^2},$$

The explicit solution of probability of fixation, or loss, of an allele by generation t, given the initial frequency p_0, can be derived by means of (1) and shown to be

$$f_0(t|p_0)=1-\frac{1-e^{-4cp_0}}{1-e^{-4c}}-\sum_{k=1}^{\infty}\frac{C_k}{\lambda_k+c^2}e^{-((\lambda_k+c^2)/4N)t}\sum_{n=0,1}^{\prime}\frac{(n+1)(n+2)}{2}f_n^k,$$

and

$$f_1(t|p_0)=\frac{1-e^{-4cp_0}}{1-e^{-4c}}-\sum_{k=1}^{\infty}(-1)\frac{C_k}{\lambda_k+c^2}e^{-((\lambda_k+c^2)/4N)t+2c}\sum_{n=0,1}^{\prime}\frac{(n+1)(n+2)}{2}f_n^k,$$

respectively. This is the solution to a genetic situation of selection without dominance complicated by random sampling. Kimura gave also an explicit solution to the problem of random drift only ($s=0$).[34] *To sum up, the neutral theory says that, through the history of life from beginning to end, random*

[34] To dig deeper, see:

- Bharucha–Reid, A.T., *Elements of the Theory of Markov Processes and Their Applications*, McGraw-Hill Book Company: New York, 1960 *statistical fluctuations have been more important than Darwinian selection in causing species to evolve. Evolution by random statistical fluctuation is called genetic drift. Kimura maintains that genetic drift drives evolution more powerfully than natural selection.*

According to **F. Dyson** *(1985) Genetic drift and natural selection are both important, and there are times and places where one or the other may be dominant. In particular it is reasonable to suppose that genetic drift was dominant in the very earliest phase of biological evolution, before the mechanism of heredity had become exact.*

We know almost nothing about the origin of life. We do not even know whether the origin was gradual or sudden. It might have been a process of slow growth stretched out over millions of years or it might have been a single molecular event that happened in a fraction of a second. As a rule, natural selection is more important over long periods of time and genetic drift is more important over short periods. If one thinks of the origin of life as being slow, one must think of it as a Darwinian process driven by natural selection. If one thinks of it as being quick, then the Kimura picture of evolution by statistical fluctuation without selection is appropriate. In reality the origin of life must have been a complicated process, with incidents of rapid change separated by long periods of slow adaptation. A complete description needs to take into account both drift and selection.

If one wishes to examine seriously the double-origin hypothesis, the hypothesis that life began and flourished without the benefit of exact replication, then it is natural to imagine that genetic drift remained strong and natural selection remained relatively weak during the early exploratory phases of evolution. But this is not to say that Darwinian selection had to wait until life learned to replicate exactly. Darwinian selection is not logically dependent on exact replication. Indeed, Darwin himself knew nothing of exact replication when he invoked the idea of natural selection. Darwinian selection would have operated to guide the evolution of living creatures even at a time when those creatures may have lacked anything resembling a modern genetic apparatus. All that is necessary for natural selection to operate is that there be some inheritance of chemical constituents from an organism to its progeny. The inheritance need not be exact. It is sufficient if a cell splitting into two daughter cells transmits to each of its daughters with a high probability a population of molecules capable of continuing its own pattern of metabolism. Statistical inheritance, as Darwin well knew, can be good enough. Darwinian selection is unavoidable as soon as inheritance begins, no matter how sloppy the mechanism of inheritance may be.

Lewontin *(1974) outlined the theoretical task for population genetics. He imagined two spaces: a "genotypic space" and a "phenotypic space". The* *challenge of a complete theory of population genetics is to provide a set of laws that predictably map a population of genotypes (G1) to a phenotype space (P1), where selection takes place, and another set of laws that map the resulting population (P2) back to genotype space (G2) where Mendelian genetics can predict the next generation of genotypes, thus completing the cycle. Even leaving aside for the moment the non-Mendelian aspects revealed by molecular genetics, this is clearly a gargantuan task. Visualizing this transformation:*

$$G_1 \rightarrow^{T_1} P_1 \rightarrow^{T_2} P_2 \rightarrow^{T_3} G_2 \rightarrow^{T_4} G_1' \rightarrow \ldots$$

T_1 represents the genetic and epigenetic laws, the aspects of functional biology, or development, that transform a genotype into phenotype. We will refer to this as the "genotype-phenotype map". T_2 is the transformation due to natural selection, T_3 are epigenetic relations that predict genotypes based on the selected phenotypes and finally T_4 the rules of Mendelian genetics.

In practice, there are two bodies of evolutionary theory that exist in parallel, traditional population genetics operating in the genotype space and the biometric theory used in plant and animal breeding, operating in phenotype space. The missing part is the mapping between the genotype and phenotype space. This leads to a "sleight of hand" (as Lewontin terms it) whereby variables in the equations of one domain, are considered parameters or constants, where, in a full-treatment they would be transformed themselves by the evolutionary process and are in reality functions of the state variables in the other domain. The "sleight of hand" is assuming that we know this mapping. Proceeding as if we do understand it is enough to analyze many cases of interest. For example, if the phenotype is almost one-to-one with genotype (sickle-cell disease) or the time-scale is sufficiently short, the "constants" can be treated as such; however, there are many situations where it is inaccurate.

(d) Panspermia and the Anthropic Principle

Panspermia is the theory that life on earth was seeded by microbial life from space. There are several variations on this theme held by many historical advocates, including the Greek philosopher, **Anaxagoras** *(500–428 BCE),* **Hermann von Helmholtz** *(1821–1894), and* **William Thomson** *(1824–1897).*

More recently, **Svante Arrhenius** *promulgated the theory of radio-panspermia, wherein microbes from space are transported by light pressure.* **Fred Hoyle** *and* **Chandra Wickramasinghe** *have advocated that DNA arrived on earth via meteorites (ballistic panspermia) or by comets (modern* *panspermia).* ***Francis Crick*** *has advocated the theory of directed panspermia, wherein RNA was transported by unmanned spaceships, or the space probes of intelligent extraterrestrial civilizations. He was led to argue for Panspermia by his belief that the chances of life accidentally originating on Earth were very low.*

Crick argued that the universality of the genetic code can only be explained by an "infective" theory of the origin of life. In this theory, life on earth would be a "clone" derived from a single set of organisms. Crick's radical theory of directed panspermia *suggests that RNA was the first replicator molecule on earth in an early biological era.*

The anthropic principle *is based on a biological argument: the minimum time required for the evolution of "intelligent observers." In this scheme, a billion years is required for the evolution of intelligence; therefore, a star must have been stable for at least that long. The anthropic timescale argument allows that the types of processes allowed in the Universe must be of such an age that "slow evolutionary processes will have had time to produce intelligent beings from non-living matter."*

Interestingly, the contemporary advocates for the existence of extraterrestrial intelligent life are primarily astronomers and physicists, while most leading experts in evolutionary biology contend that the earth is probably unique in harboring intelligence, and that human beings are alone in the universe because Darwinian evolution has told us so: intelligence was never there beforehand and can only be gotten through an incremental succession of steps involving pure random luck – an earth-based anomaly.

(E) Cosmology and Darwinian Evolution

The origin of life is one of the few scientific problems which is broad enough to make use of ideas from almost all scientific disciplines: ***E. Schrödinger*** *(1887–1961) brought his ideas from* physics; ***J. von Neumann*** *(1903–1957) from* mathematical logic; ***M. Eigen*** *(b. 1927) and* ***L. Orgel*** *(b. 1927) from* chemistry; ***L. Margulis*** *(b. 1938) from* ecology, *and* ***M. Kimura*** *(1924–1994) from* population biology.

Recently[35]*, Darwinian evolution was also linked to cosmology.*

[35] **Lee Smolin** (b. 1955) "The Life of the Cosmos", Oxford University Press, New York, 358 pp. 1997. A succinct popular account was given by John Gribbin in *Prospect* magazine, issue 19, May 1997.

If the universe began in the hot fireball of a big bang some 15 billion years ago, how did it evolve to produce galaxies and stars, planets and people?

The Big Bang theory does not provide a satisfying account for why the flying debris of the primordial blast congealed into stars and galaxies, and galaxies of galaxies – a hierarchy of structure. It is more likely that the universe would have turned out to be utterly random, a featureless fog.

More miraculous till is that the Big Bang seems to have produced a universe perfectly designed to support life. It is often argued that if gravity or electromagnetism were a little stronger or a little weaker, or if the nuclear forces were not precisely as they are, there would be no stars. And without stars, which cook hydrogen and helium into carbon and other complex atoms, there would be no chemistry, no biology, no complexity, no life. The question is "Why is the universe so interesting?", and the answer, given by Lee Smolin (1997) is:
"Since evolution so successfully explains why the biosphere is the way it is, why not apply the theory of Darwinian evolution by natural selection to the entire creation? – The universe is perfectly tuned to life because it evolved that way."

Smolin's thesis is then that the way the universe works can best be understood not simply by applying the rules of physics worked out by Newton and Einstein, but by taking account as well of the rules of evolution worked out by Darwin – the theory of natural selection.

The universe itself, and its main components (notably galaxies such as our own Milky Way) may have evolved through natural selection from a simpler state to produce the complexity we see around us. To take literally the equations of GTR, the big bang itself emerged from a point of infinite density, known as a singularity. There is, however, another place where singularities are known to occur – at the heart of black hole.

Indeed, as **Roger Penrose** *and* **Stephen Hawking** *proved in the 1960's, the expanding universe is described by exactly the same equations as a collapsing black hole, but with opposite direction of time.*

If all the complexity of galaxies, stars, planets and organic life has emerged from the singularity in which our universe was born within a black hole, could not something similar be happening to the singularities at the heart of other black holes?

The most basic view of what might happen to a collapsing singularity to turn it into the kind of expanding singularity that we see in our universe is that there is simply a "bounce," turning collapse into expansion. Unfortunately, that will not do as an explanation. A singularity forming from a collapse within our three dimensions of space and one of time cannot turn itself around and explode back outwards in the same three dimensions of space and one of time.

But in the 1980s relativists realized that there is nothing to stop the material that falls into a singularity from being shunted through a kind of spacetime warp and emerging as an expanding singularity in another set of dimensions — another spacetime.

Mathematically, this "new" spacetime is represented by a set of four dimensions (three of space and one of time), just like our own but with all of the new dimensions at right angles to all of the familiar dimensions of our own spacetime. Every singularity, on this picture, has its own set of spacetime dimensions, forming a bubble universe within the framework of some "super" spacetime, which we can refer to simply as "superspace."

One way to picture what this involves is to use the old analogy between the three dimensions of expanding space around us and the two-dimensional expanding surface of a balloon that is being steadily filled with air. The analogy is not with the volume of air inside the balloon, but with the expanding skin of the balloon, stretching uniformly in two dimensions, but curved around upon itself in a closed surface.

Imagine a black hole as forming from a tiny pimple on the surface of the balloon, a small piece of the stretching rubber that gets pinched off, and starts to expand in its own right. There is a new bubble, attached to the original balloon by a tiny, narrow throat – the black hole. And this new bubble can expand away happily in its own right, to become as big as the original balloon, or even bigger, without the skin of the original balloon (the original universe) being affected at all. There can be many bubbles growing out of the skin (the spacetime) of the original universe in this way at the same time. And, of course, new bubbles can grow out of the skin of each new universe, ad infinitum.

The dramatic implication is that many – perhaps all – of the black holes that form in our universe may be the seeds of new universes. And, of course, our own universe may have been born in this way out of a black hole in another universe. This means that the universe may not be unique. Instead, it may be one of a population of universes, interconnected by what physicists call wormholes. The key element that Smolin has introduced into the argument is the idea that every time a black hole collapses into a singularity and a new baby universe is formed, the basic laws of physics are altered slightly as spacetime it-self is crushed out of existence and reshaped. The process is analogous (perhaps more than analogous) to the way mutations provide the variability among organic life forms on which natural selection can operate. Each baby universe is, says Smolin, not a replica of its parent, but a slightly mutated form.

The original, natural state of such baby universes is to expand out to only about the Planck length, before collapsing once again. But if the random changes in the workings of the laws of physics-the mutations-happen to allow a little bit more expansion, a baby universe will grow a little larger. If it becomes big enough, it may separate into two, or several, different regions, that each collapse to make a new singularity, and thereby trigger the birth of a new universe. Those new universes will also be slightly different from their parents. Some may lose the ability to grow much larger than the Planck length and will fade back into the quantum foam. But some may have a little more inflation still than their parents, growing even larger, producing more black holes and giving birth to more baby universes in their turn. The number of new universes that are produced in each generation will be roughly proportional to the volume of the parent universe. There is even an element of competition involved, as the many baby universes are in some sense vying with one another, jostling for spacetime elbow room within superspace. Heredity is an essential feature of life, and this description of the evolution of universes works in a similar manner to living systems.

On this picture, universes pass on their characteristics to their offspring with only minor changes, just as people pass on their characteristics to their children with only minor changes. Universes that are "successful" are the ones that leave most offspring. Provided that the random mutations are indeed small, there will be a genuinely evolutionary process favoring larger and larger universes.

Once universes start to be big enough to allow stars to form, in succeeding generations of universes there will be a natural evolution, a drift in the laws of physics, to favor the production of the kinds of stars that will eventually form black holes.

The end product of this process should be not one but many universes which are all about as big as it is possible to get while still being inside a black hole, and in which the parameters of physics are such that the formation of stars and black holes is favored. Our universe exactly matches that description.

This explains the otherwise baffling mystery of why the universe we live in should be "set up" in what seems, at first sight, such an unusual way. Just as you would not expect a random collection of chemicals to suddenly organize themselves into a human being, so you would not expect a random collection of physical laws emerging from a singularity to give rise to a universe such as the one we live in.

To sum up, universes evolve in favor of the production of black holes. *It is a theory of the origin of universes by means of natural selection. Smolin's answer to the question as why our universe is the way it is (i.e, the value of its parameters) is: the parameters have the values we observe, because these make the formation of black holes much more likely than most other values. Hence, our universe is a product of mutation and selection analogous to the evolution of species described by Charles Darwin (1859).*

Cosmologists are now having to learn to think like biologists and ecologists, and to develop their ideas not within the context of a single, unique universe, but in the context of an evolving population of universes. Each universe starts from its own big bang, but all the universes are interconnected in complex ways by black hole "umbilical cords," and closely related universes share the "genetic" influence of a similar set of physical laws.

But the realization that our universe is just one among many, that it is alive and that no supernatural influences need be invoked to explain its existence, is still not the most dramatic conclusion we can draw from the new cosmology. Although it is now clear that the universe has not been set up for our benefit, and that the existence of organic life forms on Earth is a minor side effect of an evolutionary process involving universes, galaxies and stars, nevertheless it is clear that the existence of life forms such as ourselves is an inevitable side effect of those greater evolutionary processes.

The same laws of physics apply throughout our universe and throughout many other universes besides. Organic (carbon based) material occurs in profusion between the stars of a spiral galaxy such as our Milky Way. This carbon-rich material seems to be crucially involved in the processes which allow gas clouds to cool and new stars to form, so a universe that is good at making black holes will also be good at making carbon based compounds. Those compounds will undoubtedly seed any earth-like planet that forms with each new generation of stars.

Astronomers calculate that there may be as many as 10^{20} planets suitable for life forms such as ourselves in our universe. We see the components of organic life everywhere in the universe, and the chances are that most of those 10^{20} planets actually are carriers of our kind of life, in the same way that earth is a carrier of life. The birth of the living universe inevitably gave rise to the birth of living planets. Which still leaves physicists the task of explaining just how complexity arose in a hot universe expanding out of a big bang.

Note that in principle, life and Cosmological Natural Selection could be independent of each other. There are two reasons for this:

- *On the one hand there may be universes full of black holes were life as we know it couldn't evolve. For example it might be possible that there are only short-lived giant stars which collapse quickly into black holes, or that there are universes dominated either by helium or by neutrons (corresponding to the neutron/proton mass difference being either zero or negative), or that there are universes with many (more) primordial black holes and maybe without stars at all. Such universes might be very reproductive because of their giant stars or primordial black holes but are not able to produce earth-like life.*

- *On the other hand we can conceive a universe without black holes at all (if supernova lead to neutron stars only, or if there are not stars above a critical mass limit) but which could be rich in earth-like life nevertheless.*

Thus, there is not a (logically) necessary connection between black holes and life, unless, perhaps, black holes could be advantageous to life, or vice versa.

If Smolin is right, the implications for our view of nature would be enormous. Our universe would be only a grain of sand on the incredible large beach of the Multiverse. And it would be in no way special.

According to the Cosmological Principle, our universe is homogeneous and isotropic on the large scale, that is, it appears the same at all places and, from any one place, looks the same in all directions. If Smolin is right, we could accept a truly Perfect Cosmological Principle, that is, the Multiverse is literally "full" of universes like our own universe.

It has often been said that the Copernican revolution has catapulted the earth and hence, mankind, out of the center of the universe. In the twentieth century it became clear that neither the earth, nor the sun, nor the milky way, nor the local supercluster of galaxies are at the center of the universe, because there is no center at all. Nor does the baryonic matter of which we consist dominate the universe, i.e. in comparison with the assumed particles of dark matter even the material we are made of is quantitatively negligible.

This dramatic widening of our horizon and diminution of our role in the universe led to a radical cosmic expulsion. Our position in space-time is completely irrelevant, providing no evidence for a universal meaning or a cosmic value of mankind, nor a protection against contingency and absurdity.

In recent decades, some interpretations of the Anthropic Principle and the alleged fine-tuning of the physical parameters have been examples for a tendency to reverse this development. But the idea of the Multiverse made out of many different universes, which implies an Observational Selection effect, is once again sufficient to wipe out any romantic dreams of anthropocentrism.

But in this picture at least our life-bearing universe might still be special. In Smolin's account however, our universe has most properties in common with all other universes. It would be a very ordinary world indeed. Thus, if

Smolin is right, there is no reason to believe that (human) life is unique or that the constants of nature have a singular status.[36]

(F) Critique of Evolutionary Biology (1978–2008)

Darwin's theory of evolution basically states that life on earth began with single-celled life-forms, which evolved into multicellular life-forms, which over countless aeons evolved into higher life-forms, including man – all as a result of the chance process of random mutation of desirable attributes followed by natural selection, without guidance or assistance from any intelligent entity. Thus, for example, evolution holds that the human eye came into existence purely by accident.

However, Darwin knew nothing of DNA and the vastly complex systems studied by molecular biologists, such as the information processing, storage, and retrieval in DNA.

Moreover, until recently, scientists did not know what the inside of a cell looked like: the cell was a mysterious "black box".

Michael Behe[37] *(1996) used discoveries in microbiology to cast serious doubts upon Darwinism: for example, a bacterial motor, that propels bacteria*

---

[36] To dig deeper, see:

- Schrödinger, E., *What is Life*, Cambridge University Press, 1944
- Jacob, Francois, *The Possible and the Actual*, Pantheon Books: New York, 1982
- Hoyle, Fred, *The Intelligent Universe*, 1983
- Dyson, F., *Origins of life*, Cambridge University Press, 1985
- Barrow, J.D. and F.J. Tipler, *The Anthropic Cosmological Principle*, Oxford University Press: New York, 1986
- Kimura, M., *The Neutral Theory of Molecular Evolution*, 1986
- Smith, J. Maynard, *The Theory of Evolution*, 1997
- Rees, Martin, *Just Six Numbers*, Basic Book: New York, 2000

[37] **Michael J. Behe** (b. 1952, USA), biochemist. Advocates the idea that some structures are too complex at the biochemical level to be adequately explained as a result of evolutionary mechanisms and that these systems could not, even in principle, have evolved by natural selection. This is because the calculated

and sperm, called a flagellum, depends on the coordinated interaction of 30–40 complex protein parts. The absence of almost any one of the parts would render the flagellum useless. Likewise, an animal cell's whiplike oar, called a cilium, is composed of about 200 protein parts.

Now, it does not matter if 200 mutations happened a once or over a billion years. All 200 mutations would have to (1) occur, (2) be the "most fit", (3) survive long enough to exist at the same time and place, in order to (4) assemble themselves into a working cilium. The cell is as complicated as the entire city of New York. Natural selection has never been demonstrated to change anything fancier than the shape of a bird's beak.

Behe then argued that it is extremely unlikely for all 30 parts of the flagellum (or 200 parts of the cilium) to have been brought together by numerous, successive, slight modifications primed by natural selection. He then concluded that life at the molecular level is a "loud, clear, piercing cry of design".

Hence the contention that natural processes of mutation and natural selection cannot explain the complexity of living things.

In fact, the more we know about molecules, cells and DNA, the less plausible Darwin's theory of natural selection becomes. Indeed, Darwin himself

---

probabilities of mutations required for evolution to succeed are too small. He termed this concept "*irreducible complexity*" [*Darwin's Black Box*, Free Press, 1996]. He argues that the eye, or the bacterial flagellum are nanotechnological machines and cannot have evolved by any number, however large, of small mutations that each confer reproductive advantage.

Behe's claims about the irreducible complexity of key cellular structures are strongly contested by the community of Darwin followers, and his claims about "intelligent design" have been characterized as pseudoscience. His adversaries have also pointed out that he offered no design theory, or attempted to model the design process, and in general failed to offer an alternative to evolution.

However, Behe's critics themselves tend to evade his (quite reasonable) statistical arguments; Furthermore, adducing an alternative theory is not a precondition for pointing out potentially fatal flaws in a widely accepted theory or dogma. The truth is that both sides in this debate have not the slightest idea how to compute the relevant probabilities; and that the neo-Darwinian synthesis is not really a quantitative scientific theory-since it consists entirely of non-refutable, non-quantitative *scenarios* (what the Harvard scientist R. Lewontin has duffed "just-so stories").

noted the difficulty of explaining the eye in The Origin of the Species, admitting he could not do it.[38]

In his book "The Design Inference" (Cambridge University Press, 1998) **William A. Dembski** *(b.1960) added his doctrine of Specified Complexity to Behe's assertion that irreducibly complex systems cannot evolve gradually. Again, Dembski's work was strongly criticized within the scientific community, who argued that there were a number of major logical inconsistencies and evidential gaps in Dembski's hypothesis. His writings were labeled as pseudoscience.*[39]

Evolutionary biologists argue that material mechanisms suffice to account for biological complexity, while intelligent design advocates reject this claim. Both sides are trying to determine the truth of some definite matter of fact – whether life is the result of mindless material mechanisms or whether, to the contrary, it demonstrably points to a designing intelligence.

A prominent critic of Evolutionary Biology from inside the scientific establishment is **Richard Charles Lewontin** *(b. 1929, USA), Alexander Agassiz Professor at Harvard. In "Biology and Ideology" (1991), Lewontin argued that while traditional Darwinism has portrayed the organism as passive receiver of environment influences, a correct understanding should emphasize*

---

[38] Darwin hypothesized that the eye might have begun as a patch of light- sensitive cells upon which natural selection could then work its magic, making gradual improvements - creating an eye socket and slowly increasing focus and perspective and so on – until these special cells became a full – fledged eye. But this "explanation" explains nothing – it is just a story about how something might have happened: for light sensitive cells to work the cells would have to have the capacity to initiate an electric signal, a nerve capable of carrying the electric signal to a brain, and a brain capable of processing the signal and using it to emit other electric signals. No one disputes that organisms can develop small improvements on something that already exist. The interesting question is: How did the "light sensitive cells" come to exist in the first place?

[39] The mark of a *pseudoscience* is not that it is false but, in the words of physicist **Wolfgang Pauli**, that it is "*not even false*". In other words, with a pseudoscience there is no way to decide whether it is true or false.
Recently, many physicists and mathematicians have accused *string theory* of having become a pseudoscience. Psychoanalysis is doubtless in this category; and so is the so-called neo-Darwinian synthesis.
At present, the latter tells us various often-shifting, verbal, qualitative stories, albeit buttressed by results from exact sciences such as biochemistry & molecular genetics (as well as much less precise findings from fossil records). It is conceivable that one day evolution will truly merit the term "theory".

the organism as an active constructor of its environment. Niches are not preformed, empty receptacles into which organisms are inserted, but are defined and created by organisms.

Lewontin has also been a critic of traditional neo-Darwinian approaches to adaptation: he emphasized the need to give an engineering characterization of adaptation separate from measurement of number of offspring. This grew out of his recognition that the fallacies of sociobiology reflect fundamentally flawed assumptions of adaptiveness of all traits in much of the modern evolutionary synthesis.

Along with others, Lewontin has been a persistent critic of some themes in neo-Darwinism such as sociobiology and evolutionary psychology, which attempt to explain animal behavior and social structures in terms of evolutionary advantage or strategy.

Darwin *thought that "the mind of man developed from a mind as low as that possessed by the lowest animal". But if our abilities "evolved to let us get along in the cave, how can it be that they permit us to obtain deep insight into cosmology, elementary particles, molecular genetics, number theory?" asks molecular biologist and physicist* **Max Delbrück**.

None of these abstract enterprises has any direct relation to survival. Indeed, mans possession of such traits as morality, consciousness of morality, religion, or the ability to create & appreciate art – let alone his capacity to develop such intellectual constructs as relativity and quantum theory, does not seem likely to be the result of gradualistic evolution.

Darwinists and neo-Darwinists believe that humans evolved from bacteria. But for these who believe in intelligent design, this challenges our ideas of individuality, independence and the alleged uniqueness of human intelligent consciousness.

When Darwin first published The Origin of the Species (1859), his most virulent opponents were paleontologists, for there was absolutely nothing in the fossil record to support his claims. Far from showing gradual change with one species slowly giving way to another, as Darwin hypothesized, the fossil record showed vast numbers of new species suddenly appearing out of nowhere, remaining largely unchanged for millions of years, and then disappearing.

Darwin blamed the absence of fossil support for his theory on the extreme imperfection of the geological record, but was sure that paleontologists will soon produce the necessary evidence.

However, after 150 years of intense looking, the geologic record still does not yield a finely graduated chain of slow and progressive evolution!

If mutations are utterly random, as Darwinism claims, there ought to be an infinite variety of transitional animals with small mutations that eventually led to a new attribute (like a wing or a lung). But we do not have fossils connecting the extinct to the extant along fine graduated steps. What the fossil record shows is sudden bursts of all manner of animals, modest change, and then sudden and total extinction. Dinosaurs appeared, lived for 150 million years, and then disappeared, only to be quickly replaced with mammals.

We don't have fossils for the vast quantity of hapless creatures that ought to have died out in the survival-of-the fittest regime! If each one of the incremental mutations is more "fit" than what preceded it (which it had to be in order to survive), those transitional mutations should have stayed around long enough to appear in the fossil record, before mutating their way to something even better. But in the course of millions and millions of years, all we see are slight variations on the final product.

Indeed, Darwinian evolution is supposedly the completely accidental process that created butterfly wings, bat radar, the human brain, and the millions of species alive today. The theory of evolution requires millions of mutations just to create an eye. A process that is supposed to have transformed an amoeba into Wolfgang Amadeus Mozart or Albert Einstein by "random mutations" must have produced some spectacular failures. Why can't we find any of them?

For over a hundred years, evolutionists proudly pointed to the same sad birdlike animal, archaeopteryx, as their lone transitional fossil linking dinosaurs and birds. Discovered a few years after Darwin published The Origin of Species, Archaeopteryx was instantly hailed as the transitional species that proved Darwin's theory. This unfortunate creature had wings, feathers, teeth, claws, and a long, bony tail. If it flew at all, it didn't fly very well. Alas, it is now agreed that poor Archaeopteryx is no relation of modern birds. It's just a dead end. It transitioned to nothing.

But could Archaeopteryx be our one example of bad mutations eliminated by natural selection? Archaeopteryx can't fill that role either, because it seems to have no predecessors. The fossils that look like Archaeopteryx lived millions of years after Archaeopteryx, and the fossils that preceded Archaeopteryx look nothing at all like it. The bizarre bird is just an odd creation that came out of nowhere and went nowhere.

The more advances paleontologists make in uncovering the fossil record, the more absurd the evolution fable becomes.Most nettlesome for evolutionists is the Cambrian period, showing a vast quantity of plants and animals appearing on the scene in the blink of an evolutionary eye more than 500 million years ago. In a period of less than 10 million years, there is a sudden explosion of nearly all the animal phyla we have today. It is as though they were just planted there, without evolutionary history. Darwin himself referred to the great difficulty of explaining the absence of "vast piles of strata rich in fossils" before the Cambrian explosion.

In 1984, Chinese paleontologists discovered fossils just preceding the Cambrian era. The discovery showed that the dramatic transformation of life from primeval single-cell organisms to the complex multicellular precursors of modern fauna was sudden, swift and widespread within a mere 5 to 10 million years. Even the famously difficult-to-evolve eye appeared at the beginning of the Cambrian period. And there were no light-sensitive pits.

It seems now that traditional Darwinian evolution is a conjecture about how species might have arisen that is contradicted by the fossils record and by nearly everything we have learned about molecular biology since Darwin's day.

To 'save the phenomenon', modern revisionists of the Darwinian "theory" have concocted a sophisticated scheme called "punctuated equilibrium" (**Stephen Jay Gould** *and* **Niles Eldredge**, *1993): Instead of gradual change occurring by random mutation and natural selection choosing the most "fit" to survive and reproduce, evolution could also happen really fast and then stop happening at all for 150 million years – all this occurring completely by chance!*

Darwinian evolution theory also run into other difficulties. To begin with, there is not a single observable example of one species evolving into another by the Darwinian mechanism of variation and selection.

Then, there is the problem of establishing progress via mutation and natural selection: the successive appearance of more complex species does seem to show something that looks like progress. But that has nothing to do with the Darwinian mechanism of natural selection. The appearance of progress hardly establishes mutation and natural selection as the engine of change. To the contrary, the similarities, look more like the progress of a designed object than the result of a series of lucky accidents. Far from the competition of dog-eat-dog struggle to survive, we see a fossil record that reveals a rather clean, well-organized sequence.

Fossils do not reveal a parent/descendal relationship. We certainly do not know whether any particular mammal descended from any particular reptile. But more important, the apparent progress from simple animals to more sophisticated higher animals – with no transitional species – looks more like planned, deliberate progress than a series of random mutations.

In this connection, another nasty question can be asked: if all species evolved from the same single-celled organism beginning in the same little mud puddle, why hasn't the earthworm made a little more progress? Was it never, ever desirable in any of the worm's many dirt holes to mutate eyes or legs or wings or a brain? How could one clump of cells starting in the same little puddle become a human being while others never make it past the amoeba stage?

In 1835, Darwin counted 13 species of finches on the Galapagos Islands. He then hypothesized that species evolved from one species.

Today, after more than 170 years of wild variation in the environment, mutation, and "natural selection", there are still 13 species – not one more. The finches beaks have moved back and forth in shape and nothing more.

In her book "Godless" (2006), Ann Coulter succinctly summarized the many setbacks, hoaxes, pranks, fakeries and frauds that beset the theory of evolution: "Finches on Galapagos Islands with deeper beaks begin to outnumber finches with shallower beaks during a drought – and then the population of shallow beaks finches immediately rebounds after a rainy season. Bacteria develop a resistance to antibiotics and viruses develop resistance to antiviral medication – but nothing new is ever created. A bacterium remains a bacterium, a virus remains a virus, a finch remains a finch"...

"Human breeders have not been able to produce one biologically novel structure in the laboratory – much less a new animal species – even under artificial conditions. No such demonstrations exists; none has ever been provided. The fruit fly has been abused, mutilated, and stressed over the course of thousands and thousands of generations. The poor dumb creature remains what it has always been, a fruit fly in the first instance, dumb in the second. This negative result is perfectly consistent with the long history of breeding experiments, which demonstrate beyond question that species may be changed only within very narrow margins of variability. No practical breeder imagines, for example, that he will ever succeed in creating a chicken with antennae or pig with a dorsal fin.

Amid this dismal record, there have been a few exciting developments for the Darwinians. There was the discovery of a manlike ape that looked like a transitional fossil between ape and man – the long-sought after "missing link". There were drawings of embryos demonstrating that vertebrates all looked alike in the earliest stages of development. There was the peppered moth that became darker – allegedly to better camouflage itself from predatory birds – when industrial air pollution blackened the trees in England. It wasn't terribly impressive in terms of "evidence," but it filled out a few pages in biology textbooks claiming evolution was a FACT.

And then, one by one, each of these pillars of evidence for evolution was exposed as s fraud. (Ironically, each appeared to have been an intelligently designed prank.) It's difficult to imagine that any other "scientific" theory has been beset with as many hoaxes as the theory of evolution – always a good sign of a serious scientific endeavor."...

"The only time "radiocarbon dating" was used in connection with the theory of evolution was the time it was used to expose the Piltdown Man as a hoax being pawned off as proof of evolution. It was one of the greatest scientific frauds of all time, right up there with the Pepsi challenge and that commercial where ordinary laundry detergent gets red wine out of a white blouse.

For half a century, Piltdown Man constituted a major piece of evidence for Darwin's theory. After decades of being embarrassed by the fossil record's stubborn refusal to come to Darwin's aid, in 1912 the Piltdown Man miraculously appeared in a gravel pit in Sussex, England. Amateur paleontologist Charles Dawson claimed to have discovered a skull with a human-like cranium and an apelike jaw in Piltdown Quarry. It was a creature that was not quite ape, not quite man, but a transitional species between the two, rather like the actor Pauly Shore. This Pauli Shore-like fossil wouldn't have proved evolution, but it would have given evolutionists a possible link between apes and man on their imaginary "tree of life."

It was almost uncanny how precisely Piltdown Man matched what prevailing scientific theory predicted the "missing link" would look like. The New York Times *headline for the article on the Piltdown Man proclaimed, "Darwin Theory Is Proved True."*

The Piltdown fossil was "peer-reviewed" – so we know it would pass muster with the editors of Scientific American, *still flush with success after triumphantly exposing the "Ohio flight hoax." Experts confirmed the age and origin of the bones. Indeed, the Piltdown Man received the approval of Arthur smith Woodward, the leading geologist at the British Museum (Natural History).* Eoanthropus dawsoni *was born.*

Dawson was showered with praise, fame, and awards. If only Vanity Fair *had been around, Dawson could have been photographed in his Jaguar and hailed for "speaking truth to power." He was made a fellow of the Geological Society and a fellow of the Society of Antiquaries. (He was even offered a position writing editorials for* Scientific American.)

For more than forty years, the Piltdown Man was taught as scientific fact. Then, in 1953, it was exposed as a complete and utter fraud – in part through the process of radiocarbon dating."

In conclusion, anti-Darwinists claim that Darwinism is a nondisprovable conjecture, and certainly not a science. In that sense it belongs to the same category as "String Theory" and psychoanalysis.

IV. Ecology — Living Organisms and their Environment

The science of ecology – the study of the interrelationships among the biological and physical components in the natural world – has emerged as a distinct discipline only in the twentieth century. While naturalists from ancient times to the late nineteenth century noted the interdependence among organisms and the adaptations of organisms to their environment, the attempt to study those interactions as part of a larger natural system has occurred – only recently.

Ecology is a multi-disciplinary science. Because of its focus on the higher levels of the organization of life on earth and on the interrelations between organisms and their environment, ecology draws heavily on many other branches of science, especially geology and geography, meteorology, pedology, chemistry, and physics.

Thus, ecology studies the distribution and abundance of living organisms, and interactions among organisms and between organisms and their environment. The environment of an organism includes both its habitat, which can be described as the sum of local abiotic factors such as climate, and geology, as well as the other organisms that share its habitat. Ecological systems are studied at several different levels, from individuals and populations to ecosystems and the biosphere.

*Ethology studies animal behavior (particularly of social animals such as primates), and is sometimes considered a branch of zoology. Ethologists have been particularly concerned with the evolution of behavior and the understanding of behavior in terms of the theory of natural selection. In one sense the first modern ethologist was Charles Darwin, whose book '*The Expression of the Emotions in Animals and Men*' influenced many ethologists.*

Biogeography studies the spatial distribution of organisms on the earth, focusing on topics like plate tectonics, climate change, dispersal and migration, and cladistics.

Agriculture, fisheries, forestry, medicine and urban development are among human activities that would fall within the definition of ecology.

Ecology is a broad discipline comprised of many sub-disciplines. A common, broad classification, moving from lowest to highest complexity, where complexity is defined as the number of entities and processes in the system under study, is:

- *Physiological Ecology (or ecophysiology) and Behavioral ecology examine adaptations of the individual to its environment.*

- *Population ecology* (or autecology) studies the dynamics of populations of a single species.
- *Community ecology* (or synecology) focuses on the interactions between species within an ecological community.
- *Ecosystem ecology* studies the flows of energy and matter through the biotic and abiotic components of ecosystems.
- *Landscape ecology* examines processes and relationship across multiple ecosystems or very large geographic areas.

Ecology can also be sub-divided according to the species of interest into fields such as animal ecology, plant ecology, insect ecology, and so on. Another frequent method of subdivision is by biome studies, e.g., Arctic ecology (or polar ecology), tropical ecology, desert ecology, etc. The primary technique used for investigation is often used to subdivide the discipline into groups such as chemical ecology, genetic ecology, field ecology, statistical ecology, theoretical ecology, and so forth. Note that these different systems are unrelated and often applied at the same time; one could be a theoretical plant community ecologist, or a polar ecologist interested in animal genetics.

Biosphere

For modern ecologists, ecology can be studied at several levels: population level (individuals of the same species in the same or similar environment), biocoenosis level (or community of species), ecosystem level, and biosphere level.

The outer layer of planet earth can be divided into several compartments: the hydrosphere (or sphere of water), the lithosphere (or sphere of soils and rocks), and the atmosphere (or sphere of the air). The biosphere (or sphere of life), sometimes described as "the fourth envelope", is all living matter on the planet or that portion of the planet occupied by life. It reaches well into the other three spheres, although there are no permanent inhabitants of the atmosphere. Relative to the volume of earth, the biosphere is only the very thin surface layer which extends from 11,000 meters below sea level to 15,000 meters above.

It is thought that life first developed in the hydrosphere, at shallow depths, in the photic zone. (Recently, though, a competing theory has emerged, that life originated around hydrothermal vents in the deeper ocean.) Multicellular organisms then appeared and colonized benthic zones. Photosynthetic organisms gradually produced the chemically unstable oxygen-rich atmosphere that characterizes our planet. Terrestrial life developed later, after the ozone layer protecting living beings from UV rays formed.

Diversification of terrestrial species is thought to be increased by the continents drifting apart, or alternately, colliding. Biodiversity is expressed at the ecological level (ecosystem), population level (intraspecific diversity), species level (specific diversity), and genetic level. Recently, technology has allowed the discovery of the deep ocean vent communities. This remarkable ecological system is not dependent on sunlight but bacteria, utilizing the chemistry of the hot volcanic vents, are at the base of its food chain.

The biosphere contains great quantities of elements such as carbon, nitrogen, hydrogen and oxygen. Other elements, such as phosphorus, calcium, and potassium, are also essential to life, yet are present in smaller amounts. At the ecosystem and biosphere levels, there is a continual recycling of all these elements, which alternate between the mineral and organic states.

While there is a slight input of geothermal energy, the bulk of the functioning of the ecosystem is based on the input of solar energy. Plants and photosynthetic microorganisms convert light into chemical energy by the process of photosynthesis, which creates glucose (a simple sugar) and releases free oxygen. Glucose thus becomes the secondary energy source which drives the ecosystem. Some of this glucose is used directly by other organisms for energy. Other sugar molecules can be converted to other molecules such as amino acids. Plants use some of this sugar, concentrated in nectar to entice pollinators to aid them in reproduction.

Cellular respiration is the process by which organisms (like mammals) break the glucose back down into its constituents, water and carbon dioxide, thus regaining the stored energy the sun originally gave to the plants. The proportion of photosynthetic activity of plants and other photosynthesizers to the respiration of other organisms determines the specific composition of the earth's atmosphere, particularly its oxygen level. Global air currents mix the atmosphere and maintain nearly the same balance of elements in areas of intense biological activity and areas of slight biological activity.

Water is also exchanged between the hydrosphere, lithosphere, atmosphere and biosphere in regular cycles. The oceans are large tanks, which store water, ensure thermal and climatic stability, as well as the transport of chemical elements thanks to large oceanic currents.

Ecosystems

Each living organism has an ongoing and continual relationship with every other element that makes up its environment. An ecosystem can be defined as any situation where there is interaction between organisms and their environment.

These concepts lead to the idea of biomass (the total living matter in a given place), of primary productivity (the increase in the mass of plants during a given time) and of secondary productivity (the living matter produced by consumers and the decomposers in a given time).

These two last ideas are key, since they make it possible to evaluate the load capacity – the number of organisms which can be supported by a given ecosystem. In any food network, the energy contained in the level of the producers is not completely transferred to the consumers. Thus, from an energy – and environmental – point of view, it is more efficient for humans to be primary consumers (to subsist from vegetables, grains, legumes, fruit, cotton, etc.) than as secondary consumers (from eating herbivores, omnivores, or their products, such as milk, chickens, cattle, sheep, etc.) and still more so than as a tertiary consumer (from consuming carnivores, omnivores, or their products, such as fur, pigs, snakes, alligators, etc.). An ecosystem(s) is unstable when the load capacity is overrun and is especially unstable when a population doesn't have an ecological niche and overconsumers.

The productivity of ecosystems is sometimes estimated by comparing three types of land-based ecosystems and the total of aquatic ecosystems:

- *The forests (1/3 of the earth's land area) contain dense biomasses and are very productive. The total production of the world's forests corresponds to half of the primary production.*
- *Savannas, meadows, and marshes (1/3 of the earth's land area) contain less dense biomasses, but are productive. These ecosystems represent the major part of what humans depend on for food.*
- *Extreme ecosystems in the areas with more extreme climates – deserts and semi-deserts, tundra, alpine meadows, and steppes – (1/3 of the earth's surface) have very sparse biomasses and low productivity.*
- *Finally, the marine and fresh water ecosystems (3/4 of Earth's surface) contain very sparse biomasses (apart from the coastal zones).*

Humanity's actions over the last few centuries have seriously reduced the amount of the earth covered by forests (deforestation), and have increased agro-ecosystems (agriculture). In recent decades, an increase in the areas occupied by extreme ecosystems has occurred (desertification).

Dynamics and Stability

Ecological factors which can affect dynamic change in a population or species in a given ecology or environment are usually divided into two groups: abiotic and biotic.

Abiotic factors are geological, geographical, hydrological and climatological parameters. A biotope is an environmentally uniform region characterized by a particular set of abiotic ecological factors. Specific abiotic factors include:

- *Water, which is at the same time an essential element to life and a milieu*
- *Air, which provides oxygen, nitrogen, and carbon dioxide to living species and allows the dissemination of pollen and spores*
- *Soil, at the same time source of nutriment and physical support*
- *Soil pH, salinity, nitrogen and phosphorus content, ability to retain water, and density are all influential*
- *Temperature, which should not exceed certain extremes, even if tolerance to heat is significant for some species*
- *Light, which provides energy to the ecosystem through photosynthesis*
- *Natural disasters can also be considered abiotic*

Biocenose, or community, is a group of populations of plants, animals, micro-organisms. Each population is the result of procreations between individuals of same species and cohabitation in a given place and for a given time. When a population consists of an insufficient number of individuals, that population is threatened with extinction; the extinction of a species can approach when all biocenoses composed of individuals of the species are in decline. In small populations, consanguinity (inbreeding) can result in reduced genetic diversity that can further weaken the biocenose.

Biotic ecological factors also influence biocenose viability; these factors are considered as either intraspecific and interspecific relations.

Intraspecific relations are those which are established between individuals of the same species, forming a population. They are relations of co-operation or competition, with division of the territory, and sometimes organization in hierarchical societies.

Interspecific relations – interactions between different species – are numerous, and usually described according to their beneficial, detrimental or neutral effect (for example, mutualism (relation ++) or competition (relation –). The most significant relation is the relation of predation (to eat or to be eaten), which leads to the essential concepts in ecology of food chains (for example, the grass is consumed by the herbivore, itself consumed by a carnivore, itself consumed by a carnivore of larger size).

A high predator to prey ratio can have a negative influence on both the predator and prey biocenoses in that low availability of food and high death rate prior to sexual maturity can decrease (or prevent the increase of) populations of each, respectively.

Selective hunting of species by humans which leads to population decline is one example of a high predator to prey ratio in action. Other interspecific relations include parasitism, infectious disease and competition for limiting resources, which can occur when two species share the same ecological niche.

The existing interactions between the various living beings go along with a permanent mixing of mineral and organic substances, absorbed by organisms for their growth, their maintenance and their reproduction, to be finally rejected as waste. These permanent recyclings of the elements (in particular carbon, oxygen and nitrogen) as well as the water are called biogeochemical cycles. They guarantee a durable stability of the biosphere (at least when unchecked human influence and extreme weather or geological phenomena are left aside).

This self-regulation, supported by negative feedback controls, ensures the perenniality of the ecosystems. It is shown by the very stable concentrations of most elements of each compartment. This is referred to as homeostasis. The ecosystem also tends to evolve to a state of ideal balance, reached after a succession of events (for example a pond can become a peat bog).

Ecological crisis

Generally, an ecological crisis occurs when the environment of a species or a population evolves in a way unfavorable to that species survival.

It may be that the environment quality degrades compared to the species needs, after a change in an abiotic ecological factor (for example, an increase of temperature, less significant rainfalls). It may be that the environment becomes unfavorable for the survival of a species (or a population) due to an increased pressure of predation (for example overfishing). Lastly, it may be that the situation becomes unfavorable to the quality of life of the species (or the population) due to a rise in the number of individuals (overpopulation).

Ecological crises may be more or less brutal (occurring between a few months to a few million years). They can also be of natural or anthropic origin. They may relate to one unique species or on the contrary, to a high number of species.

An ecological crisis may be local (as an oil spill) or global (a rise in the sea level related to global warming).

According to its degree of endemism, a local crisis will have more or less significant consequences, from the death of many individuals to the total extinction of a species. Whatever its origin, disappearance of one or several species often will involve a rupture in the food chain, further impacting the survival of other species.

In the case of a global crisis, the consequences can be much more significant; some extinction events showed the disappearance of more than 90% of existing species at that time. However, it should be noted that the disappearance of certain species, such as the dinosaurs, by freeing an ecological niche, allowed the development and the diversification of the mammals. An ecological crisis thus paradoxically favored biodiversity.

Sometimes, an ecological crisis can be a specific and reversible phenomenon at the ecosystem scale. But more generally, the crises impact will last. Indeed, it rather is a connected series of events, that occur till a final point. From this stage, no return to the previous stable state is possible, and a new stable state will be set up gradually.

Lastly, if an ecological crisis can cause extinction, it can also more simply reduce the quality of life of the remaining individuals. Thus, even if the diversity of the human population is sometimes considered threatened, few people envision human disappearance at short span. However, epidemic diseases, famines, impact on health of reduction of air quality, food crises, reduction of living space, accumulation of toxic or non degradable wastes, threats on keystone species (great apes, panda, whales) are also factors influencing the well-being of people.

During the past decades, this increasing responsibility of humanity in some ecological crises has been clearly observed. Due to the increases in technology and a rapidly increasing population, humans have more influence on their own environment than any other ecosystem engineer.

Some usually quoted examples as ecological crises are:

- *Permian-Triassic extinction event 250 million of years ago*
- *Cretaceous-Tertiary extinction event 65 million years ago*

- *Global warming related to the greenhouse effect. Warming could involve flooding of the Asian deltas, multiplication of extreme weather phenomena and changes in the nature and quantity of the food resources.*
- *Ozone layer hole issue.*
- *Deforestation and desertification, with disappearance of many species.*
- *The nuclear meltdown at Chernobyl in 1986 caused the death of many people and animals from cancer, and caused mutations in a large number of animals and people. The area around the plant is now abandoned because of the large amount of radiation generated by the meltdown. Twenty years after the accident, the animals have returned.*

Human Ecology

Human ecology began in the 1920s, through the study of changes in vegetation succession in the city of Chicago. It became a distinct field of study in the 1970s. This marked the first recognition that humans, who had colonized all of the earth's continents, were a major ecological factor. Humans greatly modify the environment through the development of the habitat (in particular urban planning), by intensive exploitation activities such as logging and fishing, and as side effects of agriculture, mining, and industry.

Besides ecology and biology, this discipline involved many other natural and social sciences, such as anthropology and ethnology, economics, demography, architecture and urban planning, medicine and psychology, and many more. The development of human ecology led to the increasing role of ecological science in the design and management of cities.

Timelines

Timeline of Biophysics

1780–1794 **Luigi Galvani** *(1737–1798, Italy) discovered bioelectricity (causing muscular contraction in a frog's leg by application of static electricity). Pioneer of electrophysiology.*

1847–1894 **Herman von Helmholtz** *(1821–1894) First to measure the speed of nerve impulses. Studied human hearing and vision.*

1856 **Adolf Eugen Fick** *(1829–1901, Germany) Developed fundamental laws of diffusion in living organisms; discovered Fick's law of diffusion.*

Efforts to capture visions beyond the range of the human eye have long engaged scientists and engineers. By the mid-1880s **George Eastman** *had improved upon celluloid and at the turn of the 20th century used it with his new camera, the Brownie. That boxy little contraption is still remembered by many adults today, even as digital cameras record the world around us by harnessing electrons. The discovery of X-rays was only the first of many achievements leading to the development of imaging devices that today support all manner of endeavors in the medical sciences.*

1900 **George Eastman** *introduces the Kodak Brownie Camera.*

1895 **Wilhelm Conrad Röntgen** *discovered X-rays.*

1903 **Willem Einthoven** *founded electrocardiography (EKG)*

1913 **William David Coolidge** *invented the hot cathode x-ray tube, using a thermionic tube with a heated cathode electron emitter to replace the cold, or gas, tube. All modern x-ray tubes are of the thermionic type.*

1913 **Albert Solomon**, *a pathologist in Berlin, uses a conventional x-ray machine to produce images of 3,000 gross anatomic mastectomy specimens, observing black spots at the centers of breast carcinomas. Mammography, the resulting imaging, has been used since 1927 as a diagnostic tool in the early detection of breast cancer.*

1926 **Hermann J. Miller** *discovered that X-rays cause mutation in living cells.*

1928 **Georg von Békésy** *(1899–1972) expounded the physical mechanism within the Cochlea of the inner ear.*

1932 **Max Knoll** *and* **Ernst A.F. Ruska** *developed the transmission electron microscope. It can magnify objects one million times.*

1945 **Bernard Katz** *discovered how synapses work.*

1945 **Alan Lloyed Hodgkin** *(1914–1998) and* **Andrew Fielding Huxley** *(1917-) described the ionic mechanism by which neurons transmit electron pulses.*

1952 **Rosalind Franklin** *and* **Maurice Wilkins** *pioneered DNA crystallography.*

1953 *First application of positron emission tomography for medical diagnosis of brain tumor by* **Gordon Brownell** *and* **William Sweat**.

1953 **Max Perutz** *(1914–2002) and* **John Kendrew** *(1917–1997) pioneered protein crystallography (Hemoglobin).*

1954 **David Kuhl** *introduced radionuclide emission tomography.*

1950s **Russel Morgan**, **Edward Chemberlain** *and* **John W. Coltman**, *perfect a method of screen intensification that reduces radiation exposure and improves fluoroscopic vision. Their image intensifier is used in medical fluoroscopy.*

1959 **Ian Donald** *developed practical technology and applications of ultrasound as a diagnostic tool in obstetrics and gynecology. Ultrasound displays images on a screen of tissues or organs formed by echoes of inaudible sound waves at high frequencies (20 KHz or more) beamed into the body.*

1960 **Powell Richards** *and* **Walter Tucker** *invented a short halflife radionuclide generator that produces technetium-99m for use in diagnostic imaging procedures in nuclear medicine – a branch of medicine that uses radioisotopes for research, diagnosis, and treatment of disease. [Technetium-99m was discovered in 1939 by* **Emilio Segré** *and* **Glenn Seaborg***]*

1964 *Single-photon emission computerized tomography (SPECT) methods become capable of yielding accurate information similar to PET.*

1972 **Godfrey Hounsfield** and **Alan Cormack** invented *Computer Assisted Tomography (CT scanners).*

1973 **Paul Lauterbur** adapted *Magnetic Resonance Imaging (MIR) for medical purposes, using high speed computers.*

1974 **J.E. Greenleaf** and **S.A. Johnson** developed *ultrasound computer tomography.*

1974 **Mikhail Volkenshtein** *(1912–1992)* developed *Quantum Biophysics: a quantum-mechanical model of enzyme catalysis, supporting a theory that enzyme catalysis depends on quantum-mechanical effects such as tunneling.*

1977 **Peter Mansfield** developed the *Echo-planar imaging (EPI) technique to produce movie of a single cardiac cycle.*

1978 *Emission Computer tomography (ECT) by means of radionuclide imaging.*

1981 **Gerd Binnig** and **Heinrich Rohrer** designed and built the first *scanning tunneling microscope (STM) with a small tungsten probe tip, about one or two atoms wide.*

1986 **Gerd Binnig**, **Cal Quate** and **Cristoph Gerber** introduced the *atomic force microscope (AFM) which is used in microbiology and cellular biology.*

1993 *EPI is used (with functional MRI) in mapping regions of the brain responsible for thought and motor control.*

1986 **Benoit Roux** *investigated dynamics and function of biological macromolecular systems (receptors, protein kinases); functioning of biological systems at the molecular level.*

Carlos Bustamante *(1951–) used novel methods of single-molecule visualization, such as scanning force microscopy to study structure & function of nucleoprotein assemblies. Used methods of single-molecule manipulations (such as optical tweezers) to characterize the elasticity of DNA, to induce mechanical unfolding of individual protein* *molecules and to investigate the machine-like behavior of molecular motors.*

Steven Chu *(1948–) helped develop techniques for cooling and trapping atoms using laser light.*

Steven Block *(1952–)* and **Arthur Ashkin** *pioneered the use of optical tweezers to study the motion of enzymes (kinesin and RNA polymerase) at a single-molecule level.*

Howard Berg *characterized properties of bacterial chemotaxis.*

Timeline of Genetics

Ca 1300 BCE *The Bible (**Leviticus** 18, 6–17 forbids one to mate with one's relatives.)*

470–322 BCE *The Greek philosophers **Hippocratos**, **Aristotle** and **Plato** wrote about the inheritance of human traits. They observed that certain traits are passed from parent to child. Although they did not understand the exact contribution of the male and female parent to the offspring, they believed that semen is in some way responsible for passing on traits.*

1814 CE ***Joseph Adams** published "A Treatise on the Supposed Hereditary Properties of Diseases", stating therein that there is some intrafamilial correlation of diseases.*

1839 ***Matthias Schleiden** and **Theodore Schwann** suggested that cells with nuclei are the fundamental units of life.*

1855 ***Rudolph Virchow** hypothesized that new cells can only be formed by the division of existing cells.*

1859 ***Charles Darwin** published "On the Origin of Species", proposing evolution by natural selection. His key premise was that evolution occurs through the selection of inherent and transmissible (rather than acquired) characteristics between individual members of a species. Darwin did not specify the means by which characteristics are inherited and the mechanism of heredity had not been determined at that time.*

1865 ***Gregor Mendel** discovered the fundamental* laws of inheritance. *He deduced that genes come in pairs and are inherited as distinct units, one from each parent. Mendel tracked the segregation of parent genes and their appearance in the offspring as dominant or recessive traits. He recognized the mathematical patterns of inheritance from one generation to the next.*

Mendel's results were not appreciated until 1900, when they were rediscovered. Today he is widely considered founding father of modern genetics. Moreover, his work eventually helped to partly explain Darwin's concept of evolution.

1869 **Johann F. Miescher** *isolated DNA (first called "nuclein") as an acidic substance found in cell nuclei of white blood cells (puss). We now know that he had discovered the material basis of heredity, but it took another 80 years before nuclein was shown to be DNA (it become known as nucleic acid after 1874, when Miescher separated it into a protein and an acid molecule. It was suspected of exerting some function in the heredity process.)*

1819–1882 **Walther Flemming** *used new staining techniques to see tiny threads within the nucleus of cells in salamander larvae that appear to be dividing. In so doing he discovered chromosomes.*

1889 **August Weissman** *theorized that the material basis of heredity is located on the chromosomes.*

1900 • *The science of genetics was finally born when* **G. Mendel***'s work was rediscovered by* **Hugo de Vries**, **Erich von Tshchermak** *and* **Carl Correns**.

• *Major outbreaks of disease in overloaded industrial cities led to the introduction of large-scale sewage purification system based on microbial activity (first time in Manchester, England, 1911).*

• *It was first shown that key industrial chemicals (glycerol, and butanol) could be synthesized using bacteria.*

1901–1912 *Scientists suggested that food contains ingredients essential to life that are not proteins or carbohydrates:* **E. Wildiers** *(1901) discovered a new substance – a growth-factor indispensable for the development of yeast;* **Frederick Hopkins** *and* **C.A. Pekelharing** *(1906) discovered such substances in rice and citrus. In succeeding years they came to be called vitamins (***Casimir Funk***, 1912).*

1902 *The term 'immunology' first appeared.*

1902–1912 *Scientists identified chromosomes as carriers of heredity.* **Walter Sutton** *(1902) stated that chromosomes are paired and that genes are carried by chromosomes. He argued that each egg or sperm contains only one of each chromosomes which accounts for the random factor in heredity and accords with Mendel's theory. He thus pointed out the interrelationships between cytology and Mendelism, closing the gap between cell morphology end heredity.*

Theodor Boveri reached the same result independently (1903). Edmund Wilson and Nettie Stevens independently described the behavior of sex chromosomes: XX determines female; XY determines male.

1903–1909 *First experiments on quantitative traits in broad beans by* **Wilhelm Johanssen** *and in wheat by* **Herman Nilsson-Ehle**.

1905 *The word "genetic" as coined by* **William Bateson.** *proposed the idea that separate X and Y chromosome determine sex.*

1905–1908 **William Bateson** *and* **Reginald Punnett** *demonstrated that action of some genes modify action of other genes: The first time gene regulation was demonstrated.*

1908 **Hardy – Weinberg** *law was derived.*

1908–1909 **Archibald Garrod** *proposed (1908) that some human diseases are due to inborn errors of metabolism that result from lack of specific enzyme*[40]. *He is now considered the founder of biochemical genetics. In 1909 he proposed that genes dictate phenotypes through enzymes that catalyze specific processes in the cell.*

1909 **Wilhelm L. Johannsen** *coined the words gene, genotype, phenotype.*

1909–1929 **A. Levene** *discovered that the sugar ribose is found in some nucleic acids, those we now call RNA (1909). He discovered (1929) deoxyribose in nucleic acids that do not contain ribose, those known today as DNA.*

1910 **Thomas H. Morgan** *(1907–1911) proved that chromosomes have a definite function in heredity, established mutation theory, and led to a fundamental understanding of the mechanisms of heredity. He explained the separation of certain inherited characteristics that are usually linked*

[40] In 1929, **Richard Schönheimer** studied a patient with hepatomegaly due to massive glycogen storage and suggested that this disorder may by due to an enzyme deficiency. It was not until 1952 that **Cori and Cori** found glucose-6-phosphatase to be deficient in "von Gierke disease"; (glycogen storage disease type I). This observation marks the first time that an inborn error of metabolism was attributed to a specific enzyme deficiency.

as caused by breaking of chromosomes during the process of cell division, and began to map the positions of genes on chromosomes of the fruit fly. Morgan proved that the genes responsible for the appearance of a specific phenotype were located on chromosomes and that some genetically–determined traits are sex-linked. He also found that genes on the same chromosome do not always assort independently.

Morgan suggested that the strength of linkage between genes depended on the distance between them on the chromosome. The nearer two genes lie on a chromosome, the greater the chance of being inherited together. Likewise the farther away they are from each other, the more is the chance of being separated by the process of crossing-over. The genes are separated when a crossover takes place in the distance between the two genes during cell division.

One of his students, **Calvin Bridges**, *in 1913, established that genes are located on chromosomes. In the same year, another student of Morgan's,* **Alfred Sturtevant**, *determined that genes are arranged on the chromosomes in a linear fashion, much like beads on a necklace. Moreover, Sturtevant demonstrated that the gene for any specific trait is in a fixed location or locus. Yet another Morgan student,* **Herman J. Muller**, *in 1926 discovered methods for artificially producing mutants in fruit flies by ionizing radiation and other mutagens. In so doing, he discovered the origin of new genes by mutations, a theory first proposed by* **Hugo de Vries** *in the early 1900s.*

1912 **Lawrence Bragg** *discovered that X-rays can be used to study the molecular structure of simple crystalline substances. This discovery led to the development of X-ray crystallography, which made it possible to further explore the 3-dimensional structure of acids and proteins.*

While two World Wars killed millions and pushed medicine to new limits, various fields of science began to converge to explore the mechanism of reproduction – the nature and structure of the heredity-carrying materials (eventually found to be DNA molecules).

1915 **Frederick W. Twort** *discovered phages, viruses that prey on bacteria.*

1918 **R.A. Fisher** *publishes "The Correlation Between Relatives on the Supposition of Mendelian Inheritance" the modern synthesis of genetics and evolutionary biology starts. (See population genetics.)*

1920's *Nucleic acid found to be a major component of the chromosomes, but it was not considered good candidate for a carrier of genetic information.*

1927 *Physical changes in genes are called mutations.*

1928 **Frederick Griffith** *discovers that hereditary material from dead bacteria can be incorporated into live bacteria. (See Griffiths experiment.)*

1930 **W.C. Rose** *discovered essential amino acids.*

1930's *Chemical nature of nucleic acids was investigated. The ubiquitous presence of nucleic acid in the chromosome was generally explained in purely physiological or structural terms.*

1931 *Crossing over is identified as the cause of recombination.*

1933 **Arne Tiselius** *introduced electrophoresis, a new technique for separating proteins in solution.*

1935–1939 **Andrei Nicolaevitch Belozersky** *isolated DNA in its pure state for the first time (1936). Showed (1939) that both DNA and RNA are always present in bacteria.*

1937 **Frederick C. Bawden** *discovered that tobacco mosaic virus contains RNA.*

1938
- *The term 'molecular biology' was coined.*
- *Proteins and DNA were studied in various labs with the aid of X-ray crystallography.*

early 1940s *Nucleic acid still viewed as a uniform polymer unaffected by its biological source. Hereditary information commonly thought to reside in the chromosomal proteins.*

1941 **George Beadle** and **Edward Tatum** *show that genes code for proteins: they performed experiment that suggest that one gene codes for one enzyme.*

1941 *The term 'genetic engineering' was first used.*

1942 *The electron microscope was used to identify a bacteriophage*

Barbara McClintock discovered *transposable genetic elements: c.e. genes can move on chromosome and jump from one chromosome to another. Her findings were greeted with initial skepticism until the 1970's when molecular biologists confirmed the existence of an enzyme that enables jumping genes to hop around on the DNA. At age 81 (1983) she was awarded the NP. Scientists now think such transposons may be linked to some genetic disorders such a hemophilia, leukemia and breast cancer. They also think that transposons have played a crucial role in evolution.*

1944–1959 **Oswald T. Avery**, **Colin MacLeod** and **Maclyn McCarty** *continued the work of* **Frederick Griffith** *and demonstrated that DNA is the material of genes, i.e. the molecule of genetic information. Most people were skeptical of these findings until 1952.*

1946 **Joshua Lederberg** and **Edward Tatum** *showed that material can be transferred laterally between bacterial cells, c.e: bacteria can exchange genetic material directly through conjugation.*

1950 **Erwin Chargaff** *discovered a one-to-one ratio of adenine to thymine and guanine to cytosine in DNA samples from a variety of organisms. DNA rather then protein carry genetic information.*

1951 *The first animal, the tadpole was cloned by nuclear transfer (cloning in plants was known for millennia).*

1952 **Alfred Hershey** and **Martha Chase** *show in bacteriophage labeling experiments that DNA is the molecule of heredity.*

1952 **Rosalind Franklin** and **Maurice Wilkins** *perform X-ray crystallography studies of DNA, providing sharp diffraction photographs that led to the elucidation of the structure of DNA. These diffraction patterns of the DNA molecule revealed the helical structure and the location of the phosphate sugar on the DNA molecule.*

• **Jean Brachet** *suggested that RNA, a nucleic acid, plays a part in the synthesis of proteins.*

1953 **Francis Crick** *and* **James D. Watson** *proposed, on the basis of* **Franklin***'s data, the double-stranded, helical, complementary, anti-parallel model for DNA.*

They determined that deoxyribonucleic acid (DNA) is a double strand helix of nucleotides. Each nucleotide consists of deoxyribose sugar molecule to which is attached a phosphate group and one of four nitrogenous bases: two purines (adenine and guanine) and two pyrimidines (cytosine and thymine). The nucleotides are joined together by covalent bonds between the phosphate of one nucleotide and the sugar of the next, forming a phosphate-sugar backbone from which the nitrogenous bases protrude. The two strands are linked by selective hydrogen bonds: the purin adenine bonds only with the pyrimidine thymine, and the purine cytosine only with the pyrimidine guanine.

DNA replication is possible through the complementary nature of the two strands. The chemical complexity of the molecule is thought to be sufficient to store the requisite information.

The precise manner in which the information in the DNA is activated to build an organism is still very poorly understood; what is firmly demonstrated is that so-called structural genes manufacture the proteins for living tissues.

1955–1959 *Biologists work out the mechanism by which DNA functions to make protein. They hypothesized that the DNA sequence specifies the amino acid sequence in a protein and that genetic information flows only in one direction, from DNA to messenger RNA to protein. The replication mechanism of DNA is demonstrated [***Francis Crick** *and* **George Gamow***, 1957;* **Arthur Kornberg***, 1958].*

1956 **Jo Hin Tjoi** *and* **Albert Levan** *established the correct chromosome number in humans to be 46.*

1957 **Crick** *and* **Gamow** *worked out the "Central Dogma" to explain protein synthesis from DNA; the DNA sequence codes for amino acid sequences and genetic information flows in one direction — from DNA to mRNA to protein.*

1958 *The* **Meselson – Stahl** *experiment demonstrates that DNA is semiconservatively replicated.*

1961–1967 ***Marshall Nirenberg***, ***Har Gobind Khorana***, ***Heinrich Mathaei*** *and* ***Severo Ochoa*** *cracked the Genetic Code. They demonstrated that a sequence of 3 nucleotide bases, a codon, determines each of the 20 amino acids. This means that there are 64 combinations possible for 20 amino acids. Nirenberg and* ***Philip Leder*** *found that there are extra redundant codons that serve as stop signs for RNA synthesizing protein.*

1965 ***Leonard Hayflick*** *(1928–) observed that cells dividing in cell culture divided about 50 times before dying. The human limit is around 52.*[41]

1961 ***Francois Jacob*** *(1920– , France) and* ***Jacques Monod*** *(1910–1976, France) advanced the OPERON MECHANISM theory for controlling enzyme activities in the cell.*

---

[41] This is known as the *Hayflick limit*. It has been linked to the shortening of *telomeres*, a region of DNA at the end of chromosomes.
The only known way of circumventing the Hayflick limit is with the enzyme telomerase, which regenerates telomeres during DNA replication.
Stem cells, by definition, have not yet been fully differentiated, and therefore many of these cells may continue to regenerate new cells for the entire lifespan of the organism, without limit, thus constituting a notable exception to the Hayflick limit in humans and other organisms. While the manifestations of the constant regenerative effects of stem cells is most easily seen in tissues which must constantly produce replacements for existing cells, such as skin and blood cells, stem cells of one form or another are found in every tissue of the human body, even if only as dormant stem cells known as "spore-like cells".
Cancer cells constitute the other main exception to the limits on cell division. It is believed that the Hayflick limit exists principally to help prevent cancer. If a cell becomes cancerous and the Hayflick limit is approaching, the cell will only be able to divide a certain number of times. Once it reaches this limiting number of divisions, the formed tumor will no longer be able to reproduce and the cells will die off. Cancers become problems after having reactivated telomerase-encoding genes. Cells that have found a way around the limit are referred to as "immortal". Such immortal cells may still die, but the group of immortalized cells produced from cell division of an immortal cell has no limit as to how many times cell division might take place among the cells that constitute such a group of immortalized cells.
It is believed by some that some or all cancers start off as stem cells that become genetically damaged over their long lives. This would mean they already aren't limited by the Hayflick limit and can easily metastasize into the pool of cells in their final cell type destination.

The experimental system used by them was the common bacterium **E. Coli**. *They discovered in these cells a class of genes that control the activity of other genes. If for some reason, the controlling genes function improperly, the other genes may get out of control and damage the cell. This basic regulatory concept is fundamental to cellular regulation for all organisms.*

Lac operon is a DNA sequence that governs the production of enzymes for metabolizing lactose (milk sugar) in bacteria such as E. Coli. The key idea is that E. coli does not bother to waste energy making such enzymes if there is no need to metabolize lactose, such as when other sugars like glucose are available.

1967 • **Mary Weiss** *and* **Howard Green** *found a technique for combining human cells with mouse cells in one culture (somatic cell hybridization).*

• **W.M. Fitch** *and* **E. Maroliash** *set up the first evolutionary trees from protein sequences.*

1968 **Gerold Edelman** *(1929–) and* **Rudney Porter** *isolated the first DNA ligase.*

1970's *Important discoveries were made that led to modern techniques for studying genetics:*

1970 **Howard Temin** *and* **David Baltimore** *discovered how viruses affect the genes of cancer cells: Their work described how viral RNA that infects a host bacterium uses an enzyme to integrate its message to the host's DNA. This discovery allow scientists to create clones*

1970 **Hamilton Smith**, **Daniel Nathans** *and* **Kent Wilcox** *isolated first restriction enzyme that could cut DNA molecules within specific recognition sites. The restriction enzymes were discovered in studies of a bacterium and were used to cut-up foreign DNA from invading organisms such as viruses (NP 1978).*

1972 **Paul Berg** *isolated and employed a restriction enzyme to cut DNA. He also used ligase to paste two DNA strands* *together to form a hybrid circular molecule. This was the first recombinant DNA molecule*[42]*. (NP 1981)*

1973 *Scientists for the first time successfully transferred DNA molecules from one life form into another:* **Stanley H. Cohen**, **Annie Chang** *and* **Herbert Boyer** *used a restriction enzyme to cut sections of viral DNA and bacterial DNA, "spliced" them together and inserted this recombinant molecule into the DNA of the bacterium Eschericia Coli, thereby introducing the first recombinant-DNA organism; this is the beginning of genetic engineering.*

1973–1977 **Frederick Sanger** *invented a DNA sequencing technique.*

1975 **Cesar Milstein** *discovered how to fuse cells together to produce monoclonal antibodies.*

1976 • **Michael J. Bishop** *and* **Harold Varmus** *showed that oncogenes appear on animal chromosomes, and alterations in their structure can result in cancerous growth.*

• **Martin F. Gellert** *discovered the enzyme gyraze that caused DNA to form supercoils (larger helix).*

1977 • *Advent of the Age of Biotechnology: The production of the first human protein manufactured by a bacteria (human growth hormone – releasing inhibitory factor). For the first time, a synthetic recombinant gene was used to clone a protein.*

• *The first genetic engineering company (Genentech) is founded, using recombinant DNA methods to make medically important drugs.*

• **Walter Gilbert** *and* **Allan Maxam** *devised a procedure for rapidly sequencing long sequences of DNA.*

• *Bacteriophage FX-174 was the first complete genome (DNA) to be sequenced.*

---

[42] Beng realized the risks of his experiment and temporarily terminated it before the recombinant DNA molecule was added to *E.coli*, where it would have been quickly reproduced. He proposed a one-year moratorium on recombinant DNA studies while safety issues were addressed.

1978 • *Genetic engineering techniques used to produce rat insulin.*

• *Scientists show it is possible to introduce specific mutations at specific sites in a DNA molecule.*

• *Scientists successfully transplanted mammalian gene.*

•**Yuel Wai Kan**, **Andree-Marie Dozy** and **David Botstein** *discovered restriction-fragment-length polymorphisms.*

Individual humans differ one basepair in every 500 nucleotides or so. The most interesting variations for geneticists are those that are recognized by certain enzymes, called restriction enzymes. These enzymes, each of which cut DNA only when they see a specific sequence, for instance GAATTC in case of the restriction enzyme EcoR1. This sequence is called a restriction site. The enzyme will bypass the region if it has mutated to GACTTC. Thus, when a specific restriction enzyme cuts the DNA of different people, it may produce fragments of different lengths.

These DNA fragments can be separated according to size by making them move through a porous gel in an electric field. Since the smaller fragments move more rapidly than the larger ones, their sizes can be determined by examining their positions in the gel. Variations in their lengths are called restriction-fragment-length polymorphisms, or RFLPs.

1980 *Swiss researchers introduced a gene for human interferon into bacteria and then cloned millions of cells to produce an inexpensive and abundant supply of this previously rare protein. This was the first big success story in the commercial production of drugs by genetic engineering.*

1980–1986 • **Kary B. Mullis** *invented the polymerase chain reaction (PCR), a method for rapidly and easily cloning of DNA fragments: this method for multiplying DNA sequences in vitro uses heat and enzymes to make unlimited copies of genes and gene fragments. It later becomes a major tool in biotech research and product development worldwide.*

The purpose of PCR is to make a high number of copies of a specific DNA fragment, a gene for instance. This method amplifies fragments of DNA million of times to make sufficient quantities available for DNA sequence analysis.

1980 • *Researchers introduced a human gene (one that codes for the protein interferon) into a bacterium.*

1981 • *Three independent research teams announced the discovery of human oncogenes (cancer genes).*

• *Human mitochondrial DNA sequenced.*

• *Scientists produced the first transgenic animals by transferring genes from other mammals to mice.*

1982 • *Human insulin drug produced by genetically-engineered bacteria (using recombinant DNA methods) for the treatment of diabetes.*

• *A human cancer gene (isolated from bladder cancer cells) is cloned in Eschericia coli. The base sequence of the cancer gene is found to differ from the same locus in a normal cell by a single base pair, which causes a substitution of an amino acid in the resulting protein.*

1983 • *The first artificial chromosome was created.*

• *Discovery of the homeobox genes in the fruit fly.*

• *First genetic modified plant is created; a tobacco plant resistant to an antibiotic.*

1980–1993 *The work of* **Stuart Orkin** *in 1986,* **Lou Kunkel** *in 1987,* **Jim Gusella** *and* **Nancy Wexler** *in 1982 and* **Mary-Claire King** *in 1991, led to the birth of modern clinical genetics.*

1984 **Alec Jeffreys** *introduced a technique for DNA fingerprinting to identify individuals: it is based on identification of certain core sequences of DNA unique to each person. It is to be used for establishing family relationships.*

• *Scientists cloned and sequenced the entire genome of the HIV virus.*

• *The first genetically engineered vaccine was developed.*

• *successful cloning of sheep producing genetically identical animals. It is done by separating an embryo into separate cells and introducing a cell's nucleus into sheep ova that have had their nuclei removed: the altered eggs (ova) are then implanted in female sheep for development into fetuses and consequent birth.*

• British scientists mixed goat and sheep embryo cells and implanted them into a surrogate animal. This led to the birth of the first chimera, a cross between a goat and a sheep.

• ***McGinnis*** *discovered homeotic (Hox) regulatory genes, responsible for the basic body plan of most animals. In subsequent work, his team demonstrates that a single mutation in a Hox gene suffices to suppress all limb development in the thoracic region of fruit flies.*

1985 *• Genetically engineered plants resistant to insects, viruses, and bacteria were field-tested for the first time.*

• ***Lap-Chee Tsui*** *working in Massachusetts mapped the gene for cystic fibrosis to the long arm of chromosome 7.*

1986 *• Applied Biosystems introduced the first automated DNA fluorescence sequencer.*

• The Environmental Protection Agency (USA) approves the release of the first genetically engineered crop: a herbicide resistant tobacco plants.

1986 *• The first genetically engineered human vaccine – recombinant HB – was approved as a treatment for prevention of hepatatis B.*

• ***Peter Schultz*** *described how to combine antibodies and enzymes (creating "abzymes") to create pharmaceuticals.*

• Scientists and technicians at CalTech invented an automated DNA fluorescence sequencer .

1988 *• A US National Center for Biotechnology Information (NCBI) founded at NIH/NLM.*

U.S. Congress funded the Human Genome Project, a massive effort to map and sequence the human genetic code as well as the genomes of other species. All human DNA was to be mapped and sequenced by 2005 CE. The program was launched in 1990 with an estimated cost of $13 billion.

1989 *Scanning tunneling electron microscope is used to obtain, for the first time, a direct images of pure DNA. Eventually, this method may be used to observe active viruses or the action of molecules on the cell surface.*

1989 **Francis Colins** *and* **Lap-Chee Tsui** *identified the gene coding for the cystic fibrosis transmembrane conductance regulator protein (CFTR) on chromosome 7 that, when mutant, causes cystic fibrosis.*

1990 *The U.S. Human Genome Project launched at estimated cost of 13 billion dollars. The project started as a 15-year effort co-ordinated by the U.S. Department of Energy and the National Institutes of Health. The project goals were to:*

1) identify all the genes in human DNA,
2) determine the sequences of the 3 billion chemical base pairs that make up human DNA, store this information in databases,
3) improve tools for data analysis,
4) transfer related technologies to the private sector, and
5) address the ethical, legal, and social issues (ELSI) that may arise from the project.

To help achieve these goals, researchers also are studying the genetic makeup of several nonhuman organisms. These include the common human gut bacterium Escherichia coli, the fruit fly,the nematode Caenerhabditis elegans, the rat and the mouse.

1990's *DNA fingerprinting, gene therapy and genetically modified foods come onto the scene.*

1990 *The first transgenic dairy cow was created. It was used to produce human milk proteins for infant formula.*

1992 *The Institute for Genome Research (TIGR), associated with plans to exploit sequencing commercially through gene identification and drug discovery, was formed.* **Mel Simon** *introduced the use of BACs for cloning.*

1993 *The Sanger Centre, a genome research institute with the purpose to further the knowledge of genomes, was crated in Hinxton, UK*
The EMBL European Bioinformatics Institute, the center for research and service for bioinformatics was established in Hinxton, UK

1993 *FlavrSavr tomatoes, genetically engineered for longer shelf life, were marketed.*

1993 *The first gene therapy took place on a 4-year-old girl with an immune-system disorder called ADA deficiency. The therapy appeared to work.*

1993 *Scientists in Paris produced a rough map of all 23 pairs of human chromosomes.*

1993 **Dean Hamer** *and colleagues reported at least one gene related to homosexual orientation that appears to reside on the X chromosome and is inherited from the mother.*

1994 *The FDA approved the commercial use of bovine somatropin, also known as bovine growth hormone. This hormone increases the production of milk in cows, and became one of the first genetically engineered products available to farmers.*

1994 • *A multitude of genes, human and otherswise, were identified and their function determined [e.g a gene predisposing to obesity, a breast cancer susceptibility gene, a gene associated with* apoptosis *(programmed cell death)].*

• *Linkage studies identify genes for a variety of ailments including: melanoma, hearing loss, dyslexia, thyroid cancer, sudden infant death syndrome.*

1995 • *Further evidence is found to support the idea that RNA was the central molecule in the origin of life.*

• *Gene therapy, immune system modulation and genetically engineered antibodies enter the clinic in the war against cancer.*

1995 *Duke University researchers announced the transplant of hearts from genetically altered pigs into baboons.*

1995–2000 *First completed sequences of the following genomes:*

- *Bacteria* Haemophilus influenza *(1995)*
- *Yeast and* E.Coli *(1996)*
- *Soil Nematode* Caenerhabditis *(1998)*
- *Fruit fly* Drosophila melanogaster *(1999)*
- Arabidopsis *(2000)*

1996 • *Sottish scientists clone identical lambs from early embryonic sheep.*

• *Scientists sequenced the complete genome of baker's yeast – more than 12 million base pairs of DNA.*

• *The discovery of a gene associated with Parkinson's disease.*

• *Sequencing the genome of ancient microorganisms, archaea, found in inhospitable climates deep in thermal vents under the sea – a step in understanding of the evolution of life on earth.*

1996 *Genzyme Transgenics announced the birth of a goat carrying BR-96 monoclonal antibodies to be used experimentally to deliver conjugated anticancer drugs to humans.*

1997 • *Artificial human* chromosomes *created for the first time.*

• *Clock, the first gene providing the* circadian rhythm *of mammalian life, identified.*

• *Using a bit of DNA and some commonplace laboratory techniques, researchers engineered the first DNA computer "hardware" ever: logic made of DNA*[43].

• *A new DNA technique combined PCR, DNA chips, and computer programming, thus providing a new tool for the search for disease - causing genes.*

1997 **Hunt Willard** *and others working at Case Western Reserve created the first artificial human chromosome, opening the door to designer babies.*

1997 **Stanley Prusiner** *earned the Nobel Prize for his pioneering work on prion diseases such as Bovine Spongiform Encephalopathy, which is thought to have caused the outbreak of mad cow disease in Britain and Creutzfeldt-Jakob disease.*

1997 *A sheep named Dolly was cloned.*

1998 *Dolly gave birth to Bonnie, a lamb conceived by conventional means. The birth offered reassurance that cloned animals like Dolly can develop into healthy animals capable of reproducing.*

---

[43] It is not a *general purpose* computer, thought: it is tailored to solve just a single mathematical problem (the *traveling salesman* problem).

1998 *A rough draft of the human genome map was produced, showing the locations of more than 30,00 genes.*

2000 *The Human Genome Project presents its preliminary results: each of the body's 100 trillion cells contains some 3.1 billion nucleotide units. Only 1% of these are thought to be transcriptional, clustered in possibly as few as 30,000 genes. An accurate chemical map of the genome tells us surprisingly little about how it functions. Targeted experimentation is now possible.*

2002 *Presentation of human genome by Celera Genomics and the collaborating group of laboratories supported by public foundation.*

2003 *(14 April) Successful completion of Human Genome Project with 99% of the genome sequenced to a 99.99% accuracy.*

Timeline History of Structural Biology

1910 **Albrecht Kossel** *(1853–1927) discovered the amino acids Histidine (1896), thymic acid and agmatine (1910) in the cell.*

1925 **Theodor Svedberg** *(1884–1971) developed the ultracentrifuge that could spin its samples with a force of over 100,000 g.*

1932 *Invention of the electron microscope. All electron microscopes suffer from a serious drawback: since no specimen can survive under their high vacuum, they cannot show the ever-changing movements that characterize a living cell.*

1949–1956 **Christian de Duve** *(1917–),* **Albert Claude** *(1899–1983) and* **George E. Palode** *(1912–) discovered and elucidated the subcellular biochemical structure and function of organelles in biological cells.*

1953 **James D. Watson** *(1928–), and* **Francis Crick** *(1916–2004) discovered the structure of the DNA molecule.*

1955 *Microscopists and biochemists began to communicate: the microscopists discovered that the biochemists' particles matched what they had seen under their microscopes, and the biochemists learned that the microscopists could actually see what they had been analyzing. The result was an avalanche of discoveries about the world within the cells.*

1959 **Max Perutz** *(1914–2002) and* **John Kendrew** *(1917–1997) determined the molecular structure of the protein hemoglobin.*

1961 **Christian Anfinsen** *(1916–1995) first discovered the folding of protein molecules. (These molecules begin their biological activity (life) only after having folded into intricate, convoluted shapes.) Once the protein has folded, its shape determines its physical and chemical properties and particularly its active sites.*

1964 ***Cyrus Levinthal*** *(1922–1990) and* ***Robert Langridge*** *(1933–) developed the first system that showed biological molecules on a computer screen.*

1965 ***Alex Novikoff*** *(1913–1987), first to show that lysosomes (their enzymes can digest substances) exist in eukaryotic cells.*

1972 ***Michael S. Brown*** *(1941–) and* ***Joseph Goldstein*** *(1940–) elucidated the genetic defects and role of the LDL (Low Density Lipoprotein) receptor in Familial Hypercholesterolemia. These studied led to development of a drug to treat FH.*

1982 ***Thomas Coch*** *(1947–) and* ***Sydney Altman*** *(1939–) discovered that the nucleic acid RNA can act as enzyme because it can cut and splice itself.*

1985 ***Michael Rossmann*** *(1930–) mapped the structure of human common-cold virus – the first animal virus to be seen at atomic resolution.*

1989 *Researches reported using scanning electron microscopy to obtain, for the first time, direct images of pure DNA. This method is used to observe living viruses, or the action of molecules on the cell's surface, and enables scientists to see the tiniest details of cell structure and activity in ways undreamed of a few years ago.*

1995 ***J.M. Hogle***, ***David Filman*** *and* ***Marie Chow*** *reported the structure of the polio virus at atomic resolution.*

Timeline History of Immunology

1796 *Smallpox vaccination;* **Edward Jenner** *(1749–1823) discovered that cowpox induces protection against smallpox. Jenner had no knowledge of microorganisms or of immunology. This had to wait for Pasteur some 100 years later.*

1857 **Louis Pasteur** *(1822–1895) noted the existence of substances capable of exerting antimicrobial effects.*

Phase I: 1879–1910

Most of the major components of the immune response were described.

1879 **Louis Pasteur** *developed attenuated vaccines of chicken cholera, anthrax and rabies. He showed that inoculation with preparations of weakened pathogens could be used to develop immunity against infectious forms of the pathogen. (However, not until some 80 years later, did scientists gain a modern understanding of vaccines!)*

1883 **Eli Metchnikov** *(1845–1916), pioneer of immunology, established a cellular theory of vaccination, (NP 1908).*

1888 **Emil Roux** *(1853–1933) and* **Alexandre Yersin** *(1863–1943) proved that diphtheria bacilli produced a toxin which can be separated from the bacterial cells (antitoxin is the specific antibody capable of neutralizing the pathogenic toxin).*

1890 **Emil von Behring** *(1854–1917), pioneer of immunology, discovered that serum could be used to treat diphtheria (NP 1901); explained that diphtheria immunity depended on the capacity of the cell-free blood serum to neutralize the toxic substance produced by the diphtheria bacilli.*

1897 **Paul R. Ehrlich** *(1854–1915) laid the chemical theory accounting for the molecular basis of antibody–antigen reactions. Father of modern immunology, (NP 1908). Conceptualized the interactions between cells, antibodies and antigens as essentially chemical responses. Shared the 1908 NP with Metchnikov for their contributions to immunity and serum therapy.*

1901 **Karl Landsteiner** *(1868–1943) used the specific antibody–antigen reaction to identify major human blood groups A, B, AB, and O.*

Phase II: 1910–1938

Knowledge was increasingly applied in preparation of immune sera and diagnostic reagents for infection diseases: First vaccine for Diphtheria (1923), Pertussis (1926), Tetanus (1927), Yellow Fever (1935), Polio (1652, 1962); Measles (1964); Mumps (1967); Rubella (1970); Hepatitis B (1981).

1928 **Alexander Fleming** *(1881–1955) discovered Penicillin.*

1932 **Gerhard Domagk** *(1895–1964) found the sulfonamide prontosil to be effective against streptococcus. This was the first drug effective against bacterial infections. Sulfonamides became a revolutionary weapon at the time, but were later replaced by penicillin, which showed both better effects, and fewer side effects. (Kidney stones and changes in bone marrow). Domagk's work on sulfonamides eventually led to the development of the antituberculosis drugs.*

Phase III: 1938–1988*:*

Cellular and molecular aspects of the immune-system.

Immunology has benefited from the techniques and theories of molecular biology. The lymphocytes of the blood have become an important focus of study and clonal selection theory had dominant research.

1938 ***John Marrack*** *expounded the antigen-antibody binding hypothesis.*[44]

1949–1957 ***Peter Medawar*** *(1915–1987) and* ***Frank M. Burnet*** *(1899–1985) discovered how the immune system rejects or accept organ transplantation, thus creating a platform for developing methods of transplanting solid organs (Immunological Tolerance hypothesis). (NP 1960).*

1953–1978 ***Michael Heidelberger*** *(1888–1991) The father of modern immunology. Together with* ***Oswald Avery*** *showed that polysaccharides of pneumococcus are antigens, enabling him to show that antibodies are proteins.*

1956 ***Niels K. Jerne, David Talmage*** *and* ***Frank M. Burnet*** *developed clonal selection hypothesis; it proposes that before a lymphocyte ever encounters an antigen, the lymphocyte has specific receptors for that antigen on its surface.*

1956–1961 ***Baruj Benacerraf*** *(1920–),* ***Jean Dausset*** *(1916–), and* ***George D. Snell*** *(1903–1996) discovered genetically–determined structures on the cell surface that regulate immunological reactions. (NP 1980).*

1957 ***Alick Isaacs*** *(1921–1967) and* ***Jean Lindenmann*** *(1924–) discovered interferon.*[45]

---

[44] Antibodies are Y-shaped molecules, two arms of which are identical and have the ability to bind to certain proteins called antigens. The binding takes place by van der Waals forces. Because these forces are very short range, there must be a close fit between the atoms in the arm of the antibody and those in the antigen, and the bonding is very specific. An inactive part of the antibody can be tagged with fluorescein isothiocyanate. Antigens to which these tagged antibodies have become attached can then be identified by their green fluorescence when exposed to ultraviolet light. A high-energy photon is absorbed by the atom, which then loses energy in two or more successive steps. If the high-energy photon is in the ultraviolet, one or both of the lower-energy photons my be in the visible range. Typical lifetimes for the decay of the excited atoms are 10^{-9} to 10^{-7} s; if the decay is much longer than this, the phenomenon is called phosphorescence. Lifetimes can range up to hours.

[45] Its effects were noticed earlier (1954) by the Japanese virologists **Yasuichi Nagano** (1906–1998) and **Yasuhiko Kojima**. Interferons are a class of natural proteins produced by cells of the immune system of most animals in response to challenges of foreign agents such as *viruses, bacteria, parasites* and *tumor* cells. Interferons belong to the large class of glycoproteins known as *cytokines*.

1958–1962 **Gerald M. Edelman** *(1929–) and* **Rodney R. Porter** *(1917–1985) discovered human leukocyte antigens and antibody structure, thymus involvement in cellular immunity and T and B cell cooperation in immune response. (NP 1972).*

1965–1972 **Elvin A. Kabat** *(1914–2000) elucidated structure and genetic basis for specificity of antibodies. First to demonstrate that antibodies are γ-globulins.*[46]

1966 **Kimishige Ishizaka** *(1925–) discovered a new type of immunoglobulin IgE that develops allergy and elucidated the mechanism of allergy at molecular and cellular levels. Following this discovery,* **S.G.O. Johansson** *found an IgE myeloma, confirming Ishizaka's discovery. This achievement contributes to the clinical diagnosis and treatment of allergic diseases.*

1974 **Rolf M. Zinkernagel** *(1944–) and* **Peter C. Doherty** *(1940–) discovered how the immune system recognizes virus-infected cells. (NP 1996).*

1975 **Cesar Milstein** *(1927–2002),* **Georges J.F. Köhler** *(1946–1995) and* **Niels K. Jerne** *(1911–1994) developed theories concerning the specificity in development and control of the immune system and the discovery of the principle for production of monoclonal antibodies. (NP 1984). This discovery led to an enormous expansion in the exploitation of antibodies in science and medicine.*

1976 **Susumu Tonegawa** *(1939–) discovered a genetic principle for generation of antibody diversity.*[47] *(NP 1987).*

---

[46] Memory B cells are a *B cell* sub-type that are formed following primary infection. When a B cell is activated, by recognizing a specific *antigen*, it proliferates to form *antibody* producing *plasma cells* and long-lived memory B cells. The memory B cells are specific for the antigen that first stimulated their production. If this antigen is encountered again, memory B cells can recognize it and quickly proliferate. This forms a new generation of antibody-producing plasma cells. This means that the antibody response is much more rapid in subsequent infections, than primary infection, reducing the chance of symptom development. This is the principle behind *vaccination*.

[47] To achieve the diversity of antibodies needed to protect against any type of antigen, the immune system would require millions of *genes* coding for different

1983 *Discovery of HIV.*

---

antibodies, if each antibody was encoded by one gene. Instead, as **Tonegawa** showed in a series of experiments beginning in 1976, genetic material can rearrange itself to form the vast array of available antibodies. Comparing the *DNA* of *B* cells (a type of *white blood cells*) in embryonic and adult mice, he observed that genes in the mature B cells of the adult *mice* are moved around, recombined, and deleted to form the diversity of the variable region of antibodies.

Timeline History of Neurobiology

1873–1880 **Camilo Golgi** *(1843–1926), pioneer of modern neurophysiology. Discovered dendritic nerve cells called "Golgi cells" and "Golgi tendon spindle" (NP 1906)*

1889–1928 **Santiago Ramón y Cajal** *(1852–1934), pioneer of modern neurophysiology. First to formulate the neuronal theory (based on the individuality of the nerve cell. NP 1906)*

1895–1932 **Charles Scott Sherrington** *(1857–1952) formed the scientific basis of modern neurology; coined the terms neuron and synapse. Outlined the nature of communication between nerves and between nerves and muscles. (NP 1932)*

1903–1912 **Robert Barany** *(1876–1936) pioneered in the study of the human inner ear's balancing organ. (NP 1914)*

1913–1928 **Edgar Douglas Adrian** *(1889–1977). Pioneering studies of the electrophysiology of the brain and the nervous system. (NP 1932)*

1920–1929 **Otto Loewi** *(1873–1961) and* **Henry Hallett Dale** *(1875–1968) independently isolated acetylcholine, the substance released by the vagus nerve, thus providing evidence for chemical transmission of nerve impulses across the synapse. (NP 1936)*

1921–1935 **Joseph Erlanger** *(1874–1965) and* **Herbert Spencer Gasser** *(1888–1963) analyzed fundamental properties of neural conduction impulses and discovered that the velocity of the impulse was proportional to the diameter of the nerve fiber. (NP 1944)*

1928–1960 **Georg von Békésy** *(1899–1972). Discovered physical mechanisms of stimulation within the cochlea of the inner ear. (NP 1961)*

1932 **Walter B. Cannon** *(1871–1945, USA) coined the term homeostasis.*

1932–1956 **Ragnar Granit** *(1900–1991),* **Haldan K. Hartline** *(1903–1983) and* **George Wald** *(1906–1997), share the 1967 Nobel Prize for their discoveries in the neurophysiology of vision.*

1945–1952 **Alan Lloyed Hodgkin** *(1914–1998, England),* **John C. Eccles** *(1903–1997) and* **Andrew Fielding Huxley** *(1917– , England) described the ionic mechanism by which neurons transmit electrical impulses. (NP 1952)*

1945–1965 **Julius Axelrod** *(1912–2004),* **Bernard Katz** *(1911–2003) and* **Ulf Svante von Euler** *(1905–1983) share the 1970 Nobel Prize for their work on electrical simulation of nerves and the processes of neuromuscular transmission.*

1965–1977 **David Hunter Hubel** *(1926–) and* **Torsten Nils Wiesel** *(1924–), followed on from the work of Granit and Hartline to study the way in which the brain processes visual information. They demonstrated that there is a hierarchical processing pathway, of increasingly sophisticated analysis of visual information by nerve cells from the retinae to the cerebral cortex. Shared the NP in 1981.*

1968–1981 **Stanley H. Cohen** *(1917–) and* **Rita Levi-Montalcini** *(1909–) awarded the Nobel Prize (1986) for their work on the control of nerve cell growth.*

1976–1986 **Erwin Neher** *(1944–) and* **Bert Sakmann** *(1942–). Revolutionized cell physiology with the invention of the "patch–clamp" recording technique which made it possible to record the electrical activity of very small areas of membrane, by eliminating the membrane's electrical noise. This improved the sensitivity of previously available methods by a factor of a million. This was done by touching a cell membrane with the tip of a glass pipette filled with saline solution, and by applying suction through the pipette, creating a seal which isolated a small section of the membrane. This method was then applied to the study of nerve impulse propagation along axons.*

1969–1985 **Alfred G. Gilman** *(1941–) and* **Martin Rodbell** *(1925–1998), discovered G-proteins and the role of these proteins in signal transduction in cells. (G-proteins are a vital intermediary between the activation of receptors on the cell membrane and actions within the cell). NP 1994.*

Arvid Carlsson *(1923–),* **Paul Greengard** *(1925–)* and **Eric Kandel** *(1929–) share the 2000 NP for their discoveries concerning signal transduction in the nervous system.*

1973 **Solomon H. Snyder** *(1938–). Discovered the presence of opiate receptors in nervous tissues.*

Timeline History of Endocrinology

1902 **William Bayliss** *(1860–1924) and* **Ernest H. Starling** *(1866–1927) discovered secretin, the hormone secreted by the duodenum that stimulates pancreatic secretions. Previously, the process had been considered (e.g., by* **Ivan Pavlov***) to be regulated by the nervous system. Starling and Bayliss demonstrated that injected duodenal extract into dogs rapidly increased pancreatic secretions, raising the possibility of a chemical messenger.*

1905 **William B. Hardy** *(1864–1934), a Cambridge physiologist coined the word "hormone" and suggested its use to Starling.*

1911 **Bernardo Houssay** *(1887–1971) discovered the part played by the hormone of the anterior pituitary lobe in the metabolism of sugar (NP 1947)*

1921 **Frederick Grant Banting** *(1891–1941) and* **John J.R. Macleod** *discovered the hormone Insulin. (NP 1923)*

1927 **Bernard Zondek** *(1891–1966) discovered the sex hormone gonadotrophine and developed the first reliable hormone pregnancy test.*

1929 **Edward Doisy** *(1893–1986) discovered the sex hormones oestrone, oestriol and oestradiol. (NP 1943 for discovering the chemical nature of vitamin K)*

1930 **Adolf F.J. Butenandt** *(1903–1995). Discovered the sex hormones androsterone and oestriol (NP 1939;* **Zondek** *should have shared it, but at that time he was a refugee in Palestine, driven out of Nazi Germany in 1933)*

1935 **Karoly G. David** *isolated pure crystalline testosterone from testicles.*

1948 **E.C. Kendall** *(1886–1972),* **Tadues Reichstein** *(1897–1996) and* **P.S. Hench** *(1896–1965), discovered the structure and biological effects of the hormones of the adrenal cortex. (NP 1950)*

1951 **Robert Woodward** *(1917–1979) synthesized cortisone. (NP 1965)*

1951 **Charles Brenton Huggins** *(1901–1997) won the Nobel Prize (1966) for his hormonal treatment of prostate cancer.*

1951 **Frederick Sanger** *(1918–) determined the exact sequence of amino acids comprising the insulin molecule. (NP 1958)*

1960 **Rosalyn Sussman Yalow** *(1921–) received the Nobel Prize for the development of radioimmunoassays for insulin (1977).*

1960–1976 **Sune K. Bergström** *(1916–2004),* **Bengt I. Samuelsson** *(1934–) and* **John R. Vane** *(1927–2004) won the NP for 1982 for discoveries concerning prostaglandins and related biologically active substances.*

1942–1969 **Dorothy Crowfoot Hodgkin** *(1910–1994) determined the spatial conformation of the insulin molecule by means of X-ray diffraction studies.[She has been awarded earlier in 1964 the NP in chemistry for the development of crystallography in 1954].*

1970 **Roger Guillemin** *(1924–) and* **Andrew V. Schally** *(1926–) discovered the peptide hormone production of the brain. (NP 1977)*

Timeline History of Ecology

Ecology is generally spoken of as a new science, having only become prominent in the second half of the 20th Century. Nonetheless, ecological thinking at some level has been around for a long time, and the principles of ecology have developed gradually, closely intertwined with the development of other biological disciplines.

354–322 BCE *One of the first ecologists may have been* **Aristotle** *or perhaps his student,* **Theophrastos**, *both of whom had interest in many species of animals. Theophrastos described interrelationships between animals and between animals and their environment as early as the 4th century BCE.*

75–79 CE **Pliny the Elder** *(23–79 CE) writes: "Natural History"*

1555 CE **Georg Bauer** *(Agricola, 1490–1555, Germany) writes: "De re Metallica"*

Throughout the 18th and the beginning of the 19th century, the great maritime powers such as Britain, Spain, and Portugal launched many world exploratory expeditions to develop maritime commerce with other countries, and to discover new natural resources, as well as to catalog them. At the beginning of the 18th century, about twenty thousand plant species were known, versus forty thousand at the beginning of the 19th century, and almost 400,000 today.

1789–1804 *Ecology blossomed due to the new discoveries of Antoine Lavoisier (1743–1794, France) and* **de Saussure** *(1767–1845, France), notably – the nitrogen cycle.*

1789 **James Hutton** *(1726–1797) considered the earth to be a super-organism and that its proper study should be by physiology. He was thus the first to coin the term 'geo-physiology'.*

1798 **Thomas Malthus** *(1766–1834, England) writes: "Essay of the Principle of Population Theory".*

The above-mentioned expeditions were joined by many scientists, including botanists, such as the explorer:

1802–1807 **Alexander von Humboldt** *(1769–1859, Germany). He is often considered a father of ecology.*

He was the first to take on the study of the relationship between organisms and their environment. He exposed the existing relationships between observed plant species and climate, and described vegetation zones using latitude and altitude, a discipline now known as geobotany.

In 1804, for example, he reported an impressive number of species, particularly plants, for which he sought to explain their geographic distribution with respect to geological data. One of Humboldt's famous works was "Idea for a Plant Geography" (1805).

1852 **David Thoreau** *(1817–1862, USA), first to coin the term ecology.*

1858 **Charles Darwin** *(1809–1882, England) published "The Origin of species". Ecology passed from a repetitive, mechanical model to a biological, organic and hence evolutionary model.*
Alfred Russel Wallace *(1823–1913, England), contemporary and competitor to Darwin, was first to propose a "geography" of animal species. Several authors recognized at the time that species were not independent of each other, and grouped them into plant species, animal species, and later into communities of living beings.*

1866 **Ernst Haeckel** *(1834–1919, Germany) recoined the term oekologie. The word derives from the Greek oikos= "household" and logos= "study". It meant to the Greek – the study of nature.*

The word "ecology" is often used in common parlance as a synonym for the natural environment or environmentalism. Likewise "ecologic" or "ecological" is often taken in the sense of environmentally friendly.

1875 **Eduard Suess** *(1831–1914, Austria) proposed the name biosphere for the conditions promoting life, such as those on earth, which induces flora, fauna, minerals, matter cycles, etc. The need for this term followed his observation that life developed only within strict limits of each compartment that makes up the atmosphere, hydrosphere and lithosphere.*

1879–1914 **John Muir** *(1838–1914, USA), explorer, naturalist and ecologist. Campaigned for the forest conservation in the United States. Founded the Sierra Club which became a leading conservation organization.*

1899 **Henry Chandler Cowles** *(1869–1939, USA). Ecological Pioneer. Invented the concept Ecological succession as a process by which a natural community moved from a simpler level of organization to a more complex community.*

1926 **Vladimir Vernadsky** *(1863–1945, Russia). Founder of modern biogeochemistry. Detailed the idea of the biosphere in his work "The biosphere" (1926) and described the fundamental principles of the biogeochemical cycles. He thus redefined the biosphere as the sum of all ecosystems.*

First ecological damages were reported in the 18th century, as the multiplication of colonies caused deforestation. Since the 19th century, with the industrial revolution, more and more pressing concerns have grown about the impact of human activity on the environment. The term ecologist has been in use since the end of the 19th century.

Over the 19th century, botanical geography and zoogeography combined to form the basis of biogeography. This science, which deals with habitats of species, seeks to explain the reasons for the presence of certain species in a given location.

1935 **Arthur Tansley** *(1871–1955, England) coined the term ecosystem, the interactive system established between the biocoenosis (the group of living creatures), and their biotope, the environment in which they live. Ecology thus became the science of ecosystems.*

1951–1962 **Rachel Carson** *(1907–1964, USA). Zoologist and marine biologist. Her book "Silent Spring" launched the global environmental movement. In the United States it spurred a reversal in national pesticide policy.*

1953 **Eugene P. Odum** *(1913–2002, USA). Known for his pioneering work on ecosystem ecology. Explored how one natural system can interact with another. Believed that homeostasis and stability in ecosystems was a result of evolutionary processes.*

1971 *Ecology became a central part of the World's politics as UNESCO launched a research program called Man and Biosphere, with the objective of increasing knowledge about the mutual relationship between humans and nature. A few years later it defined the concept of Biosphere Reserve.*

1972 *The United Nations held the first international conference on the human environment in Stockholm. This conference was the origin of the phrase "Think Globally, Act Locally".*

1979 **James Lovelock** *(1919– , England). Researcher and environmentalist. Proposed the 'Gaia Hypothesis', in which he postulated that the earth functions as a kind of superorganism. He popularized the term "geophysiology" (foreshadowed by* **James Hutton** *(1789) and* **Huxley** *(1825–1895)). While the Gaia Hypothesis was readily accepted by many in the environmental community, it has not been fully accepted within the scientific community. Critics point out that since natural selection operates on individuals, it is not obvious how planetary-scale homeostasis can evolve.*

1992 *Earth Summit in Rio de Janeiro; The concept of the biosphere was recognized by the major international organizations, and risks associated with reduction in biodiversity were publicly acknowledged.*

1997 *The dangers the biosphere was facing were recognized from an international point of view at the conference leading to the Kyoto Protocol. In particular, this conference highlighted the increasing dangers of the greenhouse effect – related to the increasing concentration of greenhouse gases in the atmosphere, leading to global changes in climate. In Kyoto, most of the world's nations recognized the importance of looking at ecology from a global point of view, on a worldwide scale, and taking into account the impact of humans on the earth's environment.*

Life Prospects on a Hazardous Planet

Cycles of matter and energy on earth, caused both by internal and external agents, have always been accompanied by catastrophes hazardous to life. The energy sources and energy content of the various natural processes are well known: In general, the combination of the internal heat flow and gravity results in plate tectonic movements and the associated earthquakes, volcanism, mountain uplift as well as forming a few types of tsunami. The combination of the sun's energy and gravity results in all weathering, erosion and mass wasting phenomena including rain-floods, blizzards, hurricanes, tornadoes, lightning, hail, avalanches, mudflows, waves, landslides and associated tsunami.

During the early history of the earth, interior energy sources were contributed by:

- *Bolide impacts*[48]*: these helped raise the temperature of the early earth and some of that heat was left over as the internal heat of the earth.*
- *Gravity: gravitational collapse of the earth raised the interior temperatures.*

---

[48] Up to 1970's, there was little interest in asteroids, including near earth asteroids. They were considered low class astronomical objects. Indeed, the small asteroid which destroyed hundreds of square kilometers in remote Siberia in 1908 was an event little known to the general public. A small asteroid which skimmed the upper atmosphere in the 1970's, as detected by a U.S. military satellite, received little publicity.
In the 1970's, things started to change. A small but increasing number of astronomers interested in asteroids began to realize the abundance of asteroids which passed close to earth, by instituting processes to catalog asteroids accidentally seen on telescopic plates and previously not recorded.
Theoretical computer models revealed that the gravity of the planets caused a sizable number of asteroids from the Main Belt between Mars and Jupiter to cascade down into lower orbits approaching or crossing earth's. Further, a significant fraction of comets passing through the inner solar system would be diverted into orbits near earth due to gravitational encounters with the inner planets. As a result of these discoveries, the estimated number of near-earth objects dramatically expanded by about 1000 times. Finally, new telescope technology, emerging around 1900, increased the discovery rate of all asteroids. It is now estimated that there are about 300,000 near-earth asteroids which are over 100 m in diameter, and about 2000 over 1 kilometer in diameter.

- *Radioactivity: it continues to create tremendous heat in the earth, although more heat was generated earlier. Radioactivity allows for precise age dating of rock strata.*

Exterior energy sources[49] *are due to solar energy which surpasses heat flow from the interior by a factor of 5300; 30 percent of the sun's energy is reflected as short wavelength radiation, 47 percent is absorbed and re-radiated as long wavelength radiation (infrared) and 23 percent evaporates water (mostly in the tropics) to begin the hydrological cycle. The hydrological cycle in combination with gravity results in waterfalls, rivers, floods, atmospheric circulation, hurricanes, thunderstorms, lightning, tornadoes, mass movement, waves and landslide induced tsunami.*

Another exterior energy source is impact energy of asteroids and comets with solar system velocities ranging from 11 to 70 $\frac{km}{sec}$. *Temperatures of at least 10,000 degrees are produced during impact. These events have also been proposed to have produced global devastation and mass extinctions.*

Terrestrial catastrophism via bombardment episodes by 'cosmic bullets' impacted the mind of man since times immemorial: the same basic story repeated itself in many ancient cultures. The Bible, for example, tells us of darkness, pillar of cloud by day and fire by night, hail and fire dropped on earth, winged serpents in the sky, huge burning stars falling from heaven, dust from impact, earthquakes, large ecological upsets, tsunamis – all recognizable symptoms of asteroid collision with earth.

These phenomena, although faithfully observed by prehistoric people, could not have been documented prior to the invention of writing in ca 3200 BCE and certainly could not be understood before the advent of modern science in the early 16th *century.*

Natural disasters may be classified into the following categories:

---

[49] Earth receives from the sun the total of 1.75×10^{17} watt ($\frac{Joule}{sec}$) which amounts to an average flux of 1367 $\frac{watt}{m^2}$. Of this only 1.25×10^{17} watt is absorbed by the atmosphere, hydrosphere and surface (69%). The average *total solar* flux on the earth's surface amounts to 168 $\frac{watt}{m^2}$. The *non-solar* geothermal energy release (radioactive decay + primordial heat) reaches a level of 6.6×10^{13} watt $= 6.6 \times 10^{20}\ \frac{erg}{sec} = 2.07 \times 10^{28}\ \frac{erg}{year} = 5 \times 10^5\ \frac{MT}{year}$. This energy flow feeds continental drift, earthquakes, volcanoes and hot springs. Direct solar energy feeds atmospheric circulation and all accompanying meteorological processes. Appreciable global changes occur at the level of one percent of the solar input.

- *Climatic changes and ecological collapses; return of a severe glacial epoch*[50].
- *Collision of earth with asteroids, mini-planets*[51] *and comets.*
- *Volcanic eruptions and earthquakes.*
- *Floods, tsunamis, mega-tsunamis and major storms.*
- *Droughts, famines.*
- *Pandemics.*
- *Supernova explosion*[52].

Collision with Earth of Asteroids and Comets

How do asteroids arrive at earth? Detailed scenarios for the formation and transport of meteorites and other near-earth asteroids were clarified only

---

[50] The frightening aspect of this 'option' is that the sun's smallest tantrum can influence our climate significantly in spite of the vast distance of 150 million km separating it from earth. *Sunspots* correspond to changes in solar activity: times of greatest activity correspond to warm periods on earth. In the middle of the 20th century observations showed a great density of sunspots while recorded air temperatures were noticeably higher. In contrast, between 1430 to 1850, the sun seems to have gone through a calm period and sunspots were practically absent. This period coincided with a decrease in temperatures.

[51] With diameter> 1 km.

[52] A *supernova explosion* could completely extinguish life on earth by high-energy gamma-ray bursts and particles from the explosion, provided it explodes within 200 light-years from the solar system. Among the 100,000 closest stars we expect one supernova to explode in a time-scale of roughly a billion years – of the order of the time interval during which life has thus far existed on earth!
The last one seen in our galaxy was Kepler's, in 1604. In general, supernovas derive their energy from the *gravitational collapse* of the central core of a heavy enough star: the core, having a mass of order that of the sun, collapses into a neutron star and thereby liberates enormous amounts of energy (ca 10^{53} erg). Other known supernova explosions are the Crab Nebula explosion (1054 CE) and the 1987A explosion in the large Magellanic Cloud.

recently. The basic idea goes back to **Kirkwood** *(1866), who identified gaps in the distribution of asteroids at distances from the sun at which their orbital periods would be commensurable with Jupiter's period (such as 5.93 years), forming ratios of small whole numbers. Kirkwood proposed that the gaps arose as a consequence of perturbations caused by Jupiter: the effect of Jupiter's mass would be to force any asteroid that appeared in one of the asteroid-free zones into another orbit, with the result that it would immediately leave the zone (according to Kepler's law, a period of 5.93 years corresponds to a semimajor axis of 3.28 AU).*

It has been shown that such 'resonances' (as they are called) could raise asteroids' orbital eccentricities to a point that bodies swept away from the gaps could reach Mars. Scientists demonstrated that material could reach earth through chaotic dynamics at the 3:1 resonance. It was reported that numerous weak resonances and interactions among the resonances cause chaotic behavior capable of raising main-belt eccentricities to allow for a Mars crossing. Objects then break away from the resonances when they have close encounters with Mars and subsequently evolve as Mars crossers, until they become near-earth asteroids.

This scenario is sped up by a series of minor resonances that yield fairly chaotic evolution, lasting a few times 10^7 *years, during which a reasonable fraction is likely to hit earth. In short — several resonances provide the escape route from the main belt and allow bodies to reach earth in a few tens of million of years.*

The most energetic – and best-documented – encounter with the earth of an earth-crossing body produced a great fireball over the Podkamennaya–Tunguska River region of Siberia on the morning of June 30, 1908. Traveling from southeast to northwest, the bolide nearly passed over the town of Kirensk; the endpoint of the trajectory was about 60 km northwest of the remote trading post of Vanovara, over a very sparsely inhabited part of the Siberian taiga.

The bolide was observed from distances as great as 600–1000 km from the endpoint; the atmospheric shock was audible at a distance of 1270 km and heat was felt at a distance of 70 km. Trees were knocked down at distances up to 40 km from the endpoint, and circumstantial evidence suggests that dry timber was ignited by thermal radiation from the fireball at distances up to 15 km from the endpoint.

*Intensive investigation by expeditions carried out over many decades has shown that the Tunguska bolide disintegrated in the atmosphere; it deposited most of its kinetic energy at an estimated altitude of ∼8.5 km (***A. Ben-Menahem***, 1975). Only microscopic spheres of glass and magnetite, formed by ablation, reached the ground.*

Long-period atmospheric gravity and acoustic waves excited by the atmospheric shock were well recorded on weather-station barographs in Siberia and southern England; the passage of these waves in both the direct and reverse paths was recorded on a barograph at Potsdam, Germany. Coupling of the air wave to the ground near the endpoint produced seismic waves detected at Irkutsk, Tashkent, Tiflis, and Jena; local coupling of the air wave to the ground as it passed over some seismic stations was also recorded.

On the basis of a thorough study of the seismic and acoustic records, Ben-Menahem (1975) estimated the released energy at 12.5 ± 2.5 *megatons. The distance to which trees were felled near the endpoint is consistent with a* $\sim$*13-megaton explosion at an altitude of* $\sim$*8 km and scaling from the effects of nuclear weapons; the ignition of wood by the thermal radiation also suggests an energy of the order of the same yield.*

E. Shoemaker (1983) has shown that the 'best estimate' of the frequency of a 12-megaton encounter with earth is about once every 150–600 years. The explosion of the world's entire nuclear arsenal (ca 10^5 *MT ; see Table 6.12) could not match the energy released when a kilometer-size asteroid hits the earth.*

Stimulated by the renewed interest of scientists in the Tunguska explosion, Alvarez et al. (1980) hypothesized that the major dinosaur (and other fauna) extinction event at the Cretaceous-Tertiary boundary (65 My) was the result of an enormous impact of an asteroid or a comet. The heat from the impact caused forest fires, and lofted huge amounts of dust from the impact crater into the atmosphere. The smoke and ash from the fires and the dust particles together blocked the sun for at least several months. The temperature decreased since the sunlight was partially blocked from reaching the earth's surface.

Without sunlight and in the cold, plants began to die since they could not obtain nutrition. This in turn affected the plant-eating dinosaurs; lacking their source of food they too began to die. Since the plant-eating dinosaurs were dying, there was a lack of food for the meat-eating dinosaurs, and they perished as well. In other words, the effects of the impact caused the ecosystem to collapse. It is speculated that this impact led to the extinction of 75 percent of the species alive on earth during that time.

Such a 'relatively common' event would not show up much in geological records on a global scale. There have been many local tsunamis and brief climate changes in recorded history without any plausible explanation (e.g., Amos **9**, *5–7).*

If an asteroid of size 1 kilometer hit earth, its impact would wipe out life within proximity of the impact site. However, more serious is how it would affect the whole world in indirect ways. The dust and/or vapor cloud created by an impact to either the land or the ocean could be big enough to create a "nuclear winter" like mini-ice age, and disrupt climatological wind patterns, adversely affecting major food-growing regions of the world, thus straining world food supplies, governments and civilization. This is comparable to what caused the dinosaur extinction, as well as other major extinctions of smaller creatures over geologic time scales.

Earth's atmosphere gives protection against the vast majority of small asteroids which hit. Asteroids hit the atmosphere at typical speeds in excess of 10 km/sec — an average of about 20 km/sec for asteroids whose entire orbits reside within the inner solar system (with exact relative speed depending upon their angle of approach) and with speeds over 50 km/sec common for small cometary objects making a pass from the outer solar system. At such speeds, they usually break up due to severe shock pressures, and burn up due to friction with the atmosphere.

For asteroids coming in at 20 km/sec, it is generally thought that to penetrate the atmosphere and cause major damage by tsunami, an iron asteroid must be around 40 to 60 meters in diameter, and a stony asteroid 200 meters in diameter. However, a stony asteroid 60 meters in diameter can cause significant damage by airbursts. The exact damage caused by an asteroid or comet depends upon a number of factors: size, speed, composition, and whether it hits land or ocean.

For a land impact, it can be said in general that an object of roughly 75 meters in diameter can destroy a city, a 160 meter object can destroy a large urban area, a 350 meter object can destroy a small district or state, and a 700 meter object can destroy a small country.

For an ocean impact, the destruction is much greater. Smaller objects can cause far more widespread damage. The effects of an ocean impact are felt much further away than the effects of an airburst due to the more effective propagation of water waves, and the fact that human populations and assets are largely concentrated in coastal cities which historically became established due to water transport (i.e., shipping and trade) and the ports that serve it.

For example, the earthquake-induced tsunami in Chile in 1960 produced waves in Hawaii, 10,600 km away, of height up to over 10 meters, and up to 5 meters in Japan 17,000 km away with an average height of 2 meters, causing heavy damages and loss of lives.

The effects of an airburst are far more localized because the intensity of the phenomenon decreases with the inverse square of the distance, whereas the height of a water wave decreases only with the inverse of the distance, i.e., as the first inverse power, due to its circular, two-dimensional nature.

On March 23, 1989, *an asteroid with a kinetic energy of over 1000 one-megaton hydrogen bombs (i.e., about 5000 times more powerful than the bomb dropped on Hiroshima) was recorded to have passed very close to earth, using new equipment recently deployed. Named* 1989FC, *this asteroid was detected well after its point of closest approach, and astronomers found out it had passed so close only after calculating backwards its orbital path having once realized its nearness. This was a key event that brought near earth asteroids into the political arena.*

Later, the new Spacewatch Camera in Arizona, using the latest technology in electronic sensors and computerized automated detection, discovered four asteroids that came closer to the earth than the moon (actually within half the distance of the moon) in 1991–94.

It is thought that the very massive asteroid Geographos, a cigar-shaped body of 5.1 kilometers by 1.8 kilometers which passed near earth in 1969 and again in 1994, could collide with earth in the not too distant future. It is probably an iron or stony iron asteroid. It would make the Tunguska meteorite look like a trivial impact. Geographos would cause a global ice age for several years from the dust it would kick up. But it will not impact in the next few hundred years, at least. We can not project Geographos' orbit in the far future with enough precision to determine if it will impact earth or the moon, or whether it will have a close encounter which will fling it elsewhere.

Table 6.11 lists forthcoming close approaches to earth of certain minor planets and comets.

The regional and global effects from an impact of an asteroid of diameter $d > 1$km *are as follows:*

- *Massive* earthquake *with magnitude* $M \geq 13$ *(Richter scale).*
- Global winter *syndrome: dust pumped into the atmosphere blocks incoming solar radiation, producing* darkness *and* temperature depression *for months. Disruption of ecosystems by diminished photosynthesis.*
- Greenhouse gases H_2O *and* CO_2 *are formed if impact occurs in the oceans.* Warming effects *are created for many years after cooling.*
- Giant tsunamis, *1–3 km high, flood the interior of continents.*
- Shock waves *in the atmosphere produce* nitrogen oxides *(Nitrogen + Oxygen). Combined with water, nitric acid is formed, and precipitates as* Acid Rain.

Figure 6.1 exhibits the evidence concerning the dependence of the cumulative frequency distribution of earth-bolide encounter on the energy released during impact. It tells us that:

1 – year event has the energy equivalence of	*10 KT*
10 –	*100 KT*
100 –	*3 MT*
1,000 –	*50 MT*
10,000 –	*1,000 MT*
100,000 –	*20,000 MT*
1,000,000 –	6×10^5 *MT*
10,000,000 –	10^7 *MT*
100,000,000 –	10^8 *MT*

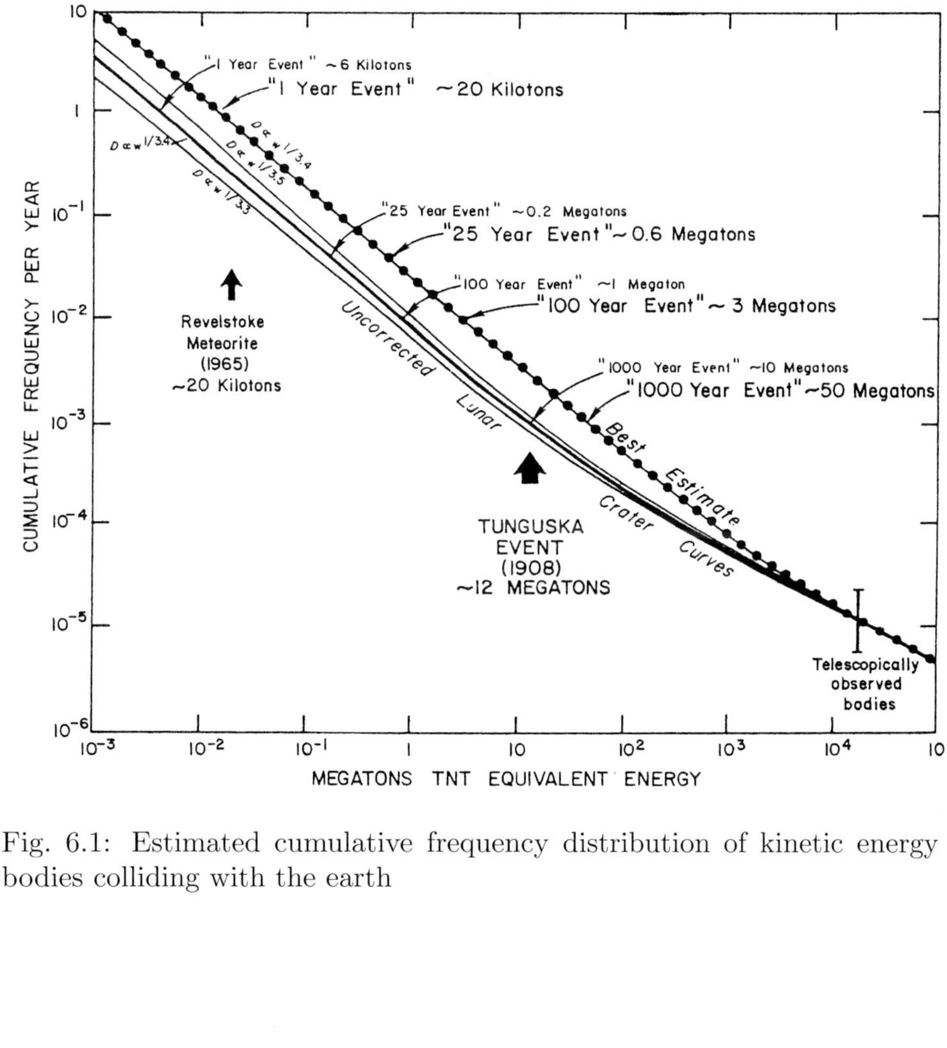

Fig. 6.1: Estimated cumulative frequency distribution of kinetic energy of bodies colliding with the earth

Table 6.11: PREDICTED MINOR-PLANET AND COMET ENCOUNTERS WITH EARTH TO WITHIN 0.05 AU (ASTRONOMICAL UNIT = CA 150 MILLION KM) DURING 2005–2031. AT 40 KM/SEC, AN OBJECT COVERS A DISTANCE OF ABOUT 3.5×10^6 KM PER DAY

Object	**Date of encounter**	**Distance (AU)**
1992 UK4	*Aug. 08, 2005*	*0.0404*
(4450) Pan	*Feb. 19, 2008*	*0.0408*
(1620) Geographos	*Mar. 17, 2008*	*0.1251*
	Aug. 12, 2026	*0.1704*
1991 VH	*Aug. 15, 2008*	*0.0458*
1998 CS1	*Jan. 17, 2009*	*0.0286*
1994 CC	*June 10, 2009*	*0.0169*
1998 HE3	*May 10, 2012*	*0.0340*
1998 KJ17	*May 27, 2012*	*0.0355*
1988 TA	*May 09, 2013*	*0.0411*
1998 FW4	*Sept. 27, 2013*	*0.0075*
(2340) Hathor	*Oct. 21, 2014*	*0.0482*
(5604) 1992 FE	*Feb. 24, 2017*	*0.0336*
(3122) Florence	*Sept. 01, 2017*	*0.0472*
1991 VG	*Feb. 11, 2018*	*0.0471*
1998 HL1	*Oct. 25, 2019*	*0.0416*
1997 BQ	*May 22,2020*	*0.0416*
1993 BX3	*Jan. 17, 2021*	*0.0473*
(3361) Orpheus	*Nov. 21, 2021*	*0.0386*
	Nov. 19, 2025	*0.0379*
(7335) 1989 JA	*May 27, 2022*	*0.0269*
(6037) 1988 EG	*Aug. 23, 2023*	*0.0407*
1998 KY26	*June 01, 2024*	*0.0309*
1997 QK1	*Aug. 03, 2025*	*0.0304*
1997 NC1	*June 27, 2026*	*0.0171*
1990 MU	*June 06, 2027*	*0.0309*
1998 ML14	*Aug. 03, 2028*	*0.0307*
1997 XF11	*Oct. 26, 2028*	*0.0064*
1994 WR12	*Nov. 26, 2030*	*0.0180*

Energetics

The energy sources and energy content of the various natural disasters are well known. Prominent features of energy conversion processes are:

- Internal *heat flow and gravity are responsible for the generation of earthquakes, volcanism and certain types of tsunamis.*

  Internal heat flow due to radioactivity, gravitational collapse *(when objects move closer to earth's center by falling, slipping, sinking – kinetic energy is released) and* bolide impact *(especially during early history) helped raise* earth's temperature.

- Exterior energy sources *are mainly due to solar energy. In conjunction with gravity and* earth's rotation *this drives the* hydrological cycle *which is responsible for floods, atmospheric circulation, hurricanes, thunderstorms, lightning, tornadoes, waves, and certain tsunami.*

- *Only 23% of total solar energy is available for evaporating water (mostly in the tropics) to drive the hydrological cycle.*

Table 6.12: ENERGY CONTENT OF SINGULAR NATURAL PHENOMENA
(1 KT $= 4.2 \times 10^{19}$ ERG; 1 MT $= 4.2 \times 10^{22}$ ERG)

(a) Single Events *(BP = Before Present, KE = Kinetic Energy)*

Thunderstorm	*1 KT*
Atomic bomb (Hiroshima)	*10 KT*
St. Helens (1980; VEI=5)	*5 MT*
Hydrogen bomb	*10 MT*
Tunguska (1908)	*10 MT*
Hurricane (average KE)	*10 MT*
Barringer asteroid (diameter=50 m)	*20 MT*
Earthquake (total seismic wave energy; $M_s = 8.1$*)*	*100 MT*
Cyclone (total KE)	*100 MT*
Krakatoa explosion (1883; VEI=6)	*200 MT*
Thera (1500/1600 BCE)	*ca 1000 MT*
Asteroid 1989 FC (close encounter in 1989), KE	1.0×10^3 *MT*
Asteroid impact $d = 570$ *m; crater diameter=10 km*	1.0×10^4 *MT*
Tambora (1815; VEI=7)	2.0×10^4 *MT*
Crater Lake eruption (4895 BCE; VEI=7)	5.0×10^4 *MT*
World nuclear arsenal (10,000 warheads; 10 MT each)	1.0×10^5 *MT*
Tuba volcano, Indonesia (73,500 BP; VEI=8)	6.0×10^5 *MT*
Total seismic energy release over the past 6000 years	7.0×10^5 *MT*
Yellowstone, Wyoming (2My BP; VEI=8)	2.0×10^6 *MT*
Yucatan (Chicxulub) asteroid ($\sim 65 \times 10^6$ *years BP)*	10^8 *MT*
Hephaistos asteroid (d$\simeq$*10 km), KE*	10^8 *MT*

Table 6.12: (Cont.)

(b) Average global energy flow

	Erg/sec	*MT/year*
Volcanic eruptions	3.3×10^{16}	*25*
Earthquakes	1.6×10^{17}	*120*
Thunderstorms	3.0×10^{18}	*2400*
Total tidal dissipation per year	3.3×10^{19}	*26,000*
Atmospheric circulation	1.0×10^{23}	*80,000,000*

(c) Available terrestrial energy (MT/year)

Geothermal	5.0×10^{5}
Solar	8.4×10^{8}

Estimated Human Death-toll

Table 6.13 renders estimated cumulative death tolls since 4000 BCE due to all natural disasters combined, on the background of the evolution of world population Table 6.13.

In this connection it is of interest to estimate the total number of people who have ever lived: One begins by determining the mean population size for a birth-death stochastic process (i.e., the average behavior of a population whose size varies stochastically, growing over time due to random occurrence of births and deaths). One then assumes a starting population of two persons 1.5 million years ago and divides the total time span into a number of smaller subintervals by using times for which estimates of world population have been made (e.g., N (8000 BCE) $= 5 \times 10^6$*; N (0 BCE)* $= 250 \times 10^6$*; N (1750 CE)* $= 800 \times 10^6$*; N (1825 CE)* $= 10^9$*; N (1930)* $= 2 \times 10^9$*; N (1960)* $= 3 \times 10^9$*; N (1980)* $= 4.4 \times 10^9$*). The total number of people who ever lived since 1.5* *million years before present is then found to range from* 50 to 100 billion (10^{11}).

In the second category of Table 6.13, flood *is by far the worst of natural disasters, claiming some 40% of all deaths caused world-wide by acts of nature (epidemics excluded).*

Table 6.13: Estimated human death toll since 4000 BCE

Earthquakes and volcanoes and their immediate aftereffects	30×10^6
Floods, storms, tsunamis and droughts	100×10^6
Epidemics	370×10^6
Total:	ca 500×10^6

Table 6.14: World population growth (in units of 10^6 person)

2000 CE	6000
1991	5400
1960	3000
1930	2000
1800	1000
1750	800
1200	400
1000	250
500	210
0	250
-500	140
-1000	120
-1500	100
-2000	50
-8000	5
-12,000	4

Conclusion

Natural disasters have been the scourge of mankind since times immemorial.

It is estimated that ca 500 million people were killed since 4000 BCE by all catastrophes combined. This amounts to about one percent of all people who have ever lived since 4000 BCE. The total fatal energy unleashed on the earth's surface against its inhabitants throughout the said time window is estimated at 2×10^7 *MT.*

The direct effect of this onslaught on man's survival on earth was manifested through

- *Decimation and mass migration of populations.*
- *Destruction of habitat, agriculture and water systems on at least a regional scale.*

However, the interaction between nature and man is certainly ambivalent and not always unidirectional.

We know that life changed a primordial reducing *atmosphere into one rich in oxygen. On the other hand, comets could have brought an abundant amount of water to the early earth, and volcanic outgassing resulted in the formation of the primitive atmosphere and oceans. Mass extinctions due to asteroid encounters changed the course of evolution, promoting the eventual emergence of Homo Sapiens.*

Even today, nature's agents of destruction are at the same time a blessing in disguise. Take hurricanes, for example – one of the most powerful engines of death and destruction on earth. These storms are products of the tropical ocean and atmosphere: powered by heat from the sea, steered by the Easterly trades and temperate Westerlies, and their own fierce energy. But they are major source of rain and may have other hidden benefits as well for the planet.

In fact, Humans are now changing the planet in such a way that is already causing a wave of extinction on a scale unparalleled since the demise of the dinosaurs.

When we use the words 'Natural disaster' we are taking the narrow anthropocentric point of view. But in the grand scheme of things what is bad for us is not necessarily bad for the earth and its environment. In this wider context we should remember that although mankind had survived so far, there is no such guarantee for the remote future.

Cosmology and Biological Evolution: A Speculative Connection[53]

At the current stage in the development of mathematics, biochemistry and computer technology, Science is still far from being able to answer even the simplest questions regarding the feasibility of Darwinian evolution. Our level of understanding of stochastic processes, and our knowledge of the chemistry of macro-molecules, are insufficient to derive realistic estimates of the probabilities involved in a population of self-replicating biotic molecules arising in any given inanimate thermodynamic environment. We posses only the vaguest theoretical insights (and almost no empirical evidence) as to how Nature "invented" the genetic code; nor do we even know whether replication and metabolism preceded the advent of this code. And if our understanding of molecular evolution is so rudimentary, it goes without saying that we have almost no clue as to how (or indeed whether) random mutations pruned by natural selection can transform one species into another, or allow organisms of a given species to change their body plan or grow new organs.

However, as pointed out by the late astronomer Fred Hoyle, such naive probability estimates as we are able to derive, indicate that the timeline of terrestrial biological evolution, as outlined by evolutionary biologists, is almost miraculously compressed, especially in the period immediately following the onset of the so-called "Cambrian Explosion" (the emergence of virtually all presently-existing organism body plans, about 0.5 Gy ago, within the space of a few million years).

Indeed, Hoyle has pointed out that not only life and its molecules, but even the existence of carbon atoms (surely a prerequisite for life as we know it on Earth to emerge) seems a priori to have an extremely low probability of occurring within the framework of the currently accepted theories of cosmic origins and stellar evolution. Indeed, Hoyle adduced the cosmic abundance of carbon as a posteriori evidence that the constants of Nature – governing

[53] This essay explains how particle physics and cosmology might play a role in enabling biological evolution, partly aided (perhaps) by intelligent design (but not necessarily). These ideas are highly speculative, of course – no one has actually SEEN the constants of particle physics mutating, or one universe spawn baby-universes through narrow spacetime "necks", etc. But then again, it is important to keep in mind that no one has actually seen evolution happen, either (in either real time, the fossil record or even mathematical simulations).

elementary particle physics, their four types of fundamental interactions and the very structure of spacetime – have somehow been "rigged" (fine-tuned) to allow life and sentient beings to emerge. Other physicists have pointed out other apparent coincidences pointing to possible fine-tuning of the constants of nature. As an example, the following three large dimensionless numbers are all roughly ten to the power 40:

(A) The ratio of the electrostatic force between a proton and electron to the corresponding gravitational force;

(B) The square-root of the number of hydrogen atoms in the observable universe; and

(C) The ratio of the age of the universe to the time it takes light to traverse a proton[54].

Another example involves the cosmological constant ("dark energy"), which governs the innate tendency of empty space to expand (a form of "antigravity"). Observations performed in the 1990s with the aid of the Hubble Space Telescope revealed that this constant, while very small (of order 10 to the negative 120th power!) on the natural scale provided by particle physics, is nonetheless not zero, and is causing a slight acceleration of the (Big-Bang caused) expansion of our universe.

It has been calculated that if this tiny constant were only ten times larger, the local "antigravity" effect it would cause would prevent galaxies from accreting, and thus possibly prevent life from emerging – although we cannot be sure about that last inference, since our knowledge about how, why and with what likelihood life emerges is so scant at present.

Two other unexplained coincidences which seem crucial to the emergence of life are: the longevity of the universe (sixty orders of magnitude longer than the Planck timescale!), and the longevity of current-generation main-sequence stars. The latter longevity is due to a fine balance (involving all four fundamental forces of Nature!) between the nuclear forces, tending to explode stars, and their self-gravitation, which tends to collapse them. A long-lived universe containing stable, long-lived stars is probably an important enabling

---

[54] One of the two mathematical relations expressing these coincidences is equivalent to the statement that the total (gravitational plus kinetic) energy of the observable universe is approximately zero; or, in other words, that the Schwarzschild radius of the observable cosmos is of the order of its actual radius. Since this is also a property of black holes, this coincidence bears an intriguing similarity to the conjecture, discussed below, that our observed universe might have been spawned by the gravitational collapse of a relatively small and light lump of mass in a "parent" universe.

condition in the long (and still poorly understood) chain that gave rise to life (at least on our Earth).

Al these coincidences need not have been "engineered" by some intelligent super-being, although we lack sufficient knowledge to rule that possibility out. There are other logical possibilities, one of which (involving a Darwinian-like evolution of multiple universes and their natural constants) we now describe. Theorists attempting to develop a theory of quantum gravity have discovered that Einstein's General Theory of Relativity, in conjunction with Quantum Mechanics, may allow one universe to spawn others via gravitational collapse – the same phenomenon that forms black holes.

Such a "daughter universe" would appear, to beings in the parent universe, as a relatively small and light black hole; yet any observer that manages to safely "glide" into the daughter universe, would find himself in a vast, new universe which expands in accordance with its own "Hubble Law", and may eventually produce its own stars, galaxies and even life and intelligence.

The spacetime manifolds of the two universes are connected through a slender "neck", and the geometry of this neck explains the curious fact that in such solutions of Einstein's field equations, the daughter universe appears to have a more or less fixed event horizon from the parent-universe side, while undergoing cosmological expansion as viewed by observers on the other side of the neck (i.e., within the daughter universe).

Theorists have argued that such "reproduction" of universes may occur naturally, but also that daughter universes could be designed by intelligent beings in the parent universe. A universe may differ from those of its parent – depending, perhaps, on physical conditions (electromagnetic fields, temperature, spatial curvature, etc.) in the region of the parent universe in which the daughter was created.

Theorist Lee Smolin as argued that this combination of ensemble of reproducing universes and (possibly random) mutations in their natural constants might amount to a form of Darwinian evolution – in which universes play the roles of living organisms, while their natural constants are analogous to genomes!

However, in order for this meta-cosmology of mutating and spawning universes to undergo evolution in the Darwinian sense, a further ingredient is of course needed – namely, natural selection.

Now, what does it mean that a given universe in this ensemble is "fit"? Smolin's answer is simple: the more fecund a given universe is at producing daughter universes, the universes will be added to the ensemble which have constants of nature close to those of the given universe (assuming that a daughter universe's constants are not very different from those of its parent).

And since universe-spawning is mediated via gravitational collapse, it follows that in Smolin's scenario, a universe in which the probability of black hole formation is higher, will on average produce more universes similar to it.

Smolin further argues that a universe's rate of black-hole formation is positively correlated with its probabilities of developing life (and possibly intelligence), although of course (as with all such arguments) the last inference is the least sound. This is simply because – as emphasized above – our existing scientific knowledge does not allow us to derive any estimates, whatever, of the local constants of nature and other environmental factors.

Peering ahead into the far future of human science, we may speculate that some day, our computing technology, in addition to advances in pure and applied mathematics and in physics and chemistry, will allow us to reliably estimate such probabilities. When that stage is reached, we may finally be able to determine whether our existence is inevitable, or whether (to the contrary) many universe had to live and die in order that one allowing the likes of us to emerge, happened to be spawned. Or, in the prescient words of the British empiricist philosopher David Hume:

"But were this world ever so perfect a production, it must still remain uncertain, whether all the excellences of the work can justly be ascribed to the workman. If we survey a ship, what an exalted idea must we form of the ingenuity of the carpenter who framed so complicated, useful, and beautiful a machine? And what surprise must we feel, when we find him a stupid mechanic, who imitated other, and copied an art, which,. through a long succession of ages, after multiplied trials, mistakes, corrections, deliberations, and controversies, had been gradually improving?

Many worlds might have been botched and bungled, throughout an eternity, ere this system was struck out; much labor lost, many fruitless trials made; and a slow, but continued improvement carried on during infinite ages in the art of world-making. In such subjects, who can determine, where the truth; nay, who can conjecture where the probability lies, amidst a greater which may be imagined?"

Epilogue

In the 20th century, a revolution occurred, one beyond the boldest dream of earlier scientists and engineers. In that century, science revolutionized technology to a degree that completely changed the daily life of man, society as a whole and in fact, the fate of mankind.

For three hundred years Western science pictured the world as a giant clock or machine, in which knowable causes produced predictable effects. It was a deterministic and totally ordered universe which, once set in motion, rendered all subsequent events meritable.

If this were an accurate description of the real world, the initial conditions of any process would determine its outcome — a machine-like universe set in motion by a Prime Mover, divine or otherwise.

If, on the other extreme, events were entirely random in an entirely accidental universe of completely random processes, there would be no law, regularity and predictability, and nature would have been nothing more than an endless series of random events, each with random consequences. It is unlikely that life, with its intricately regular processes, could ever arise – or indeed survive – in such a universe.

We now know that ours is a universe that combines both chance and necessity, chaos and order. Indeed, in accordance with the ideas of the 18th-century British empiricist philosopher David Hume, our *free will* and ability to detect causal relationships in the world around us, are both made possible by this dualistic nature of the universe we inhabit.

Appendices

- Philosophers of Nature
- Mathematicians
- Number Theorists
- Physicists
- Chemists and Biologists
- Mathematical Economists and Statisticians
- Historians of Science and Mathematics
- Earliest known use of Mathematical Symbols
- Earliest known use of Mathematical Terminology
- Photographs of celebrated scientists that flourished during 1800–1950
- List of Abbreviations and Acronyms
- The Greco – Latin Origins of Scientific Terminology
- Quotations

Table 6.15: PHILOSOPHERS OF NATURE

FL.	NAME	LIFE-SPAN
585 BCE	Thales of Miletos	(624–548 BCE)
560 BCE	Anaximander of Miletos	(611–547 BCE)
500 BCE	Heraclitos of Ephesos	(540–475 BCE)
470 BCE	Parmenides of Elea	(504–456 BCE)
450 BCE	Philolaos of Tarentum	(480–420 BCE)
420 BCE	Democritos of Abdera	(460–370 BCE)
375 BCE	Plato	(427–347 BCE)
350 BCE	Aristotle	(384–322 BCE)
50 BCE	Lucretius	(94–55 BCE)
50 CE	L.A. Seneca	(4 BCE–65 CE)
1260	Roger Bacon	(1214–1292 CE)
1610	Francis Bacon	(1561–1626 CE)
1640	René Descartes	(1596–1650 CE)
1660	Baruch Spinoza	(1632–1677 CE)
1685	Isaac Newton	(1642–1727 CE)
1770	Immanuel Kant	(1724–1804 CE)
1800	Johann Wolfgang von Goethe	(1749–1832 CE)
1810	Jean de Lamarck	(1744–1829 CE)
1860	Charles Darwin	(1809–1882 CE)
1890	Henri J. Poincare	(1854–1912 CE)
1920	Albert Einstein	(1879–1955 CE)

Table 6.16: MATHEMATICIANS

NAME	LIFE-SPAN
Pythagoras	(580–500 BCE)
Eudoxos	(408–355 BCE)
Euclid	(330–260 BCE)
Archimedes	(287–212 BCE)
Bhaskara	(1114–1185 CE)
P. Fermat	(1601–1665 CE)
I. Newton	(1642–1727 CE)
G.W. von Leibniz	(1646–1716 CE)
L. Euler	(1707–1783 CE)
J. Lagrange	(1736–1813 CE)
Laplace	(1749–1827 CE)
C. Gauss	(1777–1855 CE)
F. Abel	(1802–1829 CE)
C.G.J. Jacobi	(1804–1851 CE)
Hamilton	(1805–1865 CE)
Galois	(1811–1832 CE)
B. Riemann	(1826–1866 CE)
G. Cantor	(1845–1918 CE)
D. Hilbert	(1862–1943 CE)
Hardy	(1877–1947 CE)
Ramanujan	(1887–1920 CE)

Table 6.17: NUMBER-THEORISTS

FL.	NAME	LIFE-SPAN
535 BCE	Pythagoras	(580–500 BCE)
300 BCE	Euclid	(330–260 BCE)
250 CE	Diophantos	(206–290 CE)
1640	P. Fermat	(1601–1665 CE)
1750	L. Euler	(1707–1783 CE)
1780	J. Lagrange	(1736–1813 CE)
1800	A. Legendre	(1752–1833 CE)
1810	C. Gauss	(1777–1855 CE)
1830	C.G.J. Jacobi	(1804–1851 CE)
1860	B. Riemann	(1826–1866 CE)
1910	Hardy	(1877–1947 CE)
1917	Ramanujan	(1887–1920 CE)

Table 6.18: PHYSICISTS[55]

NAME	LIFE-SPAN	
Democritos	(460–370 BCE)	
Aristarchos	(310–230 BCE)	
Archimedes	(287–212 BCE)	
Eratosthenes	(276–194 BCE)	
Hipparchos	(180–110 BCE)	
Ptolemy	(85–165 CE)	
N. Copernicus	(1473–1543 CE)	
Galileo	(1564–1642 CE)	
J. Kepler	(1571–1630 CE)	
C. Huygens	(1629–1695 CE)	
I. Newton	(1642–1727 CE)	
C. Coulomb	(1736–1806 CE)	
J. Lagrange	(1736–1813 CE)	
W. Herschel	(1738–1922 CE)	
Laplace	(1749–1827 CE)	
Ampère	(1775–1836 CE)	
M. Faraday	(1791–1867 CE)	
Carnot	(1796–1832 CE)	
Hamilton	(1805–1896 CE)	
J. Maxwell	(1831–1879 CE)	
J.W. Gibbs	(1839–1903 CE)	
Boltzmann	(1844–1906 CE)	
M. Planck	(1858–1947 CE)	NL 1918

[55] Including greatest astronomers of antiquity and the Middle Ages.
NL = Nobel Laureate

Table 6.18: (Cont.)

NAME	LIFE-SPAN
M. Curie	(1867–1934 CE)
A. Einstein	(1879–1955 CE)
M. Born	(1882–1970 CE)
N. Bohr	(1885–1962 CE)
E. Schrödinger	(1887–1961 CE)
E. Hubble	(1889–1953 CE)
L. de-Broglie	(1892–1987 CE)
W. Pauli	(1900–1958 CE)
E. Fermi	(1901–1954 CE)
W. Heisenberg	(1901–1976 CE)
P. Dirac	(1902–1984 CE)
R. Feynman	(1918–1988 CE)

Table 6.19: CHEMISTS[56] (1665–1965)

NAME	LIFE-SPAN	
Robert Boyle	(1637–1691 CE)	
Joseph Priestly	(1733–1804 CE)	
C.W. Scheele	(1742–1786 CE)	
Nicolas Le Blanc	(1742–1806 CE)	
A.L. Lavoisier	(1743–1794 CE)	
A. Volta	(1745–1827 CE)	
C.L. Berthollet	(1748–1822 CE)	
John Dalton	(1766–1844 CE)	
A. Avogadro	(1776–1856 CE)	
Humphry Davy	(1778–1829 CE)	
J.L. Gay–Lussac	(1778–1850 CE)	
J.J. Berzelius	(1779–1850 CE)	
F. Wöhler	(1800–1882 CE)	
C. Goodyear	(1800–1860 CE)	
J. von Liebig	(1803–1873 CE)	
Thomas Graham	(1822–1869 CE)	
Louis Pasteur	(1822–1895 CE)	
Joseph Lister	(1827–1912 CE)	
A. Kekulé	(1829–1896 CE)	
D.I. Mendeleev	(1834–1907 CE)	
J.W. Hyatt	(1837–1920 CE)	
W.H. Perkin	(1838–1907 CE)	
H. Chardonnet	(1839–1924 CE)	
W.K. Röntgen	(1845–1923 CE)	NL 1901

[56] NL = Nobel Laureate

Table 6.20: (Cont.)

NAME	LIFE-SPAN	
Santiago Ramon y Cajal	(1852–1934 CE)	
Emil von Behring	(1854–1917 CE)	NL 1901
Paul Ehrlich	(1854–1915 CE)	NL 1908
Sigmund Freud	(1856–1939 CE)	
Charles S. Sherrington	(1857–1922 CE)	NL 1932
Thomas Hunt Morgan	(1868–1945 CE)	NL 1933
Karl Landsteiner	(1868–1943 CE)	NL 1930
Jules Bordet	(1870–1961 CE)	NL 1919
Leonor Michaelis	(1875–1949 CE)	
Karl von Frisch	(1886–1982 CE)	NL 1973
Selman Waksman	(1888–1973 CE)	NL 1952
Sewall Wright	(1889–1988 CE)	
Ronald Fischer	(1890–1962 CE)	
Alexander Oparin	(1894–1980 CE)	
Gerhard Domagk	(1895–1964 CE)	NL 1939
Gerta F. Cori	(1896–1957 CE)	NL 1947
Carl F. Cori	(1896–1984 CE)	NL 1947
Howard W. Florey	(1898–1968 CE)	NL 1945
Hans Adolf Krebs	(1900–1981 CE)	NL 1953
Andre Lwoff	(1902–1994 CE)	NL 1965
Gregory G. Pincus	(1903–1967 CE)	
Max Delbrück	(1906–1981 CE)	NL 1969
George Wald	(1906–1997 CE)	
Jacques Monod	(1910–1976 CE)	NL 1965

Table 6.20: (Cont.)

Name	Life-span	
Salvador Luria	(1912–1991 CE)	NL 1969
Jonas Salk	(1914–1995 CE)	
Max F. Perutz	(1914–2002 CE)	NL 1962
John C. Kendrew	(1917–1997 CE)	NL 1962
Rosalind Franklin	(1920–1958 CE)	
François Jacob	(1920– CE)	NL 1965
Christiaan Barnard	(1922–2001 CE)	
Motoo Kimura	(1924–1994 CE)	

Table 6.21: Mathematical Economists and Statisticians (S)

fl.	Name	Life-span
1817	David Ricardo	(1772–1823)
1838	Antoine Cournot	(1801–1877)
1840	Adolph Quetelet (S)	(1796–1874)
1870	William Jevons	(1835–1882)
1880	Francis Edgeworth	(1845–1926)
1880	Wilhelm Lexis (S)	(1837–1914)
1880	Leon Walras	(1824–1910)
1890	Vilfredo Pareto	(1848–1923)
1890	Alfred Marshal	(1842–1924)
1900	Carl Pearson (S)	(1857–1936)
1900	William Gosset (S)	(1876–1937)
1920	Ludwig von Mises	(1881–1973)
1920	Irving Fisher	(1867–1947)
1920	George U. Yule (S)	(1871–1951)
1925	Ronald Fisher (S)	(1890–1962)
1935	John M. Keynes	(1883–1946)
1950	Ragnar Frisch	(1895–1973)

Table 6.22: HISTORIANS OF SCIENCE AND MATHEMATICS (M)

FL.	NAME	LIFE-SPAN
325 BCE	Eudemos of Rhodes (M)	(360–300 BCE)
70 CE	Pliny the Elder	(23–79 CE)
340 CE	Pappos of Alexandria (M)	(290–350 CE)
435 CE	Maritanus Capella	(390–455 CE)
520 CE	Severinus Boethius	(480–524 CE)
1835 CE	William Whewell	(1794–1866 CE)
1835 CE	Auguste Comte	(1798–1857 CE)
1890 CE	Moritz Cantor (M)	(1829–1920 CE)
1890 CE	Agnes Mary Clarke	(1842–1907 CE)
1900 CE	Jules Tannery (M)	(1848–1910 CE)
1935 CE	George Sarton	(1884–1956 CE)
1950 CE	Otto Neugebauer (M)	(1899–1980 CE)

Table 6.23: EARLIEST KNOWN USE OF MATHEMATICAL SYMBOLS

SYMBOL	MEANING, NAME, USE	DATE INTRODUCED	INVENTOR
	Zero as a placeholder	ca 1000 BCE	Meso Americans
Greek Letters	Lettering of points, lines, planes	ca 440 BCE	Hippocrates of Chios
	Zero as blank space on counting board	ca 350 BCE	Chinese
$\parallel$	Parallelism	ca 150 CE	Heron, Pappos
o (omicron)	Zero as a placeholder	ca 150	Ptolemy
Greek Letters	As variables	ca 250	Diophantos
$\bigcirc$ (circle)	Zero, both as a *number* and a placeholder	ca 505	Hindus
e.g. $\frac{2}{3}$	Horizontal fraction bar	ca 1200	Al-Hassar
+, −	Plus and Minus	1486	J. Wideman
$\sqrt{\ }$	Square root	1525	C. Rudolff
	Multiplication by juxtaposition	1544	M. Stifel
[ ]	Brackets	1550	R. Bombelli
( )	Parentheses	1556	N. Fontana
=	Equals	1557	R. Recorde
tangent, sine, cosine	Trigonometric functions	1583	Thomas Finck(e)
.	Decimal point	1592	G.A. Magini

Table 6.23: (Cont.)

Symbol	Meaning, Name, Use	Date Introduced	Inventor
{ }	Braces	1593	F. Viéta
$>$	Greter than	1631	Thomas Harriot
$<$	Less than	1631	Thomas Harriot
$\cdot$	Multiplication (dot) [p=posthumously]	1631 (p)	Thomas Harriot
$\times$	Multiplication	1631	W. Oughtred
$\angle$	Angle	1634	Pierre Herigone
$\perp$	Perpendicularity	1634	Pierre Herigone
AB	Multiplication by juxtaposition	1637	R. Descartes
$a^1, a^2, a^3, \dots$	Exponents (positive integers only)	1637	R. Descartes
x, y, z	Letters for unknown quantities	1637	R. Descartes
a, b, c	Letters for known quantities	1637	R. Descartes
$\overline{AB}$	Line segment	1647	B. Cavalieri
log	Logarithm	1647	W. Oughtred
∞	Infinity	1655	J. Wallis
$a^{-1}, a^{1/2}, \dots$	Negative and fractional exponents	1656	J. Wallis
$\div$	The Obelus sign for division	1659	John Rahn

Table 6.23: (Cont.)

SYMBOL	MEANING, NAME, USE	DATE INTRODUCED	INVENTOR
$\therefore$	Therefore	1659	John Rahn
cot	Trigonometric cotangent	1674	J. Moore
$\int$	Integral sign	1675	G.W. von Leibniz
dx, dy, $\frac{dy}{dx}$	Infinitesimals and derivative	1675	G.W. von Leibniz
a^n	for any real number, n	1676	I. Newton
$\sim$	Geometrical similarity	1679	G.W. von Leibniz
$\simeq$	Congruence	1679	G.W. von Leibniz
:	Double-dot for both ratio and division	1698	G.W. von Leibniz
π	Ratio of circumference to diameter of circle	1706	W. Jones
/	Diagonal fraction bar	1718	Thomas Twining
e	2.718 281 828 149 045...	1728	L. Euler
A.S	Arcsine	1729	D. Bernoulli
$f(x)$	The function of x	1734	L. Euler
$\leq$	Less than or equal to	1734	P. Bouguer
$\geq$	Greater than or equal to	1734	P. Bouguer
$\neq$	Not equal	1740	L. Euler

Table 6.23: (Cont.)

SYMBOL	MEANING, NAME, USE	DATE INTRODUCED	INVENTOR
sh, ch	Hyperbolic functions	1750	Vincento Riccati
$\sum$	Summation	1755	L. Euler
Ψ'	$\frac{d\Psi}{dx}$	1770	J.L. Lagrange
∂	Partial derivation	1770	A.N. Caritat
sinh, cosh	Hyperbolic function	1771	J. Lambert
i	$\sqrt{-1}$	1777	L. Euler
$\cong$	Approximately	1777	J.F. Häseler
$\left(\begin{matrix} n \\ r \end{matrix}\right)$	Combination	1778	L. Euler
$\frac{\partial u}{\partial x}$	Partial derivative	1786	A.M. Legendre
lim	Limit	1786	S.A.J. L'Huuilier
$f'(x),\ f''(x)$	First and second derivatives of $f(x)$	1797	J.L. Lagrange
$D_x y$	Derivative of $y(x)$ w.r.t. x	1800	L.F.A. Arbogast
$\phi(m)$	Euler ϕ function	1801	C.F. Gauss
$\equiv$	Congruence of numbers	1801	C.F. Gauss
$n!$	Factorial	1804	C. Kramp
$\Gamma(n)$	Gamma function	1811	A.M. Legendre
sin^{-1}	inverse sine function	1813	J.F. Herschel

Table 6.23: (Cont.)

SYMBOL	MEANING, NAME, USE	DATE INTRODUCED	INVENTOR
$\int_a^b$	Limit of integration	1822	J.B. Fourier
$\delta - \epsilon$	In proofs	1823	A.L. Cauchy
∇	Nabla	1837	W.R. Hamilton
grad	Gradient operator	1860	J.C. Maxwell
B	Beta function	1839	P.M. Binet
$\vert\ \vert$	Determinant	1841	A. Cayley
$\vert x\vert$	Absolute value function of x	1841	K. Weierstrass
$a^2 + b^2$	Norm of $a + ib$	1842	L. Dirichlet
J	Bessel function	1843	P.H. Hansen
$y = mx + b$, $\frac{x}{a} + \frac{y}{b} = 1$	Equations of a line in analytic geometry	1848	G. Salmon
ζ	Zeta function	1857	B. Riemann
R	Rationals	1872	R. Dedekind
div	Divergence	1875	W.K. Clifford
$[x]$	Signum function (sign of x)	1878	L. Kronecker
Rho	Radian	1881	G.B. Halstead
∇^2	Laplacian operator	1883	R. Murphy
$\cap$, $\cup$	Intersection, union	1888	G. Peano
ln	Natural logarithm	1893	I. Stringham

Table 6.23: (Cont.)

Symbol	Meaning, Name, Use	Date Introduced	Inventor
$\aleph_0$	Aleph null	1893	G. Cantor
σ	Standard deviation	1894	K. Pearson
O	Big O	1894	P. Bachmann
ϵ	Membership	1894	G. Peano
N, n	Positive integer, integer	1895	G. Peano
$\exists$	Existence	1897	G. Peano
${}_nC_r$	Combination	1899	G. Chrystal
$\cdot$	Dot product	1902	J.W. Gibbs
$\times$	Vector product	1902	J.W. Gibbs
o	Little o	1909	E. Landau
$\pi(x)$	Number of primes less than x	1909	E. Landau
ϕ	Golden ratio	1914	Theodore Cook
$\oint$	Integral around closed path	1917	A. Sommerfeld
Z	The set of integers	1930	E. Landau
μ	Mean of normal distribution	1936	A. Fisher
C	Complex number	1939	N. Jacobson

Table 6.24: EARLIEST KNOWN MATHEMATICAL TERMINOLOGY

YEAR	WORD	MATHEMATICIAN
ca 1550 BCE	Pyramid	Ahmes Papyrus
ca 585 BCE	Geometry	Thales
ca 533 BCE	Odd, even and Prime numbers	Pythagoras
ca 380 BCE	Harmonic Mean	Archytas of Tarentum
ca 350 BCE	Axiom	Aristotle
ca 340 BCE	Logic, Number Theory	Xenocrates of Chalcedon
ca 300 BCE	Diameter, polygon, polyhedron, cube, trapezium	Euclid
ca 250 BCE	Helix	Archimedes
ca 230 BCE	Cylinder, asymptote, ellipse, parabola, hyperbole	Apollonios of Perga
ca 100 CE	Diagonal, Torus	Hero of Alexandria
ca 370	Analysis	Theon of Alexandria
ca 510	Radius, progression	Severinus Boethius
ca 550	Rational, irrational (numbers)	Cassiodorus
ca 630	Negative number	Brahmagupta
ca 825	Algebra	Al-Khowarizmi
ca 1145	Sinus (sine)	Robert of Chester
1202	Factor (noun), nominator, denominator, plus, minus	Fibonacci

Table 6.24: (Cont.)

YEAR, CE	WORD	MATHEMATICIAN
ca 1235	Product	Albertus Magnus
ca 1350	Zero	—
1386	Degree (angle)	G. Chaucer
1484	Natural number	N. Choquet
1543	Arithmetic and geometric progressions	M. Stifel
1557	Subtract, digit (number under 10)	Robert Recorde
1570	Vertex, discrete	John Dee
1571	Axis, orthogonal, convex	Thomas Digges
1580	Polynomial, coefficient	F. Viéta
1583	Secant, tangent (trigonometry)	Thomas Finck(e)
1594	Great circle	John Davis
1595	Trigonometry	B. Pitiscus
1596	Cosecant	Rheticus
1599	Cycloid	
1604	Focus (of an ellipse)	Johannes Kepler
1605	Pure Mathematics	Francis Bacon
1608	Operation (algebraic)	C. Clavius
1614–1617	Logarithm, decimal point	John Napier
1619	Isocahedron, dodecahedron (truncated)	Johannes Kepler

Table 6.24: (Cont.)

Year, CE	Word	Mathematician
1620	Cosine, cotangent	Edmund Gunter
1637	Real and imaginary numbers	René Descartes
1651	Spherical trigonometry	N. Mercator
1655	Continued fraction	J. Wallis
1657	Series	John Collins
1667	Convergent and divergent series	J. Gregory
1671–1672	Polar coordinates, ellipsoid	Isaac Newton
1673	Evolute	C. Huygens
1673–1694	Function (1673); Differential equation (1676); Calculus (1680); Differential calculus (1684); Algorithms (a systematic technique for solving a problem; 1684); Transcendental (1684); Differential (noun; 1690); Abscissa (1692); Coordinate (1692); Combinatorial (1694); Ordinate (1694); Variable (1694); Constant (1694)	G.W. von Leibniz
1685	Harmonic (conic)	Philippe de la Hire
1690	Integral calculus, Exponential function	Jakob Bernoulli
1694	Combination	Blaise Pascal

Table 6.24: (Cont.)

Year, CE	Word	Mathematician
1696	Normal	E. Scarburgh
1704	Vector	—
1706	Order (degree)	H. Ditton
1730	Degree	—
1738	Potential function	Daniel Bernoulli
1742	Binomial theorem	Colin Maclaurin
1753	Radius-vector (astronomy)	—
1756	Calculus of variations	L. Euler
1770	Statistics	W. Hooper
1779	Analytic geometry	S. Horley
1789	Conformal mapping	F.T. Schubert
1796	Formula	Richard Kirwan
1797	Analytic function, singular integral	J.L. Lagrange
1800	Definite integral	S-F. Lacroix
1801	Congruence (modular arithmetic)	C.F. Gauss
1811	Radix (base of a number system)	Peter Barlow
1812	Determinant	A.L. Cauchy
1814	Implicit function	—
1814	Commutativity, distributivity	F.J. Servois

Table 6.24: (Cont.)

Year, CE	Word	Mathematician
1820	Coordinate geometry	Matthew O'Brien
1821	Conjugate ($a \pm ib$)	A.L. Cauchy
1824	Algebraic number	N.H. Abel
1825	Torsion	L.T. Vallee
1825	Elliptic function	A.M. Legendre
1831	Separable variables	J.R. Young
1831–1840	Complex number, norm, potential function (1840)	C.F. Gauss
1839	Radius vector	J.R. Young
1840	Characteristic equation, absolute convergence	A.L. Cauchy
1843	Associative, Cartesian coordinate, vector, scalar, versor	W.R. Hamilton
1844	Inner product, outer product, vector space	H.G. Grassmann
1845	Linear transformation	A. Cayley
1846	X-axis	B. Peirce
1847	Topology	J.B. Listing
1848	Factor (verb)	J. Ray
1850	Geodesic	Joseph Liouville

Table 6.24: (Cont.)

YEAR, CE	WORD	MATHEMATICIAN
1851–1853	Invariant (1851); Canonical form (1851); Discriminant (1852); Jacobian (1852); Covariant (1853)	Joseph Sylvester
1854	Orthogonal matrix	Charles Hermite
1854	Bessel function	O.X. Schlömilch
1857	Order of Magnitude	I.T. Danson
1858	Field	R. Dedekind
1861–1863	Group theory, X-component etc (1863)	Joseph Sylvester
1870	Curl	J.C. Maxwell
1871	Radian	James Thomson
1871	Ideal (number theory)	R. Dedekind
1872	Limit point (accumulation point)	G. Cantor
1877	Covariant derivative	G. Ricci, T. Levi-Civita
1878	Vector product, scalar product, cross-ratio (of 4 point), divergence (of vector field)	W.K. Clifford
1878	Base of a vector space, rank of matrix	F.G. Frobenius
1878	Extremum, integral equation	P. Du Bois-Reymond
1878–1883	Graph (older sense, noun), latent root (1883)	Joseph Sylvester

Table 6.24: (Cont.)

Year, CE	Word	Mathematician
1879	Numerical analysis	S. Realis
1882	Abelian group (commutative)	H. Weber
1882	Isomorphism	W. Dyck
1883	Set (menge), closed set	G. Cantor
1884	Dyad	J.W. Gibbs
1885	Cofactor	W. Johnson
1888	Correlation, correlation coefficient	F. Galton
1890	Automorphic function	Felix Klein
1893	Circle of convergence, Integrand	John Harkness, F. Morley
1893	Standard deviation	Karl Pearson
1896	Group character	F.G. Frobenius
1896	Domain	—
1897	Gradient	Horace Lamb
1898	Tensor	W. Voigt
1900	Chi-square	Karl Pearson
1900	Class field	Teji Takagi
1901	Dot product	J.W. Gibbs
1901	Improper definite integral	E.H. Moore
1902	Adjoint of a matrix	L.E. Dickson

Table 6.24: (Cont.)

Year, CE	Word	Mathematician
1904	Kernel (as integral), Eigenvalue	David Hilbert
1904	Axiom of choice	E. Zermelo
1905	Positive definite	J. Pierpont
1906	Compactness	M.R. Fréchet
1907	Convergence in the mean	Ernst Fischer
1909	Cross product (vectors)	J.W. Gibbs
1911	Binomial distribution	G.U. Yule
1913	Hilbert space	F. Riesz
1916	Tensor Analysis	A. Einstein
1917	Stochastic	L.J. Bortkiewicz
1918	Gauge	H. Weyl
1919	Time Series	W.M. Person
1919	Central limit theorem	R. von Mises
1922	Saddle point	G.N. Watson
1922	Functional analysis (analytic properties of functionals)	Paul Levy
1923	Fourier transform	—
1927	Random number	—
1933	Convolution	N. Wiener
1934	Random variable	A. Winter
1942	Algebraic topology	Solomon Lefschetz

Photo 5: Gauss (1777–1855)

Photo 6: Hamilton (1805–1865)

Photo 7: Sylvester (1814–1897)

Photo 8: G. Stokes (1819–1903)

Photo 9: A. Cayley (1821–1895)

Photo 10: B. Riemann (1826–1866)

Photo 11: Maxwell (1831–1879)

Photo 12: Lord Rayleigh (1842–1919)

Photo 13: Hertz (1857–1894)

Photo 14: Minkowski (1864–1909)

Photo 15: Schrödinger (1887–1961)

Photo 16: Einstein (1879–1955)

A. PICCARD E. HENRIOT P. EHRENFEST Ed. HERZEN Th. DE DONDER E. SCHRÖDINGER E. VERSCHAFFELT W. PAULI W. HEISENBERG R.H. FOWLER L. BRILLOUIN
P. DEBYE M. KNUDSEN W.L. BRAGG H.A. KRAMERS P.A.M. DIRAC A.H. COMPTON L. de BROGLIE M. BORN N. BOHR
I. LANGMUIR M. PLANCK Mme CURIE H.A. LORENTZ A. EINSTEIN P. LANGEVIN Ch.E. GUYE C.T.R. WILSON O.W. RICHARDSON
Absents : Sir W.H. BRAGG, H. DESLANDRES et E. VAN AUBEL

Photo 17

Photo 18

Photo 19

List of Abbreviations and Acronyms

Å	*Ångström unit (=* 10^{-8} *cm)*
ABC	*American Broadcasting Corporation*
AC, ac	*Alternating current*
AFM	*Atomic Force Microscopy*
AFMR	*Anti-Ferromagnetic Resonance*
AI	*Artificial Intelligence*
am (AM)	*ante meridian (= before noon)*
APFIM	*Atomic Probe Field Ion Microscopy*
ATP	*Adenosine 5-TriPhosphate*
au	*astronomical unit (ca 150 million km)*
b.	*born*
BA	*Bachelor of Arts*
BBC	*British Broadcasting Cooperation*
BCE	*Before the Christian Era*
BCS	*Bardeen-Cooper-Schrieffer*
BEC	*Bose-Einstein Condensates*
BEEM	*Ballistic Electron Emission Microscopy*
BeV	*Billion electron-volt*
BMNH	*British Museum of Natural History*
BS, BSc	*Bachelor of Science*
c	*velocity of light (ca 300,000 km/sec)*
°C	*degrees Celsius*
CA	*California*
CALTECH	*California Institute of Technology*
CBI	*Cosmic Background Imager (telescope in Chile)*
CBS	*Columbia Broadcasting Service*
CBS	*Columbia Broadcasting System*
CCD	*Charged-Coupled Device*
CD	*Compact Disc*
CE	*Christian Era*
CERN	*Conseil Europeen pour Recherches Nucleaires*
c.g.s.	*centimeter, gram, second (system of units)*
cm	*centimeter*
CMB	*Cosmic Microwave Background*

CNN	*Cable News Network (Cellular Neural Network)*
COBE	*Cosmic Background Explorer*
COM	*Center of Mass*
CPU	*Central Processing Unit*
CRT	*Cathode-Ray Tube*
CT	*Computed Tomography*
d.	*died*
2D	*two-dimensional*
3D	*three-dimensional*
DALR	*Dry Adiabatic Lapse Rate*
DASI	*Degree Angular Scale Interferometer (at Antarctica)*
db	*decibel*
DBS	*Direct Broadcast Satellite*
DC, dc	*Direct Current*
deg.	*degree*
DNA	*Deoxyribo Nucleic Acid*
DVD	*Digital Video Disc*
e	*charge of the electron*
E	*East*
ECG	*Electrocardiogram*
ECT	*Emission Computed Tomography*
ed.	*editor*
EDSAC	*Electronic Delay Storage Automatic Computer*
EM	*ElectroMagnetic*
EMS	*Electron Momentum Spectroscopy*
ENIAC	*Electronic Numerical Integration and Calculation*
EPFA	*Erbium-doped Fiber Amplifier*
EPR	*Electron Parametric Resonance*
EPR	*Einstein-Podolsky-Rosen*
eq.	*equation*
eV	*electron-volt (atomic energy unit)*
$^\circ F$	*degrees Farenheit*
FC	*Fermat's Conjecture (last theorem)*
FEL	*Free Electron Laser*
FEM	*Field Emission Microscopy*
FFT	*Fast Fourier Transform*
fig	*figure*

FIM	*Field Ion Microscopy*
fl.	*flourished*
FLT	*Fermat Little Theorem*
FMR	*Ferromagnetic Resonance*
FRS	*Fellow of the Royal Society*

g	*gram (unit of mass or weight)*
G	*Newton's universal gravitational constant*
GB	*Giga Byte*
GB	*Great Britain*
gcd	*= greatest common divisor*
GMT	*Greenwich Mean Time*
Gr.	*Greek*
GTR	*General Theory of Relativity*

h	*hour*
$\hbar$	$= \frac{h}{2\pi}$ *(h = Planck's constant)*
H_0	*Present-day value of the Hubble Constant*
hcf	*= highest common factor (= gcd)*
HTML	*Hypertext Markup Language*

IBM	*International Business Machines*
ICBM	*Intercontinental Ballistic Missile*
IRT	*Infrared Reflecting Telescope*
ISS	*Ion Scattering Spectroscopy*
IT	*Information Technology*

Jr	*Junior*

$^\circ K$	*degrees Kelvin*
kg	*kilogram* (= 10^3 *gram*)
km	*kilometer* (= 10^3 *m* = 10^5 *cm*)

L	*Length*
laser	*light amplification by stimulated emission of radiation*
LCD	*Liquid Crystal Display*
lcm	*= lowest common multiplier*

LED	*Light-Emitting Diode*
LEP	*Large Electron Positron Collider*
LHC	*Large Hadron Collider (at CERN)*
l.h.s.	*left hand side (of an equation)*
LIDAR	*Light Detection and Ranging*
LIGO	*Laser Interferometer Gravitational-wave Observatory*
LINEAC	*Large Linear Electron Accelerator*
LISA	*Laser Interferometer Space Antenna*
log_{10}	*logarithm to base 10*
log_e, ln	*logarithm to base e (natural logarithm)*
LSP	*Lightest Supersymmetric Partner*
LY	$=$ *light year* $\approx 10^{13}$ *km*

m	*meter, minute*
MA	*Master of Arts*
MACHO	*Massive Compact Halo Object*
MAP	*Microwave Anisotropy Probe (satellite, 2001)*
maser	*microwave amplification by stimulated emission of radiation*
MB	*Mega Byte*
MD	*Doctor of Medicine*
MeV	*million electron-volt*
MFM	*Magnetic Force Microscopy*
mg	*milligram* $(= 10^{-3}$ *g)*
MHD	*MagnetoHydroDynamics*
MIR	*Multiple Internal Reflection*
MIT	*Massachusetts Institute of Technology*
ml	*milliliter*
mm	*millimeter* $(= 10^{-3}\ m = 10^{-1}\ cm)$
$m\mu m$	*millimicron* $(= 10^{-9}\ m = 10^{-7}\ cm = 10Å)$
mod	*modulus*
MOKE	*Magneto-Optic Kerr Effect*
MOND	*Modified Newtonian Dynamics*
MOSFET	*Metal-Oxide-Semiconductor-Field-Effect-Transistor*
MOT	*Magneto-Optical Traps*
MRI	*Magnetic Resonance Imagining*
M_s	*Surface-wave magnitude of earthquakes*
MT	*Megaton of TNT (Energy)*
M-theory	*An eleven-dimensional unification of Superstring Theory*
My	*Megayear* $(10^6$ *years)*
MYA, Mya	*millions years ago*

N	*North*
NAFS	*North Anatolian Fault System*
NASA	*National Aeronautics and Space Administration*
NBC	*National Broadcasting Corporation*
NBS	*National Bureau of Standards (renamed NIST)*
nm	nanometer (=mμm)
NMR	*Nuclear Magnetic Resonance*
NSOM	*Near-Field Scanning Optical Microscopy*
NY	*New York*
ODE	*Ordinary Differential Equation*
ORT	*Optical Reflection Telescope*
P	*Seismic Compressional Waves*
PAL	*Phase Alternating by Line*
PC	*Personal Computer*
PCR	*Polymerase Chain Reaction*
PDE	*Partial Differential Equation*
PEEM	*Photo Emission Electron Microscopy*
PEP	*Positron-Electron Project*
PET	*Positron Emission Tomography*
pH	*Potential of Hydrogen (level of acidity)*
Ph.D.	*Doctor of Philosophy*
pixel	*Picture Element*
pp.	*pages*
PST	*Pacific Standard Time*
PV	*PhotoVoltaic*
QCD	*Quantum ChromoDynamics*
QED	*Quantum ElectroDynamics*
QFT	*Quantum Field Theory*
QM	*Quantum Mechanics*
R	*Seismic Rayleigh Waves*
RADAR	*Radio Detection and Ranging*
RAM	*Random Access Memory*
RCA	*Radio Corporation of America*
ref.	*reference*
RF, rf	*radio frequency*

RHIC	*Relativistic Heavy Ion Collider (at Brookhaven N.Y.)*
r.h.s.	*right hand side (of an equation)*
RIA	*Radio Immuno Assay*
RNA	*RiboNucleic Acid*
ROM	*Read-Only Memory*
rpm	*revolutions per minute*
RSA	*Rivest-Shamir-Adelman (Public Key Cryptography)*
RT	*Radio Telescope*
R.V.	*Revised Version*

S	*South*
S	*Seismic Shear Waves*
SDI	*Space Defense Initiative*
SE	*Schrödinger's Equation*
sec.	*second (unit of time)*
SECAM	*Sequenced Color and Memory*
SEM	*Scanning Electron Microscopy*
SF	*Science Fiction*
SLAC	*Stanford Linear Accelerator Center*
SLC	*Stanford Linear Collider*
SLT	*The Second Law of Thermodynamics*
SNCF	*Societé Nationale de Chemins de Fer (France)*
SONAR	*Sonic Navigation and Ranging*
SORT	*Schmidt Optical Reflecting Telescope*
SPECT	*Single Photon Emission Tomography*
Sr	*Senior*
SRT	*Steerable Radio Telescope*
SSRL	*Stanford Synchrotron Radiation Laboratory*
STM	*Scanning Tunneling Microscopy*
S.T.P.	*Standard Temperature and Pressure*
STR	*Special Theory of Relativity*

tanh	*hyperbolic tangent*
TGV	*le Train à Grand Vitesse (France)*
TNT	*Tri-Nitro-Toluene (explosive)*
TPM	*Two-Photon Microscopy*
TV	*Television*

u *(or* amu*)*	*Atomic mass-unit (1/12 of the mass of an atom of Carbon 12)*

U	*Displacement on fault*
UFD	*Unique Factorization Domain*
UHV	*Ultra High Vacuum*
UK	*United Kingdom*
USA (US)	*United States of America*
USSR	*Union of the Soviet Socialist Republics*
UT	*Universal Time*

VCR	*Video Cassette Recorder*
VEI	*Volcanic Explosivity Index*
VLBI	*Very Long Baseline Interferometry*
VLSI	*Very Large Scale Integration*
VR	*Virtual Reality*

W	*west*
WIMP	*Weakly Interacting Massive Particle (a prime candidate for the Exotic Dark Matter in the universe)*
WKBJ	*Wentzel-Kramers-Brillouin-Jeffreys*
w.r.t.	*with respect to*
WWI	*World War I*
WWII	*World War II*

ya	*years ago*

μm	*micron* $(= 10^{-6}\ m = 10^{-4}\ cm = 10^{-3}\ mm)$
$[x]$	$=$ *greatest integer not exceeding* x *(e.g.* $[7.61] = 7$*)*

micro	=	$10^{-6} = \mu$	*mega*	=	$10^{6} = M$
nano	=	$10^{-9} = n$	*giga*	=	$10^{9} = G$
pico	=	$10^{-12} = p$	*tera*	=	$10^{12} = T$
femto	=	$10^{-15} = f$	*peta*	=	$10^{15} = P$
atto	=	$10^{-18} = a$	*exa*	=	$10^{18} = E$

The Greco–Latin Origins of Scientific Terminology

- Introduction
- The Greek Heritage
- The Latin Heritage
- Dictionary of word origins
- References

Introduction

Latin and Greek were the two common languages of Western scholars well into the 1800's. Much of this was influenced by the Catholic church which kept Latin alive in its ceremonies and in its illumination of the biblical scriptures. Ancient texts like the gospels and Greek myths were written in Greek and later translated into the Latin Vulgate by Church scholars and scribes. It was a natural place for those early scientists to go for a rich source of new descriptive words. Along with most any scholar, poets and other icons of literature up to the 20th century, were commonly schooled in Latin and Greek. They read in the great works of the ancients: Aristotle, Plato, Homer, Pythagoras, Lucretius, Marcus Aurelius in their original languages. Even well into the 20th century, a learned person was to some extent marked by his command of Latin and Greek.

Figure 6.2 shows the family tree of the Indo–European languages that includes Greek, Latin, French, English and German.

Table 6.25 lists the major historical events during 1750 BCE–1755 CE that triggered the gradual symbiosis of Greek and Latin with the English language.

Table 6.26 lists the leading progenitors of the Greco–Roman literary culture.

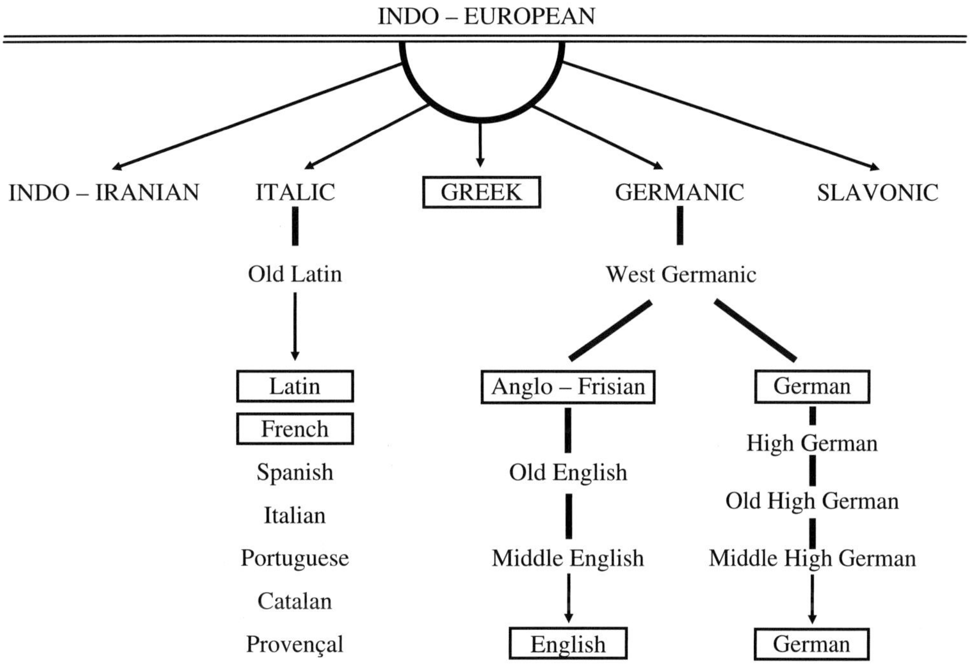

Fig. 6.2: The family of the Indo–European Languages relevant to Greek, Latin, English, German and French

Table 6.25: MAN, LANGUAGE AND HISTORY

DATE	EVENT
c.1750 BCE	*Appearance of the* SEMITIC ALPHABET
c.1450	*Earliest writings in the* HEBREW BIBLE
953	*Consecration of* **Solomon**'S TEMPLE *in Jerusalem*
776	*First* OLYMPIC GAMES *in Greece*
c.770	*Latin first brought to Italy by migrants from the north*
753	*Alleged founding of* ROME *in Latium by* ***Romulus***
750–585	*Age of* HEBREW PROPHETS: ***Amos***, ***Isaiah*** *and* ***Jeremiah***
c.750	***Homer's*** ILIAD AND ODYSSEY *in Greek writing arrived in Greece*
500–400	*The Golden Age of* ATHENIAN *democracy*
440–322	*Age of the* GREEK PHILOSOPHERS: ***Socrates***, ***Aplaton*** *and* ***Aristoteles***
323	*Death of* ***Alexander the Great***
285–250	*Jewish scholars first translated the Torah into Greek (Septuagint)*
200–100	ROME *extends its power in the Mediterranean*
146	GREECE *becomes a* ROMAN *colony*

Table 6.25: (Cont.)

Date	Event
100 BC–200 CE	*The age of "CLASSICAL LATIN" of the ROMAN LITERATI: (**Cicero, Lucretius, Cato, Vergil, Horace, Ovid, Seneca, Quintilian, Lucan, Martial, Tacitus, Marcus Aurelius, Juvenal**)*
55–54	***Julius Caesar** raids Britain*
c.7	***Jesus of Nazareth** born in the LAND of ISRAEL*
43 CE	*ROMANS conquer Britain*
122	*Hadrian's Wall built*
300–900	*The Age of "LATE LATIN"*
328	*Constantinople becomes the capital of the Eastern part of the ROMAN EMPIRE*
390–405	***Jerome** (HIERONYMUS) translated the Hebrew Bible into Latin. The translation is known as the Vulgate*
410	*ROMANS leave Britain*
476	*"Official" collapse of the Western ROMAN EMPIRE*
400–600	*The SAXONS, JUTES and ANGLES invade Britain, bringing the anglo–saxon language with them*
597	***Augustine** brings a new wave of CHRISTIANITY to Britain*
711	*The MOORS invade SPAIN, bringing their ARAB translations of GREEK books*

Table 6.25: (Cont.)

DATE	EVENT
900–1300	*Age of* "MEDIEVAL LATIN"
1066	*The* NORMANS *invade Britain, bringing Norman French with them*
1265–1374	END OF "MEDIEVAL LATIN:" *the poets* **Dante** *and* **Petrarch**
1266	**Roger Bacon** *publishes:* "OPUS MAJUS" *in Latin*
1300–1600	*Age of* "RENAISSANCE LATIN"
1453	*The Turks sack Constantinople*
1492	**Columbus** *crossed the Atlantic and heralds the colonization of the* AMERICAS
1500–1650	*Belated* "RENAISSANCE" *in Britain: rediscovery of the* ROMAN *civilization*
1543	**Andreas Vesalius** *published:* "DE HUMANI CORPORIS FABRICA"
1543	**Copernicus** *published* "DE REVOLUTIONIBUS ORBIUM COELESTIUM" *in Latin*
1600–1900	*Age of* "NEW LATIN"
1609–1621	**Kepler***'s publications:* "ASTRONOMIA NOVA" *(1609),* "HARMONICE MUNDI" *(1619) and* "MYSTERIUM COSMOGRAPHICUM" *(1621) in Latin*
1610	**Galileo***'s publication:* "SIDEREUS NUNICUS" *in Latin*

Table 6.25: (Cont.)

Date	Event
1611	*Publication of the King James version of the Hebrew Bible*
1616	*Death of* **William Shakespeare**
1620	**Francis Bacon** *published:* "NOVUM ORGANUM" *in Latin*
1644	**René Descartes** *published:* "PRINCIPIA PHILOSOPHIAE" *in Latin*
1675	**Baruch Spinoza** *completed his* "ETHICA ORDINE GEOMETRICO DEMONSTRATA" *in Latin*
1687	**Isaac Newton** *published:* "PRINCIPIA MATHEMATICA" *in Latin – the last great scientific work to be written in this language; Advent of the Scientific Revolution*
1755	**Samuel Johnson** *published his dictionary of the English language*
1776	*The motto "E Pluribus Unum" ("Out of Many — One") was proposed for the first Great seal of the United States by* **John Adams**, **Benjamin Franklin** *and* **Thomas Jefferson**. *This phrase offered a strong statement of the American determination to form a single nation from a collection of States and from people of many different backgrounds and beliefs*

Table 6.26: TIMELINE OF THE LEADING PROGENITORS AND TORCHBEARERS OF THE GRECO–ROMAN LITERARY CULTURE DURING 750 BCE–550 CE

I. Poets

1.	*Homer*	*c. 8th century BCE*	*Greek*
2.	*Hesiodos*	*fl. c. 700 BCE*	*Greek*
3.	*Archilochos*	*mid 7th century BCE*	*Greek*
4.	*Pindaros*	*c. 518–438 BCE*	*Greek*
5.	*Accius*	*c. 170–86 BCE*	*Latin*
6.	*Lucretius*	*c. 99–55 BCE*	*Latin*
7.	*Catullus*	*c. 84–54 BCE*	*Latin*
8.	*Vergil*	*c. 70–19 BCE*	*Latin*
9.	*Horace*	*c. 65–8 BCE*	*Latin*
10.	*Ovid*	*c. 43 BCE–17 CE*	*Latin*
11.	*Lucan*	*c. 39–65 CE*	*Latin*
12.	*Martial*	*c. 40–103 CE*	*Latin*
13.	*Statius*	*c. 45–96 CE*	*Latin*
14.	*Juvenal*	*c. fl. 127 CE*	*Latin*
15.	*Claudian*	*c.350–404 CE*	*Latin*

II. Thinkers (Philosophers[57], Natural philosophers, mathematicians)

1.	*Thales*	*c. 624–546 BCE*	*Greek phil.*
2.	*Anaximander*	*c. 610–546 BCE*	*Greek phil.*
3.	*Anaximenes*	*c. 585–525 BCE*	*Greek phil.*
4.	*Pythagoras of Samos*	*c. 582–507 BCE*	*Greek phil. and mathematician*
5.	*Xenophanes of Colophon*	*c. 570–480 BCE*	*Greek phil. and mathematician*
6.	*Heraclitos*	*c. 535–475 BCE*	*Greek phil. and mathematician*
7.	*Parmenides of Elea*	*c. 515–450 BCE*	*Greek phil. and mathematician*
8.	*Anaxagoras of Clazomenae*	*c. 500–428 BCE*	*Greek phil. and mathematician*
9.	*Empedocles of Acragas*	*c. 490–430 BCE*	*Greek phil. and mathematician*
10.	*Zeno of Elea*	*c. 490–430 BCE*	*Greek phil. and mathematician*
11.	*Leucippos*	*c. 480–420 BCE*	*Greek phil. and mathematician*
12.	*Antiphon*	*c. 480–411 BCE*	*Greek phil. and mathematician*
13.	*Protagoras*	*c. 480–420 BCE*	*Greek phil. and mathematician*

[57] The Hellenistic schools of thought included: *Cynicism*, *Epicureanism*, *Hedonism*, *Eclecticism*, *Neo-Platonism*, *Skepticism*, *Stoicism*, *Sophism*.
In Europe, the spread of Christianity through the Roman world marked the end of the Hellenistic philosophy and ushered in the beginning of Medieval philosophy.

II. Thinkers (cont.)

14. Hippocrates of Chios	*c. 470–410 BCE*	*Greek phil. and geometer*
15. Philolaos	*c. 480–405 BCE*	*Greek phil.*
16. Socrates	*c. 469–399 BCE*	*Greek phil.*
17. Democritos	*c. 460–370 BCE*	*Greek phil.*
18. Plato (Aplaton)	*c. 428–347 BCE*	*Greek phil.*
19. Diogenes of Sinope	*c. 400–325 BCE*	*Greek phil.*
20. Stilpo	*c. 380–330 BCE*	*Greek phil.*
21. Aristotle	*384–322 BCE*	*Greek phil.*
22. Epicuros	*341–270 BCE*	*Greek phil.*
23. Pyrrho	*c. 365–275 BCE*	*Greek phil.*
24. Zeno of Citium	*c. 333–263 BCE*	*Greek phil.*
25. Euclid	*c. 325–265 BCE*	*Greek phil.*
26. Chrysippos	*c. 287–212 BCE*	*Greek phil. and mathematician*
27. Eratosthenes	*c. 276–194 BCE*	*Hellenistic astronomer*
28. Panaetios	*c. 185–110 BCE*	*Greek stoic and Neo-Platonic phil.*
29. Posidonios	*c. 153–50 BCE*	*Stoic phil. and Historian*
30. Cicero	*106–43 BCE*	*Roman states-man and phil.*

II. Thinkers (cont.)

31.	*Philo Alexandrius*	*30 BCE–45 CE*	*Jewish philosopher*
32.	*Seneca*	*4 BCE–65 CE*	*Roman stoic phil.*
33.	*Epictetus*	*c. 55–135 CE*	*Roman phil.*
34.	*Claudius Ptolemy*	*c. 85–165 CE*	*Greek astronomer*
35.	*Marcus Aurelius*	*121–180 CE*	*Roman emperor and phil.*
36.	*Plotinus*	*c. 205–270 CE*	*Greco-Roman phil.*
37.	*Augustine of Hippo*	*354–430 CE*	*Church theologian*
38.	*Hypatia of Alexandria*	*c. 370–415 CE*	*Alexandrian mathematician*
39.	*Proclus Diadochus*	*411–485 CE*	*Greco-Roman phil.*
40.	*Boethius*	*472–524 CE*	*Christian phil.*

III. Prose authors (Historians, biographers, writers, playwrights)

a. Historians and Biographers

1.	*Herodotos*	*480–428 BCE*	*Greek hist.*
2.	*Thucydides*	*c. 460–399 BCE*	*Greek hist.*
3.	*Xenophon*	*c. 430–354 BCE*	*Greek hist.*

4.	*Cato the Elder*	*c. 234–149 BCE*	*Roman hist.*
5.	*Polybius*	*c. 230–118 BCE*	*Greek hist.*
6.	*Julius Caesar*	*c. 100–44 BCE*	*Roman Emperor and writer*
7.	*Cornelius Nepos*	*c. 100–25 BCE*	*Roman biographer*
8.	*Sallust*	*86–35 BCE*	*Roman hist.*
9.	*Strabo*	*64 BCE–c.30 CE*	*Greco-Roman geographer and hist.*
10.	*Diodoros Siculus*	*fl. 60–30 BCE*	*Sicilian Greek hist.*
11.	*Titus Livius (Livy)*	*59 BCE–17 CE*	*Roman hist.*
12.	*Josephus*	*c. 37–100 CE*	*Jewish hist.*
13.	*Plutarch*	*c. 45–128 CE*	*Greco-Roman hist.*
14.	*Tacitus*	*c. 56–117 CE*	*Roman hist.*
15.	*Pliny the Younger*	*c. 63–113 CE*	*Roman historian*
16.	*Suetonius*	*c. 71–135 CE*	*Roman biographer*
17.	*Dio Cassius*	*c. 150–235 CE*	*Roman hist.*
18.	*Ammianus Marcellinus*	4th *century CE*	*Roman hist. and biographer*
19.	*Eusebius*	*c. 260–340 CE*	*Greek-Christian hist.*

b. *Orators and Rhetoricians*

1. *Demosthenes*	*384–322 BCE*	*Greek*
2. *Cicero*	*106–43 BCE*	*Roman*
3. *Cato the Younger*	*95–46 BCE*	*Roman*
4. *Quintilian*	*35–100 CE*	*Roman*
5. *Aelius Aristeides*	*117–189 CE*	*Greek*

c. *Writers and playwrights*

1.	*Aesop*	*c. 620–560 BCE*	*Greek writer*
2.	*Aeschylus*	*c. 525–456 BCE*	*Greek playwright*
3.	*Sophocles*	*c. 496–406 BCE*	*Greek playwright*
4.	*Euripides*	*c. 480–405 BCE*	*Greek playwright*
5.	*Aristophanes*	*c. 450–385 BCE*	*Greek playwright*
6.	*Terence*	*c. 190–158 BCE*	*Roman playwright*
7.	*Varro*	*c. 116–27 BCE*	*Roman writer*
8.	*Publilius Syrus*	*first century BCE*	*Roman writer*
9.	*Tertullian*	*c. 160–240 CE*	*Roman writer*
10.	*Vegetius*	*fl. 380–400 CE*	*Roman writer*

The Greek Heritage

- Numerals
- Prefixes
- General Vocabulary
- Phrases and Epigrams

The Greek language has contributed to the English vocabulary directly as an immediate donor and, indirectly, through other intermediate languages, mainly Latin and French. In a typical English dictionary of 80,000 words, which corresponds very roughly to the vocabulary of an educated English speaker, about 5 percent of the words are borrowed from Greek directly, and about 25 percent indirectly.

Since the living Greek and English languages were not in direct contact until modern times, borrowings were necessarily indirect, coming either via Latin (through texts or various vernaculars), or from ancient Greek texts, not the living Language.[58]

More recently, a huge number of scientific, medical and technical neologisms have been coined from Greek roots.

Until the 16th *century, the few Greek words that were absorbed into English came through their Latin derivatives. Most of the early borrowings are for expressions in theology for which there were no English equivalents. In the late* 16th *century an influx of Greek words were derived directly in the new science.*

In the 19th *and* 20th *centuries a few learned words and phrases were introduced using more or less direct transliteration of Ancient Greek rather than the traditional Latin-based orthography (e.g. nous, hoi polloi).*

Many English words and word elements (roots, prefixes, suffixes) can be traced back to the ancient Greek language. For example: Demon, Lexicon, Colon, Stigma, scheme, bishop, priest, metaphor, mathematica, encyclopedia, hemoglobin, edema, dynamo, kinematica, physics, mechanics, electron, hadron, galaxy, optics, acoustics, thermodynamics – are of Greek origin.

Tables 6.27–6.29 present a glossary of Greek numerals, prefixes and general vocabulary used in the vernacular of scientists and other scholars. Table 6.30 shows some common Greek phrases and epigrams used today in the scientific and the general literature.

[58] Moreover, Greek culture and language had a direct major influence on Roman art, literature and science; Indeed, the seven Greek Muses [**Clio** (history), **Urania** (astronomy), **Calliope** (epic poetry), **Melpomene** (tragedies), **Euterpe** (harmony), **Erato** (lyric and love poetry), **Terpsichore** (dancing), **Thalia** (comedy), and **Polyhymnia** (music)] were so absorbed by the Romans that Quintilian was prompted to say: "Satura quidem tota nostra est" (At least satire is completely ours).

Table 6.27: GREEK NUMERALS

ORDINALS	CARDINALS
protos – first	*Hen – one*
deuteros – second	*Dyo – two*
tritos – third	*Treis, tria – three*
tetratos – fourth	*Tettara (tetra) – four*
	Pente – five
	Hex – six
	Hepta – seven
	Octo – eight
	Emea – nine
	Deka (deci) – ten
	Hendeka – eleven
	Dodeka – twelve
	Hekaton (hecto) – one hundred
	Chilioi (Kilo-) – one thousand
	Myrioi – ten thousand, innumerable (myriad)

Table 6.28: Greek Prefixes

Prefix	Meaning	Examples
A-, an-	*not, without, lack of*	*atom, anemia, achromatic, atypical, amoral, anesthesia, analgesic*
Amphi-	*around, about, both, in two ways*	*amphibious, amphitheater*
Ana-, ano-	*up, back, anew, again, throughout, against*	*analysis, anion, anamnesis, anabolism*
Anti	*against, opposed to, resisting*	*antibalistic,antitoxin, antiseptic, antiacid, antifreeze*
Ante	*before, in front of*	*antecedent*
Apo-	*away from, separation, lack*	*aphelion, apostasis, apogee, apologize*
Arche-, archi-	*first, chief, primitive (ancient)*	*archetype, architecture, archeology*
Cata-, Kata-, Cath-	*down, lower, under, complete, across*	*catastrophe, cathode, catarsis, catabolism, catatonia, catoptrics, catapult*
Di-	*twice, twofold, double*	*diatomic, dilemma*
Dia-	*through, across, apart, thoroughly*	*diameter, diagnosis, diabetes, diarrhea, diuretic*
Dicha-	*in two, double, asunder*	*dichotomy*
Dys-	*bad, difficult, hard, disorder, painful*	*dysentery, dystrophy, dyslexia*
Ecto-	*outer, outside*	*ectoplasm, ectozoa*
Ek-, ex-	*out from, outside, out of*	*eccentric, exit*

Table 6.28: (Cont.)

PREFIX	MEANING	EXAMPLES
En-, em-	*in, within, among*	*endemic (demos = people), embolism, enthusiasm*
Endo-, Ento-	*within*	*endocrine, entoplasm*
Epi-, eph-	*upon, on*	*ephemeral, eponym, epidemic, epidermis*
Eso-	*inward, within*	*esoteric*
Eu-	*well, good, normal, easy*	*eugenics, eulogy*
misos	*bad, hate*	*miscarriage, misanthropic, misogyny*
Exo-	*outside, outward, outer*	*exothermal, exogamy, exodus (gamos = marriage), exonerate*
Hemi-	*half, partly*	*hemisphere, hemin*
Hyper-	*above, over, excessive*	*hyperactive, hypersensitive*
Hypo-	*under, below, deficient, in bottom*	*Hippopotamus, hypothesis, hypoglycemia, hypodermic*
Meta-, meth-	*after, among, beyond, behind, change, transformation*	*metabolism, metaphysics, metamorphosis*
Pali(n)-	*back, again, once more, backwards*	*palindrome*
Para-	*by the side of, near, accessory, abnormal, prevented, beside*	*paranoia, paraxial, paragraph, paradigm*
Peri-	*all around, near*	*perimeter, periphery, pericardium, periscope*

Table 6.28: (Cont.)

PREFIX	MEANING	EXAMPLES
Pre-,pro-	*before, in front of, forward*	*prognosis, prophylactic, precede, progress*
Syn-, Sym-, Sy-	*with, together*	*sympathy, symbiosis, symphony, syndrome, synagogue, synapsis*

Table 6.29: GREEK GENERAL VOCABULARY

GREEK WORD	ENGLISH MEANING	EXAMPLES
Agora	*bringing together; assembly*	*category*
Aither	*the upper air*	*ether*
Akouein	*to hear*	*acoustic*
Algos, algesis	*pain, sense of pain*	*analgen, neuralgia*
Alos, allelon, allotrios	*other, different, external, foreign, of each other, another's*	*allergy, allotropic, parallelism*
Aner, andros	*a man, male*	*android, androgen*
Anthropos	*man, a human being*	*philanthropy, anthropoid, misanthrope*
Arthron	*joint, juncture of bones*	*arthritis*
Aster, astra	*star*	*astronomy*
Atmos	*air, breath, vapor, steam*	*atmosphere*

Table 6.29: (Cont.)

Greek Word	English Meaning	Examples
Autos	*self, by itself*	*automatic, autopsy, autism, autobiography*
Bakterion	*a little staff, rod, bacteria*	*bacteria*
Baros, barys	*weight, pressure, heavy*	*barometer, isobaric, baryon*
Bathys, bathos	*deep, inner, depth*	*bathysphere*
Bios, biosis	*a living, way of living, life*	*biology, microbe, symbiosis*
Bromos	*stench*	*bromid*
Brachys	*short*	*brachistochrone*
Breyin	*to be full, swell*	*embriology*
Chamai	*on the ground*	*chameleon*
Charakter	*to sharpen, engrave*	*character*
Chloros	*green, yellowish*	*chlorine, chlorophile*
Chole	*bile, gall, bitter, anger*	*cholic, cholera*
Choreia	*dancing*	*choreography*
Chromatos, chroma	*color of the skin, color*	*chromium, monochrome, chromosome*
Cyclo, gyro	*round*	*gyroscope, cyclone, cycloid*
Chronos	*time*	*chronic, synchronism, chronicle, chronology*
Crypto, kryptos	*hidden*	*cryptanalysis*

Table 6.29: (Cont.)

Greek Word	English Meaning	Examples
Daimon	*a divine power, demon*	*demoniac*
Demos	*country, land, people of a country*	*democracy, endemic, epidemic, demagogue*
Deuteros	*second, next*	*deuterium*
Dexios	*right (side)*	*dexterity*
Diaita	*life, way of living*	*diet*
Dosis	*a giving, dose*	*dose, antidote*
Dromos	*a running, course*	*syndrome, hypodrome*
Dynamis	*power, active force*	*hydrodynamics, dynamometer*
Eidsos, idea (id = to see)	*appearance, thought, mental impression*	*idea*
Eikon, eikonos	*image, likeness*	*icon*
Elektron	*amber, electricity*	*electron*
Enteron	*that within, the intestine*	*dysentery*
Eos	*morning red, dawn, an early age*	*eocene, eon*
Ergon	*work, functioning*	*erg, energy, allergy, synergic*
Ethnos	*race, nation*	*ethnic*
Gala, galactos	*milk*	*galaxy, galactose*
Ge	*earth*	*geode, geography, geology*

Table 6.29: (Cont.)

GREEK WORD	ENGLISH MEANING	EXAMPLES
Gen, genesis, genos	*to become, be born, coming into, being, origination, birth*	*gene, genesis, glycogen, halogen, genetics, genealogy*
Gigas, gigantos	*giant*	*giant*
Glykis, glykeros	*sweet, sugar*	*glucose, glycogen*
Gnonai, gnosis	*to know, judge, knowledge*	*gnostic, diagnosis, prognosis*
Graphein, gramma	*to draw, write, inscribe, record, something written down*	*monograph, graphic, telegram, cardiogram*
Gymnos	*naked*	*gymnasium, gymnasitcs*
Gyros	*circle, ring, turn*	*gyrate, gyroscope*
Haima, haimatos, hemo	*blood*	*anemia, leukemia, ischemia, hemoglobin*
Helios	*sun*	*helium, heliocentric*
Hemera	*day*	*ephemeral*
Heteros	*other, different from*	*heterogenous*
Hodos	*road, way, path*	*hodograph*
Holos	*entirely, all, whole*	*hologram, holistic*
Homos, homoios	*like, the same as, equal*	*homosexuality, homeopathy*
Hormon	*setting in motion, arousing, exciting*	*hormone*
Hydor, hydatos	*water, fluid*	*hydrogen, hydrostatic, hydrate*

Table 6.29: (Cont.)

Greek Word	English Meaning	Examples
Hygros	*wet, moist, fluid*	*hygrometer, hygroscopic*
Hypnos	*sleep*	*hypnosis*
Hystera	*womb, uterus*	*hysteria*
Idios	*one's own, peculiar*	*idiosyncrasy*
Isos	*equal*	*isomorphism, isomer*
Kakos	*bad, distorted, abnormal*	*cacophonia*
Kalos	*beautiful*	*calligraphy, calisthenics*
Kausos, kaustos	*burning, heat, burnt*	*caustic*
Kephale, enkephalon	*head, the brain*	*encephalic*
Kineein	*to move*	*Kinematics, cinema*
Klimax	*ladder*	*clima, climate*
Klinein, klisis, klinikos	*to bent, turn, slope, make recline, lean, lie down, pertaining to bed*	*clinic, inclination*
Kolla	*glue*	*collagen, colloid*
Koma, komatos	*deep sleep, coma*	*comatos*
Kranion	*cranium, skull*	*migrain*
Krinein, krisis	*to separate, distinguish, decide, emit, secrete, point of decision*	*crisis, critical*

Table 6.29: (Cont.)

Greek Word	English Meaning	Examples
Kryos, kristallos	*ice-cold, frost, clear ice*	*crystal, cryogen*
Kyanos	*dark blue*	*cyanamide*
Kyklos	*circle, ring, wheel*	*cyclic, cyclotron, epicycle*
Leon, leontos	*lion*	*leopard*
Leptynein	*to make thin, thin, delicate*	*lepton, lepidoma*
Lethe, lethargos	*oblivion, forgetfulness, drowsiness*	*lethe, letharge*
Leukos	*white*	*leukemia, leucite*
Lexis, legein	*to speak, speech, word, phrase, diction, reading*	*lexicon, dyslectic*
Lipos	*fat, oily*	*lipids, lipoma*
Lithos	*stone*	*Lithosphere, neolithic*
Logos, -logy	*word, speech, thought, reason, treatise, body of knowledge*	*biology, hematology, monologue, travelogue*
Makros	*long, large*	*macroscopic*
Mania	*madness, frenzy, enthusiasm*	*maniac, egomania, kleptomania*
Megas, megalou	*large*	*megalopolis, megalomaniac*
Meion	*less, smaller*	*miosis, miocardia*
Melas, melanos	*black, dark*	*melancholia, melanin*

Table 6.29: (Cont.)

Greek Word	English Meaning	Examples
Men, menos, meniskos	*month, semi-lunar*	*menopause, menstruation, meniscus*
Meros	*part, segment, partition*	*isomer,polymer*
Mesos	*middle, intermediate*	*meson, mesopotamia*
Metron	*measure, measuring instrument*	*geometry, kilometer, perimeter, parameter*
Mikros	*small, little*	*microbe, micron, microscope, microcosmos*
Mnaesthai, mneme	*to remember, memory*	*amnesia, mnenothecnic*
Monos	*alone, single, one*	*monotonic, monolith, monotheism, monograph, monosyllable, monocrome, monopoly, monoplane*
Moros	*dull, sluggish, stupid*	*moron*
Morph	*form, shape, figure*	*amorphous, morphology*
Myein	*to close, shut*	*myopia, myosis*
Nanos	*dwarf, billionth*	*nanosecond*
Narke	*numbness, stupor*	*narcosis, narcomania*
Naus, nautes	*ship, sailor*	*nausea, aeronautic*
Nekros	*dead body, dead*	*necrophil*
Neos	*new, young, recent*	*neodymium, neophyte, neonatal*
Neuron	*nerve, tendon, sinew*	*neuralgia, neurosis*

Table 6.29: (Cont.)

Greek Word	English Meaning	Examples
Nomos	*law, ordering*	*autonomic, economy, taxonomy*
Noos (nous), noema	*mind, thought*	*paranoia*
Nosos	*disease, sickness*	*nosogenic, neuronosis*
Oikos	*house, dwelling*	*economy, ecology*
Oligos	*few, small, scanty*	*oligarchy*
Onoma, onomatos	*name*	*eponymic, onomatopoeia, anonymous*
Optikos, optos (op = to see)	*pertaining to vision or the eye*	*optical, optics, cataoptrics*
Organ, ergon	*something that does work, instrument, tool, organ of the body*	*organism, organic*
Orthos	*straight, correct, normal*	*orthopedics, orthonormal, orthogonal*
Osmos	*thrusting, pushing*	*osmosis*
Oxys	*sharp, swift, quick, sour, acid*	*oxygen, oxymoron, dioxide, paroxysm, anoxia*
Palaios	*old, ancient*	*paleontology*
Pas, pantos, pan	*all, entire*	*panacea, pancreatic, pandemic, panorama, pantheon*
Pathein, pathos	*to be affected, experience, suffer feeling, disease*	*sympathy, telepathy, empathy, pathology*
Peiraein	*to attempt, try, test*	*empiric*

Table 6.29: (Cont.)

Greek Word	English Meaning	Examples
Phagein	*to eat, devour*	*bacteriophage, phagocyte*
Phain, phasma	*to bring to light, show, appear, apparition*	*fantasy, phantom*
Phanai, phon	*to speak, speech, voice, sound*	*phone, telephone, phonetics, cacophony*
Pherein, phoros	*to bear, carry, bring*	*euphoria, periphery*
Philein, philia	*to love, affinity for*	*philosophy, philanderer, philanthropy, philharmonic*
Phlegein, phlegma	*to burn, became hot, heat, flame*	*phlogiston, flame*
Phlogos	*inflammation*	*phlogosis*
Phobos	*fear, flight*	*hydrophobia, xenophobia*
Phos, photos	*light*	*phosphorous, photograph*
Phyein, physis	*to be by nature, arise, grow*	*physics, physical*
Planos, planktos	*wandering*	*plankton*
Plasscin, plastos	*to form, mold*	*plastic, plasma*
Polys	*much, many, more that usual*	*polygonal, multiply, polymer, polystyrene*
Pragma, praxis, praktikos	*a thing done, deed, fact, practical, fit for doing*	*pragmatism, practice*
Protos, proteins	*first, primitive, simple, primary*	*protocol, protein, proton, protozoan, prototype*

Table 6.29: (Cont.)

Greek Word	English Meaning	Examples
Pseudeo	*false, imaginary*	*pseudoscalar, pseudoscience, pseudonym*
Psyche	*spirit, soul, mind*	*psychic, psychotic, psychiatry*
Pyros	*fire, heat, fever*	*pyrotechnics, pyromania*
Rheein, rhema	*flow, stream, current*	*rheology, diarrhea, gonorrhea*
Rhythm	*rhythm, measure*	*arrhythmia*
Sakcharon	*sugar*	*sacchrine*
Sarx, sarkos	*flesh*	*sarcoma*
Seirein, seismos	*to shake, earthquake*	*seismology*
Sepsis, sapros	*rotting, decayed, putrid*	*asepsis, antiseptic, saprophyte*
Skopeein	*to look at, view*	*stethoscope, microscope*
Skotos	*darkness*	*scotoma*
Soma, somatos	*body*	*psychosomatic*
Sphaira	*sphere, ball, globe*	*spheroid, spherula*
Sta, stasis	*to cause to stand, set, fix, arresting*	*ecstasy, apostasis*
Steno	*narrow*	*stenographer*
Stereo	*solid*	*stereogeometry, cholesterol*
Sthenos	*strength*	*calisthenics*
Stizein, stigma	*to prick, puncture, brand, dot, mark*	*stigma*

Table 6.29: (Cont.)

Greek Word	English Meaning	Examples
Tachys	*swift, quick, rapid*	*tachion, tachycardia*
Tauto	*the same*	*tautology*
Taxis	*arrangement, order*	*syntactic, taxonomy*
Techne	*art, skill, craft*	*technic, technique*
Tele	*far off, at a distance*	*telepathy, television, telephone, telemetry*
Thenai, thesis	*to put, place, set down, a proposition*	*thesis, synthesis*
Theos	*God*	*theist, theology*
Therapeuein	*to take care of*	*therapy*
Therapeia	*to heal, treat medically*	*physiotherapy*
Thermos, therme	*hot, warm, heat*	*thermodynamics, diathermy*
Tomos, tome	*cutting, a segment*	*anatomy, atom*
Topos	*place, region, spot*	*topology, isotopes*
Trephein, trophe	*nourish, nourishment*	*atrophy*
Tropein	*to turn, bend*	*allotropic*
Xanthos	*yellow*	*xanthophyll*
Xenos, xenia	*foreign, stranger, host, guest, hospitality*	*xenophobia*
Zoon	*something living, an animal*	*zoology, protozoon*
Zone	*girdle, belt, zone*	
Zyme	*a ferment, fermentation*	*enzyme, zymase*

Table 6.30: List of Greek Phrases and Epigrams

(1) **Ageōmetrētos mēdeis eisitō**
(Let no-one without knowledge of geometry enter)
Motto over the entrance to Plato's Academy

(2) **Aei ho theos geōmetrei** *(Plato)*
(God always geometrizes)

(3) **Anthrōpos metron** *(Protagoras)*
(Man is the measure of all things)

(4) **Andrōn epiphanōn pasa gē taphos** *(Thucydides: Periclés Funeral Oration)*
(For illustrious men have the whole earth for their tomb)

(5) **Ariston men hudōr** *(Pindar)*
(Greatest however is water)

(6) **Gnōthi seauton [Noce te ipsum]**
(Know Thyself)
Inscribed in the forecourt of the Temple of Apollo at Delphi. Attributed to at least six Greek sages: Heraclitos, Chilon of Sparta, Thales of Miletos, Socrates, Phytagoras, Solon of Athens.
The saying "Know thyself" may refer by extension to the ideal of understanding human behavior, morals, and thought, because ultimately to understand oneself is to understand other humans as well. However, the ancient Greek philosophers thought that no man can ever comprehend the human spirit and thought thoroughly, so it would have been almost inconceivable to know oneself fully. Therefore, the saying may refer to a less ambitious ideal, such as knowing one's own habits, morals, temperament, ability to control anger, and other aspects of human behavior that we struggle with on a daily basis.

(7) **Diairei kai basileue**
(Divide and rule)

(8) **Dōs moi pā stō, kai tan gān kināsō** *(Archimedes)*
(Give me a place to stand and I will move the earth)

(9) **Hen oida hoti ouden oida** *(Socrates)*
(I know one thing, that I know nothing)

(10) **Heurēka!** *(Archimedes)*
(I found it)

(11) **Ē tan ē epi tas**
(Either with your shield, or upon it)
Spartan mothers to their sons before they went to battle

(12) **Lathe biōsas**
(Live in obscurity) *(Epicurean phrase)*

(13) **Métron áriston**
(moderation is the best thing)

(14) **Mē mou tous kyklous taratte** *(Archimedes)*
(Do not disturb my circles)

(15) **Mē cheiron vēltiston**
(The least bad (choice) is the best)
When there is no good option one should pick the one that does the least harm.

(16) ***Mēden agan*** *("Ne quid nimis" – st. Jerome)*
(Nothing in excess)
A carving from the temple of Apollo at Delphi

(17) ***Molōn labe!***
(Come take them!)
King Leonides of Sparta, in response to King Xerxes of Persia's demand that the Greek army lay down their arms before the battle of Thermopylae.

(18) ***Hoper edei deixai***
(Quod Erat Demonstrandum – Q.E.D.)
Used by early mathematicians including Euclid and Archimedes to signify the proof as complete.

(19) ***Outis emoi ǵ onoma***
(My name is Nobody)
Odysseus to Polyphemos when asked what his name was.
(Homer, Odyssey)

(20) ***Pistis, elpis, agapē*** *(1 Corinthians* ***13****,13)*
(Faith, hope, (and) love).

(21) ***Rhododaktulos Ēōs***
(Rosy-fingered dawn)
Occurs frequently in the Homeric Poems.

(22) ***Speude bradeōs***
(Latin: festina lente = less haste, more speed)

(23) ***Ta Panta rhei kai ouden menei*** *(Heraclitos)*
(Everything flows, nothing stands still)

(24) • *What is hard? — to know yourself.*
• *What is easy? — to advise others.*
• *What is quite common? — Hope.*
(When all is gone, there is still hope)
• *What is the fastest? — mind; it travels through all media.*
Thales

(25) **Phobou tous Danaous kai dōra pherontas**
(Quidquid id est) timeo Danaos et dona ferentes
(Beware of the Danaans (Greeks), even bearing gifts)
Virgil, "Aneid"

(26) **Khalepa ta kala** *(Plato)*
(The good things are hard to attain)

(27) *Oh stranger, tell the Spartans that here we lie, obedient to their laws.*
(Epigram by Simonides at Thermopylae)

(28) **Hoī polloī** *("Plebs urbanus" in Latin)*
(The many)

(29) **Hoi āristoi**
(The aristocracy)

(30) **Kalōs kīndyons**
(A beautiful risk)

(31) **Melēt e tō pān**
(Practice is everything; practice makes perfect)

(32) **Nike Somen**
(We shall overcome)
Said by the Greeks before battle.

(33) **Ou pollā, allā polȳ** *(Multum, non multa)*
(Not quantity, but quality)

The Latin Heritage

- History
- Prepositions and Conjunctions
- Numerals and Lengths
- Phrases and Epigrams
- Prefixes and Suffixes
- Phrases and Abbreviations
- Epigrams
- Root Groups
- Dictionary of Greco–Latin Word Origins

Latin	English
saed	*sad*
hete	*hate*
miser	*miserable*
imbibere	*imbibe*
maternus	*maternal*
radix	*root*
heah	*high*
waeccan	*wake (watch)*
drincan	*drink*
mercator	*merchant*
cantus	*chant*
taberna	*tavern*
elevare	*elevated*
observare	*observe*
accelerare	*accelerate*
descendre	*descend*
spowan	*speed*

As the Romans conquered most of Europe, the Latin language spread throughout the region.

In 1066 England was conquered by William, duke of Normandy, which is northern France. For several hundred years after the Norman invasion,

French was the language of court and polite society in England, It was during this period that many French words were borrowed into English. Linguists estimate that some 60% of our common everyday vocabulary today comes from French. Thus many Latin words came into English indirectly through French.

Many Latin words came into English directly, though, too. Monks from Rome brought religious vocabulary as well as Christianity to England beginning in the 6th century. From the Middle Ages onward many scientific, scholarly, and legal terms were borrowed form Latin.

Moreover, many of the early scientists spoke Latin or had learned it as part of their education. During the Dark Ages, the sciences and culture were segregated inside the monasteries where the spoken language was Latin. Thus, science started out using Latin as its universal language. Since the only real connections between countries through the Dark Ages were through the Roman Catholic Church, Latin became the standard lingual link between countries using different languages.

During the 17th and 18th centuries, dictionary writers and grammarians generally felt that English was an imperfect language whereas Latin was perfect. In order to improve the language, they deliberately made up a lot of English words from Latin words. For example, fraternity, from Latin fraternitas, was thought to be better than the native English word brotherhood.

Latin is a member of the family of Italic languages, and its alphabet, the Latin alphabet, emerged from the Old Italic alphabets, which in turn were derived from the Greek and Phoenician scripts (Table 6.31).

The Italic subfamily is a member of the Centum branch of the Indo-European language family. It includes the Romance languages (among others: French, Italian, Spanish, Portuguese, Romanian), and a number of extinct languages.

Latin was first brought to the Italian peninsula in the 9th or 8th century BCE by migrants from the north, who settled in the Latium region, specifically around the River Tiber, where the Roman civilization first developed. Latin was influenced by the Celtic dialects in northern Italy and the non-Indo-European Etruscan language in Central Italy, and by Greek in southern Italy.

Although surviving Latin literature consists almost entirely of Classical Latin, an artificial, highly stylized and polished literary language from the 1st century BCE, the actual spoken language of the Roman Empire was Vulgar Latin, which significantly differed from Classical Latin in grammar, vocabulary, and eventually pronunciation. Also, although Latin remained the main written language of the Roman Empire, Greek came to be the language spoken by the well-educated elite, as most of the literature studied by Romans *was written in Greek. In the eastern half of the Roman Empire, which became the Byzantine Empire, the Greek Koine of Hellenism remained current and was never replaced by Latin.*

Fig. 6.3: Approximate distribution of languages in Iron Age Italy during the sixth century BCE

We may divide the age of Latin into seven epochs:

–75 BCE	*Old Latin*
75 BCE–200 CE	*Classical Latin*
300–900	*Late Latin*
900–1300	*Medieval Latin*
1300–1600	*Renaissance Latin*
1600–1900	*New Latin*
1900–present	*Recent Latin*

Old Latin (also called Early Latin or Archaic Latin) refers to the period of Latin texts before the age of Classical Latin.

Classical Latin is the form of the Latin language used by the ancient Romans in what is usually regarded as "classical" Latin literature. Its use spanned the Golden Age of Latin literature – broadly the 1st century BCE and the early 1st century – possibly extending to the Silver Age – broadly the 1st and 2nd centuries.

What is now called "Classical Latin" was, in fact, a highly stylized and polished written literary language selectively constructed from Old Latin, of which far fewer works remain. Classical Latin is the product of the reconstruction of early Latin in the prototype of Attic Greek. Classical Latin differs from the earliest Latin literature, such as that of Cato the Elder, Plautus, and to some extent Lucretius.

The spoken Latin of the common people of the Roman Empire, especially from the 2nd century onward, is generally called Vulgar Latin. Vulgar Latin differed from Classical Latin in its vocabulary and grammar, and as time passed, it came to differ in pronunciation as well.

The golden age of Latin literature, in Latin Latinitas aurea, is a period consisting roughly of the time from 75 BCE to 14 CE, covering the end of the Roman Republic and the reign of Augustus Caesar. Many Classicists believe that this period represents the peak of Latin literature, and that its usage of the artificial and heavily stylized literary language known as Classical

Latin represents the ideal norm which other writers should follow. Classical Latin continued to be used into the Silver Age of Latin literature, 1st and 2nd centuries.

In reference to Roman literature, the Silver age covers the first two centuries CE directly after the Golden age (which was the first century BCE, and the start of the first century CE). Literature from the Silver age has traditionally, perhaps unfairly, been considered inferior to that of the Golden age. Silver Latin itself may be subdivided further into two periods: a period of radical experimentation in the latter half of the first century CE, and a renewed Neoclassicism in the second century CE.

Under the reigns of Nero and Domitian, writers like ***Seneca*** *the Younger,* ***Lucan*** *and* ***Statius*** *pioneered a unique style that has alternately delighted, disgusted and puzzled later critics. Stylistically, Neronian and Flavian literature shows the ascendance of rhetorical training in late Roman education. The style of these authors is unfailingly declamatory – at times eloquent, at times bombastic. Exotic vocabulary and sharply-polished aphorisms glimmer everywhere, though at times to the detriment of thematic coherence.*

Thematically, late 1st century literature is marked by an interest in terrible violence, witchcraft, and extreme passions. Under the influence of Stoicism, the gods recede in importance, while the physiology of emotions looms large. Passions like anger, pride and envy are painted in almost anatomical terms of inflammation, swelling, upsurges of blood or bile.

By the end of the 1st century, a reaction against this form of poetry had set in, and ***Tacitus***, ***Quintilian*** *and* ***Juvenal*** *all testify to the resurgence of a more restrained, classicizing style under Trajan and the Antonine emperors.*

Vulgar Latin (in Latin, sermo vulgaris) is a blanket term covering the vernacular dialects of the Latin language spoken mostly in the western provinces of the Roman Empire until those dialects, diverging still further, evolved into the early Romance languages – a distinction usually assigned to about the ninth century.

This spoken Latin differed from the literary language of classical Latin in its pronunciation, vocabulary, and grammar. Some features of Vulgar Latin did not appear until the late Empire. Other features are likely to have been in place in spoken Latin, in at least its basilectal forms, much earlier. Most definitions of "vulgar Latin" mean that it is a spoken language, rather than a written language, because the evidence suggests that spoken Latin broke up into divergent dialects during this period. Because no one transcribed phonetically the daily speech of any Latin speakers during the period in question, students of vulgar Latin must study it through indirect methods.

Romance languages, a major branch of the Indo-European language family, comprise all languages that descended from Latin, the language of the Roman Empire. The Romance languages have more than 600 million native speakers worldwide, mainly in the Americas, Europe, and Africa, as well as in many smaller regions scattered through the world.

All Romance languages descend from Vulgar Latin, the language of soldiers, settlers, and slaves of the Roman Empire, which was different from the Latin of the Roman literati. Between 200 BCE and 100 CE, the expansion of the Empire, coupled with administrative and educational policies of Rome, made Vulgar Latin the dominant native language over a wide area spanning from the Iberian Peninsula to the Western coast of the Black Sea. During the Empire's decadence and after its collapse and fragmentation in 5th century, Vulgar Latin began to evolve independently within each local area, and eventually diverged into dozens of distinct languages. The oversea empires established by Spain, Portugal and France after the 15th century then spread Romance to the other continents – to such an extent that about 2/3 of all Romance speakers are now outside Europe.

Medieval Latin refers to the Latin used in the Middle Ages, primarily as a medium of scholarly exchange and as the liturgical language of the medieval Roman Catholic Church. It is therefore largely synonymous with the term Ecclesiastical Latin (sometimes called Church Latin), which refers to the Latin language as used in documents of the Roman Catholic Church and in its Latin liturgies.

Renaissance Latin is a name given to the Latin written during the European Renaissance in the 14th–16th centuries, particularly distinguished by the distinctive Latin style developed by the humanist movement.

Ad fontes was the general cry of the humanists, and as such their Latin style sought to purge Latin of the medieval Latin vocabulary and stylistic accretions that it had acquired in the centuries after the fall of the Roman Empire. They looked to golden age Latin literature, and especially to Cicero in prose and Virgil in poetry, as the arbiters of Latin style. They abandoned the use of the sequence and other accentual forms of metre, and sought instead to revive the Greek formats that were used in Latin poetry during the Roman period. The humanists condemned the large body of medieval Latin literature as "gothic" – for them, a term of abuse – and believed instead that only ancient Latin from the Roman period was "real Latin".

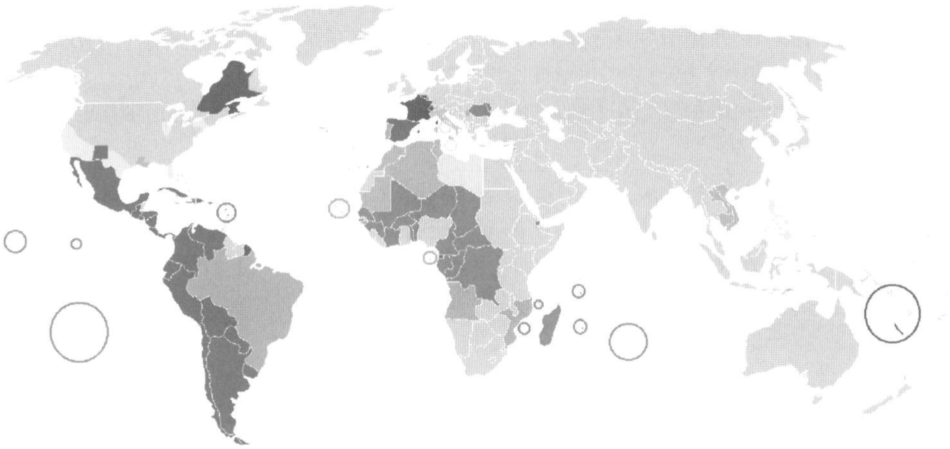

Fig. 6.4: Romance languages in the world: Blue – French; Green – Spanish; Orange – Portuguese; Yellow – Italian; Red – Romanian

The humanists also sought to purge written Latin of medieval developments in its orthography. They insisted, for example, that ae *be written out in full wherever it occurred in classical Latin; medieval scribes often wrote* e *instead of* ae. *They were much more zealous than medieval Latin writers that* t *and* c *be distinguished; because the effects of palatalization made them homophones, medieval scribes often wrote, for example,* eciam *for* etiam. *Their reforms even affected handwriting; Humanists usually wrote Latin in a script derived from Carolingian minuscule, the ultimate ancestor of most contemporary lower-case typefaces, avoiding the black-letter scripts used in the Middle Ages. Erasmus even proposed that the then-traditional pronunciations of Latin be abolished in favor of his reconstructed version of classical Latin pronunciation.*

The humanist plan to remake Latin was largely successful, at least in education. Schools now taught the humanistic spellings, and encouraged the study of the texts selected by the humanists, to the large exclusion of later Latin literature. On the other hand, while humanist Latin was an elegant literary language, it became much harder to write books about law, medicine, science or contemporary politics in Latin while observing all of the Humanists' norms about vocabulary purging and classical usage. Because humanist Latin lacked precise vocabulary to deal with modern issues, their reforms accelerated the process of turning Latin from a workday language to an object of antiquarian study. Their attempts at literary work, especially poetry, often have a strong element of pastiche. Their efforts turned Latin from a classical, but still useful language, into a truly extinct language. Latin vocabulary continued to be used by the creators of New Latin, but extensive discourses on contemporary subjects in Latin gradually ceased to be written during this period.

New Latin (or Neo-Latin) is a post-medieval version of Latin, now used primarily in International Scientific Vocabulary cladistics and systematics. The term came into widespread use towards the end of the 1890s among linguists and scientists.

Table 6.31: LATIN PREPOSITIONS AND CONJUNCTIONS

Alieni	*another's*
Aut	*either, or*
Cam	*with*
Cum	*since*
Circa	*about, approximately*
Donec	*until*
Dum	*while*
Enim, nam	*for*
Ergo, igitur	*therefore*
Et	*and*
Etiam, quoque	*also*
Ibi	*there*
Iuxta	*next to*
Modo, tantum	*only*
Neque	*nor*
Ob, propter	*because, because of*
Prope	*near*
Postquam	*after that*
Que	*and*
Quod	*because*

Si	*if*
Sed	*but*
Sine	*without*
Tanem, autem	*however*
Tandem	*at last*
Ubi	*where*
Ut	*so that*
Utinam	*may it happen*
Sui	*one's own*

Table 6.32: LATIN NUMERALS

ORDINALS	CARDINALS
Primus – first (primary)	*Unus – one (union, uniform, unit, univalent)*
Secundus – second	*Duo – two (duplex, duplicate)*
Tertius – third (tertiary, trident)	*Tres, tri – three*
Quartus – fourth	*Quattuor – four*
Quadrus – fourfold	
Quintus – fifth	*Quinque – five*
Sextus – sixth	*Sex – six*

Ordinals	Cardinals
Septimus – seventh (September)	*Septen – seven*
Octavius – eight (octavo)	*Octo – eight (octave)*
Nonus – ninth (Novembers)	*Novem – nine*
Decimus – tenth (December)	*Decem – ten*
Undecimus – eleventh	*Undecima – eleven*
Duodecimus – twelfth	*Duodecim – twelve (duodocimal)*
Vicesimus – twentieth	*Viginti – twenty*
Centesimus – hundredth	*Centum – one hundred*
Millesimus – thousandth	*Mille – one thousand*
Semi, demi – 1/2 (semicircle, semifinal)	*plus – more*
Sesqui – 3/2	*minus – less*
Bi, bis, bin – twice, twofold, double (bicycle, biped, binaural)	*equ – equal (equidistant, equilibrium)*

Table 6.33: Roman Lengths

cubit : *from the Latin cubitum (elbow). The unit represents the length of a man's forearm from his elbow to the tip of his outstretched middle finger (c. 52.35 cm)*

foot : *from the Latin pes naturalis. Based directly on the length of the human feet (31.3 cm). The modern foot (c. 30.5 cm) may have been the invention of Henry I (1100–1135)*

inch	:	*from the Latin uncia (twelfth part). Defined as the width of man's thumb at the base of the nail.*
Pace	:	*from the Latin passus (step). The distance between two successive falls of the same foot (c. 1.5 m).*
mile	:	*from the Latin milia or mille (a thousand paces): the distance a Roman legion could march in thousand paces. It's the measured distance between surviving milestones of Roman roads (close to 500 feet). Roman had set their mile equal to 8 stadia (c. 202.3 yards).*
league	:	*from the Latin leuga or leuca; it was intended to represent, roughly, the distance a person could walk in an hour (c. $1\frac{1}{2}$ Roman miles)*

Table 6.34: LATIN NEGATION PREFIXES

PREFIX	ENGLISH MEANING	EXAMPLES
In-, il-, ir-, im-	*not, beyond belief*	*infallible, impossible, immiscible, invalid*
Un-	*not*	*undo*
Non-	*not*	*non-portable, nonsense, nonstop*
Contra-	*against*	*contradict, contravene, contrarevolution*
Anti-	*against, opposed to, resisting*	*antiwar, antithesis, antisymmetrical, antibacterial, antitoxin, antiseptic*
Mis-	*badly, ill*	*miscast, misanthrope*
Mal-	*bad*	*malcontent, malpractice*
De-	*draw or remove from, down*	*devalue*

Table 6.34: (Cont.)

Prefix	English Meaning	Examples
Dis-	*apart, away from, not any*	*disport, dissect, dissonant, disbelief, discharge, disable*
Ab-	*from, away*	*abduct, abjured, aberrant*
A-, an-	*not, without, lack of*	*atom, anomaly, anoxia, amnesia, anemia, analgesic, anesthesia*
Ex-	*out of*	*export, exorbitant, ex officio, expatria*
Ob-, o-, op-	*against, in the way*	*obovate*

Table 6.35: Latin Spatial and Positional Prefixes

Prefix	English Meaning	Examples
Super-, sur-	*above, over, excessively*	*supervise, supernatural, superrealistic*
Extra-	*beyond, outside, outer, is addition to*	*extracurricular*
Dia-	*through, between*	*diameter, diaphanous*
Sub-, sup-	*under, below, near somewhat*	*submarine, suffix, subway*
Hypo-	*under*	*hypodermic, hypothesis*
De-	*down, off, from, undoing*	*descend, dementia, defrost, depart*
Epi-	*upon, beside, on*	*epidermis, epitome*

Table 6.35: (Cont.)

Prefix	English Meaning	Examples
Trans-	*across, beyond, through*	*transport, transformation*
Ambi-	*about, on all sides*	*ambition, ambidextrous, ambivalent*
Circum-	*around*	*circumnavigate, circumvent, circumpacific*
Peri-	*around*	*perimeter, perigee, pericardium*
Inter-	*between*	*intercom, internecine, interstate, intercourse, interatrial*
Ante-	*in front of, forward*	*anteroom*
Ultra-	*beyond*	*ultrasound*
Endo-	*within*	*endoscopy*
Supra-	*above, upon, upper*	*suprapelvic*
Se-	*apart, free from*	*secretion, segregation*
Retro-	*backward, behind, back*	*retrogress*
Re-	*again, back, backward*	*refraction, return, reappear, refrigerate, regurgitate*
Co-, com-, con-	*together with*	*cohesion, concert, coauthor, congress, coequal, conjugate, conspiracy, continue, cognate, colloquy, convene, concise, conduit, contravene*
A-, ad-	*at, to, toward, near*	*adhesion, admix, allegation, addict, ablate, abstract, annotate*

Table 6.35: (Cont.)

PREFIX	ENGLISH MEANING	EXAMPLES
In-, im-, ir-, il-	*in, into, on, against, onto*	*immerse, illicit, intangible, imbibe, irruption*
Infra-	*below, lower*	*infrared, infrastructure*
Intro-	*within, inward*	*introspect*
Intra-	*within, inside*	*intravenous*
Juxta-	*beside, near to*	*juxtaposition*
Pro-, pur-	*in front of, front part of, in favor of*	*prodigal, purport, protract, prospect*
Per-, pel-	*throughout, thoroughly, excessively, complete*	*pervade*
Sym-, syn-	*with, together*	*syntax, synthesis, syndicate*

Table 6.36: LATIN TEMPORAL PREFIXES

PREFIX	ENGLISH MEANING	EXAMPLES
Post-	*after, behind*	*postmortem, postpone, postnatal*
Pre-	*before*	*prepared, presume, prehistory, premeditated*

Table 6.37: Latin Suffixes

Suffix	Connotation	Examples
-or	*the agent which does*	*vector (vehera, vectum = carry)*
-io, -ura	*condition resulting from action*	*ligatio (ligare = tie), tensio (tendere = stretch)*
-men, -mentum	*action resulting from action*	*sedimentum (sedera = sit), tegmen (tegere = stand)*
-culum, -ulum	*instrument, means of action*	*curriculum, mandibula (mandare = chew)*
-ra, -itas, -or	*quality, condition, state*	*dementia (demens = out of one's senses), magnitudo (magnus = large)*
-arium, -orium, -ium	*place, apparatus, area for work*	*aquarium, sanatorium, auditorium, planetarium*
-ueas, -culus, -ellus, -illus	*small (diminutive)*	*calculus, globulus, capsula, ventriculus, flagellum, lamella*
-ilib, -bilis	*ability, capability, capacity*	*facilis, fissilis, fragilis*
-idus	*in a state of*	*fluidus, rigidus*
-eus, -ius	*made of, like, having the nature of*	*virilis, senilis*
-anus, -enus	*place, origin, belonging to*	*americanus, africanus*
-osus, -lentus	*full of*	*fibrosus, corpulentous*
-ate	*put into action, perform*	*elaborate, radiate*

Table 6.37: (Cont.)

SUFFIX	CONNOTATION	EXAMPLES
-sc	*to begin an action*	*calescence, deliquesce*
-ad	*in the direction of, toward*	*ectad, retrad*
-ase	*enzyme*	*oxidase*
-ose	*carbohydrate*	*dextrose*
-ate	*salt*	*sulphate*
-ite	*salt*	*sulphite*
-id (ide)	*a compound of two elements*	*ferric oxide*
-ol	*alcohol or phenol*	*glycerol*
-duc	*to lead, bring*	*deduce, produce, reduce*
-dict	*to say*	*contradict, edict, predict, dictate*
-gress	*to walk*	*progress, digress, transgress*
-ject	*to throw*	*eject, inject, project, reject, subject*
-pel	*to drive*	*compel, impel, repel*
-pend	*to hang*	*append, depend*
-port	*to carry*	*deport, export, import, report, support*
-tract	*to pull, drag, draw*	*attract, contract, extract, retract*
-vert	*to turn*	*convert, divert, invert, revert*

Table 6.38: LATIN TERMS, PHRASES AND ABBREVIATIONS

Mathematics is an ancient discipline, and consequently it has picked up a good deal of terminology over the years that is not commonly used in ordinary discourse. Phrases and terms from Latin make up a large part of this terminology, and reading mathematical texts - especially more advanced ones – is made easier if one is equipped with knowledge of these terms in advance.

We review below the Latin terms most commonly used in mathematics, and follow with a more extensive list of such terms and phrases as one may run into more rarely or in other contexts.

Note that when Latin or other non-English words are used in writing, they should be italicized except where they are abbreviated as single letters.

ab initio	*From the beginning.*
accessit	*Honorable mention.*
a.d.	*See anno domini.*
ad cautelem	*for safety sake: to be on the safe side.*
ad hoc	*For the immediate purpose. An ad hoc committee is appointed for some specific purpose.*
ad hominem	*"To the man." An argument is ad hominem when it attacks the opponent personally rather than addressing his arguments.*
ad infinitum	*Literally, "to infinity," indicates that a process or operation is to be carried out endlessly.*
ad nauseam	*Something continues ad nauseam when it goes on so long you become sick of it.*
ad rem	*to the point.*
a fortiori	*"With stronger reason." If every multiple of two is even, then a fortiori every multiple of four is even.*
alias	*also known as (at another time).*

alibi	*elsewhere.*
alma mater	*Your alma mater is the university or college which granted your degree.*
alumnus/alumna	*An alum, as it is sometimes shortly said, is a former member/student of a university or college. (The "us" ending is masculine, the "a" ending feminine. The plurals are alumni and alumnae, respectively.)*
agenda	*things that have to be done.*
a.m ante meridiem	*(before noon)*
anno domini	*"In the year of Our Lord." Indicates that a date is given in the Western or Gregorian calendar, in which years are counted roughly from the birth of Christ.*
annus mirabilis	*a wonderful year.*
a posteriori	*"From effect to cause." A thing is known a posteriori if it is known from evidence or empirical reasoning.*
a priori	*A thing is known a priori if it is evident by logic alone from what is already known.*
arguendo	*for the sake of argument; hypothetically.*
ars perdita	*a lost art.*
B.A.	*Baccalaureus Artium*
bona fide	*"In good faith." One's bona fides are documents or testimonials establishing one's credentials or honesty.*
B.Sc.	*Baccalaureus Scientiae*
carpe diem	*"Seize the day." A motto which says to live in the now, and/or to not waste time or opportunity.*
ceteris paribus	*other things being equal.*

cf. *See confer.*

circa *Approximately. Used with dates, abbreviated as c or ca.*

confer *"Compare." Usually abbreviated cf. and often used in footnotes, this indicates that one should compare the present passage or statement with the one referred to.*

cum laude *"With praise." Used on degree certificates to indicate exceptional academic standing.*

de facto *"In reality." Used to indicate that, whatever may be believed or legislated, the reality is as indicated here.*

de jure *"In law." Contrast to de facto.*

dixi *That settles it. Literally, "I have spoken."*

e.g. *See exempli gratia.*

emeritus *(feminine: emerita) Indicates someone who has served out his or her time and retired honorably.*

ergo *Therefore.*

erratum/errata *Literally, "error/errors," this term in fact refers to the corrections included in a paper or book after it is published to correct minor errors in the text.*

et al. *Abbreviation of et alia, meaning "and others." Used to indicate an unstated list of contributing authors following the main one, for instance.*

et cetera *And so forth.*

exempli gratia *"For example." Usually abbreviated to "e.g." and often confused with "i.e."*

ex paritate *by analogy.*

ex post facto *"From what is done afterward."*

fac simile	*make a copy (fax).*
grand nimis	*too great.*
hic of nunc	*here and now.*
ibid.	*See ibidem.*
ibidem	*"In the same place." Used in footnotes to indicate that the reference is the same as the preceding one(s).*
id est	*Literally, "that is." Usually abbreviated "i.e." and often confused with "e.g." The decision whether to use "i.e.," or "e.g." should be based on whether "that is" or "for example" is what is wanted in the sentence.*
i.e.	*See id est.*
in extenso	*in full.*
in globo	*globally.*
in re	*"In regards to." Often used to head formal correspondence. When only re is written, it should be translated as "regarding", or "concerning".*
inter alia	*Among other things.*
in toto	*Entirely.*
in vacuo	*Literally, "in a vacuum." Should be taken to mean "in the absence of other conditions or influences."*
in vitro	*taking place in the laboratory test tube, artificial.*
in vivo	*taking place in a living organism.*
ipso facto	*Literally, "by that very fact."*
lapsus calami	*a slip of the pen.*
LL.D.	*Legum Doctor*

loc. cit loco citato	*in the place quoted*
M.A.	*Magister Artium*
magna cum laude	*With great praise. See cum laude.*
magnum opus	*a masterpiece*
M.D.	*Medicinae Doctor*
modus operandi	*Manner or method of work characterizing a particular person's professional habits.*
M.Sc.	*Magister Scientiae*
mutatis mutandis	*With necessary changes. "Mutatis mutandis, this proof applies in more general cases."*
n.b.	*See nota bene*
non sequitur	*"Not following." Used to indicate a statement or conclusion that does not follow from what has gone before.*
nota bene	*Literally, "note well." Usually abbreviated "n.b.", this is a way of saying, "take note of this."*
para avis	*a rare bird.*
passim	*everywhere.*
per impossibile	*"As is impossible." Qualifies a proposition that cannot be true.*
per se	*"In and of itself." Example: "This argument does not force the conclusion per se, but with this added premise the result would follow."*
Ph.D.	*Philosophia Doctor*
plusve minusve	*more or less.*

p.m. post meridiem	*after noon*
post hoc, ergo propter hoc	*"After, therefore because of." A common fallacy in reasoning, in which causality is ascribed to preceding conditions which were in fact irrelevant to the supposed effect.*
post scriptum	*"Written after." Indicates an afterword or footnote to a main text, and is often used in written correspondence (where it is abbreviated p.s.).*
prima facie	*"On its face." Indicates that a conclusion is indicated (but not necessarily proved) from the appearance of things.*
pro forma	*"For form's sake." E.g., "It was a pro forma interview – the decision to hire her had already been made."*
Q.E.D	*See quod erat demonstrandum.*
Q.E.F	*See quod erat faciendum.*
qua	*"In the capacity of."*
quod erat demonstrandum	*"That which was to have been proved." Traditionally placed at the end of proofs, the QED is now usually indicated by a small square.*
quod erat faciendum	*"That which was to have been shown." Abbreviated QEF, it was traditionally used to mark the end of a solution or calculation. It is rarely used now.*
quod vide	*Usually abbreviated q.v., this is a scholarly way of directing the reader to a reference.*
q.v.	*See quod vide.*
R.I.P. requiescat in pace	*Rest In Peace*

semper idem	*always the same.*
sic!	*wrong, but that was how the original speaker said or wrote it.*
sine qua non	*"That without which nothing." Indicates an essential element or condition.*
stet (editorial)	*let it stand.*
summa cum laude	*With greatest praise. See cum laude.*
summum opus	*his greatest work.*
s.v. sub voce	*under that word*
tabula rasa	*"Blank Slate." Often refers to a person who has not yet formed prejudices or preconceptions on a given matter.*
temp tempore	*in the period of*
terra incognita	*unknown land.*
vade mecum	*"go with me" – a favorite book, guidebook.*
verbatim	*Word-for-word. Indicates a precise transmission of a phrase, discussion, or text.*
via	*by way of.*
videlicet	*Usually abbreviated viz., this is translated as "namely" or "in other words", or "that", or "to say".*
viz.	*See videlicet.*
vs. versus	*against*

Table 6.39: Selected Latin Epigrams, Maxims and Proverbs

While the majority of the entries in this Table date back to the classical times, some came into use in the Middle Ages.

One principle of selection was the inherent wisdom reflected in the thought. Another was the insight into a civilization implicit in a thought; Indeed, one is struck by the universality of people's problems throughout the ages and the satisfying solutions afforded, despite the often contradictory nature of these solutions. Finally, some of the entries have an inherent poetic beauty (and sadness!) of their own, reminiscent of the greatness of Biblical passages and lines from Shakespeare. To this category belong most of these epigrams.

1. ***In necessariis unitas, in dubiis libertas, in omnibus caritas.***

(In necessary things – unity;
in dubious things – liberty;
in all things – charity)

[st. ***Augustine****]*

2. ***Sic transit gloria mundi***

(at times of deep human shortcomings, we realize the transitory nature of grand projects)

*[****Thomas à Kempis****]*

3. ***Ave Imperator: morituri te salutamus!***

(Hail emperor: we who are about to die salute you!)

[quoted by ***Suetonius****]*

4. **Fame in magnis, dignitas autem in homilitate habitat.**

(Fame lives in great things, but dignity lives in humility)

*[**Cicero**]*

5. **Ingenita levitas et erudita vanitas.**

(Frivolity is inborn, conceit acquired by education)

*[**Cicero**]*

6. **Longumeque illud tempus cum non ero magis me movet quam hoc exiguum, quod mihi tamen longum videtur.**

(That long time to come when I shall not exist has more effect on me than this short present time, which nevertheless seems endless)

*[**Cicero**]*

7. **Utinam tam facile vera invenire possem quam falsa convincere.**

(I only wish I could discover the truth as easily as I can expose falsehood)

*[**Cicero**]*

8. **Universus hic mundus sit una civitas communis deorum atque hominum existimanda.**

(We must conceive of this whole universe as one commonwealth of which both gods and men are members)

*[**Cicero**]*

9. **Philosophic est ars vitae**

(Philosophy is the art of life)

*[***Cicero***]*

10. **Prospeae res et in plebem ac vilia ingenia deveniunt; at calamitates terroresque mortalium sub iugum mittere proprium magni vivi est.**

(Success comes to the common man, and even to commonplace ability; but to triumph over the calamities and terrors of mortal life is the part of a great man only)

*[***Seneca***]*

11. **Magna servitus est magna fortuna.**

(A great fortune is a great slavery)

*[***Seneca***]*

12. **Labor optimos citat.**

(Toil summons the best men)

*[***Seneca***]*

13. **Quicquid bene dictum est ab ullo, meum est.**

(Whatever is well said by anyone is mine)

*[***Seneca***]*

14. **Non potest constare liberates. Hanc si magno aestimas, omnia parvo aestimanda sunt.**

(Liberty cannot be gained for nothing. If you set a high value on liberty, you must set a low value on everything else)

*[**Seneca**]*

15. **Dum spiro spero.**

(While I breath I hope)

*[**Cicero**]*

16. **Ipsa scienta potestas est.**

(Knowledge is power)

*[**Francis Bacon**]*

17. **Rex regnat sed non gubernat.**

(The king rules but does not govern)

18. **Aut viam inveniam aut faciam.**

(I'll either find a way or make one)

19. **Dum vivimus, vivamus.**

(While we live, let us really live)

[motto of the Epicureans]

20. ***Poeta nascitur, orator fit.***
(A poet is born but an orator is manufactured)

*[**Publius Annius Florus**]*

21. ***Primum viveri diende philosophari***
(Live before you philosophize)

22. ***Forsan et hae olim meminisse iuvabit.***
(Some day perhaps it will be pleasing to remember these things too)

*[**Vergil**]*

23. ***Vivitur ingenio, caetera mortis erunt.***
(Intelligence lives on, the rest eventually dies)

*[**Vesalius**]*

24. ***Omnia mea mecum porto.***
(all that is mine, I carry with me)

25. ***Homo vitae comodatus non donatus.***
(Man is loaned to life not given to it)

*[**Publilius Syrus**]*

26. **Vivere disce, cogita mori**

(Learn to live; Remember death)

27. **Vox audita perit, littera scripta manet**

(The spoken word vanishes but the written letter remains)

28. **Verba docent, exempla trahunt***.*

(Words instruct, examples lead)

29. **Sutor, ne ultra crepidam.**

(Shoemaker, do not go further than the shoes)

[quoted by ***Pliny****]*

30. **Jacio en aleam, librumque scribo, seu praesentibus, seu posteris legendum, nihil interest; expectet ille suum lectoremper annos centum; si Deus ipse perannorum sena millia contemplatorem praestolatus est.**

(The die is cast, and I am writing this book – whether to be read by my contemporaries or by posterity matters not. Let it wait its reader for a hundred years, if God Himself has been ready for his contemplator for six thousand years)

*[***Johannes Kepler***, 1619]*

31. **Ubi materia, ibi geometria.**
(Where there is matter, there is geometry)

[**Johannes Kepler**]

32. **Non nobis solum nati sumus.**
(We are not born just for our own sake)

[**Cicero**]

33. **Qui non profit – deficit.**
(He who does not gain – loses)

34. **Non multa sed multum**
(Not many but much)

35. **Vulturum non capit muscam.**
(The eagle does not catch flies)

36. **Qui desiderat pacem, praeparet bellum.**
(Whoever desires peace, should prepare for war)

[**Varro**]

37. **Amare et sapere vix deo conceditur.**

(Even a god can scarcely love and be wise at the same time)

*[**Publilius Syrus**]*

38. **Qui usque tandem abutere, Catalina, patienta nostra?**

(How long, then, Cataline, will you abuse our patience?)

*[**Cicero**]*

39. **Vita sine libris mors est.**

(Life without books is death)

40. **Nullus est instar domus.**

(There is no place like home)

41. **Esse quam videri**

(To be rather than to seem)

42. **Docent omnia.**

(Everything teaches)

*[**Seneca**]*

43. **In silvan ne ligna feras**

(Don't carry logs into the forest)

44. ***In medio tutissimus ibis***
(You will go safest in the middle)

*[**Ovid**]*

45. ***Homo sum, humani nil a me alienum puto.***
(I am a man; nothing human is alien to me)

*[**Terrence**]*

46. ***Per angusta ad augusta.***
(Through difficulties to great things)

47. ***Quieta non movere***
(Don't move quiet things; "Let sleeping dogs lie")

48. ***Ubi bene, ibi Patria***
(Where you feel good, there is your home)

*[**Pacuvius**]*

49. ***Non semper ea sunt qua videntur.***
(Things are not always what they seem to be)

50. ***Nulla dies sine linea.***
(Not a day without something done)

*[**Pliny the Elder**]*

Table 6.40: (Cont.)

Indication	Root	English meaning	Examples
	Pellere	*push, drive*	*dispel, repel, expel, propeller, impel*
	Caedere	*cut*	*excise, cadaver, incision, incidence, decadence*
	Saltare	*leap, jump*	*insult, salient, result, resilience*
	Mittere	*send*	*emit, commit*
	Ducere	*lead*	*duct, conductor, educe, induce*
	Venere	*come*	*convened, advent*
	Tangere	*touch*	*tangent, tangible, contiguous*
Activities of the intellect			
	Anima	*mind, breath, soul, equanimity*	*animation, magnanimous*
	Cognoscere	*to know*	*cognizant, recognize*
	Quaerere	*ask, seek*	*inquest, query, acquisitive, exquisite*
	Judicare	*judge*	*judicial, judicious, dictum, adjudge*
	Credere	*believe*	*credo, credit, credulity, credentials*

Table 6.40: (Cont.)

Indication	Root	English meaning	Examples
	Scribere	*write*	*prescribe, scribe, ascribe, circumscribe*
	Nomen	*name*	*nomenclature, nominal, nominative*
	Dicere	*talk*	*diction, dictum*
	Monere	*warn*	*monitor, admonish*
	Videre	*see*	*video, envision*
	Philos	*love*	*philosophy, bibliophile*
	Cantere	*sing*	*cantor*
	Psych	*mind*	*psychology*
	Gnos	*know*	*agnostic*
	Graph	*write*	*graphology, autograph, graphite*
	Lagos	*the study of*	*logic*
Growth, Change and movement			
	Genus	*birth, kind*	*genesis, eugenics, engender*
	Natura	*nature*	*supernatural, cognate, innate*
	Novus	*new*	*novel, novice, innovation*
	Crescere	*grow, increase*	*crescendo*

Table 6.40: (Cont.)

Indication	Root	English meaning	Examples
	Stare	*stand*	*distance, static*
	Gradus	*step*	*graduate, gradation, gradient*
	Movere	*move*	*promote, mobile, motile*
	Jungere	*join*	*junction, junta, enjoined, conjugal*
	Frangere	*break*	*fracture, fragile*
	Torquere	*twist*	*torsion, torque*
	Plicere	*fold*	*complicated, complicity, imply*
	Fluere	*flow*	*effluent, affluent, confluence, fluent*
	Mortis	*dead*	*mortal, morbid*

Table 6.41: Dictionary of Greco-Latin Origins

[(*) used in modern Hebrew]

Root	Core	Meaning	English Words
acer (acris)	**acri**	*sharp, bitter*	*acrimony, acrid*

Table 6.41: (Cont.)

Root	Core	Meaning	English Words
acere	**eas, jac, ject, jet**	*to lie, to throw*	*jet, trajectory, jettisoned, adjacent, object, conjecture, reject, objective, injection, ejecta, subject, adjective, abject, deject, eject, project, projectile, projection, projector*
aedes	**edif**	*a building, temple*	*edifice*
aequus		*level, equal*	*equality, equal, equilibrium,equivalence*
aevum	**ev**	*age*	*primeval, medieval, longevity, eternity*
ager	**gri**	*field*	*peregrine, pilgrim, agriculture*
agere	**act(actum)**	*do, act, drive, set in motion, function*	*action, active, actual, actuate, enact, redact, reactor, transact, actuary, exact, agent, reaction*
agere	**gate**		*castigate, divagate, expurgate, fumigate, fustigate, navigate, objurgate, variegate, levigate, litigate, mitigate*
agere	**gen**		*cogent, agendum, agency, exigent, intransigent*
agere	**gi, gu**		*ambiguity, agility, agitate, cogitate, assay, putge*
agora *(*)*	**gor**	*marketplace*	*category, agora, agoraphobia*

Table 6.41: (Cont.)

Root	Core	Meaning	English Words
akademia	**academ**	*park near Athens*	*academy, academic*
akros	**acro**	*a point, top-most*	*acrobat, acronym, acrostic, acme*
albus	**alb**	*white*	*albatross, albumen, album, auburn, albino, albedo, albion*
Alere	**al**	*to feed, nourish, grow*	*alimentary, coalition, alimony, coalesces, albuminus*
altus	**alt**	*high*	*altitude, altimeter, alt (voice)*
allos	**alle**	*other*	*allegory, parallel, allergy*
amare	**ama, amo**	*to love*	*amateur, amour, enamored*
amicus	**ami**	*friend*	*amity, amigo, amiable, (enemy)*
amplus	**ampl**	*large, wide*	*amplifier, amplitude, ample*
angelos	**angel**	*messenger*	*angel, evangelist, archangel*
angulus	**ang**	*angle*	*triangle*
animus	**anim**	*mind, soul*	*animal, unanimous, equanimity, anima, magnanimous, animated, animosity, animalcule*
annus	**anni, annu, enni**	*year*	*annual, perennial, anniversary, centennial, millenium, annuity, annals*
anthos	**ant**	*a flower*	*anthology*

Table 6.41: (Cont.)

Root	Core	Meaning	English Words
antiquus	**antiqu**	*ancient*	*antique, antiquity, antiquarian*
aristos		*best*	*aristocrat*
arkhein	**arch**	*rule, begin, ancient, first*	*monarch, oligarchy, anarchy, archetype, archaic, patriarch, matriarch, hierarchy, archbishop, architect, archipelago, archives, archaeology, archduke*
arctos		*a bear*	*arctic ocean*
astron	**astro, aster**	*star*	*astronomy, astrology, disaster, asterix, asteroid, astronaut*
atavus	**atav**	*ancestor*	*atavism*
athlon	**athl**	*a prize*	*athlete, decathlon, pentathlon, triathlon, decathlon*
audire	**audi**	*to hear*	*audible, auditory*
aura		*air, breath*	*aura*
ballein	**blem, bol**	*to throw*	*emblem, symbol, hyperbole, parabola, parabolic, hyperbolic, matebolism, anabolism, problem, diabolic*
bassus (*)	**base, bass**	*low, short*	*bass, bassoon, base, abase, debase*
battuere	**bat**	*to beat*	*battery, battalion, battle, battlement, combat, batten, abate, debate*
bellum	**bell**	*war*	*rebel (revel), bellicose, belligerent, duell, casus belli*

Table 6.41: (Cont.)

Root	Core	Meaning	English Words
bene, bonus	**bene, beni**	*good, well*	*benediction, benevolent, benign, beneficent, benefactor, benefit*
bibere	**bib**	*to drink*	*imbibe, bibulous, beverage, beer*
biblion	**bibli, biblio**	*book, papyrus, scroll*	*bible (from Byblos, a phoenician city from which papyrus was exported), bibliophile, bibliography, bibliotheca*
bombos	**bomb**	*a booming, humming, sound*	*bomb, bombard, bombastic*
brev	**brev**	*short*	*abbreviation*
calx (calcis)	**cal**	*limestone, pebbles*	*calculus*
capere	**cip, cup**		*anticipate, emancipate, incipient, municipal, participate, percipient, participle, principal, recipient, recipes, principle, occupy, occupation, preoccupy, prince, cable*
carus	**cheri, chari**	*dear*	*cherish, charity, caress*
cavus	**cave**	*hollow*	*care, concave*
censere	**cen**	*asses, judge*	*censuse, recension, censor*
chronos	**chron**	*time*	*chronism, chronicle, chronology, chronological, chronometer*
civis	**cit, civ**	*citizen*	*civilization, civil, city, citizen, civic*

Table 6.41: (Cont.)

Root	Core	Meaning	English Words
citare	***cit***	*arouse, motivate*	*excite, excitation*
clamere	***claim, clam***	*to call out, shout*	*claim, exclaim, declaim, reclaim, proclaim, acclaim, disclaim, clamant, clamor, acclamation*
claudere	***clos, clude, clus***	*to shut, close*	*conclude, disclose, exclude, enclose, foreclose, include, occlude, preclude, recluse, secluded, closet, cloister, claustrophobia*
colare	***col***	*strain, filter*	*percolate, coulee*
coquere	***coc, cot***	*cook*	*precocious, apricot, concoct, ricotta*
corpus		*body*	*corpse, corporal, corpulent, corpuscle*
caballus	***caval***	*horse*	*cavalry, cavalier, cavalcade*
cadere	***cad, cas, cid***	*to fall*	*accident, cadaver, causality, case, cascade, cadenza, cadence, caducity, escheat, chute, parachute, coincide, decadence, incidence, occasion, occident, recidivation*
calere	***cal, cha***	*to be warm, hot*	*caldron, chafing, nonchalant, chafe, chaff, calory*
calvi	***cal, chal***	*to deceive*	*calumny, challenge*
canere	***cant, chant***	*to sing*	*chant, cantata, cantor, canto, recant, accent, enchanting, incentive*

Table 6.41: (Cont.)

Root	Core	Meaning	English Words
capere	**cap, capt**	*to seize, contain, lay hold of*	*capable, capacity, capsule, capacitate, capstan, captive, caption, capture*
capere	**cas, chas**		*case, cassette, purchase, encase, cash, cashier, caskets, cask, chase*
captere	**ceit, ceiv**		*conceive, receive, deceive, perceive, conceit*
captere	**cept**		*accept, concept, deception, exceptional, inception, intercept, percept, precept, susceptible, receptionist*
cosmos		*order, arrangement*	*cosmetics, cosmonaut, cosmopolitan*
credere	**cred**	*to believe*	*credit, credible, credulous, credulity, accredit, discredit, creed*
dare	**dat, dit**	*give*	*date, dative, datum, data, addendum, add, edit, edition, tradition, perdition, traitor, render, rent, rendezvous*
decem, deka	**dec**	*ten*	*December, decade, decimate, decathlon*
derma	**derm**	*skin*	*epiderm, dermatology*
dexter	**dext**	*on the right side*	*dexterity, ambidextrous, dexter, dextrose*
dicare	**dic, dict**	*to tell, to say*	*abdicate, benedict, addict*

Table 6.41: (Cont.)

Root	Core	Meaning	English Words
dicere			*dedication, dictate, dictator, contradict, dictum, edict, dictionary, indicator, indices, indicative, indict, interdict, judicially, judiciary, predicament, predicate, prejudice, vindicate, valediction, digit, predict*
didonai	**dos, dot**	*to give*	*antidote, dose, overdose*
digitus	**digit**	*finger, toe*	*digit, digital*
discere	**disci**	*to learn*	*discipline, disciple*
docere	**doc**	*to teach*	*doctor, doctorate, document, documentary, indoctrinate, docile, docent*
dokein	**dox**	*to seem. think*	*dogma, dogmatism, orthodox, heterodox, paradox*
dolere	**dol**	*to feel pain, grieve, suffer*	*condole, condolence, doleful, dolorous, indolent, dolor*
dominus	**domin**	*master*	*dominate, dominant, dominion, domineering, dame, madam, madonna, don (*), domino, donna, anno domini, dominique, domingo, condominium*
donare	**don**	*to give*	*pardon, condone, donation, donor*
dromein	**drome**	*to run*	*aerodrome, hippodrome, syndrome, palindrome*
dunamis	**dyna**	*force, power, strength*	*dynamite, dynamo, dynamic, dynasty*

Table 6.41: (Cont.)

Root	Core	Meaning	English Words
duo	**dou, dub**	*two*	*doubt, dubious, doublet*
	duo, dup		*duplicate, duplicity, duplex, doubloon, dual, duel, duet*
durare	**dur**	*hard, continue to exist*	*endure, durable, duration, duress, indurate, obdurate, predurate*
edere	**edi**	*to est*	*edible*
ego	**ego**	*I*	*egoist, egotist, egocentric, egomania*
eidos	**ido**	*form*	*idol, idolatrous, kaleidoscope*
eikon *(*)*	**icon**	*image, likeness*	*icon, iconoclast*
ergon	**erg, urg**	*work*	*erg, ergomania, allergy, energy, liturgy, surgeon, metallurgy*
errare	**err**	*to wander*	*err, error, erratum, aberrant, erroneous, aberration*
esse	**sen, sent**	*to be, exist*	*absence, absentia, essence, entity, essential, quintessential, present, presentation, represent*
eus	**eu**	*good, well*	*euphoria, euphony, eulogies, eugenics, euthenics, euthanasia, Eunice, Eugene*
facree	**fac, face**	*to make, do*	*fact, factory, faction, factor, manufacture, benefactor, factotum*
facere	**feat, feas**		*feature, defeat, defeasance, feasible*

Table 6.41: (Cont.)

ROOT	CORE	MEANING	ENGLISH WORDS
facere	**fect**		*affect, effect, affectionate, confection, disaffect, defect, defective, infect, infection, perfect, prefect*
facere	**feit, fit**		*comfit, profit, benefit, counterfeit, forfeit, surfeit*
facere	**fic, fice**		*aficionado, artifice, beneficiary, deficit, deficient, difficult, efficient, edifice, magnificent, munificent, office, officer*
figere	**fix**	*to fasten, pierce, fix*	*affix, fix, fixture, fixative, fixer, prefix, suffix, infix, transfix, fixation, idee fix, crucifix*
fluere	**flu**	*to flow*	*affluent, confluent, effluent, effluvia, fluorescent, fluoride, flush, influence, influenza, superflous, superfluid, fluid, flux, fluctuate, fluent*
forma	**form**	*shape*	*conform, reform, conformist, formalist, formation, deform, deformity, form, formal, format, informal, formality, formula, formation, informer, information, disinform, transform, transformer, uniform, proforma*
fortis	**fort**	*strong*	*comfort, comfortable, comforter, effort, force, enforce, reinforce, fort, fortress, fortifications, fortitude, pianoforte, fortissimo*
fortuna	**fort**	*luck*	*fortune, fortuitous, misfortune, unfortunate*

Table 6.41: (Cont.)

Root	Core	Meaning	English Words
fallere	**fal, fail, faul**	*to deceive*	*false, falsification, falsetto, fallacy, fallible, fail, failure, fault, default, faulty, faucet*
felix	**feli**	*happy, lucky*	*Felix, felicity, felicitation, felicitous*
femina	**femin**	*woman*	*feminine, femininity, femme fatale*
fendere	**fen**	*to ward off*	*offence, offensive, offend, defence, fencing, fender*
ferre, ferere	**fer**	*to bear, bring, carry, produce*	*aquifer, confer, conference, defer, different, difference, differ, indifference, differential, differentiate, fertile, infer, offer, prefer, proliferate, refer, referendum, suffer, referee, insufferable, sufferance, transfer, vociferate, vociferous, fertile, ferry*
fervere	**ferv**	*to boil, be hot*	*fervent, fervid, fervor, fervency, ferment, effervescence*
fidere	**fid**	*to trust*	*confide, fiance, affiance, confidence, defiance, diffident, fidelity, fiduciary, affidavit, infidel, perfidy, perfidious*
gamos	**gam, gamy**	*marriage*	*bigamy, monogamy, polygamy, allogamy, autogamy, cryptogram*
gemin		*twin, paired, born together*	*gemini*

Table 6.41: (Cont.)

Root	Core	Meaning	English Words
genus, (generis)	**gen**	*race, kind, class, origin*	*general, generalize, generation, engender, generate, generator, degenerate, regenerate, generous, generic, genre, genus, gender, misce generation, gene*
gignoskein	**gnos**	*to know*	*gnosis, gnosticism, gnostics, agnosticism, agnostic, prognosis, diagnosis*
gladius		*sword*	*gladiator*
globus	**glob**	*ball, sphere*	*globule, hemoglobin, globus*
glossa	**gloss**	*tongue*	*glossary, gloss*
gnoscere	**cogn, gnor, noti**	*to get to know*	*cognitive, cognizance, recognition, precognitive, cognoscenti, ignorance, ignoramus, ignore, incognito, notice, notion, noticeable, reconnoiter, notify, reconnaissance*
gradi *(*)*	**grad, gress**	*to go, step, walk*	*egress, congress, digress, ingress, progress, regress, retrogress, transgress, aggression, grade, graduate, gradual, gradation, gradient, degrade, retrograde*
grandum	**gran**	*grain, seed*	*pomegranate, granary, granules, granite, granulate*
gratus	**grac, grat, gree**	*beloved, dear, pleasing*	*grace, disgrace, gracious, grateful, gratitude, gratify, gratuitous, gratuity, ingratiate, ingrate, agree, agreement, disagree, disagreeable, gratis*

Table 6.41: (Cont.)

Root	Core	Meaning	English Words
gravis	***grav, griev***	*heavy, serious, weighty*	*gravity, gravitation, grave, aggravate, grievance, grieve, aggrieve*
gregare, grex (gregis)	***greg***	*to herd, flock*	*gregarious, congregation, segregate, aggregate, egregious, gregarious, segregation*
gustus	***gust***	*a tasting*	*disgust, gusto, gustation*
haerere (haesum)	***her, hes***	*to stick, to cling*	*adhesion, coherent*
helios	***sol, helio***	*sun*	*Helliocentric, solar, parasol, insolation, solstice, helium*
hepta	***hepta***	*seven*	*heptagon, heptathlon, heptahedron*
heres	***heir, heri***	*on heir*	*heir, inherit, heritage, inheritance, disinherit*
hodos	***od***	*way, journey*	*period, periodic, episode, method, synod, exodus*
horrere	***horr***	*to bristle, dread, shudder*	*abhor, horrendous, horrific, horrible, horror, horrid*
hostis	***hostil***	*enemy, stranger*	*hostility, hostile*
humor (humidis)	***hum***	*liquid, moisture, damp, moist*	*humidity, humoral*
identi	***identi***	*same*	*identical, identity*

Table 6.41: (Cont.)

Root	Core	Meaning	English Words
insula	**sola, sula**	*island*	*peninsula, insulation, isolate, insular, isolation*
intimus		*to be close, private, within*	*intimate*
ire (itum)	***it***	*to go*	*ambition, circuit, initiate, circuitous, initial, initiative, concomitant, exit, coition, itinerary, itinerate, obit, obituary, transitory, seditious, transit, transitive*
jugum	***jug***	*yoke*	*jugular, conjugate, subjugate*
jungere	***join, junct***	*to join*	*adjoin, join, adjunct, conjoin, conjunction, enjoin, injunction, disjoin, joint, junction, juncture, rejoin, subjoin, subjunctive, conjunctive*
jurare	***jur, jus, juris***	*to swear, law*	*adjure, conjure, injure, objurgate, perjury, juridical, jurisdiction, jurist, jury, just, justify, justice*
juvenis	***juv***	*young*	*juvenile, rejuvenation*
kaiein	**cau**	*to burn*	*caustic, holocaust*
kamara	**cam, cham**	*vault*	*chamber, camarade, camera, chambermaid, chamberlain*
kosmos	**cosm**	*order, the world, the universe*	*cosmopolitan, cosmic, cosmology. cosmonaut, cosmetic, microcosm*

Table 6.41: (Cont.)

Root	Core	Meaning	English Words
krinein	**crit**	*distinguish, separate*	*criterion, critic, criticism, critical, hypocrite, criticize*
kyklos	**cycl**	*cycle, wheel*	*cycle, recycle, cyclotron, encyclopedia, cyclone, cycloid, cyclop, cyclic, recycle, motorcycle, tricycle, unicycle*
labor	**labor**	*labor, exertion, toil*	*labor, laborious, laboratory, collaboration, elaborate, laborious*
lacuna		*small pit, gap*	*lacuna*
latus (dilatatus, lateris)	**lat**	*carried, borne, wide, broad, expanded, side*	*ablate, collate, collateral, correlate, dilate, elate, oblate, relate, relation, relative, superlative, translate, lateral, latitude, legislator, dilatation, bilateral*
laudare	**laud**	*to praise*	*laud, laudatory, cum laude*
legare	**lega, lege, legi**	*to bind, choose, send*	*alleged, allegation, allegedly, college, collegial, legate, delegation, legation, legacy, relegate*
legein	**lexi, log, logue**	*to gather, speak*	*lexicon, lexicographer, dyslexia, catalog, dialog, monolog, decalogue, analog, prolog, epilog, travelog*
levare	**lev**	*to lighten, lift, raise*	*alleviate, elevate, elevator, elevation, lever, leverage, levant, levity, levitation, levy, levigate, relevant irrelevant*
lex	**leg**	*law*	*legal, legitimate, illegal, illegitimate, legislative, legislature, privilege*

Table 6.41: (Cont.)

Root	Core	Meaning	English Words
liber	***lib***	*free*	*liberty, liberate, liberation, liberal,*
limen (liminis)		*treshold*	*elimination, subliminal*
lingua	***lingu***	*tongue, language*	*linguist, bilingual, linguistics, lingua franc.*
liquere	***liqu***	*flow, be liquid*	*liquid, liquidity, liquidate, liquify, liqueur, liquor*
lithos	***lith***	*stone*	*monolith, megalith, paleolithic, neolithic, mesolithic, lithography, lithology, lithium*
littera	***liter***	*letter of the alphabet*	*illiteracy, literally, literary, literature, alliteration, obliterate, literati*
locare (locus)	***loca***	*to place, place*	*local, locate, allocate, collocate, dislocate, location, loci, localize*
logos	***log, logy***	*word*	*logic, logical, logistics, sylogism, logo, analogous, apologetic, apologia, eulogy, biology, tautology, anthology, etymology, seismology*
longus	***long***	*long*	*longevity, longitude, prolong, elongation, oblong, longueur, longanimity*
loqui	***locu, loqu***	*to speak*	*colloquium, grandiloquence, loquacious, ventriloquist, soliloquy, somniloquy, eloquence, interlocutor, locution, obloquy*

Table 6.41: (Cont.)

Root	Core	Meaning	English Words
lucere	**lumen, luc, lum**	*to shine, light*	*luminous, lucid, elucidate, illumination, illuminati, lucubration, luminosity, lucent, luminiferous*
luna, selena	**luna**	*the moon*	*lunatic, lunacy, lunar, lunation*
magister	**mas, mis**	*a master*	*master, mastermind, masterpiece, mister, mistress, magistrate, masterwork*
magnus	**magn, maj**	*great, large, big*	*majority, majesty, major, mayor, magnate, Magna Carta, magnifico, magniloquent, magnitude, magnificent, majestic, magnanimous*
malleus	**mall**	*hammer*	*mall, mallet, pall mall*
malus	**mal, mali, male**	*bad, evil*	*malignant, malevolent, malediction*
mandare	**mand, mend**	*to order*	*commandment, commander, commandant, commando, commandeer, commend, recommend, commendation, demand, demanding, mandate, mandatory, remand*
manere	**man**	*to remain*	*mansion, manor, immanent, permanent*
manus	**man**	*hand*	*manual, manage, manager, management, manure, maneuver, manner, mannerism, mannered, emancipate, manufacture, manifest, manifesto, manipulate, manicure*

Table 6.41: (Cont.)

Root	Core	Meaning	English Words
mappa		*napkin, cloth*	*map, napkin, apron*
margo (marginis)	***marg***	*edge, border*	*margin*
mater	***mat, metro***	*mother*	*maternal, matrimony, matron, maternity*
medius	***medi***	*middle*	*mediate, immediate*
meter			*matriarch, matrix, matriculate, alma mater, metropolis, metropolitan, metrodome, metro*
merx	***merc***	*goods, trade, traffic*	*gramercy, mercy, merci, mercenary, mercantile, merchant, commerce, mercury, mercurial, mercer*
mikros	***micro***	*small*	*microwave, microphone, microscope, micrometer, microbe, microcosmos, microorganism, microfilm*
mole	***mole***	*mass*	*molecule*
movere (motum)	***motio***	*to move*	*emotion, motion, locomotion, motile*
mutare (mutatum)	***mut***	*to change*	*mutation, commute, permute*
miscere	***misc, mix***	*to mix*	*miscellany, mix, mixture, intermix, admix, promiscuous, promiscuity, miscegenation, miscible*

Table 6.41: (Cont.)

Root	Core	Meaning	English Words
mittere	***mise, miss, mit, mitt***	*send*	*admit, admissible, admission, admittance, admittedly, commit, commitment, commissar, commissary, commission, compromise, demit, demise, dismiss, demit, emit, emission, emissary, intermittent, intermission, mission, missioner, missive, missile, omit, omission, permit, permissible, permission, permissive, premise, promise, remit, remiss, remittal, remission, unremitting, submit, submissive, submission, surmise, transmit*
misein	***misa, miso***	*to hate*	*misanthrope, misandry, misogynism, misogamist, misoneism*
nasci	***nat***	*to be born*	*cognate, nature, native, naturalization, naturally, nation, nationalist, natal, international*
naus *(*)*	***nau***	*ship*	*nausea, nautical, astronaut*
negativus	***nega, negl***	*no, negative*	*negate, neglect, negative, negligible*
nekros	***necro***	*corpse*	*necrology, necrophilia, necromancy*
nepos	***neptis***	*grandson, nephew, niece*	*nepotism, nephew*
nocere	***noka, noc, nox***	*to harm*	*innocent, innoxious, innocuous, noxious, obnoxious*

Table 6.41: (Cont.)

Root	Core	Meaning	English Words
nomen, nominare	***nom***	*name, to name*	*misnomer, nomenclature, nominate, nominator, denominator, ignominy, nominal, nominee, nom de plunc*
norm	***norm***	*measure, standard, pattern*	*normal, abnormal, normalization*
novus	***nov, novi***	*new*	*novel, novella, novelist, innovator, novelty, renovate, novice, novitiate, nova, nouveau*
nox	***noct***	*night*	*nocturnal, nocturne, noctilucent, equinoctial*
numerus	***numer***	*number*	*numerate, enumerate, numeral, numerical, numeration, innumerable, numerology, innumerate*
oculus	***oc***	*eye*	*pinochle, inoculate, monocle, oculist, ocular, binoculars*
odium	***no***	*hatred*	*annoy, noisome, annoying*
offici(um)	***fic***		*official, orifice, proficient, pontificate, sacrifice, significant, specific, suffice, sufficient, superficial, beatific, honorific, pacific, soporific, prolific*
oikos	***eco***	*house*	*ecology, economy, parochial, ecumenical, economize*
omnis	***omni***	*all, every*	*omnibus, omnipotent, omnivorous*

Table 6.41: (Cont.)

Root	Core	Meaning	English Words
onoma	**onum**	*a name*	*anonym, pseudonym, eponym, antonym, toponym, anonymous, heteronym, homonym, metonymy, onomatopoeia, patronym, synonym*
orbis	**orb**	*circle, wheel*	*orbit, orbital*
ordo	**ord**	*order*	*order, disorder, ordinary, ordained, ordination, ordinarily, ordinance, ordinate, coordinate, inordinate, subordinate, insubordination*
pais	**ped**	*child*	*pediatrics, pediatrician, pedodontics, encyclopedia, pedagogue, pedant, pedantry, pederasty, pedophile*
par	**par**	*equal, peer*	*compare, par, parity, disparity, disparage, comparative, au pair, parlay, peer, at par, comparable*
parere	**par**	*give birth, come in sight*	*parents, parental, parenthood, transparent, apparent, apparition, oviparous, viviparous*
paschein	**path**	*to suffer*	*pathetic, pathos, pathology, pathogen, apathy, antipathy, sociopath, psychopath, empathy, sympathy, telepathy, homeopathy, osteopathy, allopathy*
pater	**patri, patro**	*father*	*patronize, patronage, patriot, compatriot, expatriate, pater, patricians, patrimony, patriarch, patronymic, paternity, perpetrate*

Table 6.41: (Cont.)

Root	Core	Meaning	English Words
pax	**pac, peace**	*peace*	*peace, pacification, pacifist, pacific, pacifier*
peccare	**pecca**	*stable, sin*	*peccable, peccant, impeccable*
pecus	**pecu**	*cattle, properly*	*pecuniary, impecunious, peculate, peculiar, peculation, peculiarity*
pellere	**pel, pul**	*to drive*	*appeal, appellation, appellant, compel, compulsion, compulsive, compulsory, expel, expulsion, impel, impulse, impulsive, propel, propeller, propulsion, repel, repulse, repellant, repulsion, pelt, pulsate, pulse*
pendere	**pen, pend, pens, pond**	*weigh, hang, pay*	*ponder, perpend, preponderate, ponderous, append, appendage, appendix, appendant, penthouse, appendectomy, compensate, depend, dependable, dependence, independence, dispense, dispensable, dispensary, dispensation, expend, expense, expenditure, expensive, impend, impending, pendant, pendulant, pending, penchant, propensity, pendular, pendulum, perpendicular, pension, equiponderance, spend, suspend, suspense*
penis	**pen**	*tail*	*penicilin*
penna	**pen**	*feather*	*pen, pennant*

Table 6.41: (Cont.)

Root	Core	Meaning	English Words
pes	**ped**	*foot*	*impend, expedite, impediment, expedition, expedient, pedal, pedometer, pedicure, pedestal, centipede, millipede, pedigree, bipeds, quadruped, pedestrian*
pius	**pi**	*devoted*	*pieta, pious, piety, pity, pittance*
plebes, demos	**pleb, dem, demo**	*the people*	*demagogue, plebiscite, plebeian, endemic, epidemic, pandemic, democracy, democrats, demotic*
pleetere (plexum)	**plex**	*to plait, interweave*	*complex, plexus*
plenus		*full*	*plenty, plenary, replenish*
planetes		*a wanderer*	*planet*
polis	**poli**	*city*	*politics, political, polity, politburo, politic, apolitical, acropolis, metropolis, police, megalopolis, cosmopolitan*
porcus	**por**	*hog, pig*	*porcupine, porpoise, porcelain, pork*
portare	**port**	*to carry*	*sport, sportsmanship, porter, portmanteau, portfolio, portable, comportment, deportment, comport, import, export, importance, importunate, importune, port, portly, purport, report, rapport, support, transport*
potentis, potensi	**potent**	*having power*	*potent, potential*

Table 6.41: (Cont.)

Root	Core	Meaning	English Words
pungere	***pun, punc***	*to point, stab*	*punctuate, punctilious, punctuation, punctual, pungent, compunctions, expunge, pun*
quaerere	***quest, quir, quis***	*to seek*	*quest, inquire, questionable, question, inquest, inquiry, conquest, request, acquire, acquisition, acquirement, acquisitive, disquisition, exquisite, inquisitive, inquisition, requisite, prerequisite*
qualis	***qual***	*of what kind*	*quality, qualify, qualitative*
quantus	***qualt***	*how many, how much*	*quantify, quantity, quantum, quantitative*
qui, quam, quom		*who, how, when*	*quibble, quorum, quasi, quasar, quondam*
quid, quies	***qui***	*what, something, quiet, rest*	*quid, pro quo (something for something), acquiesce, acquiescence, acquit, quiet, quiescence, quit, quite, requiem, requite*
quot		*how many*	*quota, quotient, quotidian, quotation*
radere	***ras***	*to scrape*	*rascal, rash, abrasion, abrasive, eraser, razor*
radius	***radi***	*rod, spoke, ray, beam*	*irradiation, radial, radiology, radium*
radix	***radi***	*a root*	*radix, radish, eradicate, radical*

Table 6.41: (Cont.)

Root	Core	Meaning	English Words
regere, rectum	**rect**	*keep straight*	*erectile, rectum*
rete (retis)	**ret**	*net, network*	*retina*
rex	**reg, roy**	*king*	*royal, regal, royalist, royalties, Regina, regalia, regale, viceroy, regnant, regulus, regicide, regnum*
ridere	**rid, ris**	*to laugh*	*derision, ridicule, deride, derisive, ridiculous, risible*
rigor	**rig**	*stiffness*	*rigidity, rigorous*
rodere, rosum	**ro, ros**	*to gnaw*	*erosion, corrosion, erode, corrode, rodents*
rogare	**gat, gate**	*to ask*	*Roguery, abrogate, arrogance, derogate, interrogate, perogative, supererogation, surrogate, subrogate*
rota	**rot**	*wheel*	*rotator, rotunda*
ruber (rubris)	**rub**	*red*	*bilirubin, rubella*
rumpere	**rupt**	*to break*	*abrupt, bankrupt, corrupt, disrupt, interrupt, erupt, irrupt, rupture*
salire	**sal, sult, xult**	*to leap, spring forward*	*insult, desultory, exult, result, resilience, somersault, salient, salacious, salmon, saltant, saltation*

Table 6.41: (Cont.)

Root	Core	Meaning	English Words
salus	***salu***	*health*	*salute, salud, salutation, salutary, salubrious*
salvus	***sal***	*safety, health*	*salvage, salve, salvation, savior, save, salutary, salubrious*
sanguis	***sang***	*blood*	*sanguine, sangria*
sacer	***sac***	*sacred, holy*	*sacerdotal, sacred*
sapere	***sav***	*to taste, to be wise*	*savant,unsavy, savory, savor*
scala	***scal***	*ladder*	*escalade, escalator, upscale, escalate*
scandere	***scen, scend***	*climb, leap*	*ascend, descend, ascent, ascendance, descendant, condescend, transcend*
scire (scitum)	***sci***	*to know*	*science, conscience, unconscious, self-conscious, subconscious, plebiscite, omniscient, prescient, nescient, scientific, sciolism*
scribere	***scri***	*to write*	*scribe, manuscript, scripture, script, scribbler, postscript, inscription, inscribe, transcribe, descriptive, prescript, transcript, proscribe, circumscribe, ascribe, conscript, conscription, subscription, prescribe, describe, prescription, description, nondescript, indescribable*
secare (segmentum)	***seg, sect, sex***	*to cut, part, segment*	*segment, secant, bisect, dissect, sex, section*

Table 6.41: (Cont.)

Root	Core	Meaning	English Words
sedere	**sed, sess, sid**	*to sit, settle*	*preside, president, reside, resident, residence, residue, residuum, residual, assess, assiduous, sessions, assize, obsess, possess, dispossess, repossess, sedate, sedative, sediment, sedentary, subside, subsidiary, subsidy, dissident, superseded, insidious, sedulous*
semen	**semin**	*seed*	*seminary, seminal, insemination, semination, dissemination*
senex	**sen**	*old, old man*	*senator, senior, senile, senescent*
sequor	**secu, xeco, sequ, suit**	*to follow*	*consequence, consequential, consequently, inconsequent, obsequencious, sequel, subsequently, sequence, sequential, sequitur, nonsequitur, sequested, consecutive, execute, executive, persecute, prosecute, lawsuit, suitable, unsuited, pursuit, suite*
signum	**sign**	*a mark, seal, sign, indication*	*sign, signature, signal, signify, significance, signet, insignia, assign, designate, assignment, consign, design, designate, designer, ensign, resign, resignation*
similis	**simul, sembl, simil**	*like, at the same time, alike*	*assemble, assembly, assimilate, assimilation, similar, ensemble, resemble, semblance, simulacrum, verisimilar, similitude, simile, facsimile, simulation, simulcasts, simultaneous, dissimilar*

Table 6.41: (Cont.)

Root	Core	Meaning	English Words
sinister	***sin***	*on the left*	*sinistral*
skhizein	***schiz***	*to split*	*schizophrenic*
skopein	***scope***	*see, look at*	*scope, microscope, telescope, horoscope, periscope, kaleidoscope, spectroscope*
solus	***sol***	*alone*	*solo, solitude, solist, desolate, solifidian, sole, solipsism, soliloquy, solitaire, solitary*
solvere	***solu, solv***	*to loosen*	*absolute, resolute, solution, resolution, irresolute, unresolved, resolve, dissolve, soluble, insolvent, insoluble, dissolute, absolve, absolution*
sonare (sonatum)	***son, sound***	*to make a sound*	*sonnet, sonata, sound, sonorous, resound, resonant, resonance, assonance, consonant, consonant, dissonant, sonic, supersonic, transonic*
spargo	***sper***	*to scatter besprinkle*	*aspersions, asperse, disperse, sparse, intersperse, dispersion*
spondere	***spond, spons***	*to pledge*	*respond, respondent, correspond, correspondent, irresponsible, responsibility, responsive, sponsor, despondent*
stare (statum)	***stat, stet***	*to stand*	*status, substitution, obstetrics*
stinguere	***stinct, stingu***	*pierce, quench*	*instinct, distinct, distinction, distinctive, distinguish, indistinct, indistinguishable*

Table 6.41: (Cont.)

Root	Core	Meaning	English Words
tropos	***trop***	*a turn*	*trophies, tropics, tropical, entropy, trope*
tupos	***type***	*a blow, impression*	*type, archetype, phototype, typeset, stereotype, typecast, typewriter, stenotype, logotype*
turbane (turbatum)	***turb***	*to stirr, disturb*	*turbid, turbine*
ultra	***ult***	*beyond*	*ultimate, penultimate, ultra, ulterior, ultimatum*
umbra	***umbra***	*shadow*	*umbrage, umbra, penumbra, adumbra*
unda	***und***	*wave, billow*	*abound, abundance, inundation, redundant, surround, undulation*
uti	***uti***	*to use*	*utilize, use, useful, utile, utility, abuse, misuse, peruse, abusive*
vacare (vacuus)	***vac, void***	*to be empty*	*vacant, avoid, vacous, vacuity, unavoidable, void, vacuum, vacate, devoid evacuate*
valere	***val***	*to be strong, be well*	*equivalent, valence, invalid, bivalent*
varius	***vari, vario***	*bent, changeable, crooked, diverse, manifold speckled, different*	*various, variety, vary, variance, variation, variant, variegated*

Table 6.41: (Cont.)

Root	Core	Meaning	English Words
vehere	***vec***	*to carry*	*vector, convect, vehicle, convex, convey, advect*
verbum	***verb***	*word*	*verbal, verbiage, verbose, verbatim, verb, adverb, proverb, proverbial*
venire (ventum)	***vent***	*to come*	*circumvent, intervention*
verei	***revere***	*to fear, feel awe*	*revere, reverence, reverent, irreverence*
vertere (versum)	***vert***	*to turn*	*version, vertebra, pervert, invert*
vertex (verticis)	***vert***	*summit, top*	*vertex, vertical, vertigo*
verus	***ver***	*true*	*veracity, veritable, veracious, verification, very, verify*
vestis	***vest***	*garment*	*vest, vesture, invest, investment, divest, transvestite, travesty, revest*
vestus	***vet***	*old, long-standing*	*veterinary, vet, veteran, veterate*
via	***via, vio, vium***	*way, road*	*trivia (trivium = three ways, crossroad), trivial, obviously, pervious, impervious, deviate, deviant, devious, previous, obviate, obvious, viaduct*
vicis	***vic***	*change, instead of*	*vicar, vicissitudes, vicarious, vice versa*

Table 6.41: (Cont.)

Root	Core	Meaning	English Words
videre (visum)		*to see*	*vision, television*
vigilare	**vig, veil**	*to watch*	*vigil, vigilant, vigilantes, surveillance, reveille*
villa	**villa**	*farmhouse*	*villa, village, villain*
vir	**vir**	*man, manliness, virtue, strength*	*virtue, virility, virtual, triumvirate, virtuoso, virtual*
virus	**virus**	*potent juice, poison*	*virulent, virus*
vita, vivere	**vit, via, viv**	*life, to live*	*vital, vitamin, viable, convivial, vivacious, revive, revitalize, revivify, revival, survive, victual, viviparous, vita, vitals, vitality, viva, vivid, vivacity, vivify, vivisection*
vitrum	**vit**	*glass*	*vitrina, vitriol, vitreous*
vocare	**vox, voc, vok**	*to call, voice*	*advocation, vocation, provoke, evoke, revoke, vociferous, provocative, evocative, convoke, invoke, invocation, revocation, irrevocable, advocate, equivocate, unequivocal*
volvere	**vol**	*to roll*	*volume, voluble, revolve, revolution*
vovare	**vor**	*to devour, oat*	*voracious, carnivorous, herbivorous, omnivorous, insectivorous*

Table 6.41: (Cont.)

Root	Core	Meaning	English Words
vovere	**vot, vow**	*to vow, wish, pledge*	*vote, votary, devotee, vow, votive, devote, devotion*
vulgare	**vulg**	*publish, make common*	*vulgate, vulgar, divulge*
xenos	**xeno**	*foreign, strange*	*xenophobe, xenon*
zelos	**zeal, jeal**	*ardor*	*jealous, jealousy, zeal, zealot, zealous*
zoion	**zoo**	*animal*	*zoology, zoo, zoolatry*

Quotations

1. On Nature, Universe, and Man

* *
*

"What is man, that thou art mindful of him? And the son of man, that thou visitest him? For thou hast made him a little lower than God, and hast crowned him with glory and honor. Thou madest him to have dominion over the works of thy hands..."

Psalms **8**, 5–7 (ca 1000 BCE)

* *
*

"And now, rejoicing in the prosperous gales,
With beating heart Ulysses spread his sails:
Placed at the helm he state, and mark'd the skies,
Nor closed in sleep his ever-watchful eyes.
There view'd the Pleiads, and the Northern Team,
And great Orion more refulgent beam,
To which, around the axle of the sky,
The Bear, revolving, points his golden eye:
Who shines exalted on the etheral plain,
Nor baths his blazing forehead in the main.
Far on the left those radiant fires to keep
The nymph directed, as he sail'd the deep"

Homer, Odyssey **V**, 270–276 (ca 800 BCE)

* *
*

Nature does not hurry, yet everything is accomplished

Lao Tzu, 4th century BCE

* *
*

And the Lord answered Job out of the Whirlwind, and said,
"...
Where wast thou when I layed the foundation of the earth?... Who hath laid the measured thereof... Whereupon are the foundations thereof fastened? Or who laid the corner stone thereof?... Or who shut up the sea with doors, when it brake forth...? Hast thou commanded the morning...? Where is the way where light dwelleth?... Canst thou bind the sweet influences of Pleiades, or loose the bands of Orion? Canst thou bring forth Venus in his season? Or canst thou guide Arcturus with his sons?..."

(Job **38**), ca 600 BCE

* *
*

"What profit is there in my blood,
When I go down the pit?
Shall the dust praise thee?
Shall it declare thy truth?"

Psalms **30**, 10

* *
*

"It is the glory of God to conceal a thing; but the honor of king is to search out a matter."

Proverbs **25**, 2

* *
*

"He hate made everything beautiful in his time: *also he hate set the world in their heart, so that no man can find out the work that God maketh from the beginning to the end."*

Ecclesiastes, **3**:11

* *
*

"Opinions say hot and cold, but the reality is atoms and empty space."

"Everything existing in the Universe is the fruit of chance and necessity."

Democritos (ca 460–370 BCE)

* *
*

"Even God cannot change the past."

Agathon (448–400 BCE)

* *
*

"Do you think it a matter worthy of lamentation that when there is such a vast multitude of them (universes), we have not yet conquered one?"

Alexander the Great (356–323 BCE)
Wept when told by Anaxarchos of Abdera
that there exist an infinity of worlds.

* *
*

"Nature does nothing uselessly."

Aristotle (384–322 BCE)

* *
*

"The laws of nature are but the mathematical thoughts of God."

Euclid (ca 300 BCE)

* *
*

"The universe is not bounded in any direction. If it were, it would have a limit somewhere. But clearly a thing cannot have a limit unless there is something outside to limit it... It makes no odds in which part of it you make take your stand: whatever spot anyone may occupy, the universe stretches away from him just the same in all directions without limit."

Lucretius (ca 99–55 BCE)

* *
*

"Nothing which we can imagine about nature is incredible."

Pliny, the Elder (23–79 AD)

* *
*

"Nothing is difficult for nature, especially when she rushes to destroy herself. At the beginning of things she uses her strength sparingly and apportions herself out in imperceptible increases. For destruction she comes suddenly with all her violence. A long time is needed so that a child, once conceived, may come to be born. The tender infant is reared only with great toil. The frail body finally develops only with diligent nurture. But how with no effort it is all undone!
It takes an age to establish cities, an hour to destroy them. A forest grows for a long time, becomes ashes in a moment. Great safeguards may exist and all things may be flourishing, but quickly and suddenly they all fall apart."

"The day will come when diligent research over long periods will bring to light the mysteries of nature which now lie hidden. A single lifetime, even though totally devoted to the sky, would not be enough for the investigation of so vast a subject... And so this knowledge will be unfolded only through long successive ages. The day will yet come when our descendants will be amused that we did not know things that are so plain to them... Many discoveries are reserved for ages still to come, when memory of us will have been perished. Our universe is a sorry little affair unless it has in it something for every age to investigate...
Nature does not reveal her mysteries once and for all."

"Any deviation by nature from the existing state of the Universe is enough for the destruction of mankind."

"What else is nature but God?"

Lucius Annaeus Seneca (4 BCE–65)

* *
*

"The rational soul goes about the whole universe and the void surrounding it and traces its plan and stretches forth into the infinitude of time."

Marcus Aurelius, "Meditations", **11**,1,2

* *
*

"Opened Rabbi Tanhuma and said: "He hate made everything beautiful in his time" – the world was created at the right time, for it was unworthy to be created earlier. Followed Rabbi Abbahu and said: Hence that the Lord was creating universes and destroying them, recreating and destroying them again, until he made the present universe."

(Genesis Rabba, **9** ca 450 CE)

* *
*

"Thou art wise, and from Thy wisdom
Thou hast set apart Thy appointed purpose,
Like a craftsman and an artist
To draw up the films of Being from Nothingness
As light is drawn that darteth from the eye:
Without bucket from the fountain of light
Thy workman drawn it up,
And without tool hath he wrought,
Hewing, graving, cleansing, refining.
Calling into the void – and it was cleft,
And unto existence – and its was urged,
And to the universe – and it was spread out."

Shlomo Ibn-Gabirol (1021–1070)

* *
*

Do there exist many worlds, or is there but a single world? This is one of the most noble and exalted questions in the study of nature.

Albertus Magnus (ca 1200–1280)

* *
*

If a man in the heavens ... could see the earth distinctly ... it would appear to him that the earth was moving in daily motion, just as to us on earth it seems as though the heavens are moving.
...One could then believe that the earth moves and not the heavens.

Nicole Oresme (ca 1325–1382)

* *
*

"Nature never breaks her own laws."

"The genius of man will never discover a more beautiful, a more economical, or a more direct one that nature's, since in her inventions nothing is wanting and nothing is superfluous."

"Oh God, you sell all things to men at the cost of their effort, but life is short for this kind of commerce!"

"Natura semper agit per vias brevissimas."

Leonardo da Vinci (1452–1519)

* *
*

"A dark flame issued forth from the innermost hiddenness, from the mystery of the infinite..."

The *Zohar* (13th century)

* *
*

"Pluritas non est pondera sine necessitate." ('It is in vain to do with more than can be done with less.')

William of Ockham (1285–1349)

* *
*

"Let us permit nature to have her way. She understands her business better than we do."

Michael de Montaigne (1533–1592)

* *
*

"The laws of nature ultimately are more important then nature."

René Descartes (1596–1650)

* *
*

"Time and space are of the same essence."

Yehudah Liwa of Prague (1512–1609)

* *
*

"Man can only conquer nature by obeying her."

"Nature is often hidden, sometimes overcome, seldom extinguished."

"God never wrought miracle to convince atheism, because his ordinary works convince it."

Francis Bacon (1561–1626)

* *
*

"Ubi materia – ibi geometria."

Johannes Kepler (1571–1630)

* *
*

"One touch of nature makes the whole world kin."

"Thou, Nature, art my goddess;
To thy law my services are bound."

William Shakespeare, *'King Lear', Act I, Scene II* (1564–1616)

"I could be bounded in a nut-shell, and count myself a king of infinite space."

William Shakespeare, '*Hamlet*' (1564–1616)

* *
*

"Nature hath no goal though she hath law."

"A fountain breaks out in the wilderness, but that fountain cares not, whether any man may come to fetch water, or no; a fresh and fit gale blows upon the sea, but it cares not whether the mariners hoist sail or no; a rose blows in your garden, but it calls you not to smell it."

John Donne (1572–1631), ca 1620

* *
*

"Beautiful are the things we see; More beautiful those we understand; But the most beautiful are those we do not understand."

Niels Steno (1638–1685)

* *
*

"Nature has set no end before herself, and all final causes are nothing but human fictions."

"Space and matter are not really different."

"I believe that a triangle, if it could speak, would say that is eminently triangular, and a circle, that the divine nature is eminently circular; and thus would every one ascribe his attributes to God."

Baruch Spinoza (1632–1677)

* *
*

"All nature is but art unknown to thee;
All chance, direction which thou canst not see;
All discord, harmony not understood;
All partial evil, universal good;
And, spite of pride, in erring reason's spite,
One truth is clear, Whatever IS, is RIGHT."

Alexander Pope (1688–1744)

* *
*

"The enormous and the minute are interchangeable manifestations of the eternal."

"To see a World in Grain of Sand
  And a Heaven in a Wild Flower,
Hold Infinity in the palm of your hand,
  And Eternity in an hour."

William Blake (1757–1827)

* *
*

"Give me matter, and I will construct a world out of it."

Immanuel Kant (1724–1804)

* *
*

"Nature has neither kernel nor shell; she is everything at once."

"Nature goes her own way, and all that to us seems an exception is really according to order."

"Nature! We are surrounded by her and locked in her clasp: powerless to leave her, and powerless to come closer to her. Without asking us or warning us she takes us up into the whirl of her dance, and hurries on with us until we are weary and fall from her arms."

"Nature alone knows what she wants."

"Every moment Nature starts on the longest journey, and every moment she reaches her goal."

"The difficulty in nature is to see the law where it is concealed from us, and not to be misled by phenomena that contradict our senses. For in nature there is much that contradict our senses and nevertheless true. That the sun stands still, that he does not rise and set, but that the earth performs a diurnal revolution with incredible swiftness, contradicts to senses as much as anything; but yet no well-informed person doubts that this is the case."

"I love him who yearns for the impossible."

Johann Wolfgang von Goethe, '*Faust II*' (1749–1832)

* *
*

"Space is the stature of God."

Joseph Joubert (1754–1824)

* *
*

"After your death you will be what you were before your birth."

Arthur Schopenhauer (1788–1860)

* *
*

"Deep into the darkness peering, long I stood
there, wondering, fearing,
Doubting, dreaming dreams no mortal ever
dared to dream before."

Edgar Allan Poe (1809–1849)

* *
*

"When Man thinks, it is Nature which is thinking itself."

Arthur Schopenhauer (1788–1860)

* *
*

"Nature is visible thought."

"So we keep asking, over and over,
Until a handful of earth
Stops our mouth –
But is that an answer?"

"Like a great poet, Nature knows how to produce the greatest effects with the most limited means."

Heinrich Heine (1797–1856)

* *
*

"Nature is not embarrassed by difficulties of analysis."

Augustin Fresnel (1788–1827)

* *
*

"We are a sign that is not read."

Friedrich Hölderlin (1770–1843)

* *
*

"God hides in the minute details of Nature but is manifested in its grand design."

Alphonse de Lamartine (1790–1869)

* *
*

"I believe in the incomprehensibility of God."

Honore de Balzac (1799–1850)

* *
*

"I would rather believe that God did not exist than believe that He was indifferent."

George Sand (1804–1876)

* *
*

"The decisive events of the world take place in the intellect."

Henri Frédéric Amiel (1821–1881)

* *
*

"The chess-board is the world; the pieces are the phenomena of the universe; the rules of the game are what we call the laws of Nature. The player on the other side is hidden from us. We know that his play is always fair, just, and patient. But also we know, to our cost, that he never overlooks a mistake, or makes the smallest allowance for ignorance."

T.H. Huxley (1825–1895)

* *
*

"Nature teaches more than she preaches. There are no sermons in stones. It is easier to get a spark out of a stone that a moral."

"Nature is not benevolent; Nature is just, gives pound for pound, measure for measure, makes no exceptions, never tempers her decrees with mercy, or winks at any infringement of her laws."

John Burroughs (1837–1921)

* *
*

"When we try to pick out anything by itself, we find it is tied to everything else in the universe."

John Muir (1838–1914)

* *
*

"The God whom science recognizes must be a God of universal laws exclusively, a God who does a wholesale, not a retail business. He cannot accommodate his processes to the convenience of individuals."

William James (1842–1910)

* *
*

"Nature, in her indifference, makes no distinction between good and evil."

"Chance is the pseudonym of God when he did not want to sign."

Anatole France (1844–1924)

* *
*

"The true mystery of the world is the visible, not the invisible."

Oscar Wilde (1854–1900)

* *
*

"Repetition is the only form of permanence that nature can achieve."

George Santayana (1863–1952)

* *
*

"To believe in God is to desire His existence, and what is more, to act as though He existed."

Unamuno, y Jugo de Miguel (1864–1936)

* *
*

"The bee intentionally seeks for a method of economizing wax."

D'Arcy Wentworth Thompson (1860–1948)

* *
*

A man said to the universe:
"Sir, I exist".
"However", replied the universe,
"The fact has not created in me
A sense of obligations".

Stephen Crane (1871–1900)

* *
*

"The history of thought may be summed up in these words: it is absurd by what it seeks, great by what it finds."

"God made everything out of nothing. But the nothingness shows through."

Paul Valéry (1871–1945)

* *
*

"The Universe is full of magical things patiently waiting for our wits to get sharper."

"God, why did you make the evidence for your existence so insufficient."

Bertrand Arthur William Russell (1872–1970)

* *
*

"The true return to nature is the definitive return to the elements – death."

André Gide (1869–1951)

* *
*

"Seek simplicity, and distrust it."

"The aim of science is to seek the simplest explanation of complex facts."

Alfred North Whitehead (1861–1947)

* *
*

" Say something to us we can learn
By heart and when alone repeat.

Say something! And it says "I burn".
But say with great degree of heat.

Talk Fahrenheit, talk centigrade.
Use language we can comprehend.
Tell us what elements you blend.
It gives us strongly little aid,
But does tell something in the end."

Robert Lee Frost (1874–1963)

* *
*

"Rerum natura nullam nobis dedit cognitionem finitum." ('Nature has given to us no knowledge of the end of things.')

Winston Churchill (1874–1965)

* *
*

"In nature there are neither rewards nor punishments – there are consequences."

Robert G. Ingersoll (1839–1899)

* *
*

"Who speaks of victory, survival is all."

"What will you do, When I die?
When I, your pitcher, broken lie?
When I, your drink, go stale or dry?
I am your garb, the trade you ply,
You lose your meaning, losing me."

Rainer Maria Rilke (1875–1925)

* *
*

"The universe is not hostile, nor yet is it friendly. It is simply indifferent."

John Haynes Holmes (1879–1964)

* *
*

"Everything should be made as simple as possible, but not simpler."

"The eternal mystery of the world is its comprehensibility. The fact that it is comprehensible is a miracle."

"The most incomprehensible thing about the universe is that it is comprehensible."

"Nature hides her secrets because of her essential loftiness, but not by means of ruse." ("Die Natur verbirgt ihr Geheimnis durch die Erhabenheit ihres wesens, aber nicht durch List.")

"Die Natur hat es nicht angelegen sein lassen, uns die Auffindung ihrer Gesetze bequem zu machen."

"Nature did not deem it her business to make the identification of her laws comfortable for us."

"Out yonder there is a huge world, which exists independent of us human being and which stands before us like a great, eternal riddle, at least partially accessible to our inspection and thinking. The contemplation of this world beckons like a liberation."

"God is subtle but he is not malicious."

"What I'm really interested in is whether God could have made the world in a different way; that is, whether the necessity of logical simplicity leaves any freedom at all."

"For the rest of my life I will reflect on what light is!"

Albert Einstein (1879–1955)

* *
*

"The God within creates the God in nature."

Arthur Stanley Eddington (1882–1944)

* *
*

"Today there is a wide measure of agreement ... that the stream of knowledge is heading towards a non-mechanical reality; the universe begins to look more like a great thought than a great machine."

"The history of physical science in the twentieth century is one of progressive emancipation from the purely human angle of vision."

"The Universe can be pictured, although still very imperfectly and inadequately, as consisting of pure thought, the thought of what for want of a wider word, we must describe as a mathematical thinker."

James Jeans (1877–1946)

* *
*

— *" 'Never will you draw the water out of the depths of this well'.*
What water? What well?
— *'Who is asking?'*
Silence.
— *'What silence?' "*

"Man's fundamental weakness lies by no means in the fact that he cannot achieve victory, but in the fact that he cannot exploit his victory."

Franz Kafka (1883–1924)

* *
*

"I respect the idea of God too much to hold it responsible for a world as absurd as this one is."

George Duhamel (1884–1966)

* *
*

"Many people find that modern science is far removed from God. I find, on the contrary, that it is much more difficult today for the knowing person to approach God from history, from the spiritual side of the world, and from morals; for there we encounter the sufferings and evil in the world which it is difficult to bring into harmony with an all-merciful and all-mighty God. In this domain we have evidently not yet succeeded in raising the veil with which our human nature covers the essence of things. But in our knowledge of physical nature we have penetrated so far that we can obtain a vision of the flawless harmony which is in conformity with sublime reason."

Hermann Weyl (1885–1955)

* *
*

"To believe in God means to see that life has a meaning."

Ludwig Witgenstein (1889–1951)

* *
*

"The universe is not only stranger than we imagine; it is stranger than we can imagine."

John Burdon Sanderson Haldane (1892–1964)

* *
*

"The universe was dictated but not signed."

Christopher Morley (1890–1957)

* *
*

"Everything you've learned in school as "obvious" becomes less and less obvious as you begin to study the universe. For example, there are no solids in the universe. There's not even a suggestion of a solid. There are no absolute continuums. There are no surfaces. There are no straight lines."

"God is a verb."

"Humanity is acquiring all the right technology for all the wrong reasons."

"Nature is trying very hard to make us succeed, but nature does not depend on us. We are not the only experiment."

Richard Buckminster Fuller (1895–1983)

* *
*

"Nature always has the last word."

John Stewart Collis (1900–1984)

* *
*

"Atoms are not things.... When we get down to the atomic level, the objective world in space in time no longer exists, and the mathematical symbols of theoretical physics refer merely to possibilities, not to facts."

Werner Heisenberg (1901–1976)

* *
*

"The notion of structure is comprised of three key ideas: the idea of wholeness, *the idea of* transformation, *and the idea of self-regulation."*

Jean Piaget (1896–1980)

* *
*

"The Lord is a weak left-hander, but he still appears to be left-right symmetric when he expresses himself strongly."

Wolfgang Pauli (1900–1976)

* *
*

"By getting to smaller and smaller units, we do not come to fundamental units, or indivisible units, but we do come to a point where division has no meaning."

Werner Heisenberg (1901–1976)

* *
*

"The views of space and time which I wish to lay before you have sprung from the soil of experimental physics, and therein lies their strength. They are radical. Henceforth, space by itself, and time by itself, are doomed to fade away into mere shadows, and only a kind of union of the two will preserve an independent reality."

Herman Minkowski (1864–1909), 1908

* *
*

"Human destiny is bound to remain a gamble, because at some unpredictable time and in some unforeseeable manner nature will strike back."

Rene Dubos (1901–1982)

* *
*

"I am not interested in proofs, only what nature does."

Paul Dirac (1902–1984)

* *
*

"Laws of nature could not exist without principles of invariance."

Eugene Wigner (1902–1995)

* *
*

"We must expect the ice that retreated some 10,000 years ago to come back again."

George Gamow (1904–1968)

* *
*

"The universe is like a safe to which there is a combination. But the combination is locked up in the safe."

Peter de Vries (1910–1993)

* *
*

"Matter has reached the point of beginning to know itself . . . [man is] a star's way of knowing about stars."

"It would be a poor thing to be an atom in a universe without physicists, and physicists are made of atoms. A physicist is an atom's way of knowing about atoms."

George Wald (1906–1997)

* *
*

"In man's brain the impressions from outside are not merely registered; they produce concepts and ideas. They are the imprint of the external world upon the human brain. Therefore, it is not surprising that, after a long period of searching and erring, some of the concepts and ideas in human thinking should have come gradually closer to the fundamental laws of the world, that some of our thinking should reveal the true structure of atoms and the true movements of the stars. Nature, in the form of man, begins to recognize itself."

Victor Frederick Weisskopf (1908–2002)

* *
*

"Biology occupies a position among the sciences at once marginal and central. Marginal because — the living world constituting but a tiny and very "special" part of the universe — it does not seem likely that the study of living beings will ever uncover general laws applicable outside the biosphere. But if the ultimate aim of the whole science is indeed, as I believe, to clarify man's relationship to the universe, then biology must be accorded a central position..."

"Man at last knows that he is alone in the unfeeling immensity of the universe ... Neither his destiny nor his duty have been written down."

Jacques Monod (1910–1976)

* *
*

"Cosmologists are often in error, but never in doubt."

Yakov Borisovich Zeldovich (1914–1987)

* *
*

"We could *have lived in a Universe with different laws in every province, but we do not. We might have lived in a Universe in which nothing could be understood by a few simple laws, in which Nature was complex beyond our abilities to understand, in which the laws that apply on earth are invalid on a distant quasar. But the evidence proves otherwise; Luckily for us we live in a Universe in which much* can *be "reduced" to a small number of comparatively simple laws of Nature. Otherwise we might have lacked the intellectual capacity to comprehend the world."*

Carl Sagan (1934–1996)

* *
*

"Basic laws are simple only in their first approximation."

Martin Gardner (1914–)

* *
*

"My suspicion is that the universe is not only queerer than we suppose, but queerer than we can *suppose."*

J.B.S. Haldane (1892–1964), 1927

* *
*

"Man is a complex thing: he makes desert bloom and lakes die."

Gil Stern

* *
*

"The best (scientific) data we have are exactly what I would have predicted had I nothing to go on but the five books of Moses, the Psalms, the Bible as a whole. ... What we have ... is an amazing amount of order; and when we see order, in our experience it normally reflects purpose."

Arno Penzias (1933–)

* *
*

"The ability to reduce everything to simple fundamental laws does not imply the ability to start from those laws and reconstruct the universe."

Philip W. Anderson (1923–)

* *
*

"There is no reason that the universe should be designed for our convenience."

John D. Barrow (1952–)

* *
*

"Every snowflake is an avalanche pleads not guilty."

Stanislaw Lec (1909–1966)

* *
*

"Atoms of our body, were once part of stars."

"As astronomers you can't say anything except that here is a miracle. ... Can you go the other way, back outside the barrier and finally find the answer to the question why is there something rather than nothing? No, you cannot, not within science."

Allan Sandage (1926–)

* *
*

"Why is it that the world is structured in such a way that we can know something without knowing everything? If we could not understand limited parts of the universe without understanding the whole, science would be a hopeless enterprise."

"If science demonstrated an infinitely old *universe, science could disprove a Creator, because an infinitely old universe would never have been created! On the other hand, in a universe of* infinite extent, *anything that is possible must happen somewhere by pure chance. Obviously,* we *will find ourselves just where that fantastic happening has occurred."*

"As the universe is slowly choked to death by its own entropy, will God die too?
The alternative — gravitational collapse to a singularity, resulting in the total obliteration of the physical universe — seems even less promising.
Only a universe of the cyclic or steady-state varieties, would appear to offer scope for a natural God to be both infinite and eternal."

"A universe that came from nothing in the big bang will disappear into nothing in the big crunch, its glorious few zillion years of existence not even a memory."

Paul Davies (1946–)

* *
*

"If you see a formula in the Physical Review that extends over a quarter of a page, forget it. It's wrong. Nature isn't that complicated."

Matthias Berndt (1918–1980)

* *
*

"No purported inconsistency with the General Theory's predictions has ever stood the test of time. No logical inconsistency in its foundations has ever been detected. No acceptable alternative has ever been put forward of comparative simplicity and scope."

John Archibald Wheeler (1911–2008)

* *
*

"Gods are born and die, but the atom endures."

"Walking in space, man has never looked more puny or more significant."

Alexander Chase

* *
*

"As we look out into the Universe and identify the many accidents of physics and astronomy that have worked together to our benefit, it almost seems as if the Universe must in some sense have known that we were coming."

"If it should turn out that the whole of physical reality can be described by a finite set of equations, I would be disappointed. I would feel that the Creator had been uncharacteristically lacking in imagination."

Freeman Dyson (1923–)

* *
*

"I do not pretend to have all the answers. But the questions themselves are worth talking about."

"There is no reason to assume that the universe has the slightest interest in intelligence – or even in life. Both may be random accidental by-products of its operations like the beautiful patterns on a butterfly's wings. The insect would fly just as well without them."

"Sometimes I think we're alone in the universe, and sometimes I think we're not. In either case, the idea is quite staggering."

Arthur C. Clarke (1917–2008)

* *
*

"The world began without man, and it will end without him."

Claude Lévi-Strauss (1908–)

* *
*

"Einstein's postulate that the laws of nature should appear the same to all freely moving observers was the foundation of the theory of relativity, so called because it implies that only relative motion is important."

"Why does the universe go to all the bother of existing?"

"We may now be near the end of the search for the ultimate laws of nature."

Stephen Hawking (1942–)

* *
*

“Astronomers discover God!”

Headlines in a well-known periodical (1982)

* *
*

“The Strong Anthropic Principle
Because there appear to exist such a large number of remarkable and apparently disconnected ‘coincidences’ which conspire to allow life to be possible in the universe, the universe must *give rise to observers at some stage in its history (i.e. the universe* must *be such as to admit the creation of observers within it at some stage).”*

“The Weak Anthropic Principle
The observed values of all physical and cosmological quantities take on values restricted *by the requirement that there exist sites where carbon-based life can evolve and by the requirement that the universe be old enough for it to have already done so (i.e. what we* can *expect to observe must be* restricted *by the conditions necessary for our presence as observers).”*

“The universe must be such as to admit the creation of observers within it.”

Brandon Carter (1942–)

* *
*

“Has it a clock? Or is it a clock?”

Colin S. Pittendrigh (1918–1996), 1957

* *
*

"The improbable is bound to happen one day."

(Anon)

* *
*

"The familiar idea of a god who is omniscient (someone who knows everything) does not immediately ring alarm bells in our brains; it is plausible that such a being could exist. Yet, when it is probed more closely one can show that omniscience of this sort creates a logical paradox and must, by the standards of human reason, therefore be judged impossible or be qualified in some way. To see this consider this test statement:

THIS STATEMENT IS NOT KNOWN TO BE TRUE BY ANYONE.

Now consider the plight of our hypothetical Omniscient Being ('Big O'). Suppose first that this statement is true and Big O does not know it. Then Big O would not be omniscient. So, instead, suppose our statement is false. This means that someone must know the statement to be true; hence it must be true. So regardless of whether we assume at the outset that this statement is true or false, we are forced to conclude that it must be true! And therefore, since the statement is true, nobody (including Big O) can know that it is true. This shows that there must always be true statements that no being can know to be true. Hence there cannot be an Omniscient Being who knows all truths. Nor, by the same argument, could we or our future successors, ever attain such a state of omniscience. All that can be known is all that can be known, not all that is true."

John D. Barrow (1952–)

* *
*

"Why is there something rather than nothing at all?"

(Anon)

* *
*

"It is not so much that you are in the universe, *as that universe is in you."*

Meher Baba (1894–1969)

* *
*

"We can never tell whether the hand of God was at work in the moment of creation. ... In the searing heat of the first moment, all the evidence needed for a scientific study of the cause of the great explosion was melted down and destroyed."

Robert Jastrow (1925–2008)

* *
*

"It is the most persistent and greatest adventure in human history, this search to understand the universe, *how it works and where it comes from. It is difficult to imagine that a handful of residents of a small planet circling an insignificant star in a small galaxy have as their aim a complete understanding of the entire* universe, *a small speck of creation truly believing it is capable of comprehending the whole."*

Murray Gell-Mann (1929–)

* *
*

"The more the universe seems comprehensible, the more it seems pointless."

"Maybe nature is fundamentally ugly, chaotic and complicated. But if it's like that, then I want out."

Steven Weinberg (1933–)

* *
*

"Laws of science (physical laws), are at best approximation of the truth; they are inaccurate. Laws of Nature are some other laws (statements, principles), doubtless more complex (which are literally true), which govern the natural phenomena of the world. These are factual truths, not logical ones. They are true for every time and every place in the universe, are universal or statistical claims."

(Anon)

* *
*

"The whole of science is nothing more than a refinement of everyday thinking."

(Anon)

* *
*

"Only daring speculation can lead us further, and not accumulation of facts."

(Anon)

* *
*

"Scientific knowledge is the most reliable and useful knowledge that human being possess."

(Anon)

* *
*

"Nature does not care for the survival of individuals, only survival of species."

(Anon)

* *
*

"The bluebird carries the sky on his back."

Henry David Thoreau (1817–1862)

* *
*

"In order to see birds it is necessary to become a part of the silence."

Robert Lynd (1879–1949)

* *
*

"A bird thinks nothing of its flying or it would fall."

Leslie Sahler (1952–)

2. *The Quantum World*

* *
*

"Without mysticism man can achieve nothing great."

André Gide (1869–1951)

* *
*

"We have seen the truth, and the truth makes no sense."

Gilbert Keith Chesterton (1874–1936)

* *
*

"Anyone who is not shocked by Quantum Theory has not understood it."

"There is no quantum world. There is only an abstract quantum physical description. It is wrong to think that the task of physics is to find out how nature is. Physics concerns what we can say about nature."

Niels Bohr (1885–1962)

* *
*

"The electron is not as simple as it looks."

Lawrence Bragg (1890–1971)

* *
*

"Undoubtedly a piece of definitive truth – but not the whole truth, let alone the definitive truth."

"Gott würfelt nicht." (God casts the die, not the dice)

"The more success the quantum theory has, the sillier it looks."

"Quantum physics formulates laws governing crowds and not individuals."

"The statistical interpretation of quantum theory has led to important successes, but I still believe in the possibility of producing a model of reality which shall represent events themselves and not merely the probability of, their occurrence."

"There is no doubt that quantum mechanics has seized hold of a beautiful element of truth and that it will be a touchstone for a future theoretical basis in that it must be deducible as a limiting case from that basis, just as electrostatics is deducible from Maxwell equations of the electromagnetic field or as thermodynamics is deducible from statistical mechanics. I do not believe that quantum mechanics will be the starting point in the search for this basis, just as one cannot arrive at the foundations of mechanics from thermodynamics or statistical mechanics."

"All these fifty years of conscious brooding have brought me no nearer to the answer to the question 'what are light quanta?' Nowadays every Tom, Dick and Harry thinks he knows it, but he is mistaken."

Albert Einstein (1879–1955)

* *
*

"Complementarity is a thoughtless slogan. If I were thoroughly convinced that Bohr is honest and really believes in the relevance of his – I do not say theory, but – sounding word, I shall call it intellectually wicked."

Erwin Schrödinger (1887–1961)

* *
*

"Only metaphysics can inspire the hard work of theoretical physics."

"I don't like it, and I'm sorry I ever had anything to do with it."

Erwin Schrödinger (1887–1961)

* *
*

"By getting to smaller and smaller units, we do not come to fundamental units, or indivisible units, but we do come to a point where division has no meaning."

Werner Karl Heisenberg (1901–1976)

* *
*

"The theory of quantum electrodynamics describes nature as absurd from the point of view of common sense. And it agrees fully with experiment. So I hope you can accept nature as she is – absurd!"

"If people say they understand quantum mechanics they're lying."

"If you believe that atoms are like little solar systems, then you are back in 1910."

"I think it is safe to say that no one understands quantum mechanics. Do not keep saying to yourself: 'But how can it be like that?' because you will go into a blind alley from which nobody has yet escaped. Nobody knows how it can be like that."

Richard Phillips Feynman (1918–1988)

* *
*

"All of modern physics is governed by that magnificent and thoroughly confusing discipline called quantum mechanics invented more than fifty years ago. It has survived all tests. We suppose that it is exactly correct. Nobody understands it but we all know to use it and how to apply it to all problems: and so we have learned to live with the fact that nobody can understand it."

"Niels Bohr brainwashed a whole generation of physicists into believing that the problem of the interpretation of quantum mechanics had been solved fifty years ago."

Murray Gell-Mann (1929–)

* *
*

"When we are dealing with things as small as atoms and electrons, the observer or experimenter cannot be excluded from the description of nature. The laws of subatomic physics cannot even be formulated without some reference to the observer."

Freeman Dyson (1923–)

* *
*

"I dreamt I died and went to heaven, and Saint Peter led me into the presence of God. And God said 'You won't remember me, but I took your Quantum Mechanics Course in Berkeley in 1947'."

Robert Serber (1909–1997)

* *
*

"God not only plays dice. He also sometime throws the dice where they cannot be seen."

Stephen William Hawking (1942–), 1975

* *
*

"The basic concept and methods of quantum field theory are becoming more and more mathematical."

N.N. Bogoliubov (1909–1992)

* *
*

"Scientists have come to accept some pretty bizarre notions. The most precise theory ever created, quantum electrodynamics, starts with one of its fundamental postulates being a statement of uncertainty. We deal with concepts of curved space, of a vacuum that is rich in physical properties, and of pointlike particles that have a radius equal to zero but which, without embarrassment, carry spin, electric charge, mass and a plethora of other endowments."

Leon M. Lederman (1922–)

* *
*

"Most of the mathematics needed for quantum mechanics was forged in the 19th century, but quantum mechanics itself was never predicted."

Shahar Ben-Menahem

3. *Life, Evolution and Reductionism*

* *
*

"The tree of life in the garden and the tree of knowledge of good and evil."

(Genesis **2**, 9)

* *
*

"It is remarkable that the stupidest ape differs so little from the wisest man, that the surveyor of nature has yet to be found who can draw the line between them."

Carl Linnaeus (1707–1778)

* *
*

"As many more individuals of each species are born than can possibly survive... it follows that any being, if it vary ever so slightly in a manner profitable to itself... will have a better chance of survival, and thus be naturally selected."

Thomas Malthus (1766–1834)

* *
*

"Life, as manifested to us, is a function of the asymmetry of the universe and of the consequences of this fact. The universe is asymmetrical; for, if the whole of the bodies which compose the solar system moving with their individual movements were placed before a glass, the image in the glass would not be superposed upon the reality. Even the movement of solar light is asymmetrical... Terrestrial magnetism, the opposition which exists between the north and the south poles in a magnet and between positive and negative electricity, are but resultants of asymmetrical actions and movements... Life is dominated by asymmetrical actions. I can even imagine that all living species are primordially, in their structure, in their external functions, functions of cosmic asymmetry."

Louis Pasteur (1822–1895), 1874

* *
*

"To suppose that the eye with all its inimitable contrivances for adjusting the focus to different distances, for admitting different amounts of light, and for the correction of spherical and chromatic aberration, could have been formed by natural selection, seems, I confess, absurd in the highest degree."

Charles Darwin (1809–1882)

* *
*

"The difference between a piece of stone and an atom is that an atom is highly organized, whereas the stone is not. The atom is a pattern, and the molecule is a pattern, and the crystal is a pattern; but the stone, although it is made up of these patterns, is just a mere confusion. It's only when life appears that you begin to get organization on a larger scale. Life takes the atoms and molecules and crystals; but instead of making a mess of them like the stone, it combines them into new and more elaborate patterns of its own."

"The great end of life is not knowledge but action."

Thomas Henry Huxley (1825–1895)

* *
*

"Nothing in life is to be feared; It is to be understood."

Marie Curie (1867–1934)

* *
*

"Analogies from chemical experience will not, of course, any more than the ancient comparison of life with fire, give a better explanation of living organisms than will the resemblance, often mentioned, between living organisms and such purely mechanical contrivances as clockworks. An understanding of the essential characteristics of living beings must be sought, no doubt, in their peculiar organization, in which features that may be analyzed by the usual mechanics are interwoven with typically atomistic traits in a manner having no counterpart in inorganic matter."

Niels Bohr (1885–1962)

* *
*

"An organism has an astonishing gift of concentrating a 'stream of order' on itself and thus escaping decay into atomic chaos – of 'drinking orderliness' from a suitable environment."

Erwin Schrödinger (1887–1961)

* *
*

"The amoeba and the paramecium are potentially immortal. From time to time each divides itself into two, but... no new individual is ever produced-only fragments of the original individuals, whose life has thus been continuous back to the time when life itself was first created."

Joseph Wood Krutch (1893–1970)

* *
*

"It is selection, not mutation, that determines the direction of evolution... No mutant gene has the slightest chance of maintaining itself against even the faintest degree of adverse selection."

Gavin de Beer (1899–1972)

* *
*

"Nothing in Biology makes sense except in light of evolution."

T.G. Dobzhansky (1900–1975), 1973

* *
*

"To us as mammals, the insects seem to belong to some topsy-turvy world almost outside the reach of our understanding. They have the skeleton on the outside of the body, the main nervous system below the digestive tract... and use blood (body fluid) only for the transport of food materials."

Marston Bates (1906–1974)

* *
*

"Life is a cosmic event – so far as we know, the most complex state of organization that matter has achieved in our cosmos."

George Wald (1906–1997)

* *
*

"The apparent cause of illnesses is often an infection, an intoxication, nervous exhaustion, or merely old age. But actually a break-down of the hormonal mechanism seems to be the most common ultimate cause of death in man."

Hans Selye (1907–1982)

* *
*

"Strange though it may sound, it was a combination of Judeo-Greek ideas, amalgamated within the medieval church itself, which were to form part of the foundation out of which finally arose, in the eighteenth and nineteenth centuries, one of the greatest scientific achievements of all time: the recovery of the lost history of life."

Loren Eiseley (1907–1977)

* *
*

"Elements and stars, Planets and time, air and water – what would these things be without intelligent life to illuminate them with perception and insight?"

Preston Cloud (1912–1991)

* *
*

"The programming of a plant's form is based on the speed and direction of growth. To take a very simple example, the apple is round because growth continues equally in all directions as it matures; but a pear grows faster along its long axis than its radial one."

Anthony Huxley, 1974

* *
*

"Scrutiny of the organization of shells of many viruses with the electron microscope proves that their protein molecules are assembled according to well-known principles of solid geometry, the same ones employed by roof builders to construct quasi-spherical shells of maximum strength using uniform building elements. The shells of viruses bear close resemblance to Buckminster Fuller's domes."

"The perfect geometric shape of virus shells is in its way as remarkable as the symmetrical shape of a starfish or a sea urchin. But the shape of these animals and of all complex organisms is achieved through an elaborate process of development, involving cellular interactions whose complex mechanism is not yet understood. The shape of a virus is simply the outcome of the assembly of protein molecules tending, like all molecular structures, to reach a state of minimal energy."

Salvador Luria (1912–1991)

* *
*

"As far as the meaning of life in general, or in the abstract, as far as I can see, there is none. If all of life were suddenly to disappear from earth and anywhere else it may exist, or if none had ever formed in the first place, I think the Universe would continue to exist without perceptible change. However, it is always possible for an individual to invest his own life with meaning that he can find significant. He can so order his life that he may find as much beauty and wisdom in it as he can."

Isaac Asimov (1920–1992)

* *
*

"I feel that it is the very process of creating so many species which leads to evolutionary progress... Without speciation, there would be no diversification of the organic world, no adaptive radiation, and very little evolutionary progress."

"Attempts at a 'reduction' of purely biological phenomena or concepts to laws of the physical sciences has rarely, if ever, led to any advance in our understanding. Reduction is at best a vacuous, but more often a thoroughly misleading and futile, approach... System always have the peculiarity that the characteristics of the whole cannot (not even in theory) be deduced from the most complete knowledge of the components, taken separately or in other partial combination. The appearance of new characteristics in wholes has been designated emergence*: Species, competition, territory, migration, and hibernation are examples of organismic phenomena for which a purely physical description is at best incomplete and usually biologically irrelevant."*

Ernst Mayer (1901–1952)

* *
*

"Biology can neither reduced to physics, nor do without it."

"It is not inconceivable that in the future the thousands of chemical species contained in the bacterial cell may be synthesized one by one. But there is no chance of seeing all these compounds being assembled correctly and a bacterium emerging fully armed from a test-tube."

"It is natural selection that gives direction to changes, orients chance, and slowly, progressively produces more complex structures, new organs, and new species. Novelties come from previously unseen association of old material. To create is to recombine."

Francois Jacob (1920–), 1977

* *
*

"The greatest mystery is why there is something instead of nothing, and the greatest something is this thing we call life."

Allan R. Sandage (1926–)

* *
*

"Large, widespread, and successful species tend to be especially stable. Humans fall into this category, and the historical record supports such a prediction. Human body form has not altered appreciably in 100,000 years."

Stephen Jay Gould (1941–2002)

* *
*

"Consider the difference in size between some of the very tiniest and the very largest creatures on Earth. A small bacterium weighs as little as 0.00000000001 gram. A blue whale weighs about 100,000,000 grams. Yet a bacterium can kill a whale... Microbes, not macrobes, rule the world."

Bernard Dixon

* *
*

"Darwinism is not a theory of random chance. It is a theory of random mutation plus non-random cumulative natural selection."

"There is enough information capacity in a single human cell to store the Encyclopedia Britannica, all 30 volumes of it, three or four times over."

Richard Dawkins (1941–)

4. Time, History and Man

* *
*

"Very few things happen at the right time, and the rest do not happen at all; the conscientious historian will correct these defects."

Herodotos (ca 484–425 BCE)

* *
*

"The sun also ariseth, and the sun goeth down, and hasteth to his place where he arose. The wind goeth toward the south, and turneth about unto the north; it whirleth about continually, and the wind returneth again according to his circuits... The thing that hath been, it is that which shall be; and that which is done is that which shall be done; and there is no new thing under the sun... There is no remembrance of former things; neither shall there be any remembrance of things that are to come..."

Ecclesiastes **1**, 5–11

* *
*

"And I heard, but I understood not:
then said I, 'O my Lord, what shall be
the end of these things?'
And he said, 'Go thy way, Daniel:
for the words are closed up and sealed
till the time of the end'".

Daniel **12**, 8–9, (ca 165 BCE)

* *
*

"To be ignorant of what occurred before you were born is to remain always a child. For what is the worth of human life, unless it is woven into the life of our ancestors by the records of history?"

Cicero (46 BCE)

* *
*

"Nothing is ours except time."

Seneca (4 BCE–65)

* *
*

"What, then, is time? If no one asks me, I know; if I want to explain it to someone who does ask me, I do not know."

"The world was made, not in time, but simultaneously with time."

Aurelius Augustinus (354–430)

* *
*

"The moving Finger writes; and, having writ,
Moves on: nor all thy Piety nor Wit
Shall lure it back to cancel half a Line,
Nor all thy Tears wash out a Word of it."

Omar Khayyam (ca 1100)

* *
*

"The most precious thing in life is its uncertainty."

Yoshida Kenko (ca 1340)

* *
*

"What's past is prologue."

William Shakespeare (1564–1616), '*The Tempest*'

"When I do count the clock that tells the time
And see the brave day sunk in hideous night,
When I behold the violet past prime
And sable curls all silvered o'er with white,
When lofty trees I see barren of leaves,
Which erst from heat did canopy the herd,
And summer's green, all girded up in sheaves,
Borne on the bier with white and bristly beard:
Then on thy beauty do I question make
That thou among the wastes of Time must go,
Since sweets and beauties do themselves forsake
And die as fast as they see others grow;

And nothing 'gainst Time's scythe can make defense
Save breed, to brave him when he takes thee hence."

William Shakespeare (1564–1616), Sonnet **12**

"O time! thou must untangle this, not I;
It is too hard a knot for me to untie!"

William Shakespeare (1564–1616), 'Twelfth Night' **3**, (II)

* *
*

"Coming events cast their shadow before."

Thomas Campbell (1777–1844)

* *
*

"Prior to the creation of the universe, time did not exist. Time started at the moment of creation."

Jonathan Eibshitz (1690–1764)

* *
*

"To prevent an effect from occurring at all requires a force equal to the cause of that effect, but to give it a new direction often requires only something very trivial."

George Christoph Lichtenberg (1742–1799)

* *
*

"Greek civilization depended essentially on slave-labor but could not progress without the harnessing of natural forces to labor-saving machines. Only the free man, not a slave, has a disposition and interest to improve implements or to invent them. Accordingly, in the devising of a complicated machine, the workmen employed upon it are generally co-inventors. The eccentric *and the* governor, *most important part of the steam-engine, were devised by laborers. The improvement of established industrial methods by slaves, themselves industrial machines, is out of question."*

Justus von Liebig (1803–1873)

* *
*

"In analyzing history, do not be too profound, for often the causes are quite superficial."

Ralph Waldo Emerson (1803–1882)

* *
*

"Ideas have consequences."

F. Dostoyevski (1821–1881)

* *
*

"There is one thing stronger than all the armies in the world: and that is an idea whose time has come."

Victor Hugo (1802–1885)

* *
*

"Biography is the only history: The history of the world is but the biography of great men."

"Speech is of time, silence is of eternity."

Thomas Carlyle (1795–1881)

* *
*

"Anybody can make history. Only a great man can write it."

Oscar Wilde (1854–1900)

* *
*

"An idea isn't responsible for the people who believe in it."

Don Marquis (1878–1937)

* *
*

"If you want work well done, select a busy man – the other kind has no time."

Elbert Hubbard (1856–1915)

* *
*

"Time is neutral; but it can be made the ally of those who will seize it and use it to the full."

Winston Churchill (1874–1965)

* *
*

"Time is the stuff life is made of."

Benjamin Franklin (1706–1790)

* *
*

"With me goes your meaning too."

"Das ist der Sinn von allem, was einst war,
dass es nicht bleibt mit seiner ganzen Schwere,
dass es zu unserm Wesen wiederkehre,
in uns verwoben, tief und wunderbar."

"How can the least thing happen,
unless all future fullness,
time's completed sum,
move to meet us half-way?"

"The free animal
has its decease perpetually behind it
and God in front, and when it moves, it moves
into eternity, like running springs."

"The future enters into us, in order to transform itself in us, long before it happens."

Rainer Maria Rilke (1875–1925)

* *
*

"Nothing has really happened until it has been recorded."

Virginia Woolf (1882–1941)

* *
*

"The oldest of the old follows behind us in our thinking, and yet it comes to meet us."

Martin Heidegger (1889–1976)

* *
*

"We physicists work with time every day, but don't ask me to tell you what time is; it is too complicated to be thinking about it."

Richard Phillips Feynman (1918–1988)

* *
*

"Everything is a matter of chronology."

"In theory one is aware that the earth revolves, but in practice one does not perceive it. So it is with time in one's life."

Marcel Proust (1871–1922)

* *
*

"For the tribal man space was the uncontrollable mystery. For technological man it is time that occupies the same role."

Herbert Marshall McLuhan (1911–1980), (1951)

* *
*

"Time is in fact the hero of the plot. What we regard as impossible on the basis of human experience is meaningless here. Given so much time, the 'impossible' becomes the possible, the possible probable, and the probable virtually certain. One has only to wait: time itself performs the miracles."

George Wald (1906–1997), (1954)

* *
*

"Money-making, social achievements, family and posterity are nothing but plain nature, not culture. Culture lies outside the purpose of nature."

"The essence of culture is continuity and conservation of the past."

C.G. Jung (1875–1961)

* *
*

"Space and time are modes of our thinking and not conditions of our life."

"I never think of the future. It comes soon enough."

"Michael left this strange world just before me. This is of no importance. For us, devout physicists, the distinction between past, present and future signifies only an obstinate illusion."

Albert Einstein (1879–1955), (1955)

* *
*

"Time is a fluid condition which has no existence except in the momentary avatars of individual people."

William Faulkner (1897–1962), 1958

* *
*

"The invention of the mechanical clock was one of a number of major advances that turned Europe from a weak, peripheral, highly vulnerable outpost of Mediterranean civilization into a hegemonic aggressor."

David Landes (1924–), 1980

* *
*

"Our knowledge of time and space owes more to the labors of mathematicians and physicists than to those of professed philosophers."

Charles Broad

* *
*

"History is the study of the bones of civilizations that failed, as the pterodactyl and the dinosaur failed."

Colin Wilson (1931–)

* *
*

"Civilizations die of suicide not murder."

Toynbee (1889–1975)

* *
*

"Ideas shape the course of history."

John Maynard Keynes (1883–1946)

* *
*

"In the last 3421 years of recorded history only 268 have seen no war."

Will Durant (1885–1981)

* *
*

"A study of history shows that civilizations that abandon the quest for knowledge are doomed to disintegration."

Bernard Lovell (1913–)

* *
*

"Great events do not necessarily have great causes."

"All changes in history, all advance, comes from the non-conformists. If there had been no troublemakers, no dissenters, we should still be living in caves."

A.J.P. Taylor (1906–1990)

* *
*

"There is no such thing as history, there are only historians."

Walter Benjamin (1892–1940)

* *
*

"Civilizations rise and fall on ideas.Thus, the "Sturm und Drang" philosophy of Goethe, Kant, Hegel, Rousseau ad Blake, which was essentially a reaction to rationalism, later ruined Europe and is now undermining the United States. The Liberal-Left understands this, but the Right has nothing to offer except religion."

Alan Bloom (1930–1992)

* *
*

"When the anonymous masses enter history, it is chiefly to be slaughtered in battle, to die of famine or privation – to illustrate the failures of their betters. ... We have the mighty pyramids, but no firsthand account of the feelings of the wretches who built them."

Herbert J. Muller (1905–1980)

* *
*

"Man is the only animal to be troubled by time, and from that concern comes much of his finest art, a great deal of his religion, and almost all his science. First was the temporal regularity of nature — the rising of sun and stars, the slower rhythm of seasons — which led to the concept of law and order and in turn to astronomy, the first of all sciences. Changeless environments like in the deep ocean or the cloud-wrapped surface of Venus provide no stimulus to intelligence and in such places it may never be able to arise."

Arthur C. Clarke (1917–2008)

* *
*

"The genius of Einstein leads to Hiroshima."

Pablo Picasso (1881–1973)

* *
*

"There is never time to do it right, but there is always time to do it over."

(Anon)

* *
*

"You are never given enough time or money."

(Anon)

* *
*

"What happens first is not necessarily the beginning."

(Anon)

* *
*

"Throughout all great adventures of men, the cowards stayed behind, the weak perished on the way, only the strong survived and reached their destination."

(Anon)

* *
*

"The Protestant families which were exiled from the Catholic countries of Europe during the sixteenth and seventeenth centuries and even during the eighteenth, have given birth to an extraordinarily high number of distinguished scientists. This is not to be wondered at. These people who preferred the misery of exile to moral servitude were certainly above the average in their conscientiousness and earnestness."

(Anon)

5. On Mathematics, Zero and Infinity

* *
*

"All things are numbers."

Pythagoras (580–506 BCE)

* *
*

"When the Greek philosophers found that the square root of 2 is not a rational number, they celebrated the discovery by sacrificing 100 oxen."

Herodotos (ca 484–424 BCE)

* *
*

"In fact, everything that can be known has number. For it is not possible to conceive of, or to know, anything that has not."

Philolaos of Croton (ca 480 BCE)

* *
*

"The mathematical sciences particularly exhibit order, symmetry and limitation; and these are the greatest forms of the beautiful."

Aristotle (384–322 BCE)

* *
*

"God ever geometrizes."

"He is unworthy of the name of man who is ignorant of the fact that the diagonal of a square is incommensurable with its side."

"How did Thales measure the height of the Pyramide of Cheops? — he measures the length of its shadow when a man's shadow was equal to his height."

" Let no one ignorant of geometry enter here."(Inscribed over the door of the Academy, ca 387 BCE)

Plato (427–347 BCE)

* *
*

"If we can approach the Divine only through symbols, then it is most suitable that we use mathematical symbols, for these have an indestructible certainty."

Nicolas of Cusa (1401–1464)

* *
*

"Nothing can be created from nothing."

Lucretius (ca 99–55 BCE)

* *
*

"Beauty — the adjustment of all parts proportionately so that one cannot add or subtract or change without impairing the harmony of the whole."

Leon Battista Alberti (1404–1472)

* *
*

"There is no royal road to Geometry."

"The laws of nature are but the mathematical thoughts of God."

Euclid (323–283 BCE)

* *
*

"But let us remember that we are dealing with infinities and indivisibles, both of which transcend our finite understanding: the former on account of their magnitude, the latter because of their smallness. In spite of this, men cannot refrain from discussing them, even if it must be done in a roundabout way."

"The universe stands continually open to our gaze but it cannot be understood unless one first learns to comprehend the language and interpret the characters in which it is written. It is written in the language of mathematics."

"The book of Nature is written in mathematical symbols."

Galileo Galilei (1564–1642)

* *
*

"No human investigation can be called real science if it cannot be demonstrated mathematically."

Leonardo da Vinci (1452–1519)

* *
*

"Ubi materia — ibi geometria."

Johannes Kepler (1571–1630), 1609

* *
*

"The mathematicians who are merely mathematicians reason correctly, but only when everything has been explained to them in terms of definitions and principles. Otherwise they are limited and insufferable, for they only reason correctly when they are dealing with very clear principles."

"All that transcends geometry transcends our comprehension."

Blaise Pascal (1623–1662)

* *
*

"The imaginary number is a fine and wonderful recourse of the divine spirit, almost an amphibian between being and not being."

"Music the pleasure the human mind experiences from counting without being aware that it is counting."

"Datis ordinatis etiam sunt ordinata."

Gottfried Wilhelm von Leibniz (1646–1716)

* *
*

"Eadem mutata resurgo."

Engraved on the tombstone of Jakob Bernoulli (1654–1705)

* *
*

"I could be bounded in a nutshell and count myself a king of infinite space."

William Shakespeare (1564–1616), "*Hamlet*"

* *
*

"What is man in nature? Nothing in relation to the infinite, everything in relation to nothing, a mean between nothing and everything."

"When I consider the small span of my life absorbed in the eternity of time, or the small part of space which I can touch or see engulfed by the infinite immensity of spaces that I know and that know me not, I am frightened and astonished to see myself here instead of there... now instead of then."

Blaise Pascal (1623–1662)

* *
*

"An infinitesimal is the spirit of a departed quantity."

George Berkeley (1685–1753)

* *
*

"The enormous and the minute are interchangeable manifestations of the eternal."

William Blake (1757–1827)

* *
*

"Nature is not embarrassed by difficulties of analysis."

Augustin Fresnel (1788–1827)

* *
*

"The profound study of nature is the most fertile source of mathematical discoveries."

Jean Baptiste Joseph Fourier (1768–1830)

* *
*

"It is difficult to say which of Fourier results is most to be praised: their uniquely original quality, their transcendently intense mathematical interest, or their perennially important instructiveness for physical science."

"Mathematics is the only good metaphysics."

Lord Kelvin (1824–1907)

* *
*

"It is not knowledge but the act of learning, not possession but the act of getting there, which grants the greatest enjoyment. When I have clarified and exhausted a subject, then I turn away from it in order to go into darkness again."

"Important propositions, with the impress of simplicity on them, are often easily discovered by induction, and yet are of so profound a character that we cannot find the demonstrations till after many vain attempts; and even then, when we do succeed, it is often by some tedious and artificial process, while the simple methods may long remain concealed."

"Pauca Sed Matura."

"All the measurements in the world are not the equivalent of a single theorem that produces a significant advance in our greatest of sciences."

Carl Friedrich Gauss (1777–1855)

* *
*

"Fourier's book is a great mathematical poem."

J.C. Maxwell (1831–1879)

* *
*

The success of Fourier "It was, no doubt, partially because of his very disregard for rigor that he was able to take conceptual steps which were inherently impossible to men of more critical genius."

Rudolph Langer

* *
*

"Looking back, we can see Fourier's memoir as heralding the surge of new mathematical methods and results which were to mark the new century. His ideas are built into the commonsense of our society."

T.W. Körner, 1988

* *
*

"Fourier's book was of paramount importance in the history of mathematics and pure analysis perhaps owed it even more than applied mathematics."

Poincare (1854–1912), 1895

* *
*

"I have had my results for a long time: but I do not yet know how I am to arrive at them."

"I protest against the use of an infinite quantity as an actual entity; this is never allowed in mathematics. The infinite is a manner of speaking, in which one properly speaks of limits to which certain ratios can come as near as desired, while others are permitted to increase without bound."

"Mathematics is concerned only with the enumeration and comparison of relations."

"The higher arithmetic offers an inexhaustible wealth of interesting truths, which are not isolated but stand in intimate relationship with each other, and which reveal, as the science develops, ever new, even unexpected, connexions."

Carl Friedrich Gauss (1777–1855)

* *
*

"Between Analysis and Number Theory, which have long been held to be quite separate disciplines, one has discovered, in recent times, with increasing frequency, interconnections which are often unexpected. A rich source of such connections, and one which will long remain unexhausted, is the analysis of elliptic functions."

"The theory of elliptic functions is a vast subject of research, which in the course of its development embraces almost all algebra, the theory of definite integrals, and the science of numbers."

"The deduction of these arithmetical theorems from developments in Analysis not only increases the supply of methods of proof in arithmetic, but the theorems themselves are found to acquire a new and striking form."

"God ever arithmetizes."

"Man muss immer generalisieren."

"For Gaussian rigor, we have no time."

"Problems of number theory are just as important as problems from the real world. The honor of the human spirit is at stake."

C.G.J. Jacobi (1804–1851)

* *
*

"There is no study in the world which brings into more harmonious action all the faculties of the mind then mathematics. It seems to raise them, by successive steps of initiation, to higher and higher states of conscious intellectual being."

"The object of pure physics is the unfolding of the laws of the intelligible world; the object of pure mathematics is that of unfolding of the laws of human intelligence."

James Joseph Sylvester (1814–1897)

* *
*

"The moving power of mathematical invention is not reasoning but imagination."

Augustus de Morgan (1806–1871)

* *
*

"Time was when all the parts of the subject were dissevered, when algebra, geometry, and arithmetic either lived apart or kept up cold relations of acquaintance confined to occasional calls upon one another; but that is now at an end; they are drawn together and are constantly becoming more and more intimately related and connected by a thousand fresh ties, and we may confidently look forward to a time when they shall form but one body with one soul.
The world of ideas which it [mathematics] discloses or illuminates, the contemplation of divine beauty and order which it induces, the harmonious connection of its parts, the infinite hierarchy and absolute evidence of the truths with which it is concerned, these, and such like, are the surest grounds of the title of mathematics to human regard, and would remain unimpeached and unimpaired were the plan of the universe unrolled like a map at our feet, and the mind of man qualified to take in the whole scheme of creation at a glance."

J.J. Sylvester (1814–1897), 1869

* *
*

"I see it but I do not believe it."

"In re mathematica ars proponendi questionem pluris facienda est quam solvendi."

Georg Cantor (1845–1918)

* *
*

"The teacher should let science develop before the eyes of his pupil. As it develops and takes form in the mind of the mature thinker, out of his fundamental ideas, so shall he present it, merely adjusting it to the youthful power of understanding."

"It is true that a mathematician who is not also something of a poet, will never be a perfect mathematician."

Karl Theodor Wilhelm Weierstrass (1815–1857), 1883

* *
*

"The integer numbers have been made by God, everything else is the work of man."

Leopold Kronecker (1823–1891)

* *
*

"In eternity, I plan to spend eight million years on mathematics."

Mark Twain (1835–1910)

* *
*

"In most sciences one generation tears down what another has built, and what one has established another undoes. In mathematics alone, each generation builds a new story to the old structures."

Hermann Hankel (1839–1873)

* *
*

"A mathematician may say anything he pleases, but a physicist must be at least partially sane."

"Mathematics is a language."

Josiah Willard Gibbs (1839–1903)

* *
*

"Among the minor, yet striking characteristics of mathematics, may be mentioned the fleshless and skeletal build of its propositions; the peculiar difficulty, complication, and stress of its reasonings; the perfect exactitude of its results; their broad universality; their practical infallibility."

"The one [the logician] studies the science of drawing conclusions, the other [the mathematician] the science which draws necessary conclusions."

"...mathematics is distinguished from all other sciences except only ethics, in standing in no need of ethics. Every other science, even logic, especially in its early stages in danger of evaporating into airy nothingness, degenerating, as the Germans say, into an arachnoid film, spun from the stuff that dreams are made of. There is no such danger for pure mathematics; for that is precisely what mathematics ought to be."

Charles Sanders Peirce (1839–1914)

* *
*

"One cannot escape the feeling that these mathematical formulae have an independent existence and intelligence of their own, that they are wiser than we are, wiser even than their discoverers, that we get more out of them than was originally put into them."

Heinrich Rudolf Hertz (1857–1894)

* *
*

"Geometrical properties are characterized by their invariance under a group of transformations."

"When I was a student, Abelian functions were, as an effect of the Jacobian tradition, considered the uncontested summit of mathematics and each of us was ambitious to make progress in this field. And now? The younger generation hardly knows Abelian functions. How did this happen? In mathematics, as in other sciences, the same process can be observed again and again. First new questions arise, for internal or external reasons, and draw researchers away from the old questions. And the old questions, just because they have been worked on so much, need ever more comprehensive study for their mastery. This is unpleasant, and so one is glad to turn to problems that have been less developed and therefore require less foreknowledge — even if it only a matter of axiomatics, or set theory, or some such thing."

Felix Christian Klein (1849–1925)

* *
*

"I know mathematical processes that I have used with success for a very long time, of which neither I nor anyone else understands the scholastic logic. I have grown into them, and so understand them that way."

"Shall I refuse my dinner because I do not fully understand the process of digestion? No, not if I am satisfied with the result."

"Euclid is the worst. It is shocking that young people should be addling their brains over mere logical subtleties, trying to understand the proof of one obvious fact in terms of something equally obvious."

Oliver Heaviside (1850–1925)

* *
*

"With the introduction of the infinitely small and infinitely large, mathematics, usually so strictly ethical, fell from grace. The virgin state of absolute validity and irrefutable proof of everything mathematical was gone forever; the realm of controversy was inaugurated, and we have reached the point where most people differentiate and integrate not because they understand what they are doing, but from pure faith, because up to now it has always come out right."

Friedrich Engels (1820–1895)

* *
*

"...without mathematical infinity*, there would be no science at all, because there would be nothing general."*

"It has been said that geometry is the art of applying good reasoning to bad diagrams. This is not a joke but a truth worthy of serious thought. What do we mean by a poorly drawn figure? It is one where proportions are changed slightly or even markedly, where straight lines become zigzag, circle acquire incredible humps. But none of this matters.
An inept artist, however, must not *represent a closed curve as if it were open, three concurrent lines as if they intersected in pairs, nor must he draw an unbroken surface when the original contains holes."*

"One geometry cannot be more true than another; it can only be more convenient. Geometry is not true, it is advantageous."

"Mathematics is the art of giving the same name to different things."

"If we wish to foresee the future of mathematics, our proper course is to study the history and present conditions of the science."

Henri Jules Poincaré (1854–1912)

* *
*

"The mathematician, carried along on his flood of symbols, dealing apparently with purely formal truths, may still reach results of endless importance for our description of the physical universe."

Karl Pearson (1857–1936)

* *
*

"The mathematician has reached the highest rung of the ladder of human thought."

Havelock Ellis (1859–1939)

* *
*

"The perfection of mathematical beauty is such...that whatsoever is most beautiful and regular is also found to be most useful and excellent."

D'Arcy Thompson (1860–1948)

* *
*

"The first comment I would make after waking at the end of a thousand year sleep would be: 'Is the Riemann hypothesis established yet'?"

"As long as a branch of science offers an abundance of problems, so long is it alive; a lack of problems foreshadows extinction or the cessation of independent development."

"From time immemorial the infinite *has stirred men's emotions more than any other question. Hardly any other idea stimulated the mind so fruitfully. Yet no other concept needs clarification more than it does."*

David Hilbert (1862–1943)

* *
*

"Do not work within two hours of a substantial meal; blood cannot be in two places at once."

"Any identity, once verified, is trivial."

John Edensor Littlewood (1885–1977)

* *
*

"Oriental mathematics may be an interesting curiosity, but Greek mathematics is the real thing. The Greeks first spoke a language which modern mathematicians can understand; as Littlewood said to me once, they are not clever schoolboys or 'scholarship candidates', but 'Fellows of another college'. So Greek mathematics is 'permanent', more permanent even than Greek literature. Archimedes will be remembered when Aeschylus is forgotten, because languages die and mathematical ideas do not. 'Immortality' may be a silly word, but probably a mathematician has the best chance of whatever it may mean."

"Here is to pure mathematics, may it never find an application."

"The 'seriousness' of a mathematical theorem lies, not in its practical consequences, which are usually negligible, but in the significance of the mathematical ideas which it connects. We may say, roughly, that a mathematical idea is 'significant' if it can be connected, in a natural and illuminating way, with a large complex of other mathematical ideas. Thus a serious mathematical theorem, a theorem which connects significant ideas, is likely to lead to important advances in mathematics itself and even in other sciences."

"Sometimes one has to say difficult things, but one ought to say them as simply as one knows how."

Godfrey Harold Hardy (1877–1947)

* *
*

'If insight is the essential element in intelligent problem-solving, fixation is its archenemy. Fixation is overcome and insight attained by a sudden shift in the way the problem or the objects involved in it are viewed. The work described in this article has pointed to some of the factors that necessitate this sudden shift, but precisely what brings it about is still unknown. It remains the central problem of problem-solving."

George Polya (1887–1985)

* *
*

"Mathematics is the most powerful technique for the understanding of patterns and for the analysis of the relations of patterns."

"In the year 1500 Europe knew less than Archimedes who died in the year 212 BCE."

" I will not go so far as to say that to construct a history of thought without profound study of the mathematical ideas of successive epochs is like omitting Hamlet from the play which is named after him... But it is certainly analogous to cutting out the part of Ophelia. This simile is singularly exact. For Ophelia is quite essential to the play, she is very charming – and a little mad."

"The science of pure mathematics may claim to be the most original creation of the human spirit."

Alfred North Whitehead (1861–1947)

* *
*

"Deepest interrelationship in analysis are of an arithmetical nature."

Hermann Minkowski (1864–1909)

* *
*

"Rigor merely sanctions the conquests of the intuition."

Jacques Solomon Hadamard (1865–1963)

* *
*

"A barber is presumed to shave each man in his town who does not shave himself, but not to shave anyone else. Does the barber shave himself?"

"Physics is mathematical not because we know so much about the physical world, but because we know so little: it is only its mathematical properties that we can discover."

"It must have required many ages to discover that a brace of pheasants and a couple of days were both instances of the number two."

"Mathematics, rightly viewed, possesses not only truth, but supreme beauty — a beauty cold and austere, without appeal to any part of our weaker nature, without the gorgeous trappings of painting or music, yet sublimely pure, and capable of a stern perfection such as only the greatest art can show."

"A good notation has a subtlety and suggestiveness."

"Mathematics is the science in which we do not know what we are talking about, and do not care whether what we say about it is true."

"Mathematics is, I believe, the chief source of the belief in eternal and exact truth, as well as in a super-sensible intelligible world."

Bertrand Russell (1872–1970)

* *
*

"Mathematics may explore the forth dimension but the Czar can be overthrown only in the third dimension."

Vladimir Ilyich Lenin (1870–1924)

* *
*

"I don't believe in mathematics."

"Mathematics offers the exact natural sciences certain measure of security which, without mathematics, they could not attain."

"Pure mathematics is, in its way, the poetry of logical ideas."

"Mathematics are well and good but nature keeps dragging us around by the nose."

"Since the basic equations of physics are nonlinear, all the mathematical physics will have to be done over again."

"How can it be that mathematics, being after all a product of human thought, independent of experience, is so admirably adapted to the objects of reality?"

"As far as the laws of mathematics refer to reality, they are not certain; and as far as they are certain, they do not refer to reality."

"God does not care about our mathematical difficulties. He integrates empirically."

Albert Einstein (1879–1955)

* *
*

"Empirical evidence can never establish mathematical existence — nor can the mathematician's demand for existence be dismissed by the physicist as useless rigor. Only a mathematical existence proof can ensure that the mathematical description of a physical phenomenon is meaningful."

Richard Courant (1888–1972)

* *
*

"When feeling burdened or downcast,... the human mind will gladly turn to the realms of Mathematics, where a lucid and precise grasp of objectivities is obtained and insight is gained so pleasantly through appropriate concept formation. Here the human spirit feels at home."

Paul Bernays (1888–1977)

* *
*

"Mathematics itself is only a particular formulation of the mathematical."

Martin Heidegger (1889–1976)

* *
*

"To be a scholar of mathematics you must be born with talent, insight, concentration, taste, luck, drive and the ability to visualize and guess."

Paul Halmos (1916–2006), 1985

* *
*

"When I am working on a problem, I never think about beauty. I think only how to solve the problem. But when I have finished, if the solution is not beautiful, I know it is wrong."

Richard Buckminster Fuller (1895–1983)

* *
*

"Greek pure mathematics covers a very short time indeed, beginning in the time of Plato (Theaetetos and Eudoxos, ca 400 BCE), condensed in the Elements *and appearing for the last time in the works of Archimedes and Apollonios (200 BCE). The main reason for this early interruption of pure mathematics can be found in the purely geometric type of expression which was adopted in order to gain the higher degree of generality which the geometrical magnitudes represent, in contrast to the field of rational numbers, which was the exclusive concern of oriental mathematics and astronomy. This geometrical language, however, very soon reached such a degree of complication that development beyond the theory of conic sections was practically impossible. As a result, the development of theoretical mathematics ended two centuries after its beginning.*
I think that the influence of this Greek pure mathematics on the general standard of mathematics in antiquity has been very much overestimated. Even Euclid's own works, other than the Elements*, are on a very different level; this can be simply explained by the remark that the* Elements *are concerned with very special group of problems, mainly concentrated on the theory of irrational numbers where the exactitude of definitions and conclusions is the essential point of the discussion.*
The main part of mathematical literature, however, was less rigorous and represented the direct continuation of Babylonian and even Egyptian method. The Babylonian influence is, for instance, mainly responsible for the general character of other groups of Greek mathematical literature, as, e.g., the work of Diophantos (ca 300 CE*). This situation in the field of mathematics corresponds very much to the general character of the Hellenistic culture, with its mixture of very contradictory elements from all parts of the ancient world."*

Otto Neugebauer (1899–1990)

* *
*

"The mathematician plays a game in which he himself invents the rules while the physicist plays a game in which the rules are provided by Nature, but as time goes on it becomes increasingly evident that the rules which the mathematician finds interesting are the same as those which Nature has chosen."

"In science one tries to tell people, in such a way as to be understood by everyone, something that no one ever knew before. But in poetry, it's the exact opposite."

"If there is a God, he's a great mathematician."

"Sensible mathematics involves neglecting a quantity when it is small – not neglecting it because it is infinitely great and you do not want it!"

"If there is a God, he's a great mathematician."

"It seems to be one of the fundamental features of nature that fundamental physical laws are described in terms of mathematical theory of great beauty and power, needing quite a high standard of mathematics for one to understand it. You may wonder: Why is nature constructed along these lines? One can only answer that our present knowledge seems to show that it is so constructed. We simply have to accept it. One could perhaps describe the situation by saying that God is a mathematician of a very high order, and He used very advanced mathematics in constructing the universe. Our feeble attempts at mathematics enable us to understand a bit of the universe, and as we proceed to develop higher and higher mathematics we can hope to understand the universe better."

"I am not interested in proofs, only what nature does."

"Mathematics is the tool specially suited for dealing with abstract concepts of any kind and there is no limit to its power in this field."

Paul Adrien Maurice Dirac (1902–1984)

* *
*

"If people do not believe that mathematics is simple, it is only because they do not realize how complicated life is."

"In mathematics you don't understand things. You just get used to them."

"It is only proper to realize that language is largely a historical accident. The basic human languages are traditionally transmitted to us in various forms, but their very multiplicity proves that there is nothing absolute and necessary about them. Just as languages like Greek or Sanskrit are historical facts and not absolute logical necessities, it is only reasonable to assume that logics and mathematics are similarly historical, accidental forms of expression. They may have essential variants, i.e. they may exist in other forms than the ones to which we are accustomed. Indeed, the nature of the central nervous system and of the message systems that it transmits indicate positively that this is so. We have now accumulated sufficient evidence to see that whatever language the central nervous system is using, it is characterized by less logical and arithmetical depth than what we are normally used to. The following is an obvious example of this: the retina of the human eye performs a considerable reorganization of the visual image as perceived by the eye. Now, this reorganization is effected on the retina, or to be more precise, at the point of entry of the optic nerve by means of three successive synapses only, i.e. in terms of three consecutive logical steps. The statistical behavior of the message system used in the arithmetics of the central nervous system and its low precision also indicate that the degeneration of precision cannot proceed very far in the message system involved. Consequently, there exist here different logical structures from the ones we are ordinarily used to in logic and mathematics. They are characterized by less logical and arithmetical depth than we are used to under otherwise similar circumstances. Thus logic and mathematics in the central nervous system, when viewed as languages, must structurally be essentially different from those languages to which our common experience refers."

John von Neumann (1903–1957)

* *
*

"The zero is the most important digit. It is a stroke of genius, to make something out of nothing by giving it a name and inventing a symbol for it."

Bartel Leendert van der Waerden (1903–1996)

* *
*

"The problem is, when we try to calculate all the way down to zero distance, the equation blows up in our face and gives us meaningless answers – things like infinity. This caused a lot of trouble when the theory of quantum electrodynamics first came out. People were getting infinity for every problem they tried to calculate."

"Mathematics is a tool for reasoning. It is not a science from our point of view, in the sense that it is not a natural science. The test of its validity is not experiment."

"I love only nature, and I hate mathematicians."

Richard Feynman (1918–1988)

* *
*

"God exists since mathematics is consistent, and the Devil exists since we cannot prove it."

Andre Weil (1906–1998)

* *
*

"The danger of the mathematician making mistakes is an unavoidable corollary of his power to hit sometimes upon an entirely new method. This seems to be confirmed by the well-known fact that the most reliable people will not usually hit upon new methods."

Alan Mathison Turing (1912–1954)

* *
*

"When the answer to your question is 'yes' then you have asked the wrong question."

Fritz John (1910–1994)

* *
*

"The genesis of the majority of mathematical theories is obscure and difficult. Often, today's presentation of a classical topic, will be much more accessible and concise than it could even have been when it was developed. Any scientist's work, can only be understood within its contemporary scientific framework."

Walter K. Bühler (1944–)

* *
*

"Probability is a simple subject; all the answers are between zero and one."

Byron Goldstein

* *
*

"From long experience, all working mathematicians know that there is a preliminary period of rapid advancement in ideas without worrying about exact definitions and proofs, after which there is very hard work to go from that level of accuracy to finished mathematics, where the bugs in definitions and proofs are gone, and concepts are quite clear. A lot of things change in the process. This is the essence of finishing mathematical work."

Anil Nerode

* *
*

"Mathematics is the science of order. Its object is to find, describe, and understand the order that underlies apparently complex situations. The principal tools of mathematics are concepts which enable us to describe order. Precisely because mathematicians have been searching for centuries for the most efficient concepts for describing obscure instances of order, their tools are applicable to the outside world: for the real world is the very epitome of a complex situation in which there is a great deal of order."

Andrew Gleason (1921–)

* *
*

"It has been said that World War I was the chemist's war, World War II was the physicist's war, World War III (may it never come) will be the mathematicians' war."

(Anon)

* *
*

"The infinite is the mathematician's paradise."

(Anon)

* *
*

Q: What's the difference between a mathematician and a physicist?
A: A mathematician thinks that two points are enough to define a strait line while a physicist wants more data.

(Anon)

* *
*

"The equation $e^{\pi i} = -1$ *has been called the eutectic point of mathematics, for no matter how you boil down and explain this equation, which relates four of the most remarkable numbers of mathematics, it still has a certain mystery about it that cannot be explained away."*

(Anon)

* *
*

"An applied mathematician loves the theorem. A pure mathematician loves the proof."

(Anon)

* *
*

"The trouble with integers is that we have examined only the very small ones. Maybe all the exciting stuff happens at really big numbers, ones we can't even begin to think about in any very definite way. Our brains have evolved to get us out of the rain, find where the berries are, and keep us from getting killed. Our brains did not evolve to help us grasp really large numbers or to look at things in a hundred thousand dimensions."

Ronald L. Graham (1935–)

* *
*

"Mathematics has beauties of its own – a symmetry and proportion in its results, a lack of superfluity, an exact adaptation of means to ends, which is exceedingly remarkable and to be found only in the works of the greatest beauty. When this subject is properly ... presented, the mental emotion should be that of enjoyment of beauty..."

J.W.A. Young

* *
*

"Hiding between all the ordinary numbers was an infinity of transcendental numbers whose presence you would never have guessed unless you looked deeply into mathematics."

Carl Sagan (1934–1996)

* *
*

"Nature does not count nor do integers occur in nature. Man made them all, integers and all the rest, Kronecker to the contrary notwithstanding."

Percy William Bridgman (1882–1961)

* *
*

"The problem with linear theory is that it is not nonlinear."

John A. Adam (1735–1826)

* *
*

"Zero is powerful because it is infinity's twin. They are equal and opposite, yin and yang. They are equally paradoxical and troubling. The biggest questions in science and religion are about nothingness and eternity, the void and the infinite, zero and infinity. The clashes over zero were the battles that shook the foundations of philosophy, of science, of mathematics, and of religion. Underneath every revolution lay a zero – and an infinity.
Zero was at the heart of the battle between East and West. Zero was the center of the struggle between religion and science. Zero became the language of nature and the most important tool in mathematics. And the most profound problems in physics – the dark core of a black hole and the brilliant flash of the big bang, are struggles to defeat zero.
Yet, through all its history, despite rejection and exile, zero has always defeated those who opposed it. Humanity could never force zero to fit its philosophies. Instead, zero shaped humanity's view of the universe – and of God."

Charles Seife

* *
*

"The mathematical description of the world depends on a delicate interplay between continuity and discontinuous, discrete phenomena. The latter are perceived first. 'Functions, just like living beings, are characterized by their singularities'. Singularities, bifurcations and catastrophes are different terms for describing the emergence of discrete structures from smooth, continuous ones."

Vladimir Igorevich Arnol'd (1937–)

* *
*

"Mathematics is the grammar of science and order."

Lancelot T. Hogben (1895–1975)

* *
*

"We should now place the operational calculus with Poincaré discovery of automorphic functions and Ricci's discovery of the tensor calculus as the three most important mathematical advances of the last quarter of the 19th century. Applications, extension, and justifications of it constitute a considerable part of the mathematical activity of today."

Edmund Taylor Whittaker (1873–1956), 1928

* *
*

"The whole form of modern mathematical thinking was created by Euler. It is only with the greatest difficulty that one is able to follow the writings of any author preceding Euler, because it was not yet known how to let the formulae speak for themselves. This art Euler was the first to teach.."

F. Rudio (1857–1929)

* *
*

"Mathematics is an exciting, dynamic field that continues to generate provocative ideas and novel concepts. Some notions quickly find their way into applications; others rest on their intrinsic beauty; still others occupy a modest place in the growing structure of mathematics itself."

Ivar Peterson

* *
*

"The mathematician is simultaneously a revolutionary and a conservative; a deep–rooted skeptic and an avowed optimist."

Max Dehn (1878–1952)

* *
*

"When, after a thousand-year stupor, European thought shook off the effect of the sleeping powders so skillfully administered by the Christian Fathers, the problem of infinity was one of the first to be revived."

"In the history of culture, the discovery of zero will always stand out as one of the greatest single achievements of the human race."

Tobias Dantzig (1884–1956)

* *
*

"Perhaps the most surprising thing about mathematics is that it is so surprising. The rules which we make up at the beginning seem ordinary and inevitable, but it is impossible to foresee their consequences. These have only been found out by long study, extending over many centuries. Much of our knowledge is due to a comparatively few great mathematicians such as Newton, Euler, Gauss, Cauchy or Riemann; few careers can have been more satisfying than theirs. They have contributed something to human thought even more lasting than great literature, since it is independent of language."

E.C. Titchmarsh (1899–1963)

* *
*

"I like mathematics because it is not human and has nothing particular to do with this planet or with the whole accidental universe - because like Spinoza's God, it won't love us in return."

Bertrand Russell (1872–1970), 1912

* *
*

"Geometry, however, supplies sustenance and meaning to bare formulas. Geometry remains the major source of rich and fruitful intuitions, which in turn lend creative power to mathematics. Most mathematicians think in terms of geometric schemes, even though they leave no trace of that scaffolding when they present the complicated analytical structures. One can still believe Plato's statement that geometry draws the soul toward truth.'
Mathematics is a marvelous invention, but the marvel lies in the human mind's capacity to construct understandable models of complex and seemingly inscrutable natural phenomena and thereby give man some enlightenment and power. Mathematics may be the queen of the science and therefore entitled to royal prerogatives, but the queen who loses touch with her subjects may lose support and even be deprived of her realm. Mathematicians may like to rise into the clouds of abstract thought, but they should, and indeed they must, return to earth for nourishing food or else die of mental starvation. They are on safer and saner ground when they stay close to nature. As Wordsworth put it "Wisdom oft is nearer when we stoop than when we soar."

"The tantalizing and compelling pursuit of mathematical problems offers mental absorption, peace of mind amid endless challenges, repose in activity, battle without conflict, 'refuge from the goading urgency of contingent happenings,' and the sort of beauty changeless mountains present to sense tried by the present-day kaleidoscope of events."

Morris Kline (1908–1992)

* *
*

"Mathematics was born and nurtured in a cultural environment. Without the perspective which the cultural background affords, a proper appreciation of the content and state of present-day mathematics is hardly possible."

R.L. Wilder (1896–1982)

* *
*

"After having spent years trying to be accurate, we must spend as many more in discovering when and how to be inaccurate."

Ambrose Gwinett Bierce (1842–1914)

* *
*

"Occupation with the study of mathematics is the best remedy against the lusts of the flesh."

Thomas Mann (1875–1955)

* *
*

"I had a feeling once about mathematics — that I saw it all. Depth beyond Depth was revealed to me — the Byss and the Abyss. I saw a quantity passing through infinity and changing its sign from plus to minus. I saw exactly how it happened and why the tergiversation was inevitable — but it was after dinner and I let it go."

Winston Churchill (1874–1965)

* *
*

"The solution of problems is one of the lowest forms of mathematical research, ...yet its educational value cannot be over estimated. It is the ladder by which the mind ascends into higher fields of original research and investigation. Many dormant minds have been aroused into activity through the mastery of a single problem."

Benjamin Finkel (1865–1947), 1894

* *
*

"Mathematics is the language of the possible."

(Anon)

* *
*

"Topologist – a man who cannot tell the difference between a cup and a doughnut."

(Anon)

* *
*

"Mathematics knows no races or geographic boundaries; for mathematics, the cultural world is one country."

(Anon)

* *
*

"Part of the charm in solving a differential equation is in the feeling that we are getting something for nothing. So little information appears to go into the solution that there is a sense of surprise over the extensive results that are derived."

(Anon)

* *
*

"Black holes result from God dividing the universe by zero."

(Anon)

* *
*

"The infinite! No other question has ever moved so profoundly the spirit of man."

(Anon)

* *
*

"The true spirit of delight, the exaltation, the sense of being more than Man, which is the touchstone on the highest excellence is to be found in mathematics as surely as in poetry."

(Anon)

* *
*

"The Theory of Groups is a branch of mathematics in which one does something to something and then compares the result with the result obtained from doing the same thing to something else, or something else to the same thing."

(Anon)

* *
*

"I'm sorry to say that the subject I most disliked was mathematics. I have thought about it. I think the reason was that mathematics leaves no room for argument. If you made a mistake, that was all there was to it."

(Anon)

* *
*

"What is it that constitutes value or importance in mathematical research? To put it another way: Wherein lies the difference between valuable research and laborious trifling? Or, again, it may be asked: How is valuable research to be distinguished from the construction of examination questions or from mathematical recreations? The distinction is recognized to exist. It is recognized that some kinds of work are better worth doing than others... It is easy to note certain qualities by which valuable research work is characterized.

One of these qualities is novelty. In any valuable research this element of creation or novelty cannot be absent. The new thing, or the created thing, may be a new idea, or a new method, or a new result, or a new proof of a known result... It seems that the progress of mathematics needs many kinds of work. A worker who introduces a new idea may be compared with the exploring prospector who discovers that there is gold in a country; one who invents a new method, with the mechanician who devises the processes and perfects the tools by which the gold can be extracted; one who obtains new results, with the miner who extracts the gold; one who obtains new proofs of known results, with the metallurgist who refines the gold and uses it for making beautiful objects. Few of us can hope to play the part of the prospector or the mechanician, but their efforts would be fruitless without the work of the miner and the metallurgist.

After this all-important character of novelty, or artistic creation, we may note other qualities which valuable research must possess. One of these, which is not very easy to define, I propose to call "relevancy". A piece of work, to be valuable, must be a branch of the tree of knowledge; it must stand in a proper relation to the state of mathematical knowledge existing at the time when it is produced. If it is isolated or has no such relation, it is irrelevant. A proposition may be new and true and difficult to prove and yet it may be irrelevant. Let me give an example. Prof. Hobson, in his little book on the Squaring of the Circle, *has given us a series of most interesting surveys of the state of knowledge in regard to this problem existing at various periods. Anyone who should now spend time on developing new series for calculating approximate values for π, after the fashion of Gregory's series for the inverse tangent, or Newton's series for the inverse sine, would be doing work that might have been valuable in the seventeenth century but would be irrelevant now. This example is rather extreme, but the quality of relevancy or irrelevancy attaches in greater or less degree to all original work in mathematics.*

Another quality which characterizes valuable original research may be named "definiteness". A piece of research work should aim at giving a definite answer to a definite question. For example, the most famous work of Galois aimed at answering the question: What algebraic equations can be solved by means of radicals? We observe in regard to this question that when asked it was supremely relevant...

We have noted three qualities as characteristic of valuable research in mathematics: novelty, relevancy, definiteness, I would add to this catalogue a fourth: generality. The best work is never parochial, it is never restricted to a narrow outlook. The quality of generality may seem to be opposed to the quality of definiteness, but generality must not be confused with vagueness. As an example of a piece of work which shows conspicuously the mark of generality, I would cite Gauss' famous memoir on the hypergeometric series. At the time when this was published it could be said of it that it included the theory of almost all the functions which up to that time had been investigated by analysts. But there is nothing indefinite or vague about Gauss' work... Is there then such a thing as excessive generality? A story is told concerning a certain variety of roses which were in great demand among the makers of bouquets, not only on account of their beauty, but especially because the stalks were very long and stiff. The growers took steps to increase the length of the stalk, and were very successful, producing blooms with stalks as much as seven feet long. But unfortunately as the stalk lengthened the bloom dwindled, indeed most of the very long stalks bore no flowers at all. It may be treading on dangerous ground to suggest that there is such a thing as excessive generality, though even so convinced an analyst as Picard is not without misgivings on the subject. Yet it can sometimes be wished that writers who develop general theories at great length would pause to inquire how far they are available for the solution of special problems...

Finally, to revert to the aspect of mathematics as a creative art, I would urge that an essential element in the equipment of an investigator is a literary education, or, if you prefer it, a training in the means of expression. It is necessary to be articulate, but more than this is desirable. It is desirable to be mathematically articulate, to be able to express mathematical ideas in such a way that they can be comprehended easily by those who have the requisite training. Some great work is marred by obscurity. This charge has been brought against even so great an originator as **Abel**. *Others, such as Laplace, are models of lucidity. There is such a thing as style in mathematics, and it is worth cultivating. The mathematician is an artist; and every artist, we have been told by Mr. Bernard Shaw, must grow his own style out of himself. But there are points of style to which it is desirable to attend, such as clearness, arrangement, rigor, avoidance of haste, conciseness, notation. It* *is desirable to say exactly what one means, neither less nor more. It is desirable to introduce new ideas, or new relations, one at a time, so that each one seems to arise naturally just at the place where it makes its appearance in a piece of written work. No trouble is too great to secure rigor, if it can be secured. We have all heard how Newton kept back the publication of the work, which was ultimately embodied in the* Principia, *until he had obtained a conclusive proof that spheres attract as if their masses were condensed at their centers. If absolute rigor has occasionally to be sacrificed, it should be made perfectly clear at what points it is absent. A memoir should not bear marks of hurry; the argument should be developed in a straightforward fashion from the premises to the conclusion. On the other hand, it should not waste time, as, for instance, by undue restriction of conditions in the main argument, with the object of excluding exceptional cases, or by overloading the main argument with details; it should always be possible to distinguish the wood from the trees. The choice of notation is not to be despised; it may make all the difference to the ease with which a piece of work can be assimilated, or a new idea applied to new questions. It may be necessary to rewrite a memoir more than once or twice if these advantages are to be secured. It is worth while."*

Augustus Edward Hough Love (1863–1940), 1914

6. *Truth, Error, Probability and Statistics*

* *
*

"It is never possible to step twice into the same river."

Heraclitos

* *
*

"Everything existing in the Universe is the fruit of chance and necessity."

Democritos of Abdera (ca 460–370 BCE)

* *
*

"The probable is what usually happens."

Aristotle (384–322 BCE)

* *
*

"Probability is the very guide of life."

"Probabilities direct the conduct of the wise man."

Marcus Tullius Cicero (ca 50 BCE)

* *
*

"The only certainty is that there is nothing certain."

Pliny the elder (23–79 CE)

* *
*

"Time and chance happeneth to them all."

Ecclesiastes **9**, 11

* *
*

"Truth emerges more readily from error than from confusion."

"If a man will begin with certainties he shall end in doubts; but if he will be content to begin with doubts he shall end in certainties."

Francis Bacon (1561–1626)

* *
*

"When it is not in our power to determine what is true, we ought to follow what is probable."

René Descartes (1596–1650)

* *
*

"One can have three principle objects in the study of truth: to discover it when one searches for it, to prove it when one possesses it and to distinguish it from falsity when one examines it".

"Contradiction is not a sign of falsity, nor the lack of contradiction a sign of truth."

Blase Pascal (1623–1662)

* *
*

"Almost all human life depends on probabilities."

Voltaire (1694–1778)

* *
*

"The finite mind cannot attain to the full truth about things through similarity. For the truth is neither more or less, but rather indivisible. What is itself not true can no more measure the truth than what is not a circle can measure a circle. Hence reason, which is not the truth, can never grasp the truth so exactly that it could not be grasped infinitely more exactly. Reason stands in the same relation to the truth as the polygon to the circle; yet even when the number of vertices grows infinite, the polygon never becomes equal to a circle, unless it becomes a circle in its true nature. The real nature of what exists, which constitutes its truth, is therefore never entirely attainable. It has been sought by all the philosophers but never really found."

Nicolas of Cusa (1401–1464), 1440

* *
*

"I cannot get over my amazement as to the mental inertia of our astronomers in general who, like credulous women, believe what they read in the books, tablets, and commentaries as if it were the divine and unalterable truth; they believe the authors and neglect the truth. It is necessary to keep the stars doggedly before one eyes, and to rid posterity from ancient tradition."

Regiomontanus (1436–1476)

* *
*

"There is no isolated truth."

Jean-Francois Millet (1814–1875)

* *
*

"It is easier to perceive error than to find truth, for the former lies on the surface and is easily seen, while the latter lies in the depth, where few are willing to search for it."

Johann Wolfgang von Goethe (1749–1832)

* *
*

"The most important questions of life are, for most part, really only problems of probability."

"But since he has unnecessarily complicated the method by the considerations of diagrams, I shall present it here in its simplest analytical form."

"It is remarkable that a science which began with the consideration of games of chance should have become important object of human knowledge."

Pierre Simon de Laplace (1749–1827)

* *
*

"Everyone wishes to have truth on his side, but not everyone wishes to be on the side of truth."

Richard Whately (1787–1863)

* *
*

"All the compliments that I have received from Arago, Laplace and Biot never gave me so much pleasure as the discovery of a theoretical truth, or the confirmation of a calculation by experiment."

Augustin Jean Fresnel (1788–1827)

* *
*

"In the field of observation, chance favors the prepared mind."

Louis Pasteur (1822–1895)

* *
*

"When you can measure what you are speaking about, and express it in numbers, you know something about it."

Lord Kelvin (1824–1907)

* *
*

"It is the customary fate of new truths to begin as heresies and to end as superstitions."

Thomas Henry Huxley (1825–1895)

* *
*

"Truth is stranger than fiction, but it is because fiction is obliged to stick to probabilities; Truth isn't."

Mark Twain (1835–1910)

* *
*

"Don't be consistent, but be simply true."

Oliver Wendell Holmes (1841–1935)

* *
*

"Great men's errors are to be venerated as more fruitful than little men's truths."

Nietzsche (1844–1900)

* *
*

"If a million people believe a foolish thing, it is still a foolish thing."

Anatole France (1844–1924)

* *
*

"He uses statistics as a drunken man uses lamp posts — for support rather than illumination."

Andrew Lang (1844–1912)

* *
*

"The Gaussian distribution — physicists think that it is a mathematical theorem while mathematicians believe that physicists have verified it experimentally."

Gabriel Lippman (1845–1921)

* *
*

"Man can believe the impossible, but can never believe the improbable."

"The pure and simple truth is rarely pure and never simple."

Oscar Wilde (1854–1900)

* *
*

"A new scientific truth does not triumph by convincing opponents and making them see the light, but rather because its opponents eventually die, and a new generation grows up that is familiar with it."

Max Planck (1858–1947)

* *
*

"If your experiment needs statistics, you ought to have done a better experiment."

Ernest Rutherford (1871–1937)

* *
*

"You can only find truth with logic if you have already found truth without it."

G.K. Chesterton (1874–1936)

* *
*

"Man occasionally stumble over the truth, but most of them pick themselves up and hurry off as if nothing happened."

Winston Spencer Churchill (1874–1965)

* *
*

"Faith — an illogical belief in the occurrence of the impossible."

Henry Louis Mencken (1880–1956)

* *
*

"The conception of chance enters into the very first steps of scientific activity in virtue of the fact that no observation is absolutely correct. I think chance is more fundamental concept than causality; for whether in a concrete case, a cause-effect relation holds or not can only be judged by applying the laws of chance to the observation."

Max Born (1882–1970)

* *
*

"Truth exists, only falsehood has to be invented."

Georges Braque (1882–1963)

* *
*

"To call the statistician after the experiment is done may be no more than asking him to perform a postmortem examination: he may be able to say what the experiment died of."

R.A. Fisher (1890–1962)

* *
*

"If an effect is really there, it shouldn't take a statistician to bring it out."

Harold Jeffreys (1891–1989)

* *
*

"Facts do not cease to exist because they are ignored."

Aldous Huxley (1894–1963)

* *
*

"The progress of science implies not only the accumulation of knowledge, but its organization, its unification, and this involves the periodical invention of new syntheses, coordinating existing knowledge, and of new hypotheses, which give us methods of approaching the unknown. Science is essentially a system, but instead of being, as it was for the schoolmen, a closed system, it is never closed, but always subject to revision or even to complete discard. The true scientist considers his theories not as perfect and permanent, but as essentially incomplete and precarious; he is ever ready to abandon any part or the whole of them, should new experimental facts make it necessary. There are scientific methods; there are no scientific dogmas; there is no scientific orthodoxy. Of course this does not mean that there are no scientific doctrines; there are at any time many of them which are binding as long as they have not been shown to be erroneous, but not a moment longer; an orthodoxy as mobile as that is not a real orthodoxy as theologians understand it. Nor does it mean that men of science are never dogmatic, for being men, they are necessarily frail. In this sense, we might say that men of science are essentially heterodox. Their heterodoxy is not restricted to this or that doctrine; it extends potentially to every doctrine. We would never think of saying that a genuine man of science is loyal to this or that theory as he would be loyal to his church or country; he knows no such scientific loyalty; his only loyalty is to truth,a loyalty which causes him to abandon his most cherished opinions as soon as they are invalidated. This has been proved repeatedly during the last thirty years, for a number of revolutionary discoveries have put his scientific conscience into the crucible. For example, the discovery of radioactivity, the introduction of quanta, and the theories of relativity have obliged every physicist and chemist to change radically his ways of thinking about essential things; they have done it as soon as their conviction was completed, if not without reluctance (it is not easy, especially for older men, to change the intellectual habits of a lifetime), at least without struggle and without rancor. As for the younger men, the more revolutionary the theories, the more exhilarating; it gave them the impression of being the witnesses of a new revelation, of a new beginning. Science is not a being, but a becoming. *Thus the love of science is not the love of this or that system, which is bound sooner or later to be superseded by a better one, but simply the love of truth."*

George Sarton (1884–1956)

* *
*

"It must always be kept in mind that "true" values are unknown and must remain unknown; so that the errors being deviations from an unknown value, are likewise unknown. True values must be estimated by appropriate substitutes, namely, "best" or optimal values, and errors by the deviations of the observed from the optimal values."

Alexander Craig Aitken (1895–1967)

* *
*

"It is not so much important to be rigorous as to be right."

A.N. Kolmogorov (1903–1987)

* *
*

"Error can often be fertile but perfection is always sterile."

A.J.P. Taylor (1906–1990)

* *
*

"There is no part of mathematics that is intimately connected with everyday experiences than the theory of probability, and recent developments in mathematical physics have emphasized the importance of this theory in every branch of science. Knowledge of probability is required in such diverse fields as quantum mechanics, kinetic theory, the design of experiments, and the interpretation of data. Operations analysis applies probability methods to questions of traffic control, allocation of equipment, and the theory of strategy. Cybernetics, another field of recent origin, uses the theory to analyze problems in communication and control."

I.S. Sokolnikoff

* *
*

"Statisticians have an understandable penchant for viewing the whole of the history of science as revolving around measurement and statistical reasoning. This view, which stops very short of insisting that science is only *measurement, is not entirely parochial."*

Stephen M. Stigler

* *
*

"Statistics are like a bikini. What they reveal is suggestive, but what they conceal is vital."

A. Levenstein

* *
*

"An error doesn't become a mistake until you refuse to correct it."

Orlando A. Battista

* *
*

"If you don't make mistakes, you are not working on hard enough problems, and that is a big mistake."

Frank Wilczek (1951–)

* *
*

"The human being is an incurable optimist: he believes he has a pretty good chance to win a lottery prize, but that there is scarcely the slightest chance of his getting killed in a traffic accident."

(Anon)

* *
*

"Everyone believes that the Gauss distribution describes the distribution of random errors: Mathematicians — because they think physicists have verified it experimentally, Physicists — because they think mathematicians have proved it theoretically."

(Anon)

* *
*

"Every scientific truth is an approximation."

(Anon)

* *
*

"Truth is not determined by majority vote."

(Anon)

Some vital statistics:

- Only one out of 10,000 women have the classical measures of 35–25–35 or 36–25–36.
- 200,000 babies are born daily in the world (1993).
- A man of 60 has spent 20 years in bed and over 3 years in eating.
- Mathematicians publish more than 200,000 theorems every year.
- 40,000 persons die yearly of snake bites (mostly in India, from Cobra bites); only 12 them in the USA (1993).
- There are 37 million dogs in American homes. 5 million of them are turned each year to be annihilated (1993).
- 2 million people in the USA are in prison.
- More then a million American teen-agers get pregnant each year.
- About 100,000 people in the USA die of physician's errors each year.
- 1.2 billion people in the world are overweight and 1.2 billion are underweight.
- Close to 17,00 stores are selling and renting video cassettes in the USA. The x-rated biz has become a home and hearth industry. It finally took sex out of the theaters and put it back in the bedroom, where it belongs.
- There are 30 million singles over 40 in the US (2006).
- There are 100,000 people older then 100 years in the world (2000).
- An average US family earns $ 35,000 per year, to be spent as follows: Household (9,000), Insurance (4,800), Taxes (6,500), Payments (13,500), Savings (1,200).
- There are 27 million functionally illiterate Americans.
- An estimated 10–30 percent of American and British babies are illegitimate. The seducer usually proves to be the male next door.
- There are about 500,000 physicians in the US.
- In the US, a woman is raped every 6 minutes.
- There are 8,000 different words in King's John version of Bible, but 34,000 words in Shakespeare.

- There are 60,000 homeless people in New York city. They sleep in the streets. Many die in winter.
- There are one million millionaires in the US; every 39 minutes another millionaire is made there.
- In American homes, television sets are on an average of 42.7 hours per week.
- 5,000 young people commit suicide every year in the US.

7. On Philosophy, Philosophers and Man

* *
*

"Composite things decay. Drive diligently"

Buddha (ca 563–483 BCE)

* *
*

"All is foreseen and free will is given."

Rabbi Akiva Ben Yosef (ca 50–ca 135 CE)

* *
*

In 330 BCE ***Aristotle*** *taught as that a human being is a creature whose distinguishing feature and chief survival mechanism is its ability to consider the world rationality. In the end reason will sway emotion*

* *
*

Immanuel Kant *declared science incapable of solving the three fundamental problems of metaphysics: God, freedom of will, and immortality. He thus contended that physics can never determine if God exists, if we have free will, or if God will grant us immortal life.*

* *
*

***David Hume** maintained that since divinity and metaphysics do not contain any abstract reasoning concerning quantity or number it does not contain any experimental reasoning concerning matter of fact.*

* *
*

"Man is measure of all things; of those which are – that they are, of those which are not – that they are not."

Protagoras of Abdera (ca. 490–420 BC)

* *
*

"The unexamined life is not worth living".

Socrates (470–399 BCE)

* *
*

"The ultimate purpose of philosophy is to lead man to the good life of true happiness."

Epicurus (341–270 BCE)

* *
*

"Philosophy is the science which considers truth".

Aristotle (384–322 BCE)

* *
*

"Of all existing things – some are in our power, and other are not. In our power are: thought, desire, will to chose and will to avoid, and, in a word, everything which is our own doing. Things not in our power include the body, property, reputation, office, and, in a word, everything which is not our own doing. Things in our power are by nature free, unhindered, untrammeled; things not in our power are weak, servile, subject to hindrance, dependent on others."

"All philosophy *lies in two words,* sustain *and* abstain.*"*

Epictetos (ca. 55–ca. 135)

* *
*

"Astra inclinat, non trahunt"
(Stars impel, but they do not compel)

Latin maxim

* *
*

"Mankind must pray to God for fortune but obtain wisdom *for themselves."*

Marcus Tullius Cicero (106–43 BCE)

* *
*

"Philosophy – molds and constructed the soul, guides our conduct, shows us what we should do and what we should leave undone; it sits at the helm and directs our course as we waver amid uncertainties. Without it, no one can live fearlessly or in peace of mind".

Lucius Annaeus Seneca (4 BCE–65 CE)

* *
*

“But there was never yet philosopher that could endure the toothache patiently.”

“Much Ado About Nothing”

“Hang up philosophy!
Unless philosophy can make a Juliet,
Displant a town, reverse a prince’s doom,
It helps not, it prevail not, talk no more.”

“Romeo and Juliet” **3**, 3

“There are more things in heaven and earth, Horatio, than are dreamt of in your philosophy.”

William Shakespeare (1564–1616), “Hamlet” **1**-5, 191-2

* *
*

“Things do not pass for what they are but for what they seem.”

Baltasar Gracián

* *
*

“Esse is percipi”
(nothing exists unless it is perceived by some mind)

George Berkeley

* *
*

*"Is God willing to prevent evil, but not able? – Then he is impotent.
Is he able but not willing? – Then he is malevolent.
Is he both able and willing? – Whence then is evil."*

David Hume (1711–1776)

* *
*

"Know thyself."? If I knew myself, I'd run away."

J.W. von Goethe (1749–1832)

* *
*

*"There are two kinds of truths: those of reasoning and those of facts. The truths of reasoning are necessary and their opposite is impossible.
The truths of fact are contingent and their opposites are possible."*

Leibniz (1646–1716)

* *
*

"Life can only be understood backwards, but it must be lived forwards."

Søren Kierkegaard (1813–1855)

* *
*

"The philosopher *is not interested in truth, but only in "my truth"."*

"There is more wisdom in your body than in your deepest philosophy.*"*

"All ethics begins when the individual is taken to be of infinite importance *– in contrast to nature, which behaves cruelly and playfully toward the individual."*

F.W. Nietzsche (1844–1900)

* *
*

*"*Physics *is what you know,* philosophy *is what you don't know."*

Bertrand Russel (1872–1970)

* *
*

"Life is at the same time freedom and fatality: we accept the fatality and within it we decide on a destiny."

Jose Ortega y Gasset (1883–1955)

* *
*

"Philosophy is not a theory but an activity."

Ludwig Wittgenstein (1889–1951)

* *
*

"If Napoleon had been as intelligent as Spinoza, he would have lived in a garret and written four books."

Anatole France (1844–1924)

* *
*

"Chemistry emerged from alchemy as astronomy from astrology and physics from philosophy."

"Every man of science somewhat of a cynic*, because he does not accept words and conventions at their face value, and of a* skeptic*, because he refuses to believe anything without adequate proof."*

George Sarton (1884–1956)

* *
*

"Where there is the necessary technical skill to move mountains, there is no need for the faith that move mountains."

Eric Hoffer (1902–1983)

* *
*

"Philosophy: unintelligible answers to insoluble problems."

Henry B. Adams (1803–1873)

* *
*

"It is easy to build a philosophy. It doesn't have to run."

C.F. Kettering (1876–1958)

* *
*

"Die Logic ist zwar unerschüterlich, aber einem menschen, der leben will, widersteht sie nicht."

Franz Kafka (1883–1924)

* *
*

"Faith – an illogical belief in the occurrence of the impossible."

"There is no record in human history of a happy philosopher."

H.L. Mencken (1880–1956)

* *
*

"To believe in God is to desire his existence, and what is more, to act as though he existed."

Miguel de Unamuno (1864–1936)

* *
*

"Logic does not apply to the real world."

M.L. Minsky

* *
*

*"That which enters the mind through reason can be corrected.
That which is admitted through faith – hardly ever."*

Santiago Ramón y Cajal (1852–1934)

* *
*

"'Thinkers' are people who re-think; who think that what was thought before was never thought enough."

Paul Valéry (1871–1945)

* *
*

"Truth is the object of philosophy, but not always of philosophers."

John Churton Collins (1848–1908)

* *
*

"What pride to discover that nothing belongs to you – what a revelation."

Emile M. Cioran (1911–1995)

* *
*

"Nothing matters very much, and very few things matter at all."

A.J. Balfour (1848–1930)

* *
*

"Idealism is what precedes experience; cynicism is what follows."

David T. Wolf (1943–)

* *
*

"Once there were philosophers – today there are only professors of philosophy."

(Anon)

* *
*

"Without discipline there is no knowledge and without philosophy there is no purpose."

(Anon)

* *
*

"No real problem has a solution."

(Anon)

* *
*

"If God is dead, everything is permitted."

Jean-Paul Sartre (1905–1980)

* *
*

Napoleon: "Why, in your entire treatise Celestial Mechanics, *you had not once mentioned God?"*
Laplace: "Sir, I had no need for that hypothesis."

* *
*

"The philosopher-mathematician Bernard Russell was once asked why he did not believe in God.
He replied: "Not enough evidence".

* *
*

"In his congressional testimony (1987), Steven Weinberg asked for money to build the SSC, a $10 billion device to be constructed in Texas (funding has since been cut off). A congressman asked Weinberg if the SSC would enable us to find God, and Weinberg (unlike Laplace) declined to answer."

8. On Science, Engineering and Technology[59]

* *
*

"It is easy to distinguish those who argue from fact and those who argue from notions... The principles of every science are derived from experience: thus it is from astronomical observations that we derive the principles of astronomical science."

Aristotle (384–322 BCE)

* *
*

"Science liberates man from the terror of the gods."

Lucretius (ca 99–55 BCE)

* *
*

"Entia non sunt multiplicanda praeter necessitatem." (Entities should not be multiplied unnecessarily.)

"Pluritas non est ponendra sine necessitate."

William of Ockham (1285–1349)

[59] *Technology*: from the Greek *techne* – practical skill. The Greeks relegated skills to a lower sphere; the ideal of a free man was leisure (*schole*), and the pursuit of wisdom which it permitted. Only in the modern world was *techne* made into a prodigious instrument for scientific investigation and material progress.

* *
*

"In dealing with a scientific problem, I first arrange several experiments, since my purpose is to determine the problem in accordance with experience, and then to show why the bodies are compelled so to act. That is the method which must be followed in all researches upon the phenomena of Nature... We must consult experience in the variety of cases and circumstances until we can draw from them a general rule that is contained in them. And for what purpose are these rules good? They lead us to further investigations of Nature and to creation of art. They prevent us from deceiving ourselves, or others, by promising results to ourselves which are not to be obtained."

"There is no certainty in science where one of the mathematical sciences cannot be applied."

"Science is the observation of things possible."

Leonardo da Vinci (1452–1519)

* *
*

"Science is not a belief to be held but a work to be done."

Francis Bacon (1561–1626)

* *
*

"In matters of science, the authority of a thousand does not stand against the humble opinion of one."

Galileo Galilei (1564–1642)

* *
*

"It (my book) may wait a century for a reader, as God has waited six thousand years for an observer."

Johannes Kepler (1571–1630)

* *
*

"The world is my country, science is my religion."

Christiaan Huygens (1629–1695)

* *
*

"I wish to direct all sciences to one end and aim, so that we may attain to have supreme human perfection which we have named; and, therefore, whatsoever in the sciences does not serve to promote our object will have to be rejected as useless."

Baruch Spinoza (1632–1677)

* *
*

"The first man of science was he who looked into a thing, not to learn whether it furnished him with food, or shelter, or weapons, or tools, or armaments, or playwiths but who sought to know it for the gratification of knowing."

Samuel Taylor Coleridge (1772–1934)

* *
*

"I have had my results for a long time: but I do not yet know how I am to arrive at them."

Carl Friedrich Gauss (1777–1855)

* *
*

"Science says the first word on everything, and the last word on nothing."

Victor Hugo (1802–1885)

* *
*

"In science one must search for ideas. If there are no ideas, there is no science. A knowledge of facts is only valuable in so far as facts conceal ideas: facts without ideas are just the sweepings of the brain and the memory."

Vissarion Belinskii (1811–1848)

* *
*

"Only a misguided mind tries to introduce religion into science. More misguided still, is he who attempts to introduce science into religion, because he entertains greater respect for the scientific method."

Louis Pasteur (1822–1895)

* *
*

"Accurate and minute measurements seem to the non-scientific imagination a less lofty and dignified work than looking for something new. But nearly all the grandest discoveries of science have been but the rewards of accurate measurements and patient long-continued labors in the minute sifting of numerical results.
I often say that when you can measure what you are speaking about, and express it in numbers, you know something about it; but when you cannot express it in numbers, your knowledge is of a meager and unsatisfactory kind: it may be the beginning of knowledge, but you have scarcely, in your thoughts, advanced to the stage of science, whatever the matter may be."

"In science there are no paradoxes."

Lord Kelvin (1824–1907)

* *
*

"Science is meaningless because it gives no answer to our question, the only question important for us: 'What shall we do and how shall we be?'"

Lev Nikolayevich Tolstoy (1828–1910)

* *
*

"Science is for those who learn; poetry for those who know."

Joseph Roux (1834–1886)

* *
*

"There is something fascinating about science. One gets such wholesale returns of conjecture out of such a trifling investment of fact."

Mark Twain (1835–1910)

* *
*

"Do you believe that the sciences would ever have arisen and become great if there had not beforehand been magicians, alchemists, astrologers and wizards, who thirsted and hungered after abscondite and forbidden powers?
Indeed, infinitely more had to be promised *than could ever be fulfilled in order that anything at all might be fulfilled in the realms of knowledge."*

Friedrich Nietzsche (1844–1900), 1886

* *
*

"In science the credit goes to the man who convinces the world, not to the man to whom the idea first occurs."

Francis Darwin (1848–1925), 1914

* *
*

"Science is always simple and always profound. It is only the half-truths that are dangerous."

George Bernard Shaw (1856–1950)

* *
*

"Thus I saw that most men only care for science so far as they get a living by it, and that they worship even error when it affords them a subsistence."

Johann Wolfgang von Goethe (1749–1832)

* *
*

"Science is the knowledge of many orderly and methodically digested and arranged so as to become attainable by one."

A.F.W. Herschel

* *
*

"Culture is one thing and varnish is another."

Ralph Waldo Emerson (1803–1882)

* *
*

"Culture is to know the best that has been said and thought in the world."

Mattew Arnold (1822–1888)

* *
*

"There are no such things as applied science, only applications of science."

"In the field of observation, chance favors the prepared mind."

Louis Pasteur (1822–1895), 1872

* *
*

"Science has founded the only true religion. Science is the only redemption of this world."

Robert G. Ingersoll (1833–1899), 1906

* *
*

"An experiment is a question which science poses to Nature, and a measurement is the recording of Nature's answer."

"A new scientific truth does not triumph by convincing opponents and making them see the light, but rather because its opponents eventually die, and a new generation grows up that is familiar with it."

"Scientists don't change their minds, they just die."

"Science cannot solve the ultimate mystery of nature. And that is because, in the last analysis, we ourselves are part of nature and therefore part of the mystery we are trying to solve."

"Science is a mountain from which we can see far and wide into the surrounding terrain, but the mountain itself, is not visible."

Max Planck (1858–1947)

* *
*

"Every great advance in science has issued from a new audacity of imagination."

John Dewey (1859–1952)

* *
*

"Scientific principles and laws do not lie on the surface of nature. They are hidden, and must be wrested from nature by an active and elaborate technique of inquiry."

John Dewey (1859–1952), 1920

* *
*

"One machine can do the work of fifty ordinary men. No machine can do the work of one extraordinary man."

Elbert Hubbard (1856–1915), 1923

* *
*

"Science is nothing but trained and organized common sense."

T.H. Huxley (1825–1895)

* *
*

"It is of the highest importance in the art of detection to be able to recognize out of a number of facts which are incidental and which are vital. ...I would call your attention to the curious incident of the dog in the night-time." "The dog did nothing in the night-time." "That was the curious incident."

"It is a capital mistake to theorize before one has data. Insensibly one begins to twist facts to suit theories instead of theories to suit facts."

Arthur Conan Doyle (1859–1930)

* *
*

"The aims of scientific thought are to see the general in the particular and the eternal in the transitory."

Alfred North Whitehead (1861–1947)

* *
*

"Science is a cemetery of dead ideas."

Miguel de Unamuno (1864–1936)

* *
*

"In science we must be interested in things, not in people."

Marie Curie (1867–1934)

* *
*

"...the problem of all natural philosophy is to drive out qualitative conceptions and to replace them by quantitative relations."

Robert Millikan (1868–1953)

* *
*

"Nothing is more vulnerable and ephemeral than scientific theories, which are mere tools and not everlasting truths."

"Science is not the summa *of life. It is only one of the forms of human thought."*

Carl Gustav Jung (1875–1961)

* *
*

"Although this may seem a paradox, all exact science is dominated by the idea of approximation."

"In art nothing worth doing can be done without genius; in science, even a very moderate capacity can contribute to a supreme achievement."

"Science is what you know, philosophy is what you don't know."

"Man has existed for about a million years. He has possessed writing for about 6000 years, agriculture somewhat longer, but perhaps not much longer. Science, as a dominant factor in determining beliefs of educated men, has existed for about 300 years. In this brief period it has proven itself as an incredibly powerful revolutionary force. When we consider how recently it has risen to power, we find ourselves forced to believe that we are at the very beginning of its work in transforming human life."

"The science have developed in an order the reverse of what might have been expected. What was most remote from ourselves was first brought under the domain of law, and then, gradually, what was nearer: first the heavens, nest the earth, then animal and vegetable life, then the human body, and last of all (as yet very imperfectly) the human mind."

Bertrand Russell (1872–1970)

* *
*

"Scientists should be on top, but not at the top."

"Science, which now offers us a golden age with one hand, offers at the same time with the other the doom of all that we have built up inch by inch since the Stone Age and the dawn of any human annals. My faith is in the high progressive destiny of man. I do not believe we are to be flung back into abysmal darkness by those fiercesome discoveries which human genius has made. Let us make sure that they are servants, but not our masters."

"The Dark Ages may return on the gleaming wings of Science."

Winston Spencer Churchill (1874–1965)

* *
*

"The most beautiful thing we can experience is the mysterious. It is the source of all true art and science."

"There are no eternal theories in science, nearly every great advance in science arises from a crisis in the old theory, through an endeavor to find a way out of the difficulties created."

"No amount of experimentation can ever prove me right; a single experiment can prove me wrong."

"Science is the attempt to make the chaotic diversity of our sense-experience correspond to a logically uniform system of thought."

" Once the Teutonic barbarians had destroyed Europe's ancient culture, a new and finer cultural life slowly began to flow from two sources that had somehow escaped being altogether buried in the general havoc — the Jewish Bible and Greek philosophy and art. The union of these two sources, so different one from other, marks the beginning of our present cultural epoch, and from that union, directly or indirectly, has sprung all that makes the true values of our present-day life."

"Science without religion is lame, religion without science is blind."

"A theory is more impressive the greater the simplicity of its premises is, the more different kinds of things it relates, and the more extended its area of applicability."

"The whole of science is nothing more than a refinement of everyday thinking."

"No great discovery was ever made in science except by one who lifted his nose above the grindstone of details and ventured on a more comprehensive vision."

"The grand aim of all science is to cover the greatest number of empirical facts by logical deduction from the smallest number of hypotheses or axioms."

Albert Einstein (1879–1955)

* *
*

"The scientist describes what is: the engineer creates what never was."

Theodor von Kármán (1881–1963)

* *
*

"The moment man cast off his age-long belief in magic, Science bestowed upon him the blessings of the Electric Current."

Jean Giraudoux (1882–1944)

* *
*

*"*Culture *is the system of vital ideas by which the age lives. It borrows from* science *what is vitally necessary for the interpretation of our existence. There are entire portions of science which are not culture, but pure scientific technique. And vice versa, culture requires that we posses a complete concept of the world and of man: it is not for culture to stop, with science, at the point where the methods of absolute rigor happen to end. Life cannot wait until science may have explained the universe scientifically. We cannot put off living until we are ready. The most salient characteristic of life is its coerciveness: it is always urgent, 'here and now' without any possible postponement. Life is fired at us point-blank. And culture, which is but its interpretation, cannot wait any more than can life itself. Science is not something by which we live. The internal conduct of science is not a vital concern; that of culture is. Science is indifferent to the exigencies of our life, and follows its own necessities. Accordingly, science grows constantly more diversified and specialized without limit, and is never completed. But culture is subservient to our life here and now, and it is required to be, at every instant, a complete, unified, coherent system — the plan of life."*

José Ortega y Gasset (1883–1955)

* *
*

“Science is ‘the only human activity which is truly cumulative and progressive’.”

“Science is more than an accumulation of facts; it is not simply positive knowledge, but systematized positive knowledge; it is not simply unguided analysis and haphazard empiricism, but synthesis; it is not simply a passive recording, but constructive activity.”

George Alfred Léon Sarton (1884–1956)

* *
*

“Modern science is not relevant to the search for the underlying metaphysical and moral truths by which one lives. They must be intuitively, almost mystically arrived at.”

“Every second of our lives is saturated with the physical consequences of science or, as we could say, with excrements from the progress of research.”

“Modern science may be as far from revealing the underlying laws of the natural universe as was the science of ancient Greece.”

Erwin Schrödinger (1887–1961)

* *
*

“Physical models are as different from the world as a geographical map is from the surface of the earth.”

Leon Brillouin (1889–1969)

* *
*

"Don't confuse hypothesis and theory. The former is a possible explanation; the latter, the correct one. The establishment of theory is the very purpose of science."

"Facts are not science — as the dictionary is not literature."

Martin H. Fischer (1879–1962)

* *
*

"The fantastic advance in the field of electronic communication constitute a greater danger to the privacy of the individual."

Earl Warren (1891–1974)

* *
*

"Science is not technology, it is not gadgetry, it is not some mysterious cult, it is not a great mechanical monster! Science is an adventure of the human spirit. It is essentially an artistic enterprise, stimulated largely by the universe, served largely by disciplined imagination, and based largely on faith in the reasonableness, order, and beauty of the universe of which man is part."

Warren Weaver (1894–1978)

* *
*

"It is my belief that if a scientist cannot talk simply about his subject, he has not got to the bottom of it himself."

Edward Victor Appleton (1892–1965)

* *
*

"Technology: the invention, manufacture, and use of tools."

Arnold J. Toynbee (1889–1975), 1961

* *
*

"In science the credit goes to the man who convinces the world, not to the man to whom the idea first occurs."

William Osler (1849–1919)

* *
*

"Science is out of the reach of morals, for her eyes are fixed upon eternal truthes. Art is out of the reach of morals, for her eyes are fixed upon things beautiful and immortal and ever-changing. To morals belong the lower and less intellectual spheres."

Oscar Wilde (1854–1900), 1891

* *
*

"The simplest schoolboy is now familiar with truths for which Archimedes would have sacrificed his life."

Ernest Renan (1823–1892)

* *
*

"Science is built of facts the way a house is built of bricks; but an accumulation of facts is no more science than a pile of bricks is a house."

Henry Poincaré (1854–1912), 1905

* *
*

"In chess, if you make one wrong *move you are finished whereas in science if you can do one thing* right*, you are famous."*

Norbert Wiener (1894–1964)

* *
*

"Watch out for the engineers — they begin with sewing machines and end up with the atomic bomb."

Marcel Pagnol (1895–1974)

* *
*

"Whether we choose to call it pure or applied, the story of science is not something apart from the common life of mankind. What we call pure science only thrives when the contemporary social structure is capable of making full use of its teaching, furnishing it with new problems for solution and equipping it with new instruments for solving them. Without printing there would have been little demand for spectacles, without spectacles neither the telescope nor microscope, without these the finite velocity of light, the annual parallax of the stars, and the micro-organisms of fermentation processes and disease would never have been known to science. Without the pendulum clock and the projectile there would have been no dynamics nor theory of sound. Without the dynamics of the pendulum and projectile, no Principia*. Without deep-shaft mining in the sixteenth century, when abundant slave labor was no longer to hand, there would have been no social urge to study pressure, ventilation and explosion. Balloons would not have been invented, chemistry would have*

barely surpassed the level reached in the third millennium and the conditions for discovering the electric current would have been lacking."

Lancelot Hogben (1895–1975)

* *
*

"I do not hesitate to assert that I consider astronomy as the most important force in the development of science since its origin sometime around 500 BCE."

Otto Neugebauer (1899–1990)

* *
*

"Atomic energy bears that same duality that has faced man from time immemorial, a duality expressed in the Book of Books thousands of years ago: "See, I have set before thee this day life and good and death and evil ... therefore choose life."

David E. Lilienthal (1899–1981)

* *
*

"The danger of the past was that men became slaves. The danger of the future is that men may become robots."

Eric Fromm (1900–1980)

* *
*

"A theory is not an absolute truth but a self-consistent analytical formulation of the relations governing a group of natural phenomena."

Julius Adams Stratton (1901–1994)

* *
*

"In science one tries to tell people, in such a way as to be understood by everyone, something that no one ever knew before. But in poetry, it's the exact opposite."

Paul Dirac (1902–1984)

* *
*

"It is his intuition, his mystical insight into the nature of things, rather than his reasoning which makes a great scientist."

"Every truly scientific theory necessarily has to be couched in a form capable of being disproved."

Karl Raimund Popper (1902–1994)

* *
*

"The spiritual and intellectual decline which has overtaken us in the last thirty years ... [may be due] to the diversion of all the best brains to technology."

Kenneth Clarke (1903–1983)

* *
*

"To ask in advance for a complete recipe would be unreasonable. We can specify only the human qualities required: patience, flexibility, intelligence."

John von Neumann (1903–1957)

* *
*

"In the first-instance the work of science is co-operative; a scientist takes his colleagues as judges, competitors and collaborators."

Robert Oppenheimer (1904–1967)

* *
*

"Many civilizations invented different technologies, but science was invented only once."

Loren Eiseley (1907–1977)

* *
*

"Modesty befits the scientist, but not the ideas that inhabit him and which he is under the obligation of upholding."

"In science, self-satisfaction is death. Personal self-satisfaction is the death of the scientist. Collective self-satisfaction is the death of the research. It is restlessness, anxiety, dissatisfaction, agony of mind that nourish science."

Jacques Monod (1910–1976)

* *
*

"There is not a 'pure' science. By this I mean that physics impinges on astronomy, on the one hand, and chemistry and biology on the other. And not only does each support neighbors, but derives sustenance from them. The same can be said of chemistry. Biology is, perhaps, the example par excellence today of an 'impure' Science."

Melvin Calvin (1911–1997)

* *
*

"Technology — the knack of so arranging the world that we don't have to experience it."

Max Frisch (1911–1991)

* *
*

"No scientist is admired for failing in the attempt to solve problems that lie beyond his competence. The most he can hope for is the kindly contempt earned by the Utopian politician. If politics is the art of the possible, research is surely the art of the soluble. Good scientists study the most important problems they think they can solve. It is, after all, their professional business to solve problems, not merely to grapple with them."

Peter Medawar (1915–1987)

* *
*

"I am sorry to say that there is too much point to the wisecrack that life is extinct on other planets because their scientists were more advanced than ours."

John Fitzgerald Kennedy (1917–1963)

* *
*

"This is one of the obvious things that take a very long time to notice ..."

"In scientific practice, the onus of the proof is always on the advocate of the more complicated hypothesis."

H. Jeffreys (1891–1989)

* *
*

"If it keeps up, man will atrophy all his limbs but the push-button finger."

Frank Lloyd Wright (1867–1958)

* *
*

"Space isn't remote at all. It's only an hour's drive away if your car could go straight upwards."

Fred Hoyle (1915–2001), 1979

* *
*

"Science may be described as the art of systematic over-simplification."

Karl R. Popper (1902–1994), 1982

* *
*

"The engineer is the key figure in the material progress of the world. It is his engineering that makes a reality of the potential value of science by translating scientific knowledge into tools, resources, energy and labor to bring them into the service of man ... To make contributions of this kind the engineer requires the imagination to visualize the needs of society and to appreciate what is possible as well as the technological and broad social age understanding to bring his vision to reality."

Sir Eric Ashby (1904–1992), 1958

* *
*

"The scientist is not a person who gives the right answers, he's one who asks the right questions."

Claude Levi-Strauss (1908–), 1964

* *
*

"It is a medium of entertainment which permits millions of people to listen to the same joke at the same time, and yet remain lonesome."

T.S. Eliot (1888–1965)

* *
*

"Science has promised us truth – an understanding of such relationships as our minds cam grasp; it has never promised us either peace or happiness."

Gustav Le Bon (1841–1931)

* *
*

"Technology means the systematic application of scientific or other organized knowledge to practical tasks."

"We are becoming the servants in thought, as in action, of the machine we have created to serve us."

John Kenneth Galbraith (1908–2006)

* *
*

" Both science and art have to do with ordered complexity."

Lancelot Law White (1896–1972)

* *
*

"The planet and mankind are in grave danger of irreversible catastrophe... wars of mass destruction, overpopulation, pollution, and the depletion of resources."

Richard A. Falk

* *
*

"The true men of action in our time, those who transform the world, are not the politicians and statesmen, but the scientists. Unfortunately, poetry cannot celebrate them, because their deeds are concerned with things, not persons and are, therefore, speechless."

Wystan Hugh Auden (1907–1973)

* *
*

"The medium is the message. This is merely to say that the personal and social consequences of any medium ... result from the new scale that is introduced into our affairs by each extension of ourselves, or by any new technology."

Marshall McLuhan (1911–1981)

* *
*

"The danger always exists that our technology will serve as a buffer between us and nature, a block between us and the deeper dimensions of our own experience."

Rollo May (1909–1994)

* *
*

"The ideal engineer is a composite ... He is not scientist, he is not a mathematician, he is not a sociologist or a writer; but he may use the knowledge and techniques of any or all of these disciplines in solving engineering problems."

N.W. Dougherty, 1955

* *
*

"A first-rate theory predicts; a second-rate theory forbids; and a third-rate theory explains after the event."

A.I. Kitaigorodskii (1914–1985)

* *
*

"Modern technology
Owes ecology
An apology."

Alan M. Eddison

* *
*

"Almost all aspects of life are engineered at the molecular level, and without understanding molecules we can only have a very sketchy understanding of life itself."

Fracis Crick (1916–2004)

* *
*

"What was once thought can never be unthought."

Friedrich Dürrenmatt (1921–1990)

* *
*

"Any sufficiently advanced technology is indistinguishable from magic."

Arthur C. Clarke (1917–2008)

* *
*

"Science is wonderfully equipped to answer the question "how?" but it gets terribly confused when you ask the question "why?"."

Erwin Chargaff (1905–2002)

* *
*

"It is scientific only to say what is more likely and what is less likely: the existence of flying saucers was not impossible, just very unlikely."

Richard Phillips Feynman (1918–1988)

* *
*

"That which today calls itself science gives us more and more information, an indigestible glut of information, and less and less understanding."

Edward Abbey (1927–1989)

* *
*

"Science is far more exciting than science fiction, far more intricate, far more subtle, and science has the additional virtue of being true."

Carl Sagan (1934–1996)

* *
*

"The probability of success is difficult to estimate, but if we never search, the chance of success is zero."

Giuseppe Cocconi (1914–) and Philip Morrison (1915–2005)

* *
*

"Electronic machines can solve problems which the man who made them cannot solve; but no government subsidized commission of engineers and scientists could create a worm."

Joseph Krutch (1893–1970)

* *
*

"I've given up trying to be rigorous. All I'm concerned about is being right."

Stephen J. Hawking (1942–)

* *
*

"Science is really the search for simplicity. William of Occam, a fourteenth-century philosopher made the dictum... "Entities should not be multiplied beyond necessity." This principle of parsimony ... means that no more forces or causes should be postulated than are necessary to account for the phenomenon observed."

Claude A. Villee (1917–2003)

* *
*

"The development of science is unlike the smooth and monotonically progressive path which one would hope for. Gains are so easily followed by loss; and what one generation builds, another tears down. And yet all is not lost. Man still by nature desires to know; and if he will but appreciate that the preservation *of knowledge is just as important as its* acquisition, *the spiral course of science may yet be directed always upward."*

Carl B. Boyer (1906–1976)

* *
*

"If nature attempts to conceal her tiny secrets, science bares them publicly with magnifications of five million fold. Neither the porcupine nor the mosquito can keep its love life to itself any longer. Scientists peep into their private familiarities and delight in detailed descriptions in lectures, papers and books. If God's molecular gifts are too bulky for human utility, science chops them into little pieces of useful chemicals. If the natural bits are too small, science joins them together into larger units. If the Thanksgiving turkey is too large, a small one is bred. If seeds are not wanted in fruits, seedless varieties are developed. Not satisfied with man's mundane three-dimensional world, science conjures up four- and six-dimensional phantasms. The curiosity of science and her bent on innovation seem uncontrollable. She pries into every heavenly nook and earth cranny. She respects neither the ancient sanctity of tombs nor the caressing intimacies of boudoirs. Science speeds on unabashed!"

R.G.H. Siu

* *
*

"Innovation occurs for many reasons, including greed, ambition, conviction, happenstance, acts of nature, mistakes, and desperation. But one force above all seems to facilitate the process. The easier it is to communicate, the faster changes happens. Every time there is an improvement in the technology with ideas and people come together, major changes ensues:

- *The Greek alphabet have birth to philosophy, logic and the democratic process.*
- *The printing press generated the entire Scientific Revolution.*
- *The telegraph brought modern business methods into existence and held empires together.*

Today, supercomputers and fiber-optic networks, with their ability to make unimaginable amounts of data instantly accessible to millions of people, are accelerating the process of change by many orders of magnitude. The tidal wave of the Information Age will soon break upon us."

James Burke (1936–)

* *
*

"No good *model ever accounted for* all *the facts since some data was bound to b misleading if not plain wrong."*

James Watson (1928–)

* *
*

"It would be a poor thing to be an atom in a universe without physicists, and physicists are made of atoms. A physicist is an atom's way of knowing about atoms."

George Wald (1906–1997)

* *
*

"Western society has accepted as unquestionable a technological imperative that is quite as arbitrary as the most primitive taboo: not merely the duty to foster invention and constantly to create technological novelties, but equally the duty to surrender to these novelties unconditionally, just because they are offered, without respect to their human consequences."

Lewis Mumford (1895–1990)

* *
*

"Science is not one success after another; it is one success in a desert of failure."

"You must swim ahead; science goes where you imagine it."

Yehuda Folkman (1933–2008)

* *
*

"Despite the dazzling success of modern technology and the unprecedented power of modern military systems, they suffer from a common and catastrophic fault. While providing us with a bountiful supply of food, with great industrial plants, with high-speed transportation, and with military weapons of unprecedented power, they threaten our very survival."

Barry Commoner (1917–)

* *
*

"Contrary to popular belief, what is important in science is as much its spirit as its product: it is as much the open-mindedness, the primacy of criticism, the submission to the unforeseen, however upsetting, as the result, however new that may be. Ages ago, scientists gave up the idea of an ultimate and intangible truth, the exact image of a "reality" waiting around the corner to be unveiled. They now know they must be satisfied with the incomplete and the temporary. This approach goes against the natural inclination of the human mind, which calls for unity and coherence in its representation of the world under the most varied aspects. In fact, this conflict between the universal and the local, the eternal and the temporary, reappears at regular intervals in certain controversies — for example, in the debate between the advocates of creation and those of evolution, where arguments already used more than a hundred years ago between Huxley and Wilberforce, Agassiz and Gray, are being used again.

Scientists have come under increased attack in recent years. They are accused of being heartless and conscienceless, of not caring about their fellow humans, even of being dangerous people who do not hesitate to discover new means of destruction and coercion and to use them. That is giving them too much credit. In any population sample there is a constant proportion of stupid people and of crooks, be it among scientists or insurance agents, writers or peasants, priests or politicians. And in spite of Dr. Frankenstein and Dr. Strangelove, catastrophes in history have been caused more often by priests and politicians than by scientists.

For people do not kill each other only for material benefit but also for reasons of dogma. Nothing is more dangerous than the certainty that one is right. Nothing is potentially so destructive as the obsession with a truth one considers absolute. All crimes in history have been the result of fanaticism of one type or another. All massacres have been carried out in the name of virtue, of true religion, of legitimate nationalism, of proper policy, of right ideology: in short, in the name of the fight against somebody else's truth, of the fight against Satan. The coldness and objectivity so often held against scientists are perhaps more suitable than fervor and subjectivity when it comes to dealing with some human matters. For scientific ideas do not generate passion. It is rather passion that exploits science to support its cause. Science does not lead to racism and hatred. It is rather hatred that calls upon science to justify its racism. One can hold against scientists the ardor with which they sometimes champion their ideas. But no genocide has yet been committed for the triumph of a scientific theory. At the end of the twentieth century, it should be clear to each of us that no single system will ever explain the world in all its aspects and detail. The scientific approach has helped to destroy the idea of an intangible and eternal truth. This is not the least of its titles to fame."

Francois Jacob (1920–)

* * *

"Science deals with ideas and is a curiosity-driven, abstract, cultural activity. Technology deals with tools and other things that people use. Faraday's laws of electricity and magnetism are science. Marconi invented wireless technology. Clausius contributed to the science of thermodynamics. Watt invented the steam engine. It was science that clarified the nature of nuclear binding energy; but it was technology that, in an astonishingly short time, converted it to a weapon of unimaginable power. Whereas the first clear records of scientific concerns date back to about 600 BCE, the history of technology is much older. There is evidence that toolmaking goes back as far as one million years. Invention that produced technology did not require scientific reasoning until relatively modern times. The progress in technology was an important component of natural selection as our human ancestors learned to cope with recurrent ice ages and other conditions hardly conductive to creative contemplation. By the time of the apparently abrupt appearance of scientific thought, a considerably sophisticated technology was at hand: fire; metal working; agriculture; weights and measures; elements of arithmetic, algebra and geometry; an astronomical data base; navigation; land surveying; medicine and surgery; calendars. It is a matter for detailed historical scholarship to determine how much of the technology that was available in the 7th *century BCE came about by curiosity and those drives that we now associate with pure research. If modern experience is any guide, a substantial component did come from the dreamers; it is hard to believe that stellar navigation, for example, was discovered by lost seamen desperately trying to find their way home on a dark night in the Aegean.*

Twentieth-century technology is essentially all derived from the results of science. There is an intimate interweaving and mutual enhancement of these two disciplines that, in the past century, accounts for the ever-escalating pace of both: science begets technology, science uses technology to create more science. More science begets more technology.

This litany deserves illustration:

1. *Science begets technology; for example, the quantum theory of solids leads to the transistor.*
2. *Science uses technology to create more science. The transistor provides fast, low-cost digital computers and electronic circuitry for data acquisition, controls and analysis. Electronic controls and computer guidance vastly improve the performance of electron accelerators and not only do research in particle physics but also produce synchrotron radiation.*
3. *More science begets more technology. The synchrotron X rays are used for lithographic etching of integrated circuits for more powerful application of transistors to the making of super-computers."*

Leon M. Lederman (1922–)

* *
*

"One has to look out for engineers – they begin with sewing machines and end up with the atomic bomb."

Marcel Pagnol (1895–1974)

* *
*

"What is the origin of the urge, the fascination that drives physicists, mathematicians, and presumably other scientists as well? Psychoanalysis suggests that it is sexual curiosity... This explanation is somewhat irritating and therefore probably basically correct."

David Ruelle (1935–)

* *
*

"Science cannot solve the ultimate mystery of nature. And that is because, in the last analysis, we ourselves are part of nature and therefore part of the mystery that we are trying to solve."

(Anon)

* *
*

"Impossible only means that you haven't found the solution yet."

(Anon)

* *
*

"Scientists do the work of God, engineers do the work of man."

(Anon)

* *
*

"The interactions between science and religion have often had an aggressive character. There has been, most of the time, a real warfare. But, as a matter of fact, it is not a warfare between science and religion *— there can be no warfare between them — but between* science and theology*. It is true that the man in the street does not easily differentiate between religious feelings and faith, on one side, and dogmas, rites and religious formalism, on the other. It is true also that the theologians, while affecting that religion itself was aimed at when they alone were criticized, have not ceased from aggravating these misunderstandings."*

(Anon)

* *
*

"In the 18th century, science expounded the dynamics of ordered simplicity*.*
In the 19th century, statistical mechanics expounded the science of disordered complexity*.*
In the 20th century, foundations were laid for the science of ordered complexity*."*

(Anon)

* *
*

"Until the Scientific Revolution of the seventeenth century, meaning flowed from ourselves into the world; afterward, meaning flowed from the world to us."

(Anon)

* *
*

"Technology is the science of arranging life so that one need not experience it."

(Anon)

* *
*

"Through the telephone you loose privacy *and the charm of distance."*

(Anon)

* *
*

"Here men from the planet Earth first set foot on the moon, July 1969 A.D. We came in peace for all mankind."

Anonymous (American). Plaque marking the spot on the moon where the historic event took place

* *
*

"In Scientific practice, the onus of the proof is always on the advocate of the more complicated hypothesis."

(Anon)

* *
*

"Experimentalists observe things that cannot be explained and theoreticians explain things that cannot be observed."

(Anon)

* *
*

"Television brought the brutality of war into the comfort of the living room."

(Anon)

* *
*

"A theory can be proved by experiment; but no path leads from experiment to the birth of a theory."

(Anon)

* *
*

"What is now proved was once only imagined."

(Anon)

Ideas, Discoveries and Inventions

* *
*

"I haven't failed; I've found 10,000 ways that don't work."

Benjamin Franklin (1706–1790)

* *
*

"There is one thing stronger than all the armies in the world; and that is an idea whose time has come."

Victor Hugo (1802–1885)

* *
*

"The origin of an original work is always the pursuit of a fact which does not fit into accepted ideas."

Claude Bernard (1813–1878)

* *
*

"After previous investigation of the problem in all directions, happy ideas come unexpectedly without effort, like an inspiration."

Hermann von Helmholtz (1821–1894)

* *
*

"Ideas have consequences."

Dostoevsky (1821–1881)

* *
*

"In the field of observation, chance favors the prepare mind."

Louis Pasteur (1822–1895)

* *
*

"Choose one definite objective and drive ahead toward it. You may never reach your goal, but you will find something on the way."

Felix Klein's advice to a perplexed student (1849–1925)

* *
*

"Say what you know, do what you must, come what may."

Sonja Kovalevsky (1850–1891)

* *
*

"An idea that is not dangerous is unworthy of being called an idea at all."

Oscar Wilde (1854–1900)

* *
*

"It is through science that we prove, but through intuition that we discover."

Henri Poincare (1854–1912)

* *
*

"In the realm of thought, momentous discoveries and solution of problems are possible only to an individual, working in solitude."

Sigmund Freud (1856–1939)

* *
*

"The earth is the cradle of the mind, but one cannot live forever in a cradle. To set foot on the soil of the asteroids, to lift by hand a rock from the moon, to observe Mars from a distance of several tents of kilometers, to land on its satellite or even on its surface, what can be more fantastic?
From the moment of using rocket devices, a new great era will begin in astronomy: the epoch of the more intensive study of the firmament."

Konstantin Eduardovitch Ziolkowski (1857–1935)

* *
*

"It requires a very unusual mind to undertake the analysis of the obvious."

Alfred North Whitehead (1861–1947)

* *
*

"Young man, you can be grateful that my invention is not for sale, for it would undoubtedly ruin you. It can be exploited for a certain time as a scientific curiosity, but apart from that it has no commercial value whatsoever."

Auguste Lumière (1862–1954). French chemist. On the motion-picture camera he invented in 1895

* *
*

"It is obvious that invention or discovery, be it in mathematics or anywhere else, takes place by combining ideas."

Jacques Hadamard (1865–1963)

* *
*

"Learning the secret of flight from a bird was a good deal like learning the secret of magic from a magician. Once you know the trick and know what to look for, you see things that you did not notice."

Orville Wright (1871–1948)

* *
*

"It isn't that they can't see the solution. It is that they can't see the problem."

G.K. Chesterton (1874–1936)

* *
*

"Reason is man's faculty for grasping the world by thought. Intelligence is man's ability to manipulate the world with the help of thought."

Erich Fromm (1900–1980)

* *
*

"Don't worry about people stealing your ideas. If your ideas are any good, you'll have to ram them down people's throats."

Howard Aiken (1900–1973)

* *
*

"The machine, the genie that man has thoughtlessly let out of its bottle and cannot put back again."

George Orwell (1903–1950)

* *
*

"What counts ... in science is not so much the first as the last."

Erwin Chargaff (1905–2002)

* *
*

"Scientific discoveries and ideas are produced by the intuition, creativeness and genius of a man. Dollars of themselves don't produce this any more than they could be expected to produce another Mona Lisa."

Harry Hess (1906–1969)

* *
*

"The technological breakthroughs leading to great inventions usually come from totally unrelated areas. For instance, if a queen of ancient Crete had launched a Minoan Manhattan Project to achieve mass literacy through improved printing, she would never have thought to emphasize research into cheese, wine and olive presses — but those presses furnished prototypes for Gutenberg's most original contribution to printing technology. Similarly, American military planners trying to build powerful bombs in the 1930's would have laughed at suggestions that they finance research into anything so arcane as transuranic elements.

We picture inventors as heroes with the genius to recognize and solve a society's problems. In reality, the greatest inventors have been tinkerers who loved tinkering for its own sake and who then had to figure out what, if anything, their devices might be good for. The prime example is Thomas Edison, whose phonograph is widely considered to be his most brilliant invention. When he built his first one, in 1877, it was not in response to a national clamor for hearing Beethoven at home. Having built it, he wasn't sure what to do with it, so he drew up a list of 10 uses, like recording the last words of dying people, announcing the time and teaching spelling. When entrepreneurs used his invention to play music, Edison thought it was a debasement of his idea.

Our widespread misunderstanding of inventors as setting out to solve society's problems causes us to say that necessity is the mother of invention. Actually, invention is the mother of necessity, by creating needs that we never felt before. (Be honest: did you really feel a need for your Walkman CD player long before it existed?) Far from welcoming solutions to our supposed needs, society's entrenched interests commonly resist inventions. In Gutenberg's time, no one was pleading for a new way to churn out book copies: there were hordes of copyists whose desire not to be put out of business led to local bans on printing.

The first internal-combustion engine was built in 1867, but no motor vehicles came along for decades, because the public was content with horses and railroads. Transistors were invented in the United States, but the country's electronics industry ignored them to protect its investment in vacuum-tube products; it was left to Sony in bombed-out postwar Japan to adapt transistors to consumer-electronics products. Manufacturers of typing keyboards continue to prefer our inefficient qwerty *layout to a rationally designed one.*

All these misunderstandings about invention pervade our science and technology policies. Every year, officials decry some areas of basic research as a waste of tax dollars and urge that we instead concentrate on "solving problems": that is, applied research. Of course, much applied research is necessary to translate basic discoveries into workable products — a prime example being the Manhattan Project, which spent three years and $2 billion to turn Otto Hahn and Fritz Strassman's discovery of nuclear fission into an atomic bomb. All too often, however, the world fails to realize that neither the solutions to most difficult problems of technology nor the potential uses of most basic research discoveries have been predictable in advance. Instead, penicillin, X-rays and many other modern wonders were discovered accidentally — by tinkerers.

So forget those stories about genius inventors who perceived a need of society, solved it single-handedly and transformed the world. There has never been such a genius; there have only been processions of replaceable creative minds who made serendipitous or incremental contributions. If Gutenberg himself hadn't devised the better alloys and inks used in early printing, some other tinkerer with metals and oils would have done so. For the best invention of the millennium, do give Gutenberg some of the credit — but not too much."

Jared Diamond (1937–)

* * *

"A theory has only the alternative of being right or wrong. A model has a third possibility: it may be right, but irrelevant."

Manfred Eigen (1927–)

* * *

"The greatest discovery of the 19th century was that the equations of nature were linear, and the great discovery of the 20th century is that they are not."

T.W. Körner

* *
*

"After developing retractable landing gear, flaps, navigation devices, autopilots, and active controls to provide stability for unstable vehicles, people realized that birds had been doing these things for 100 million years."

Paul MacCready (1925–2007)

* *
*

"To invent the automobile is easy; a really good science fiction would predict the traffic jam."

Frederick Pohl (1889–1991)

* *
*

"The four 'laws' of discovery:

- *Discoveries are rarely attributed to the correct person.*
- *Nothing is ever discovered for the first time.*
- *Everything of importance has been said before by someone who did not discover it.*
- *All great discoveries are made by mistake."*

Arthur Bloch (1948–)

* *
*

"Nothing was ever achieved by a reasonable man."

Anon

* *
*

"If we can really understand the problem, the answer will come out of it, because the answer is not separate from the problem."

Anon

9. *Mind, Brain and the Computer*

* *
*

"Generally, and device that can perform numerical calculations – even an adding machine, an abacus, or a slide rule – may be called a computer. Currently, however, the term usually refers to an electronic device that can use a list of instructions, called a program, to perform calculations or to store, manipulate, and retrieve information."

L.R. Shannon

* *
*

"One ought to know that on the one hand pleasure, joy, laughter, and games, and on the other, grief, sorrow, discontent, and dissatisfaction arise only from the brain. It is especially by it that we think, comprehend, see, and hear, that we distinguish the ugly from the beautiful, the bad from the good, the agreeable from the disagreeable..."

Hippocrates of Cos (460–377 BCE)

* *
*

"The Analytical Engine has no pretensions whatsoever to originate anything. It can do whatever we know how to order it to perform."

Augusta Ada Byron (1788–1824)

* *
*

"As machine get to be more and more like men, men will come to be more and more like machines."

Joseph Wood Krutch (1893–1970)

* *
*

"The complexity of the nerve-structures for vision is even in the insect something incredibly stupendous... The intricacy of the connections defies description. Before it the mind halts, abased."

"As long as our brain is a mystery, the universe – the reflection of the structure of the brain – will also be a mystery."

Santiago Ramón y Cajal (1852–1934)

* *
*

"The brain is wider than the sky
For, put them side by side,
The one the other will contain
With ease, and you beside.

The brain is deeper than the sea,
For, hold them, blue to blue,
The one the other will absorb,
As sponges, buckets do.

The brain is just the weight of God,
For, heft them pound for pound,
And they will differ, if they do,
As syllable from sound."

Emily Dickinson (1830–1886)

* *
*

"Computers are useless. They can only give you answers."

Pablo Picasso (1881–1973)

* *
*

"How could a mechanism composed of some 10^{10} unreliable components function reliably while computers with 10^4 components regularly fail."

John von Neumann (1903–1957)

* *
*

"Just as the electron does not have a precise position and motion, I believe that consciousness has no location."

George Wald (1906–1997)

* *
*

"Computers are circumscribed by their binary, yes or no, mode of operation, by their 'two-bit wit'. Subtleties such as yes and no, are beyond them."

Marshall McLuhan (1911–1980)

* *
*

"The Christian notion of the possibility of redemption is incomprehensible to the computer."

Vance Packard (1914–1996)

* *
*

"The real danger is not that computers will begin to think like men, but that men will begin to think like computers."

Sydney J. Harris (1917–1986)

* *
*

"Computers make it easier to do a lot of things, but most of the things they make easier to do don't need to be done."

Andy Rooney (1919–)

* *
*

"Part of the inhumanity of the computer is that, once it is competently programmed and working smoothly, it is completely honest."

Isaac Asimov (1920–1992)

* *
*

"No, I'm not interested in developing a powerful brain. All I'm after is just a mediocre brain, something like the president of the American Telephone and Telegraph Company."

"I propose to consider the question, 'Can Machine Think?'"

Alan Mathison Turing, 1936

* *
*

"You, your joys and sorrows, your memories and your ambitions, your sense of personal identity and free will are in fact no more than the behavior of a vast assembly of nerve cells and their associated molecules."

Francis Crick (1916–2004)

* *
*

"There is nothing that I can see in the physical laws that says the computer elements cannot be made enormously smaller than they are now."

Richard Feynman (1918–1988)

* *
*

"My fundamental premise about the brain is that its workings – what we sometimes call 'mind' – are a consequence of its anatomy and physiology and nothing more."

Carl Sagan (1934–1996)

* *
*

"Can mental states and processes be reduced *to brain (neurobiological) states and processes? Can one be a reductionist?"*

Patricia Churchland (1943–), 1986

* *
*

"I regard a human as nothing but a particular type of a machine, the human brain as nothing but an information processing device, the human soul as nothing but a program being run on a computer called the brain. Further, all possible types of living things, intelligent or not, are of the same nature, and subject to the same laws of physics as constrain all information processing devices. A human being is a quantum mechanical object which can be exactly described by a computer program coding 10^{45} *bits of information."*

Frank J. Tipler (1947–), 1994

* *
*

"The brain is a three-pound mass you can hold in your hand that can conceive of a universe a hundred-billion light years across."

Marian Diamond

* *
*

"One can search the brain with a microscope and not find the mind and can search the stars with a telescope and not find God."

Gustav J. White

* *
*

"I'm struck by the insidious, computer-driven tendency to take things out of the domain of muscular activity and put them into the domain of mental activity. The transfer is not paying off. Sure, muscles are unreliable, but they represent several million years of accumulated finesse."

Brian Eno (1948–), 1999

* *
*

"The question of whether computers can think is just like the question of whether submarines can swim."

Edsger W. Dijkstra (1930–2002)

* *
*

"There is no way that an unassisted human brain, which is nothing more than a dog's breakfast, three and a half pounds of blood-soaked sponge, could have written Beethoven's Ninth Symphony."

Kurt Vonnegut (1922–2007)

* *
*

"The great thing about a computer notebook is that no matter how much you stuff into it, it doesn't get bigger or heavier."

"E-mail is a unique communication vehicle for a lot of reasons. However e-mail is not a substitute for direct interaction."

Bill Gates (1955–)

* *
*

" 'Out of sight, out of mind', when translated into Russian (by computer), then back again into English, became 'invisible maniac' ".

Arthur Calder–Marshall (1908–1992)

* *
*

"There is no reason for any individual to have a computer in their home."

Ken Olsen (1926–), 1977

* *
*

"If you can make a machine that contains the contents of your mind, then that machine is you... Even if it doesn't last forever, you can always dump it onto tape and make backups... Everyone would like to be immortal... I'm afraid, unfortunately, that I am the last generation to die."

Gerald Jay Sussman

* *
*

"Mankind is a catalyzing enzyme for the transition from a carbon-based to a silicon-based intelligence."

Gérard Bricogne

* *
*

"The brain and the satellite glands have now been probed to the point where no possible site remains that can reasonably be supposed to harbor any physical mind."

Edmund O. Wilson (1929–)

* *
*

"If the brain were so simple we could understand it, we would be so simple we couldn't."

Lyall Watson (1939–)

* *
*

"Video won't be able to hold onto market it captures after the first six months. People will get tired of staring at a plywood box every night."

Darryl F. Zanuck (1902–1979), 1940s

* *
*

"The computer is no better than its program."

Elting E. Morison (1910–1995), 1966

* *
*

"Nobody predicted the computer – not even Joule Verne."

"The fact that prior to 1945 physical scientists did not have computers at their disposal compelled them to delve deeper into matters, and then come up with the 'pearl in the midst of the oyster', i.e. to identify underlying operative principles; whereas, had they lived after 1945, they might be tempted to simulate rather than cogitate."

Shahar Ben-Menahem

* *
*

"I think there is a world market for maybe five computers."

Thomas J. Watson (1874–1956)

* *
*

"The intelligence of the computer is syntactic *whereas human intelligence is* semantic *(i.e. assigns meaning to concepts)."*

(Anon)

* *
*

"Where a calculator on the ENIAC is equipped with 18,000 vacuum tubes and weighs 30 tons, computers in the future may have only 1,000 vacuum tubes and perhaps weigh 1.5 tons."

(Anon)

* *
*

"Someday, computers may not only be able to beat human beings at chess, but also at tennis and ice hockey and volleyball. Someday, computers may be able to marry Brooke Shields."

(Anon)

* *
*

"Computers have lots of memory but no imagination."

(Anon)

* *
*

"• *Brain cause minds.*

• *Syntax is not sufficient for semantics.*

• *Computer programs are entirely defined by their formal, or syntactical, structure.*

• *Minds have mental contents; specifically, they have semantic contents.*

Conclusion: *No computer program by itself is sufficient to give a system a mind. Programs are not minds, and they are not by themselves sufficient for having minds."*

"The computer, in contradistinction to the brain, has no self-consciousness, self-awareness – it does not know that it is a computer, it cannot think."

"For a long time many biologists and philosophers thought it was impossible, in principle, to account for the existence of life on purely biological grounds. They thought that in addition to the biological processes some other element must be necessary, some élan vital *must be postulated in order to lend life to what was otherwise dead and inert matter. It is hard today to realize how intense the dispute was between vitalism and mechanism even a generation ago, but today these issues are no longer taken seriously. Why not? I think it is not so much because mechanism won and vitalism lost, but because we have come to understand better the biological character of the processes that are characteristic of living organisms. Once we understand how the features that are characteristic of living beings have a biological explanation, it no longer seems mysterious to us that matter should be alive. I think that exactly similar considerations should apply to our discussions of consciousness. It should seem no more mysterious, in principle, that this hunk of matter, this grey and white oatmeal-textured substance of the brain, should be conscious than it seems mysterious that this other hunk of matter, this collection of nucleo-protein molecules stuck onto a calcium frame, should be alive. The way, in short, to dispel the mystery is to understand the processes. We do not yet fully understand the processes, but we understand their general character,*

we understand that there are certain specific electro-chemical activities going on among neurons or neuron-modules and perhaps other features of the brain and these processes cause consciousness."

"By 'mind' I just mean the sequence of thoughts, feelings and experiences, whether conscious or unconscious, that go to make up our mental life.But the use of the noun 'mind' is dangerously inhabited by the ghosts of old philosophical theories."

"On the traditional account of the brain, the account that takes the neuron as the fundamental unit of brain functioning, the most remarkable thing about brain functioning is simply this. All of the enormous variety of inputs that the brain receives – the photons that strike the retina, the sound waves that stimulate the ear drum, the pressure on the skin that activates nerve endings for pressure, heat, cold, and pain, etc. – all of these inputs are converted into one common medium: variable rates of neuron firing. Furthermore, and equally remarkably, these variable rates of neuron firing in different neuronal circuits and different local conditions in the brain produce all of the variety of our mental life. The smell of a rose, the experience of the blue of the sky, the taste of onions, the thought of a mathematical formula: all of these are produced by variable rates of neuron-firing, in different circuits, relative to different local conditions in the brain."

John Searle, 1984

* * *

"One machine can do the work of fifty ordinary men. No machine can do the work of one extraordinary man."

Albert Hubbard, 1948

10. Science and Scientists — the Lighter Side

* *
*

Justus von Liebig *(1803–1873) was approached one day by his assistant who all excited informed him that he had just discovered a universal solvent. Liebig asked: "And what is a universal solvent?" Assistant: "One that dissolves all substances." Liebig: "Where are you going to keep that solvent, then?"*

* *
*

At a session of the Academy of Sciences of the (former) USSR, the notorious agronomist Lysenko gave a talk on the inheritance of acquired traits. When his report was over, the famous physicist **Landau** *asked: – So, you argue that if we will cut off the ear of a cow, and the ear of its offspring, and so on, sooner or later the earless cows will start to be born? – Yes, that's right. – Then, how can you explain that virgins are still being born?*

* *
*

When asked for his latest results, **Edison** *said: "Results! Why, man, I have gotten a lot of results. I know several thousand things that won't work."*

* *
*

On a paper submitted by a physicist colleague, **Pauli** *said: "This isn't right. This isn't even wrong."*

* *
*

The 67th Mersenne number $2^{67}-1$, claimed by Mersenne to be prime, was proven to be non-prime in 1903 by **F.N. Cole** *(1861–1927). In the October meeting of the AMS, Cole announced a talk "On the Factorization of Large Numbers".*
He walked up to the blackboard without saying a word, calculated by hand the value of 2^{67}, carefully subtracted 1. Then he multiplied two numbers (which were 193707721 and 761838257287). Both results written on the blackboard were equal. Cole silently walked back to his seat, and this is said to be the first and only talk held during an AMS meeting where the audience applauded. There were no questions.
It took Cole about 3 years, each Sunday, to find this factorization.

* *
*

Theorem:
All positive integers are interesting.

Proof:
Assume the contrary. Then there is a lowest non-interesting positive integer. But, hey, that's pretty interesting! A contradiction.
QED

(Anon)

* *
*

Dirac *was working on an equation on the board. Turning around to a silent audience he asked for any questions. A person in audience raised a hand and said "I do not understand such-and-such equation". To which Dirac replies, "That's not a question, it's a statement."*

* *
*

Ernst Eduard Kummer *(1810–1893) was rather poor at arithmetic. Whenever he had occasion to do simple arithmetic in class, he would get his students to help him. Once he had to find 7×9. "Seven times nine," he began, "Seven times nine is er – ah – ah – seven times nine is..." "Sixty-one," a student suggested. Kummer wrote 61 on the board. "Sir," said another student, "it should be sixty-nine." "Come, come, gentlemen, it can't be both," Kummer exclaimed. "It must be one or the other."*

Another version:
Kummer said to himself: "Hmmm, the product cannot be 61, because 61 is prime, it cannot be 65, because 65 is a multiple of 5, 67 is a prime, 69 is too big – Only 63 is left."

In one of his lectures, ***Richard Feynman*** *said:*
"This is the third of four lectures on a rather difficult subject – the theory of quantum electrodynamics – and since there are obviously more people here tonight than there were before, some of you haven't heard the other two lectures and will find this lecture incomprehensible. Those of you who have heard the other two lectures will also find this lecture incomprehensible, but you know that that's all right: as I explained in the first lecture, the way we have to describe Nature is generally incomprehensible to us."

* *
*

What is set theory?

When there are five people in the room and seven are leaving, two have to enter the room, so it is empty.

Paul Erdös

* *
*

"Fast cars, fast women, fast algorithms... what more could a man want?"

Joe Mattis

* *
*

An MIT student cornered the famous **John von Neumann** *in the hallway:*

Student: "Er, excuse me, Professor von Neumann, could you please help me with a calculus problem?"

John: "Okay, sonny, if it's real quick – I'm a busy man."

Student: "I'm having trouble with this integral."

John: "Let's have a look." (insert brief pause here) "Alright, sonny, the answer's two-pi over 5."

Student: "I know that, sir, the answer's in the back – I'm having trouble deriving it, though."

John: "Okay, let me see it again." (another pause) "The answer's two-pi over 5."

Student (frustrated): "Uh, sir, I know the answer, I just don't see how to derive it."

John: "Whaddya want, sonny, I worked the problem in two different ways!"

* *
*

"Two trains 200 miles apart are moving toward each other; each one is going at a speed of 50 miles per hour. A fly starting on the front of one of them flies back and forth between them at a rate of 75 miles per hour. It does this until the trains collide and crush the fly to death. What is the total distance the fly has flown?

The fly actually hits each train an infinite number of times before it gets crushed, and one could solve the problem the hard way with pencil and paper by summing an infinite series of distances. The easy way is as follows: Since the trains are 200 miles apart and each train is going 50 miles an hour, it takes 2 hours for the trains to collide. Therefore the fly was flying for two hours. Since the fly was flying at a rate of 75 miles per hour, the fly must have flown 150 miles. That's all there is to it.

When this problem was posed to ***John von Neumann****, he immediately replied, "150 miles."*

"It is very strange," said the poser, "but nearly everyone tries to sum the infinite series."

"What do you mean, strange?" asked Von Neumann. "That's how I did it!""

* *
*

A student in ***Rutherford****'s lab was very hard-working. Rutherford had noticed it and asked one evening:*
– Do you work in the mornings too?
– Yes, – proudly answered the student sure he would be commended.
– But when do you think? – amazed Rutherford.

* *
*

"The traditional mathematics professor of the popular legend is absent-minded. He usually appears in public with a lost umbrella in each hand. He prefers to face the blackboard and to turn his back to the class. He writes a, he says b, he means c; but it should be d. Some of his sayings are handed down from generation to generation.
"In order to solve this differential equation you look at it till a solution occurs to you."
"This principle is so perfectly general that no particular application of it is possible."
"Geometry is the science of correct reasoning on incorrect figures."
"My method to overcome a difficulty is to go round it."
"What is the difference between method and device? A method is a device which you used twice."

George Polya (1887–1985)

* *
*

***Niels Bohr** once commented on someone's lecture:*
"His theory is crazy... but it's not crazy enough to be true."

* *
*

"I aim at the stars, but sometimes I hit London."

Werner von Braun (1912–1977)

* *
*

David Hilbert *was invited to give a talk on any subject he liked during the early days of air travel. His subject:*

The Proof of Fermat's Last Theorem

Needless to say, his talk was eagerly anticipated. The day arrived, the talk was given, and it was brilliant – but it had nothing at all to do with Fermat's Last Theorem.
After the talk, someone asked Hilbert why he had picked a title that had nothing to do with the talk. His answer: "Oh, that title was just in case the plane crashed."

* *
*

Hilbert *has accepted an invitation to deliver a keynote address to a large engineering convention. The organizers subsequently learned that Hilbert was known for rather acerbic attitude towards engineering. Greatly concerned they decided to go back and talk to him.*

After beating around the bush for a while they managed to convey to him that they are worried that he may offend some people, and if he could sort of hold back during his speech.

When Hilbert realized what they were asking he grinned broadly and said, "You don't have to worry about that at all. How could I possibly offend anyone, for mathematics and engineering have absolutely nothing in common".

* *
*

__J.E. Littlewood__ (1885–1977) read in the proof sheets of Hardy on Ramanujan: "As someone said, each of the positive integers was one of his personal friends." His reaction was, "I wonder who said that; I wish I had." In the next proof-sheets he read (what now stands), "It was Littlewood who said..."

* *
*

Hungarian mathematician __Frigyes Riesz__ needed two assistants for his lectures: one was reading aloud his (Riesz's) book, the second one was writing everything on the board, while Riesz was standing next to the board nodding.

* *
*

When __Edmund Landau__ was asked for a testimony to the effect that Emmy Noether was a great woman mathematician, he said:
"I can testify that she is a great mathematician, but that she is a woman, I cannot swear."

* *
*

In the period that __Einstein__ was active as a professor, one of his students came to him and said: "The questions of this year's exam are the same as last years!" "True," Einstein said, "but this year all answers are different."

* *
*

"Do you realize the tremendous strategic importance of machines of this sort? They could be an effective straightjacket for that noisy shopkeeper Harry Truman. We must go ahead with it, comrades. The problem of the creation of transatlantic rockets is of extreme importance to us."

Premier I.V. Stalin, at a meeting of the Politburo (1947)

* *
*

"Give me a one-handed economist! All my economists say, on the one hand... on the other."

Harry S. Truman (1884–1972)

* *
*

"During the Paris peace negotiations between the Allies and some Eastern European countries, a Hungarian woman journalist came to me and wanted an interview. She asked me what I thought was the greatest progress in aviation in the last decade. I told her: 'Propulsion by reaction'. She said to me: 'Professor, could you express this in some other way? I cannot write in a progressive paper that progress is accomplished by reaction!'"

"Young students with athletic ability may have brilliant minds and still believe that to jump two inches farther than anybody else is an important contribution to human progress."

Theodore von Kármán (1881–1963)

* *
*

Theodore von Kármán *was a showman in the higher sense of the natural raconteur and purveyor of honest entertainment. his story-telling skills were enhanced by a rich and mysterious Hungarian accent that was so thick it provoked the late Dr.Hugh Dryden, Deputy Administrator of NASA, to comment jokingly that it must have been fabricated for "commercial reasons." Von Kármán's charming accent, coupled with hand gestures that carried extraordinary editorial comment, were enough to captivate any audience. He knew how to make a lasting effect. Like Bernard Baruch, von Kármán also carried an ornate hearing aid which he surreptitiously turned off whenever the conversation grew boring. He once confided to a friend that he thought his deafness since youth had been one of his most important assets because it enabled him to concentrate.*

He was also a master of the "squelch", which he used at times with withering effect. On one early occasion at the University of Aachen, the owner of a small tool factory came to him expressing concern about a problem of vibration. It seemed that one of the machines was in danger of shaking itself to destruction. Nobody could find the cause. Would the eminent Herr Professor take a look?

Von Kármán agreed and found the trouble in a few minutes. It was a small dislocation and he suggested turning a gear ninety degrees. When this was done, the vibration miraculously disappeared. The factory owner was overjoyed.

However, a few days later, he again sought out the professor. This time it was von Kármán's bill. "How can I pay so much money," the owner cried, "for making just a ninety-degree turn of the gear?"

"Very well," said von Kármán, "turn the gear ninety degrees back and I will tear up the bill."

* *
*

There is a story about the student who wanted to know, "Can one prove the **Mandelstam** *representation from field theory?" He went to* **Weisskopf** *who responded, "Field theory, what is field theory?" Then he sought out Wigner who said, "Mandelstam, who is Mandelstam?" Finally, our persistent student found his way to Chew, repeated the question, and heard, "Proof, what is proof?"*

Scientists Reflect on the World and Themselves

* *
*

"If I have been able to see further, it was because I stood on the shoulders of giants."

"I do not know what I may appear to the world; but to myself I seem to have been only like a boy playing on the seashore, and diverting myself in now and then finding a smoother pebble, or a prettier shell than ordinary; whilst the great ocean of truth lay all undiscovered before me.."

Isaac Newton (1643–1727)

* *
*

"When the storm rages and the state is threatened by shipwreck, we can do nothing more noble then to lower the anchor of our peaceful studied into the ground of eternity."

Johannes Kepler (1571–1630)

* *
*

"I do not feel obliged to believe that the same God who has endowed us with sense, reason, and intellect has intended us to forego their use."

Galileo Galilei (1564–1642)

* *
*

"I am truly a lone traveler and have never belonged to my country, my home, my friends, or even my immediate family, with my whole heart; in the face of all these ties, I have never lost a sense of distance and a need for solitude."

"Great spirits have always encountered violent opposition from mediocre minds."

"The trite objects of human efforts – possessions, outward success, luxury — have always seemed to me contemptible."

A. Einstein (1879–1955)

"The works of Archimedes are without exception, monuments of mathematical exposition; the gradual revelation of the plan of attack, the masterly ordering of the propositions, the stern elimination of everything not immediately relevant to the purpose, the finish of the whole, are so impressive in their perfection as to create a feeling akin to awe in the mind of the reader."

Thomas Heath (1861–1940), 1921

* *
*

"One cannot read Archimedes' complicated accounts of his quadratures and cubatures without saying to oneself, "How on earth did he imagine those expedients and reach those conclusions?"

George Sarton (1884–1956), 1952

"The greatest change in the axiomatic basis of physics — in other words, of our conception of the structure of reality — since Newton laid the foundation of theoretical physics, was brought about by Faraday's and Maxwell's work on electromagnetic phenomena.

Before Maxwell people conceived of physical reality as material points, whose changes consist exclusively of motions, which are subject to total differential equations. After Maxwell they conceived physical reality as represented by continuous fields, not mechanically explicable, which are subject to partial differential equations. This change in the conception of reality is the most profound and fruitful one that has come to physics since Newton."

Albert Einstein[60] (1879–1955), 1931

* *
*

"From a long view of the history of mankind there can be little doubt that the most significant event of the 19th century will be judged as Maxwell's discovery of the laws of electrodynamics. Even the American Civil War, will pale into provincial insignificance before this more powerful event of the 1860's."

Richard Feynman (1918–1998), 1964

* *
*

In a letter written near the end of his life (dated February 24, 1918) he recalled how he first learned of Maxwell's work one day while in a library:
"I remember my first look at the great treatise of Maxwell's when I was a young man. Up to that time there was not a single comprehensive theory, just a few scraps; I was struggling to understand electricity in the midst of a great obscurity. When I saw on the table in the library the work that had just been published (1873), I browsed through it and I was astonished! I read the preface chapter, and several bits here and there; I saw that it was great, greater and greatest, with prodigious possibilities in its power. I was determined to master the book and set to work. I was very ignorant. I had no knowledge of mathematical analysis (having learned only school algebra and trigonometry which I had largely forgotten) and thus my work was laid out for me. It took me several years before I could understand as much as I possibly could. Then I set Maxwell aside and followed my own course. And I progressed much more quickly."

Oliver Heaviside (1850–1925)

[60] Maxwell died in 1879, the year that Einstein was born.

* *
*

Von Kármán's *fabulously rich and varied life would be hard to duplicate in today's age of superspecialization. He was not modest or humble about his accomplishments. He followed Goethe's advice: "Nur die Lumpen sind Bescheiden" ("Only the loafers are modest").*

In science he saw himself among the immortals. Once he was asked to rate himself among the great scientists of the century. "If you define a great scientist as a man with great ideas," he replied, "then you will have to rate Einstein first. He had four great ideas. In the history of science perhaps only Sir Isaac Newton is ahead of Einstein, because he had five or six ideas. All the other major scientists of our age associated with just one, or at the most two, great ideas. In my case I have had three great ideas. Maybe more. Yes, perhaps three and a half great ideas."

Historians of science may assess von Kármán on a different scale. Obviously even "great ideas" are not "great" on the same level. Some scientists of our century, such as Bohr, Planck, Fermi, Dirac, and Schrödinger, have opened up vistas in physics whose impact on scientific thought may be far longer lasting, far more responsible for changing the future of civilization, than von Kármán's contribution to the conquest of air and space. But this impact is in part related to the social use being made of scientific discovery, such as the control of nuclear energy. In any evaluation of scientific genius as it deals in an inspired way with the apparently insoluble problems of the universe, there is no doubt that von Kármán rates a place among the first ten scientific minds of this century.

* *
*

"One never noticed what has been done; one can only see what remains to be done..."

Marie Curie (1867–1934), 1894

They Died Unconvinced

* *
*

"Inventions have long since reached their limit, and I see no hope for further development."

Julius Sextus Frontinus (40–103 CE)

* *
*

"The new astrologer wants to prove that the earth moves and revolves, instead of the sky, the sun and the moon. Just as if somebody moving in a carriage might hold that he was sitting still, at rest, while the earth and trees walked and moved. Ridiculous!"

Martin Luther (on Copernicus) (1483–1546)

* *
*

"To tell if a given number of 15 to 20 digits is prime or not, all time would not suffice for the test."

Marin Mersenne (1588–1648)

* *
*

"Alchemy is the only Art which might be able to complete and bring to light not only medicine but also a universal Philosophy."

Isaac Barrow (1630–1677)

* *
*

"There is no likelihood man can ever tap the power of the atom."

Robert Millikan (1868–1953), 1923

* *
*

"That one body may act on another through a vacuum, without the mediation of anything else, by and through – which their action and force may be conveyed from one to another, is to be so great an absurdity, that I believe no man who has in philosophical matters a competent faculty of thinking can ever fall into it."

Isaac Newton (1643–1727)

* *
*

"Atoms are the figment of a weak imagination."

Leibniz (1646–1716)

* *
*

"What, sir, you would make a ship sail against the wind and currents by lighting a bonfire under her decks. I pray you excuse me. I have no time to listen to such nonsense."

Napoleon, to Robert Fulton, who unsuccessfully tried
to interest the emperor in his idea to build a steamship, 1800

* *
*

"Paltry and unsubstantial papers... destitute of every species of merit... His theory have no other effect than to check the progress of science and renew all those wild phantoms of the imagination which Newton put to flight from her temple."

The London Royal Society, regarding Thomas Young's
epoch-making papers describing the wave nature of light; 1802

* *
*

"I could more easily believe that two Yankee professors would lie than that stones would fall from heaven."

Thomas Jefferson, on a report on a meteorite shower
which fell in Weston, Connecticut in 1807

* *
*

"The laws of mechanics are identical with those of Nature."

René Descartes (1596–1650)

* *
*

"It is quite impossible that the noble organs of human speech could be replaced by ignoble, senseless metal."

Jean Bouillaud, member of the French Academy, before
viewing a demonstration of Edison's phonograph; 1877.
After the demonstration, Bouillaud called Edison's
invention a fake and attributed the demonstration he had
seen to "ventriloquism"

* *
*

"Such startling announcements as these should be deprecated as being unworthy of science and mischievous to its true progress."

William Siemens, England's most distinguished electrical engineer on Edison's electric light bulb; 1879. Siemens had been unsuccessfully working on electric light bulbs for a decade.

* *
*

"His claims are so manifestly absurd as to indicate a positive want of knowledge of the electric circuit and the principles governing the construction and operation of electrical machines."

Edwin Weston, a respected specialist in arc lighting on Edison's electrical light bulb; 1879.

* *
*

"Edison's ideas are good enough for our transatlantic friends... but unworthy of the attention of practical or scientific men."

Committee set up by the British Parliament to look into Edison's work on the incandescent lamp, c. 1878.

* *
*

"If the whole of the English language could be condensed into one word, it would not suffice to express the utter contempt those invite who are so deluded as to be disciples of such an imposture as Darwinism."

Francis Orpen Morris, British ornithologist (1810–1893)

* *
*

"Heavier-than-air flying machines are impossible."

1895

"X-rays are an elaborate hoax."

1896

"We find something at every turn to show the utter futility of Darwin's philosophy."

William Thomson (Kelvin) (1824–1907)

* *
*

"I am tired of all this thing called science... We have spent millions on that sort of thing for the last few years, and it is time it should be stopped."

Senator Simon Cameron, demanding that the funding of the Smithsonian Institution be cut off, 1861

* *
*

"...in a few years, all great physical constants will have been approximately estimated, and the only occupation which will be left to men of science will be to carry these measurements to another place of decimals."

James Clerk Maxwell (1831–1879)

* *
*

"How can he [Thomas Edison] call it a wonderful success when everyone acquainted with the subject will recognize it as a conspicuous failure, trumpeted as a wonderful success. A fraud upon the public."

Henry Morton, Professor of Physics and President of the Stevens Institute of Technology, on Edison's Incandescent light bulb, in *The New York Herald*, December 18, 1879

* *
*

"The abdomen, the chest, and the brain will be forever shut from the intrusions of the wise and humane surgeon."

John Erichsen (1818–1896)

* *
*

"All true scientific progress ceased around 1900."

Joseph Larmor (1857–1942)
(steadfastly opposed the new relativity and quantum theories)

* *
*

"Attempting to fly a heavier-than-air aircraft is simply absurd."

Rear-Admiral George Melville, chief engineer, US Navy, 1902

* *
*

"By 1940 the relativity theory will be considered a joke."

George Francis Gillette, 1929

* *
*

"The energy produced by the atom is a very poor kind of thing. Anyone who expects a source of power from the transformation of these atoms is talking moonshine."

Ernest Rutherford, *The New York Herald Tribune*,
September 12, 1933

* *
*

"Professor Goddard does not know the relation between action and reaction and the need to have something better than a vacuum against which to react. He seems to lack the knowledge ladled out daily in the high schools."

editorial in The New York Times, 1921, dismissing Robert Goddard,
who proposed that someday rockets could reach the moon.

* *
*

"There is no likelihood man can ever tap the power of the atom... Nature has introduced a few foolproof devices into the great majority of elements that constitute the bulk of the world, and they have no energy to give up in the process of disintegration."

Robert Andrews Millikan (1868–1953), 1923

* *
*

"We can surely never hope to see the craft of surgery made much more perfect than it is today. We are at the end of a chapter."

Berkeley George Moynihan (1865–1936), 1930

* *
*

"There is not the slightest indication that [nuclear] energy will ever be obtainable. It would mean that the atom would have to be shattered at will."

Albert Einstein (1879–1955), 1932

* *
*

"Fortunately; uranium bombs cannot at once be adapted for war, as the apparatus needed is very heavy and also very delicate, so it cannot at present be dropped from an airplane"

J.B.S. Haldane (1892–1964), 1940

* *
*

"Real mathematics has no effect on war: No one has yet discovered any warlike purpose to be served by the theory of numbers or relativity, and it seems very unlikely that anyone will do so for many years."

G.H. Hardy (1877–1947), 1940

* *
*

"That is the biggest fool thing we have ever done... The (atomic) bomb will never go off, and I will speak as an expert on explosives."

Admiral William Leahy to President Harry Truman, 1945

* *
*

"Man will never reach the moon regardless of all future scientific advances."

Lee De Forest (1873–1961), 1957

* *
*

"Further investigation and experimentation have confirmed the findings of Isaac Newton in the 17th century, and it is now definitely established that a rocket can function in a vacuum as well as in an atmosphere. The Times *regrets the error."*

editorial in *The New York Times*, July 17, 1969

* *
*

"There is no reason for any individual to have a computer in their home."

Ken Olsen, President of Digital Equipment Corporation, USA, 1980

* *
*

"If one believes in science, one must accept the possibility – even the probability – that the great era of scientific discovery is over... Further research may yield no more great revelations or revolutions, but only incremental, diminishing returns."

John Morgan (1806–1871), 1996

* *
*

"The use of small quantities [of uranium], sufficient, say, to operate a car or an airplane, so far is impossible, and one cannot predict when it will be achieved. No doubt, it will be achieved, but nobody can say when."

Albert Einstein (1879–1955), 1945

* *
*

"I think we are not quite yet fit for Flying Machines and therefore there will be none."

Ralph Waldo Emerson (1803–1882), 1843

* *
*

"I do not hesitate to forecast that atomic batteries will be commonplace long before 1980."

David Sarnoff (1891–1971), 1955

* *
*

"The best way to teach that one should be suspicious of everything one holds dear is through the study of brilliant people of the past, who by modern standards are so wrong, and where it is easy to see that their errors were the result of cultural biases of their day."

Stephen Jay Gould (1941–2002), 1993

* *
*

"The ideas of Freud were popularized by people who only imperfectly understood them, who were incapable of the great effort required to grasp them in their relationship to larger truths, and who therefore assigned to them a prominence out of all proportion to their true importance."

Alfred North Whitehead (1861–1947)

* *
*

"It may be that the stars of heaven appear fair and pure simply because they are so far away from us, and we know nothing of their private life."

Heinrich Heine (1797–1856), 1833

* *
*

"The certain proof that intelligent life exists elsewhere in the Universe is that no one has bothered yet to make contact with us."

(Anon)

* *
*

"You know why there are so many whitefish in the Yellowstone River? Because the Fish and Game people have never done anything to help them."

Russell Chatham (1939–), 1978

* *
*

"With classical thermodynamics, one can calculate almost everything crudely; with kinetic theory, one can calculate fewer things, but more accurately; with statistical mechanics, one can calculate nothing."

Eugene Wigner (1902–1995)

* *
*

"First law of thermodynamics: You cannot win.
Second law of thermodynamics: You cannot break even.
Third law of thermodynamics: You cannot get out of the game."

(Anon)

* *
*

"Astronomers discover God!"

Headlines in a well-known periodical (1982)

* *
*

"The question is not so much whether there is life elsewhere in the universe as whether it will continue to be possible to live on earth."

(Anon)

* *
*

"If you want to see comets that are comets, you've got to get outside our solar system – where there's room for them."

"There is something fascinating about science. One gets such wholesale returns of conjecture out of such a trifling investment of fact."

Mark Twain (1835–1910)

* *
*

"Comets are the nearest thing to nothing that anything can be and still be something."

(Anon)

* *
*

"We owe a lot to Thomas Edison – if it wasn't for him, we'd be watching television by candlelight."

Milton Berle (1908–2002)

* *
*

"No one has ever complained of a parachute not opening."

(Anon)

* *
*

"Statistics prove that 50 per cent of the married people in the United States are women."

(Anon)

* *
*

"My great-grandfather looked at him severely. 'My man,' he said, 'don't you know that very few people ever die after the age of 99? Statistics prove it!'"

(Anon)

* *
*

"Never try to walk across a river just because it has an average depth of four feet."

Martin Friedman (1962–)

* *
*

"Amoebas at the start
Were not complex;
They tore themselves apart
And started Sex."

Arthur Guiterman (1871–1943)

* *
*

"When you have excluded the impossible, whatever remains, however improbable, must be the truth."

"You know my method. It is founded upon the observance of trifles."

"It has long been an axiom of mine that the little things are infinitely the most important."

"You see, but you don't observe."

'Is there any point to which you would wish to draw my attention?'
'To the curious incident of the dog in the night-time.'
'The dog did nothing in the night-time.'
'That was the curious incident,' remarked Sherlock Holmes.

Arthur Conan Doyle (1856–1930)

* *
*

"That theory is worthless. It isn't even wrong!"

Wolfgang Pauli (1900–1958)

* *
*

"The guy who invented the first wheel was an idiot. The guy who invented the other three, he was a genius."

Sid Caesar (1922–)

* *
*

"If it weren't for Philo T. Farnsworth, inventor of television, we'd still be eating frozen radio dinners."

Johnny Carson (1925–2005)

* *
*

Louise: "how did you get there?"
Johnny: "Well, basically, there was this little dot, right? And the dot went bang and the bang expanded. Energy formed into matter, matter cooled, matter lived, the amoeba to fish, to fish to fowl, to fowl to frog, to frog to mammal, the mammal to monkey, to monkey to man, amo amas amat, quid pro quo, memento mori, ad infinitum, sprinkle on a little bit of grated cheese and leave under the grill till Doomsday."

* *
*

"No committee could ever come up with anything as revolutionary as a camel — anything as practical and as perfectly designed to perform effectively under such difficult conditions."

Laurence J. Peter (1919–1990)

* *
*

"Physics is becoming so unbelievably complex that it is taking longer and longer to train a physicist. It is taking so long, in fact, to train a physicist to the place where he understands the nature of physical problems that he is already too old to solve them."

Eugene Wigner (1902–1995)

* *
*

"An expert is someone who knows some of the worst mistakes that can be made in his subject and how to avoid them."

Werner Heisenberg (1901–1976)

* *
*

"An expert is a man has made all the mistakes, which can be made, in a very narrow field."

Niels Bohr (1885–1962)

Name Index (quotations)

A

B

C

D

E

F

G

H

I

J

K

L

M

N

O

P

R

S

T

U

V

W

Y

Z

Bibliography

* *
*

In creating this treatise I have borrowed information and data from various source-media such as books, articles and Internet data bases. References to these sources have been placed in four distinct locations:

- *In the text proper (name of books, articles, historical events, quotations, etc.)*
- *Footnotes at the bottom of the pages.*
- *At the end of each chapter.*
- *A comprehensive bibliography (some 900 entries) at the end of chapter 6.*

The following bibliography is divided here into ten categories:

1. Source–media for Biographies, History and Chronologies (Encyclopedias, Lexicons, Dictionaries, Timetables, Archives, Atlases and Internet Data Bases)

2. History of Science and Technology (and Biographies of Scientists, Engineers, Inventors and Explorers)

3. Mathematics, Logic and Philosophy

4. Physics and Cosmology

5. Chemistry

6. Biology and Medicine

7. Earth–Sciences

8. Space–Science, Astronomy and Astrophysics

9. Social Sciences

10. Quotations (of Thinkers, Sages, Savants, and Scholars throughout the ages)

1. *Source–media for Biographies, History and Chronologies*

Avneyon, E. (Editor), *The Encyclopedic Dictionary of the Bible*,Itaab Press, Israel, 2003 (In Hebrew).

Bahn, P.G. (Editor), *100 Great Archaeological Discoveries*, Barnes & Noble Books: New York, 1995, 256 pp.

Barnes, I., *Historical Atlas of the Bible*,Chartwell Books, Edison: New York, 2006, 399 pp.

Bacon, J., *The Illustrated Atlas of Jewish civilization* , Quantum Books, 2005, 224 pp.

Britannica, The Encyclopaedia, Eleventh Edition, 29 Volumes, Cambridge University Press: Cambridge, 1910–1911.
(a publication now is the public domain)

Bunch, B., and A. Hellemans, *The Timetables of Technology*, Simon and Schuster: New York, 1994, 490 pp.

Bynum, W.F., et al. Eds. *Dictionary of the History of Science*, Princeton University Press: Princeton N.J., 1981, 494 pp.

Calkins, C.C. (Editor), *The story of America*, The Readers Digest Association: New York, 1975, 527 pp.

Cary, M., et al. Eds, *The Oxford Classical Dictionary*, The Clarendon Press: Oxford, 1957, 971 pp.

Cicero, *Selected Papers*,Penguin Books: England, 1976, 272 pp.

Craig, G.A. *Europe Since 1815*, The Dryden Press: Hinsdale, IL, 1974, 620 pp.

Daiches, David (Editor), *Penguin Companion to English Literature*, McGraw-Hill Book Company: New York, 1971, 576 pp.

Debus, A.G. (Editor), *World's Who's Who in Science: A Biographical Dictionary of Notable Scientists from Antiquity to the Present*, Western Publishing Company: Hannibal, MO, 1968, 1855 pp.

Derry, T.K. and T.I. Williams, *A Short History of Technology*, Dover Publications: New York, 1993, 782 pp.

Desmond, K., *A Timetable of Inventions and Discoveries* (From Pre-History to the Present Day), M. Evans and Company: New York, 1986.

Eckermann, J.P., *Conversation with Goethe*, E.P. Dutton and Company: New York, 1971, 448 pp.

Encyklopädie der Mathematischen Wissenschaften, Edited by the Academies of Göttingen, Leipzig, Münich and Vienna; 24 volumes, Teubner, 1898–1934.

Eves, H., *An Introduction to the History of Mathematics*, Holt, Reinhart and Winston: New York, 1976.

Flusser, D., *Jewish Sources in Early Christianity*: Sifriat Poalim, Tel-Aviv, 1979.

Franklin, F. (Editor), *History's Timeline*,Crescent Books: New York, 1981, 239 pp.

Gillispie, C.C. (Editor), *Dictionary of Scientific Biography*, 15 Volumes, Charles Scribner's Sons: New York, 1970.

Grun, B., *The Timetables of History*, The New Third Revised Edition, A touchstone Book, Simon and Shuster: New York, 1982, 724 pp.

Hadas, M., *Imperial Rome*, Time–Life International: Nederland, 1966, 190 pp.

Hall, J.W. (Editor), *History of the World*, JG Press, World Publication Group: North Dighton, MA, 2005, 915 pp.

Hart–Davis, A., (Editor), *History – the Definitive Visual Guide*, DK Publications: New York, 2007, 612 pp.

Harvey, D. and D.E., Fehrenbacher, *The Illustrated Biographical Dictionary*, Dorset Press: New York, 1990, 256 pp.

Hellemans, A. and B. Bunch, *The Timetables of Science* (A Chronology of the Most Important People and Events in the History of Science), Simon and Schuster: New York, 1988, 656 pp.

Hornblower, Simon and Spawforth, T. (Editors) *Who's who in the Classical World* , Oxford University Press, 2000, 440 pp.

Ito, Kiyosi (Editor), *Encyclopedic Dictionary of Mathematics*, Second Edition, MIT Press, 1993, 2148 pp.

Jacobus, M.W. (Editor), *A New Standard Bible Dictionary*,Funk and Wagnalls Company: New York and London, 1926, 965 pp.

Jewish Encyclopaedia, 12 Volumes, New York, 1901–1906.

Johnson, Paul, *Modern Times*, Harper Perenial: New York, 1992, 870 pp.

Johnson, Paul, *A History of the Jews*, Harper Perenial: New York, 1988, 644 pp.

Kline, M., *Mathematical Thought from Ancient to Modern Times*, Oxford University Press: New York, 1972, 1238 pp.

Konner, M., *Unsettled – An Anthropology of the Jews*, Viking Compass, the Penguin Group: New York, 2003, 500 pp.

Kramer, S.N., *History Begins at Summer*, Doubleday Anchor Books: New York, 1959, 247 pp.

Lang, D.M. and D.R. Dudley (Editors), *Penguin Companion to Classical Literature*, McGraw-Hill Book Company: New York, 1969, 361 pp.

Langer, W.L. (Editor), *An Encyclopedia of World History*, Houghton Mifflin Company: Boston, 1963, 1243 pp.

Maccaoby, H., *The Pariah People /*, Constable: London, 1988, 236 pp.

Meyers Universal Lexicon, 15 Volumes, Bibliographisches Institut Manheim: Zürich, 1985.

National Geographic Visual History of the World, National Geographic: Washington D.C. 2005, 656 pp.

Obe, P.M., *The Concise Atlas of the Universe*, Rand McNally and Company: New York, 1970, 189 pp.

Parkinson, L.C., *Breakthroughs* (A Chronology of Great Achievements in Science and Mathematics, 1200–1930), G.K. Hall and Company: Boston, 1985, 576 pp.

Poggendorff, J.C. (1796–1877), *Biographisch-Literarisches Handwörterbuch zür Geschichte der Exacten Wissenschaften*, Vols. I–II, 1863, Vol. III, 1898, Vol. IV, 1904.

Porter, Roy (Editor), *The Biographical Dictionary of Scientists*, Oxford University Press: New York, 1994, 891 pp.

Rich, N., *The Age of Nationalism and Reform (1850–1890)*, W.W. Norton and Company: New York, 1977, 270 pp.

Roth, C., *The Jewish Contribution to Civilization*, The Phaidon Press: Oxford, 1943, 369 pp.

Rundle, D. (Editor), *The Hutchinson Encyclopedia of the Renaissance*, Helicon Publishing, 1999, 434 pp.

Sarton, G., *Introduction to the History of Science*, 3 Volumes, Williams and Wilkins Company: Baltimore, 1927–1948.

Schweizer Lexikon, 2 Volumes, Encyclios-Verlag AG: Zürich, 1949, 936–984 pp.

Spengler, O., *The Decline of the West* (Abridged Edition by A. Knopf), New York, 1962, 415 pp.

Talmon, J.L., *Romanticism and Revolt (Europe 1815–1848)*, Harcourt, Brace and World: New York, 1970, 216 pp.

Taton, René, *History of Science*, volumes I-IV, Basic Books Inc: New York (translated from French); I. From the beginning to 1450, 551 pp., 1963; II. From 1450–1800, 665 pp., 1964; III. 19th century, 623 pp., 1965; IV. 20th century, 638 pp., 1966.

Teeple, John B. *Timelines of World History*, DK, New York, 2002, 666 pp.

Thorlby, Anthony (Editor), *Penguin Companion to European Literature*, McGraw-Hill Book Company: New York, 1969, 908 pp.

Van Loon, H.W., *The Arts*, Simon and Schuster: New York, 1937, 677 pp.

Von Hagen, V.W., *Realm of the Incas*, Mentor Books: New York, 1961, 223 pp.

Von Hagen, V.W., *The World of the Maya*, Mentor Books: New York, 1962, 224 pp.

Von Hagen, V.W., *The Aztec: Man and Tribe*, Mentor Books: New York, 1962, 224 pp.

Webster Biographical Dictionary, G. and C. Marriam Company, Springfield, MA, 1951, 1697 pp.

Welch, C. *Rise and fall of the Nazis*, Magpie, London, 2008, 314 pp.

Wittle, T., *The World of Classical Greece*, William Heinemann Medical Books: London, 1971.

World Book Encyclopedia, 22 Volumes, Field Enterprises Educational Corporation: Chicago, 1977.

The MacTutor History of Mathematics archive
http://www-history.mcs.st-andrews.ac.uk/history/index.html

Wikipedia, the free encyclopedia
http://en.wikipedia.org

2. *History of Science and Technology*

Adams, F.D., *Birth and Development of the Geological Sciences*, Dover Publications: New York, 1954, 606 pp.

Asimov, I., *Biographical Encyclopedia of Science and Technology*, Doubleday and Company: Garden City, New York, 1964, 662 pp.

Bagrow, Leo, *History of Cartography*, Harvard University Press: Cambridge MA, 1966, 312 pp.

Ball, W.W.R., *A Short Account of the History of Mathematics*, Dover Publications: New York, 1960, 522 pp.

Beckmann, P., *A History of π (Pi)*, St. Martins Press: New York, 1971, 200 pp.

Berlin, I., *Against the Current*, Penguin Books: New York, 1979, 394 pp.

Bleiler, E.F., *Science Fiction – The Early years*, The Kent State University Press, 1990.

Bleiler, E.F., Ed. *Science-Fiction Writers*, Charles Scribnu's Sons: New York, 1982.

Bowler, P. J., *Evolution: The History of an Idea*, University of California Press, 2003

Bowra, C.M., *Classical Greece*, Time-Life International: Nederland, 1970, 192 pp.

Boyer, C.B., *A History of Mathematics*, John Wiley & Sons: New York, 1968, 717 pp.

Brooks, C.E.P., *Climate Through the Ages*, Dover Publications: New York, 1970, 395 pp.

Bunt, L.N., H. Jones and J.D. Bedient, *The Historical Roots of Elementary Mathematics*, Prentice-Hall: Englewood Cliffs, NJ, 1976, 299 pp.

Burtt, G., *The Metaphysical Foundations of Modern Science*, London, 1924.

Butterfield, H., *The Origins of Modern Science*, Free Press: New York, 1965, 255 pp.

Butler, A. (Editor), *Everyman's Dictionary of Dates*, E.P. Dutton and Company: New York, 1964, 455 pp.

Bynum, W.F., et al. Eds. *Dictionary of the History of Science*, Princeton University Press: Princeton N.J., 1981, 494 pp.

Cajori, F.A., *A History of Mathematics*, Macmillan and Company: New York, 1919, 516 pp.

Cajori, F., *A History of Mathematical Notation*, Dover publications, Inc: New York, 1993, volume I (451 pp.) and volume II (365 pp.).

Camusso, L., *The Voyages of Columbus*, Dorset Press: New York, 1991, 80 pp.

Carruccio, E., *Mathematics and Logic in History and in Contemporary Thought*, Faber and Faber: London, 1964, 398 pp.

Cavalli-Sforza, L.L. and F. Cavali-Sfroza, *The Great Human Diasporas*, Addison-Wesley: New York, 1995, 300 pp.

Clark, G., *Space, Time and Man*, A Prehistoric Review, Cambridge University Press, Canto edition, 1994, 165 pp.

Cohen, M.R. and I.E. Drabkin, *A Source Book in Greek Science*, McGraw-Hill Book Company: New York, 1948.

Cooke, R., *The Mathematics of Sonya Kovalevskaya*, Springer-Verlag: New York, 1984, 234 pp.

Coolidge, J.L., *The Mathematics of Great Amateurs*, Oxford University Press: London, 1949, 211 pp.

Dales, R.C., *The Scientific Achievement of the Middle Ages*, University of Pennsylvania Press: Philadelphia, 1973, 182 pp.

Dauben, J.W., *George Cantor: His Mathematics and Philosophy of the Infinite*, Harvard University Press: Cambridge, 1979, 404 pp.

David, R.A., *The Egyptian Kingdoms*, Elsevier: Phaidon, 1975, 152 pp.

De Camp, L.S., *The Ancient Engineers*, Ballantine Books: New York, 1963, 450 pp.

Derry, T.K. and T.I. Williams, *A Short History of Technology*, Dover Publications: New York, 1993, 782 pp.

De Solla, P.D., *Science Since Babylon*, Yale University Press: New Haven, CT, 1975, 215 pp.

Desmond, K., *A Timetable of Inventions and Discoveries* (From Pre-History to the Present Day), M. Evans and Company: New York, 1986.

Dick, A. (Editor), *Emmy Noether*, Birkhäuser Verlag: Basel, 1970, 72 pp.

Diamond, J., *Guns, Germs, and Steel*, W.W. Norton and Company: New York, 1997, 480 pp.

Dieudonné, Jean (editor), *Abrégé d'histoire des mathématiques 1700–1900*, vol II, Hermann: Paris, 1978, 392+469 pp.

Discover Science Almanac, Stonesong Press Books: New York, 2003, 790 pp.

Dugas, R., *A History of Mechanics* , Dover Publications: New York, 1988, 622 pp.

Einstein, A., *Letters to Solovine*, Citadel Press, 1993, 159 pp.

Eldredge, N., *The Monkey Business, A Scientist Looks at Creationism*, Washington Square Press: New York, 1982, 157 pp.

Eves, H., *Great Moments of Mathematics*, The Mathematical Society of America, 1983.

Farber, E., *The Evolution of Chemistry, A History of its Ideas, Methods, and Materials*, Ronald Press Company: New York, 1952, 349 pp.

Flatow, I., *They All Laughed*, Harper Perrenial: New York, 1992, 238 pp.

French, A.P. (Editor), *Einstein*, A Centennial Volume, Harvard University Press: Cambridge MA, 1979, 332 pp.

Galilei, G., *Dialogues Concerning Two New Sciences (1638)*, Dover Publications: New York, 1954, 300 pp.

Galilei, G., *Discorsi e Dimostriazioni Matematiche Intorno a due Nuovo Scienze*, Elzevir: Leiden, 1638. English translation by H. Crew and A. de Slavio, *Dialogues Concerning Two New Sciences*, Macmillan and Company: New York, 1914; Reprinted at Dover Publications: New York, 1954.)

Gauss, C.F., *Disquisitiones Arithmeticae (1801)*, Yale University Press: New Haven, CT, 1966, 472 pp.

Gellert, W., H. Küstner, M. Hellwich and H. Kästner (Editors), *The VNR Concise Encyclopedia of Mathematics*, Van Nostrand Reinhold Company: New York, 1975, 760 pp.

Gillispie, C.C. and M. Dewachter (Editors), *Monuments of Egypt*, The Complete Archaeological Plates from the Description De L'Egypt. Princeton Architectural Press, 1994.

Gladstone, I., *Human Nutrition. Historic and Scientific*, International University Press: New York, 1960

Goldstine, H.H., *The Computer from Pascal to von Neumann*, Princeton University Press: Princeton, NJ, 1982, 178 pp.

Grant, E. (Editor), *A Source Book in Medieval Science*, Harvard University Press: Cambridge, 1974, 864 pp.

Guillemin, V., *The Story of Quantum Mechanics*, Charles Scribner's Sons: New York, 1968, 332 pp.

Hart, I.B., *Makers of Science, Mathematics, Physics, Astronomy*, Oxford University Press: London, 1940, 320 pp.

Heath, T.L., *A History of Greek Mathematics*, 2 Volumes, Oxford University Press: London, 1921, I, 446 pp; II, 586 pp.

Heath, T.L. (Editor), *Apollonius of Perga: Treatise on Conic Sections*, W. Heffer and Sons: Cambridge, 1961, 254 pp.

Hellman, H., *Great Feuds in Medicine*, John Wiley & Sons: New York, 2001, 237 pp.

Hogben, L., *Science for the Citizen*, George Allen and Unwin: London, 1951, 1146 pp.

Hogben, L., *Mathematics for the Million*, George Allen and Unwin: London.

Ihde, A.J., *The Development of Modern Chemistry*, Harper and Row: New York, 1964, 851 pp.

Joseph, G.G., *The Crest of the Peacock*, Princeton University Press: Princeton NJ, 2000, 455 pp.

Judson, H. F., *The Eighth Day of Creation*, Cold Spring Harbor Press: New York, 1996

Keegan, J., *Intelligence in War*, Vintage Books New York, 2002, 387 pp.

Kline, M., *Mathematical Thought from Ancient to Modern Times*, Oxford University Press: New York, 1972, 1238 pp.

Kline, M., *Mathematics – the Loss of Certainty*, Oxford University Press: New York, 1980, 366 pp.

Kolmogorov, A.N., and A.P. Yushkevich Eds, *Mathematics of the 19th Century*, Birkhäuser Verlag: Boston, 2001, 308 pp.

Kramer, E.E., *The Nature and Growth of Modern Mathematics*, 2 Volumes, Fawcett Publications: New York, 1970, 527+512 pp.

Kuhn, T.S., *The Structure of Scientific Revolutions*, University of Chicago Press, 1970.

Leake, C. D. and Charles C., Thomes *A Historical Account of Pharmacology to the Twentieth Century*, Springfield IL, 1975

Leicester, H.M. and H.S. Klickstein, *A Source Book in Chemistry*, McGraw-Hill Book Company: New York, 1952.

Leicester, H.M., *The Historical Background of Chemistry*, Dover Publications: New York, 1971, 260 pp.

Lindberg, D.C. (Editor), *Science in the Middle Ages*, University of Chicago Press: Chicago, 1978, 549 pp.

Magie, W.F., *A Source Book in Physics*, McGraw-Hill Book Company: New York, 1935.

Mason, S.F., *A History of the Sciences*, Philosophical Library: New York, 1965, 185 pp.

Serres, M., *La Naissance de la Physique dans le Texte de Lucréce*, Minuit: Paris, 1977.

Seife, C., *Zero – The Biography of a Dangerous Idea*, Penguin Books, 2000, 238 pp.

Shapley, H. and H.E. Howarth, *A Source Book in Astronomy*, McGraw-Hill Book Company: New York, 1929.

Singer, C., *A Short History of Science to the Nineteenth Century*, Oxford, 1941.

Siu, R.G.H., *The Tao of Science*, The MIT Press: Cambridge MA, 1957, 180 pp.

Smith, D.E., *History of Mathematics*, 2 Volumes, Ginn and Company: Boston, 1951–1953, I, 596 pp; II, 725 pp.

Smith, D.E., *A Source Book in Mathematics*, Dover Publications: New York, 1959.

Stillwell, J., *Mathematics and its History*, Springer-Verlag: New York, 1989, 371 pp.

Stone, I., *Clarence Darrow for the Defense*, A Signet Book: New York, 1971, 640 pp.

Stubbe, H., *History of Genetics: From Prehistoric Times to the Rediscovery of Mendel's Laws*, MIT Press

Swetz, F.J. and T.I. Kao, *Was Pythagoras Chinese?* The Pennsylvania State University Press: Pennsylvania, 1977, 75 pp.

Temple, G., *100 Years of Mathematics*, Springer-Verlag: New York, 1981.

Thomas, H., *Human origins*, Abrams publishers: New York, 1994, 159 pp.

Todhunter, I., *A History of the Progress of the Calculus of Variations in the Nineteenth Century*, Chelsea Publications: New York.

Todhunter, I. and K. Pearson, *A History of the Theory of Elasticity from Galileo to Lord Kelvin*, 2 Volumes, Dover Publications: New York, 1960.

Van der Waerden, B.L., *Science Awakening*, P. Noordhoff: Groningen, Holland, 1974, 306 pp.

Van der Waerden, B.L., *A History of Algebra From Al-Khwarizmi to Emmy Noether*, Springer-Verlag: Berlin, 1980.

Von Braun, W. and F.I. Ordway, *History of Rocketry and Space Travel*, 1967.

White, A.D., *A History of the Warfare of Science with Theology*, 2 Volumes, London, 1914.

Whittaker, E.T., *A History of the Theories of Aether and Electricity: From the Age of Descartes to the Close of the Nineteenth Century*, Longmans, Green and Company: London, 1910, 475 pp.

Whittaker, E.T., *A History of the Theories of Aether and Electricity: The Modern Theories 1900–1926*, Philosophical Library: New York, 1954, 319 pp.

Williams, L.P., *The Origins of Field Theory*, Random House: New York, 1966, 148 pp.

Wilson, A.M., *The Infinite in the Finite*, Oxford University Press: New York, 1995, 524 pp.

Wolpert, L., *The Unnatural Nature of Science*, Harvard University Press: Cambridge MA, 1993, 191 pp.

Woster, D., *Nature's Economy: A History of Ecological Ideas*, Cambridge University Press, 1994

Wuckel, D. and B. Cassiday, *The Illustrated History of Science Fiction*, Ungar: New York, 1989, 251 pp.

Biographies

Adams, A.B., *Eternal Quest: The Story of the Great Naturalists*, Isaac Putnam's Sons: New York, 1969, 509 pp

Bell, E.T., *Men of Mathematics*, Simon and Schuster: New York, 1937, 592 pp.

Bühler, W.K., *Gauss: A Biographical Study*, Springer-Verlag: Berlin, 1981, 208 pp.

Calinger, R., *Gottfried Wilhelm Leibniz*, Rensselaer Polytechnic Institute: Troy, NY, 1976, 102 pp.

Cheney, M., *Tesla: Man Out of Time*, A Laurel Book, Dell Publishing: New York, 1981, 320 pp.

Dolan, S., *Juan Ponce de León*, Chelsea House Publishers, New York, 1995, 111 pp.

Drake, S., *Galileo at Work, His Scientific Biography*, University of Chicago Press: Chicago, 1978, 536 pp.

Dunnington, G.W., *C.F. Gauss Titan of Science*, New York, 1955.

Eiseley, L., *The Man Who Saw Through Time*, Charles Scribner Sons: Ne York, 1973, 125 pp.

Everitt, C.W.F., *James Clerk Maxwell, Physicist and Natural Philosopher*, Charles Scribner's Sons: New York, 1975, 205 pp.

Fölsing,A., *Albert Einstein*, Penguin Books: New York, 1997, 882 pp.

Gadol, J., *Leon Battista Alberti, Universal Man of the Renaissance*, University of Chicago Press: Chicago, 1973, 266 pp.

Gillispie, C.C., *Lazzare Carnot-Savant*, Princeton University Press: New Jersey, 1971, 359 pp.

Gleick, J., *Genius — The Life and Science of Richard Feynman*, Pantheon Books: New York, 1992, 531 pp.

Hall, T., *Carl Friedrich Gauss*, Massachusetts Institute of Technology Press: Cambridge, 1970, 176 pp.

Hardy, G.H., *Ramanujan*, Chelsea Publishing Company: New York, 1940, 236 pp.

Hart, I.B., *The World of Leonardo da Vinci: Man of Science, Engineer and Dreamer of Flight*, Viking Press: New York, 1961, 374 pp.

Hodges, A., *Alan Turing, The Enigma*, Simon and Schuster: New York, 1983, 587 pp.

Hoffmann, P., *The Man Who Loved Only Numbers*, Fourth Estate: London, 1999, 302 pp.

Kanigel, R., *The Man Who Knew Infinity* (*A life of the genius Ramanujan*), Charles Scribner's Sons: New York, 1991, 438 pp.

Kennedy, D.H., *Little Sparrow: A Portrait of Sophia Kovalevsky*, Ohio University Press: Athens, OH, 1983, 341 pp.

Koestler, A., *The Sleepwalkers*, Macmillan and Company: New York, 1959, 623 pp.

Kosmodemianskii, A.A., *Konstantin Tsiolkovsky: His Life and Work*, Translated from the Russian by X. Danko, Foreign Language Pub. House: Moscow, 1956, 101 pp.

Lanouette, W., *Genius in the Shadows*, University of Chicago Press, 1992, 587 pp.

Mahoney, M.S., *The Mathematical Career of Pierre de Fermat*, Princeton University Press: Princeton, NJ, 1973.

Marchand, P., *Marshall McLuhann, The Medium and the Messenger*, Ticknor and Fields: New York, 1989, 320 pp.

Marquardt, Martha, *Paul Ehrlich*, William Heinemann Medical Books: London, 1949, 255 pp.

Moore, W., *Schrödinger*, Cambridge University Press: Cambridge, 1989, 513 pp.

Nahin, P.J., *Oliver Heaviside: Sage in Soltitude*, IEEE Press: New York.

Ore, O., *Cardano the Gambling Scholar*, Dover Publications: New York, 1965, 249 pp.

Osen, L.M., *Women in Mathematics*, Massachusetts Institute of Technology Press: Cambridge, 1988, 185 pp.

Pais, A., '*Subtle is the Lord...*', *The Science and the Life of Albert Einstein*, Oxford University Press: Oxford, 1983, 552 pp.

Parikh, C., *The Unreal Life of Oscar Zariski*, Academic Press: New York, 1991, 264 pp.

Reid, C., *Hilbert*, Springer-Verlag: New York, 1970, 290 pp.

Steinmann, J., *Pascal*, Harcourt, Brace and World: New York, 1966, 304 pp.

Vallentin, A., *Leonardo da Vinci, The Tragic Pursuit of Perfection*, Viking Press: New York, 1938, 561 pp.

Westfall, R.S., *Never at Rest: A Biograph of Isaac Newton*, Cambridge University Press: Cambridge, 1980, 908 pp.

Williams, L.P., *Michael Faraday, A Biography*, Basic Books: New York, 1965, 531 pp.

Wittle, T., *Leonhard Euler 1707–1783: Beiträge zu Leben und Werke*, Gedenkband des Kantons Basel-Stadt, Birkhäuser: Basel, 1983.

Wolf, E., The life and work of Christiaan Huygens, in Blok, H., H.A. Ferwerda and H.K. Kuiken (Editors), *Huygens Principle: 1690–1990: Theory and Applications*, Elsevier Science Publishers, 1992.

Moorehead, A., *Darwin and the Beagle*, Penguin Books: England, 1971, 280 pp.

3. Mathematics, Logic and Philosophy

Alexandrov, A.D., A.N. Kolmogorov and M.A. Lavrentév (Editors), *Mathematics, Its Content, Method, and Meaning*, Volumes I–III, The M.I.T. Press: Cambridge, 1969, Vol. I, 377 pp., Vol. II, 395 pp., Vol. III, 372 pp.

Allenby, R.B.J.T. and E.J. Redfern, *Introduction to Number Theory with Computing*, Edward Arnold: London, 1989, 309 pp.

Ames, W.F., *Numerical Methods for Partial Differential Equations*, Academic Press: Boston, 1992.

Apostol, T.M., *Modular Functions and Dirichlet Series in Number Theory*, Springer-Verlag: New York, 1990, 204 pp.

Aris, R., *Vectors, Tensors, and the Basic Equations of Fluid Mechanics*, Dover Publications: New York, 1962, 286 pp.

Arnold, B.H., *Intuitive Concepts in Elementary Topology*, Prentice-Hall: Englewood Cliffs, NJ, 1963, 182 pp.

Bailey, N.T.J., *The Elements of Stochastic Processes* (With Applications to Natural Sciences), John Wiley & Sons: New York, 1964, 249 pp.

Ball, W.W.R., *Mathematical Recreations and Essays*, Macmillan and Company: London, 1944, 418 pp.

Bass, J., *Exercises in Mathematics*, Academic Press: New York, 1966, 459 pp.

Batschelet, E., *Introduction to Mathematics for Life Sciences*, Springer-Verlag: Berlin, 1979, 643 pp.

Beiler, A.H., *Recreations in the Theory of Numbers* (The Queen of Mathematics Entertains), Dover Publications: New York, 1964, 349 pp.

Bell, E.T., *Development of Mathematics*, McGraw-Hill Book Company: New York, 1945, 637 pp.

Bender, C.M. and S.A. Orszag, *Advanced Mathematical Methods for Scientists and Engineers*, McGraw-Hill Book Company: New York, 1978, 593 pp.

Berndt, B.C., *Ramanujan's Notebooks*, Springer-Verlag: New York, 1985–1991, *Part I*, 357 pp., *Part II*, 510 pp., *Part III*, 510 pp., *Part IV*, 451 pp., *Part V*, 624 pp.

Berg, H.C., *Random Walks in Biology*, Princeton University Press: Princeton, NJ, 1983, 142 pp.

Berman, G. and K.D. Fryer, *Introduction to Combinatorics*, Academic Press, 1972, 300 pp.

Bharucha-Reid, A.T., *Elements of the Theory of Markov Processes and Their Applications*, McGraw-Hill Book Company: New York, 1960, 468 pp.

Blatner, D., *The Joy of* π Walker, 1999, 124 pp.

Brand, L., *Vector and Tensor Analysis*, John Wiley & Sons: New York, 1948, 439 pp.

Brauer, F. and J.A. Nohel, *The Qualitative Theory of Ordinary Differential Equations*, Dover Publications: New York, 1989, 314 pp.

Budak, B.M. and S.V. Fomin, *Multiple Integrals, Field Theory and Series* (An Advanced Course in Higher Mathematics), Mir Publications: Moscow, 1978, 640 pp.

Budden, F.J., *The Fascination of Groups*,Cambridge: Cambridge University Press, 1972.

Burnside, W.S. and A.W. Panton, *The Theory of Equations*, Dover Publications: New York, Vol. I, 286 pp., Vol. II, 318 pp.

Carnahan, B., H.A. Luther and J.O. Wilkes, *Applied Numerical Methods*, John Wiley & Sons: New York, 1969, 604 pp.

Cartan, E., *The Theory of Spinors*, Dover Publications: New York, 1981, 157 pp.

Casti, J., *Reality Rules: Picturing the World in Mathematics*, I – The Fundamentals, Wiley, New York, 1992.

Casti, J.L., *Complexification*, Abacus Books, 1994, 320 pp.

Conway, J., E. Berlekamp and R. Guy, *Winning Ways for Your Mathematical Plays*, Academic Press, New York, 1982.

Conway, J.H. and Steve Sigur, *The Triangle Book*, AK Peters, 400 pp.

Courant, R., *Dirichlet's Principle, Conformal Mapping and Minimal Surfaces*, Interscience-Wiley: New York, 1950, 330 pp.

Courant, R. and H. Hilbert, *Methods of Mathematical Physics*, Vol. I, Interscience Publishers: New York, 1953, 561 pp.

Courant, R. and H. Robbins, *What is Mathematics* (An Elementary Approach to Ideas and Methods), Oxford University Press: London, 1953, 521 pp.

Crowe, M.J., *A History of Vector Analysis: Evolution of the Idea of a Vectorial System*, University of Notre Dame Press: South Bend, 1967, 270 pp.

Cushing, J.T., *Applied Analytical Mathematics for Physical Scientists*, John Wiley & Sons: New York, 1975, 651 pp.

Danielson, D.A., *Vectors and Tensors in Engineering and Physics*, Addison-Wesley: Redwood City CA, 1992, 280 pp.

Davis, H.T., *Introduction to Nonlinear Differential and Integral Equations*, Dover Publications: New York, 1962, 566 pp.

De Koninck, J.-M. and A. Mercier, *1001 Problems in Classical Number Theory*, AMS, 2007, 336 pp.

Deans, S.R., *The Radon Transform and Some of Its Applications*, John Wiley & Sons: New York, 1983, 289 pp.

Dehn, E., *Algebraic equations*, Dover Publications Inc: New York, 1960, 208 pp.

Demidovich, B.P. and I.A. Maron, *Computational Mathematics*, Mir Publications: Moscow, 1987, 688 pp.

Demidovich, B., *Problems in Mathematical Analysis*, Mir Publishers: Moscow, 1989, 496 pp.

Dennery, P. and A. Krzywicki, *Mathematics For Physicists*, Dover, New York, 1993, 384 pp.

Dickson, L.E., *Modern Algebraic Theories*, Ben J.H. Sanborn and company: Chicago, 1930, 276 pp.

Dilke, O.A.W., *Mathematics and Measurements*, The British Museum Publications: London, 1987, 63 pp.

Dörrie, H., *100 Great Problems of Elementary Mathematics, Their History and Solutions*, Dover Publications: New York, 1965, 393 pp.

Drazin, P.G. and R.S. Johnson, *Solitons: An Introduction*, Cambridge University Press: Cambridge, 1990, 226 pp.

Durell, C.V. and A. Robson, *Advanced Trigonometry*, G. Bell and Sons: London, 1943, 335 pp.

Dutta, M. and L. Debnath, *Elements of the Theory of Elliptic and Associated Functions* (With Applications), The World Press Private, 1965, 290 pp.

Eagle, A., *The Elliptic Functions as They Should Be*, Gallaway and Porter: Cambridge, England, 1958, 508 pp.

Edwards, H.M., *Riemann's Zeta Function*, Academic Press: New York, 1974, 315 pp.

Ekeland, I., *Mathematics and the Unexpected*, Univ. of Chicago Press, 1988, 145 pp.

Ellison, W. and F. Ellison, *Prime Numbers*, John Wiley & Sons: New York, 1985, 417 pp.

Elsgolts, L., *Differential Equations and the Calculus of Variations*, Mir Publishers (English translation): Moscow, 1977, 440 pp.

Erdélyi, A. (Editor), *Higher Transcendental Functions*, 3 Volumes, McGraw-Hill Book Company: New York, 1953–1955.

Fikhtengol'ts, G.M., *The Fundamentals of Mathematical Analysis*, 2 Volumes, Pergamon Press: Oxford, 1965, I, 491 pp; II, 516 pp.

Gamow, G., *One Two Three... Infinity*, Mentor Books: New York, 1953, 318 pp.

Gantmacher, F.R., *The Theory of Matrices*, 2 Volumes, Chelsea Publishing Company: New York, 1960, I, 374 pp; II, 276 pp.

Gardner, M., *Wheels, Life, and other Mathematical Amusements*, Freeman, New York, 1983.

Gause, G.F., *The Struggle for Existence*, Williams and Wilkins Company: New York, 1934.

Gelfand, I.M., R.A. Minlos and Z.Y. Shapiro, *Representation of the Rotation and Lorentz Groups and their Applications*, Pergamon Press: New York, 1963.

Gershenfeld, N., *The Nature of Mathematical Modeling*, Cambridge University Press: New York, 1999, 344 pp.

Ghyka, M., *The Geometry of Art and Life*, Dover Publications: New York, 1977, 174 pp.

Goertzel, G. and N. Tralli, *Some Mathematical Methods of Physics*, McGraw-Hill Book Company: New York, 1960, 300 pp.

Goldsmith, D., *Nemesis*, Berkeley Books: New York, 1985, 166 pp.

Goldstine, H.H., *The Computer from Pascal to von Neumann*, Princeton University Press: Princeton, NJ, 1982, 178 pp.

Golub, G.H. and C.F. Van Loan, *Matrix Computations*, John Hopkins University Press: Baltimore, MD, 1996.

Goursat, E., *A Course in Mathematical Analysis*, Vol. 1, Ginn and Company: Boston, 1904, 548 pp.

Grassman, H.D., *Die Linear Ausdehnungslehre*, Leipzig, 1844.

Grindrod, P., *Patterns and Waves*, Clarendon Press: Oxford, 1991, 239 pp.

Gradshteyn, I.S. and I.M. Ryzhik, *Tables of Integrals, Series and Products*, Academic Press: New York, 1965, 1086 pp.

Hagerdon, P., *Non-Linear Oscillations*, Clarendon Press: Oxford, 1988, 311 pp.

Hall, G.G., *Applied Group Theory*, American Elsevier Publishing Company: New York, 1967, 128 pp.

Hall, N. Ed., *Exploring Chaos*, W.W. Norton & Company: New York, 1993, 223 pp.

Hardy, G.H., *A Course of Pure Mathematics*, Cambridge University Press: Cambridge, 1967, 509 pp.

Hardy, G.H. and E.M. Wright, *An Introduction to the Theory of Numbers*, Oxford University Press: Oxford, 1989, 426 pp.

Heading, J., *Matrix Theory for Physicists*, Longmans, Green and Company: London, 1958, 241 pp.

Heading, J., *An Introduction to Phase-Integral Methods*, Methuen and Company: London, 1962, 160 pp.

Hildebrand, F.B., *Introduction to Numerical Analysis*, McGraw-Hill Book Company: New York, 1956, 511 pp.

Hildebrand, S. and A. Tromba, *Mathematics of Optimal Forms*, Scientific American Books: New York, 1985, 215 pp.

Holden, A.V. (Editor), *Chaos*, Princeton University Press: Princeton, NJ, 1986, 324 pp.

Hopf, L., *Introduction to the Differential Equations of Physics*, Dover Publications: New York, 1948, 154 pp.

Hoskins, R.F., *Generalized Functions*, Ellis Horwood: Chichester, England, 1979, 192 pp.

Hunter, J.A.H. and J.S. Madachy, *Mathematical Diversions*, Dover Publications: New York, 1975, 178 pp.

Huntly, I. and R.M. Johnson, *Linear and Nonlinear Differential Equations*, Ellis Horwood: Chichester, England, 1983, 190 pp.

Inch, E.L., *Ordinary Differential Equations*, Dover Publications: New York, 1956, 558 pp.

Isham, C.J., *Modern Differential Geometry for Physicists*, World Scientific.

Jacobs, H.R., *Mathematics — A Human Endeavor*, W.H. Freeman and Company: New York, 1982, 649 pp.

Jeffreys, H. and B.S. Jeffreys, *Methods of Mathematical Physics*, Cambridge University Press: Cambridge, 1956, 718 pp.

Jolley, L.B.W., *Summation of Series*, Dover Publications: New York, 1961, 251 pp.

Jones, D.S. and B.D. Sleeman, *Differential Equations and Mathematical Biology*, George Allen and Unwin: London, 1983, 339 pp.

Jordan, D.W. and P. Smith, *Nonlinear Ordinary Differential Equations*, Oxford University Press: Oxford, 1977, 360 pp.

Kamke, E., *Gewöhnliche Differentialgleichungen*, Chelsea Publishing Company: New York, 1959, 666 pp.

Kappraff, J., *Connections*, McGraw-Hill Book Company: New York, 1990, 471 pp.

Khinchin, A., *A Course of Mathematical Analysis*, Hindustan Publishing Corporation: Delhi, India, 1960, 668 pp.

Klambauer, G., *Aspects of Calculus*, Springer-Verlag: New York, 1986, 515 pp.

Knopp, K., *Theory and Applications of Infinite Series*, Blackie & Son: London, 1951, 563 pp.

Kreyszig, E., *Differential Geometry*, Dover Publications: New York, 1991, 352 pp.

Lamport, L., *LATEX*, Addison-Wesley Publishing Company: Reading, MA, 1986, 242 pp.

Larsen, R.J. and M.L. Marx, *An Introduction to Mathematical Statistics and its Applications*, Prentice-Hall: Englewood Cliffs, NJ, 1981, 596 pp.

Lass, H., *Vector and Tensor Analysis*, McGraw-Hill Book Company: New York, 1950, 347 pp.

Lavrentjev, M.A. and B.W. Shabat, *Methoden Der Komplexen Funktionentheorie*, Deutscher Verlag der Wissenschaften: Berlin, 1967.

Lawden, D.F., *Elliptic Functions and Applications*, Springer-Verlag: New York, 1989, 334 pp.

Leduc, S.A., *Linear Algebra*, Wiley, 1996, 318 pp.

Levy, H. and L. Roth, *Elements of Probability*, Oxford University Press: London, 1951, 200 pp.

Levy, S., *Artificial Life*, Pantheon, 1992.

Lie, S. and F. Engel, *Theorie der Transformationsgruppen*, Vol. I, Teubner: Leipzig, 1888.

Lie, S. and G. Scheffers, *Vorlesungen über Continuierliche Gruppen*, Teubner: Leipzig, 1893, 810 pp.

Littlewood, D.E., *The Skeleton Key of Mathematics*, Harper Torchbooks: New York, 1960, 138 pp.

Littlewood, D.E., *A University Algebra*, Dover Publications: New York, 1970, 324 pp.

Locher, J.L. (Editor) *The Magic of M.C. Escher*, Barnes & Noble: New York, 2001, 196 pp.

Loney, S.L., *The Elements of Coordinate Geometry*,Macmillan & Co: London, 1952.

Lovelock, D. and H. Rund, *Tensors, Differential Forms and Variational Principles*, John Wiley & Sons: New York, 1975, 364 pp.

MacMillan, W.D., *The Theory of the Potential*, Dover Publications: New York, 1958, 469 pp.

Magnus, W., F. Oberhettinger and R.P. Soni, *Formulas and Theorems for the Special Functions of Mathematical Physics*, Springer-Verlag: Berlin, 1966, 508 pp.

Mandelbrot, B., *The Fractal Geometry of Nature*, W.H. Freeman and Company: New York, 1983, 468 pp.

Mangulis, V., *Handbook of Series for Scientists and Engineers*, Academic Press: New York, 1965, 133 pp.

Marsden, J.E. and A.J. Tromba, *Vector Calculus*, W.H. Freeman and Company: New York, 1988, 655 pp.

Masterson-Gibbons, M., *A Concrete Approach to Mathematical Modeling*, Addison-Wesley Publishing Company: Redwood City, CA, 1989, 597 pp.

McConnell, A.J., *Application of Tensor Analysis*, Dover Publications: New York, 1957, 318 pp.

Milankovitch, M., *Canon of Insolation and the Ice-Age Problem*, Translated from German by the Israel Program for Scientific Translations, Jerusalem, 1969, 484 pp.

Milne, W.E., *Numerical Solutions of Differential Equations*, John Wiley & Sons: London, 1953, 275 pp.

Mirsky, L., *An Introduction to Linear Algebra*, Oxford University Press: London, 1955, 433 pp.

Mishchenko, A. and A. Fomenko, *A Course of Differential Geometry and Topology*, Mir Publications: Moscow, 1988, 455 pp.

Monastyrsky, M., *Modern Mathematics in the Light of the Fields Medal*, AK Peters LTD: Wellesley, Massachusetts, 1997.

Moon, F.C., *Chaotic Vibrations*, John Wiley & Sons: New York, 1987, 309 pp.

Moreau, R., *The Computer Comes of Age: The People, the Hardware, and the Software*, Massachusetts Institute of Technology Press: Cambridge, 1984, 227 pp.

Mosteller, F., *Fifty Challenging Probability Problems*, Dover Publications: New York, 1965, 88 pp.

Murray, J.D., *Mathematical Biology*, Springer-Verlag: Berlin, 1989, 767 pp.

Nahin, P.J., *Dr Euler Fabulous Formula*, Princeton University Press.

Nicolis, G. and I. Prigogine, *Self-Organization in Nonequilibrium Systems* (From Dissipative Structures to Order Through Fluctuations), John Wiley & Sons: New York, 1977, 488 pp.

Oberhettinger, F. and W. Magnus, *Anwendung Der Elliptischen Functionen in Physik und Technik*, Springer-Verlag: Berlin, 1949, 126 pp.

Ogilvy, C.S. and J.T. Anderson, *Excursions in Number Theory*, Dover Publications: New York, 1988, 168 pp.

Olds, C.D., *Continued Fractions*, Random House: New York, 1963, 162 pp.

O'Shea, D., *The Poincare Conjecture*, Walker and Com., New York, 2008, 292 pp.

Parzen, E., *Modern Probability Theory and its Applications*, John Wiley & Sons: New York, 1960, 464 pp.

Pavlidis, T., *Biological Oscillators*, Academic Press: New York, 1973, 207 pp.

Pearson, Carl E., *Handbook of Applied Mathematics*, Van Nostrand: New York, 1990.

Peterson, I., *The Mathematical Tourist*, Freeman, New York, 1983.

Phillips, J.L., *How to Think About Statistics*, W.H. Freeman and Company: New York, 1988, 198 pp.

Phillips, E.P., *A Course of Analysis*, Cambridge University Press: Cambridge, 1962, 372 pp.

Pickover, C.A., *Computers, Pattern, Chaos and Beauty* (Graphics From an Unseen World), St. Martin's Press: New York, 1990, 391 pp.

Pierce, J.R., *An Introduction to Information Theory*, Dover Publications: New Work, 1980, 305 pp.

Posamentier, A.S. and I. Lehmann, *The Fabulous Fibonacci Numbers*, Prometheus Books, New York, 2007, 385 pp.

Poundstone, W., *The Recursive Universe*, William Morrow and Comp., New York, 1985.

Press, W.H. *et al.*, *Numerical Recipes in C, The Art of Scientific Computing*, Cambridge University Press: Cambridge, 1988, 735 pp.

Ramanujan, S., *Collected Papers*, Cambridge University Press: Cambridge, 1927, 355 pp.

Ribenboim, P., *The New Book of Prime Records*, Springer-Verlag: New York, 1995.

Richeson, D.S, *Euler's Gem–The Polyhedron Formula and The Birth of Topology*, Princeton Univ. Press, 2008, 317 pp.

Riesel, H., *Prime Numbers and Computer Methods of Factorization*, Birkhauser: Boston, 1994.

Rozanov, Y.A., *Probability Theory*, Dover Publications: New York, 1969, 148 pp.

Saaty, T.L., *Modern Nonlinear Equations*, Dover Publications: New York, 1981, 471 pp.

Salomaa, A., *Public-Key Cryptography*, Springer-Verlag: Berlin, 1990, 245 pp.

Sattinger, D.H. and O.L. Weaver, *Lie Groups and Algebras with Applications to Physics, Geometry and Mechanics*, Springer-Verlag: New York, 1986, 215 pp.

Saunders, P.T., *An Introduction to Catastrophe Theory*, Cambridge University Press: Cambridge, 1982, 144 pp.

Schroeder, M.R., *Number Theory in Science and Communication* (With Applications in Cryptography, Physics, Biology, Digital Information, and Computing), Springer-Verlag: Berlin, 1984, 324 pp.

Schroeder, M.R., *Fractals, Chaos, Power Laws*, W.H. Freeman and Company: New York, 1990, 429 pp.

Schwartz, M., S. Green and A.W. Ruthledge, *Vector Analysis*, Harper and Brothers: New York, 1960.

Shatten, A., *Cellular Automata*, Vienna University of Technology, Vienna, 1997.

Smith, D.E. and Y. Mikami, *A History of Japanese Mathematics*, Dover Publications: New York, 2004

Sneddon, I.N., *Fourier Transforms*, McGraw-Hill Book Company: New York, 1951, 542 pp.

Sokolnikoff, I.S., *Tensor Analysis* (Theory and Applications to Geometry and Mechanics of Continua), John Wiley & Sons: New York, 1964, 361 pp.

Spivak, M.D., *The Joy of TEX*, American Mathematical Society: Providence, RI, 1986, 290 pp.

Strang, G., *Introduction to Applied Mathematics*, Wellesly – Cambridge Press: Wellesley, MA, 1986.

Struik, D.J., *Lectures on Classical Differential Geometry*, Dover Publications: New York, 1988, 232 pp.

Tannery, J. and J. Molk, *Elements De La Théorie Des Fonctions Elliptiques*, Chelsea Publishing Company: New York, 1972, Vol. I, 416 pp., Vol. II, 510 pp.

Titchmarsh, E.C., *The Theory of Functions*, Oxford University Press: London, 1939, 452 pp.

Titchmarsh, E.C., *Introduction to the Theory of the Fourier Integrals*, Oxford University Press: Oxford, 1948, 391 pp.

Titchmarsh, E.C., *The Theory of the Riemann Zeta-Function*, Oxford University Press: Oxford, 1951, 346 pp.

Turnbull, H.W., *The Theory of Determinants, Matrices and Invariants*, Dover Publications: New York, 1960, 374 pp.

Turnbull, H.W. and A.C. Aitken, *An Introduction to the Theory of Canonical Matrices*, Blackie & Son: London, 1952, 200 pp.

Van Der Pol, B. and H. Bremmer, *Operational Calculus*, Cambridge University Press: Cambridge, 1959, 409 pp.

Vilenkin, N.Ya., *Combinatorics*, Academic Press: New York, 1971, 296 pp.

Volterra, V., *Theory of Functionals*, Blackie & Son: London, 1930, 225 pp.

Volterra, V., *Theorie Mathématique de la Lutte pour la Vie*, Gauthier-Villars: Paris, 1931.

Von Bertalanffy, L., *Theorie generale des systems*, Dunod, Paris, 1973.

Von Neumann, J., *Theory of Self-Reproducing Automata*, Urbana, Ill, University of Illinois Press, 1966.

Walace, P.R., *Mathematical Analysis of Physical Problems*, Dover Publications: New York, 1984, 616 pp.

Watson, G.N., *A Treatise on the Theory of Bessel Functions*, Cambridge University Press: Cambridge, 1966, 804 pp.

Wax, N. (Editor), *Selected Papers on Noise and Stochastic Processes*, Dover Publications: New York, 1954, 337 pp.

Weatherburn, C.E., *Differential Geometry of Three Dimensions*, Cambridge University Press: Cambridge, 1939, 268 pp.

Weaver, W., *Lady Luck: The Theory of Probability*, Doubleday Anchor Books: New York, 1963, 392 pp.

Whittaker, E.T. and G.N. Watson, *A Course of Modern Analysis* (An Introduction to the General Theory of Infinite Processes and of Analytic

Functions; With an Account of the Principal Transcendental Functions), Cambridge University Press: London, 1962, 608 pp.

Wiener, N., *The Fourier Integral and Certain of its Applications*, Dover Publications: New York, 1958, 201 pp.

Wolfram, S., *Mathematica* (A System for Doing Mathematics by Computer), Addison-Wesley Publishing Company: Redwood City, CA, 1988, 749 pp.

Wolfram, S. (ed), *Theory and Applications of Cellular Automata*, World Scientific Press, Singapore, 1986.

Woodhouse, R., *A Treatise on Isoperimetrical Problems, and the Calculus of Variations*, Chelsea Publications: New York.

Yaglom, A.M. and I.M. Yaglom, *Challenging Mathematical Problems with Elementary Solutions*, Vols. I–II, Dover Publications: New York, 1964, 231 pp.; 1967, 214 pp.

Zeldovich, Ya.B., *Higher Mathematics for Beginners and its Applications to Physics*, Mir Publishers: Moscow, 1973, 494 pp.

Zuse, K., *Calculating Space*, MIT Press, Cambridge, Mass., 1970.

Zwillinger, D., *Handbook of Differential Equations*, Academic Press: Boston, 1989, 673 pp.

Philosophy

Armstrong, K., *A History of God*, Ballantine Books: New York, 1994, 496 pp.

Cambridge Dictionary of Philosophy, Cambridge University Press, 1999.

Craig, E., *A Shorter Routledge Encyclopedia of Philosophy*, 2005, 1077 pp.

Craig, E., *Routledge Encyclopedia of Philosophy*, 10 volumes, 1998.

Durant, W., *The Story of Philosophy*, Pocket Books: New York, 543 pp.

Frost, S.E., *Basic Teachings of the Great Philosophers*, Anchor Books, Doubleday: New York, 1989, 301 pp.

Honderich, T. (Editor), *The Oxford Guide to Philosophy*, , Oxford University Press, 2005.

Kiernan, T., *Who's who in the History of Philosophy*, Philosophical Library: New York, 1965, 185 pp.

Magee, B., *The Story of Philosophy*, DK Publishing, Inc.: New York, 1998, 240 pp.

Morris, T., *Philosophy for Dummies*, IDG Books: New York, 1999, 361 pp.

Nietzsche, F., *Human, All Too Human*, Cambridge University Press, 1996, 430 pp.

Oretga y Gasset, J., *What is Philosophy*, Norton and Company: New York, 1964, 252 pp.

Oxford Companion to Philosophy, Oxford University Press, 1995.

Runes, D.D., *Pictorial History of Philosophy*, Bramhall House: New York, 1954, 406 pp.

Russel, B., *Wisdom of the West*, Crescent Books, 1977, 319 pp.

Sahakian, W.S., *History of Philosophy*, Barnes and Noble: New York, 1968, 366 pp.

Scruton, R., *An Intelligent Person's Guide to Philosophy*, Penguin Books: New York, 1999, 168 pp.

Thompson, M., *Teach Yourself Philosophy*, McGraw–Hill: New York, 2007, 289 pp.

Turnbull, N., *Get a Grip on Philosophy*, Time–Life Books: New York, 1998, 192 pp.

4. *Physics and Cosmology*

Aharoni, J., *The Special Theory of Relativity*, Dover Publications: New York, 1965, 331 pp.

Alonso, M. and E.J. Finn, *Fundamental University Physics*, 3 Volumes, Addison-Wesley Publishing Company: Reading, MA, 1976, 1563 pp.

Andrew, E.R., *Nuclear Magnetic Resonance*, Cambridge University Press: Cambridge, 1956, 265 pp.

Asimov, I., *The Neutrino*, Discus Books: New York, 1966, 188 pp.

Bailey, N.T.J., *The Elements of Stochastic Processes* (With Applications to Natural Sciences), John Wiley & Sons: New York, 1964, 249 pp.

Balibar, F. and J.P. Maury, *How Things Fly*, Barrons: New York, 1987, 76 pp.

Banach, S., *Mechanics*, Publication of the Polish Mathematical Society: Warszawa, Poland, 1951, 546 pp.

Baeyer, H.C. von, *Taming the Atom*, Dover Publications: New York, 2000, 239 pp.

Baker, J., *50 Physics Ideas You Realy Need to Know*, Quercus Pub., London, 2007, 208 pp.

Baldwin, G.C., *Nonlinear Optics*, Plenum Press: New York, 1971, 155 pp.

Ball, Philip, *Designing the Molecular World*, Princeton University Press: Princeton NJ, 1994, 379 pp.

Barrow, J.L. and F.J. Tipler, *The Anthropic Cosmological Principle*, Oxford University Press: Oxford, 1986, 706 pp.

Bartusiak, M. (editor), *Archives of the Universe*, Vintage Books: New York, 2004, 695 pp.

Becker, R., *Quantum Theory of Atoms and Radiation*, Blackie & Son: London, 1964, 403 pp.

Bergmann, P.G., *Introduction to the theory of Relativity*, Prentice-Hall: New York, 1950, 287 pp.

Bethe, H. and E.E. Salpeter, *Quantum Mechanics of One and Two-Electron Atoms*, Plenum Publishing Corporation: New York, 1977, 369 pp.

Bharucha-Reid, A.T., *Elements of the Theory of Markov Processes and Their Applications*, McGraw-Hill Book Company: New York, 1960, 468 pp.

Bonometto et al (editors), *Modern Cosmology*, Institute of Physics Publishing, 2002,480 pp.

Born, M., *Optik*, Edwards Brothers: Ann Arbor, MI, 1943, 591 pp.

Born, M., *Einstein's Theory of Relativity*, Dover Publications: New York, 1962, 376 pp.

Born, M. and E. Wolf, *Principles of Optics*, Pergamon Press: London, 1964, 808 pp.

Brown, A.F., *Statistical Physics*, Edinburgh University Press: Edinburgh, 1968, 307 pp.

Chang, R., *Physical Chemistry with Applications to Biological Sciences*, Collier Macmillan International: New York, 1977, 538 pp.

Chaikin, P.M., and T.C. Lubensky, *Principles of Condensed Matter Physics*, Cambridge University Press, 1997, 699 pp.

Contopoulos, G. and D. Kostakis, *Cosmology*, Springer–Verlag: Berlin, 1987, 235 pp.

Crighton, D.G. et all, *Modern Methods in Analytical Acoustics*, Springer Verlag: Berlin, 1992, 738 pp.

Cronin, J.A. *et al.*, *Graduate Problems in Physics (with Solutions)*, University of Chicago Press: Chicago, 1967, 263 pp.

Croxton, C.A., *Introductory Eigenphysics*, John Wiley & Sons: London, 1974, 275 pp.

D'inverno, R., *Introducing Einstein's Relativity*, Clarendon Press: Oxford, 1992, 383 pp.

Davies, F.C.W., *The Search for Gravity Waves*, Cambridge University Press: Cambridge, 1980, 144 pp.

DeVelis, J.B. and G.O. Reynolds, *Theory and Applications of Holography*, Addison-Wesley Publishing Company, Reading, MA, 1967, 196 pp.

Dickinson, R.E., *The Makers of Modern Geography*, Praeger Publishers, New York, 1969.

Eddington, A.S., *The Mathematical Theory of Relativity*, Cambridge University Press: Cambridge, 1963, 270 pp.

Einstein, A., *Investigations on the Theory of the Brownian Movement*, Dover Publications: New York, 1956, 119 pp.

Feynman, R.P., *Quantum Electrodynamics*, Addison-Wesley Publishing Company: Reading, MA, 1961, 198 pp.

Feynman, R.P., *The Character of the Physical Law*, BBC Corporation: London, 1965, 173 pp.

Feynman, R.P., *QED: The Strange Theory of Light and Matter*, Princeton University Press: Princeton, NJ, 1985, 158 pp.

Feynman, R.P., R.B. Leighton and M. Sands, *The Feynman Lectures on Physics*, 3 Volumes, Addison-Wesley Publishing Company: Reading, MA, 1963–1965.

Ferris, T., *The Red Limit* (The Search for the Edge of the Universe), Quill: New York, 1983, 286 pp.

Finkelnburg, W., *Structure of Matter*, Springer-Verlag: Berlin, 1964, 511 pp.

Finlay–Freundlich, E., *Cosmology*, University of Chicago Press: Chicago, 1951, 59 pp.

Flügge, S., *Practical Quantum Mechanics*, Springer-Verlag: New York, 1974, 287 pp.

Frauenfelder, H. and E.M. Henley, *Subatomic Physics*, Prentice-Hall: Englewood Cliffs, NJ, 1974, 554 pp.

Genz, H., *Nothingness-the science of empty space*, Basic Books, 2001, 340 pp.

Gibilisco, S., *Understanding Einstein's Theories of Relativity*, Dover Publications: New York, 1991, 200 pp.

Glass, L. and M.C. Mackey, *From Clocks to Chaos, The Rhythms of Life*, Princeton University Press: Princeton, NJ, 1988, 248 pp.

Goodstein, D.L., *States of Matter*, Dover Publications: New York, 1975, 500 pp.

Gribbin, J. and M. Rees, *Cosmic Coincidences* (Dark Matter, Mankind and Anthropic Cosmology), Bantam Books: New York, 1989, 302 pp.

Hartree, D.R., *The Calculations of Atomic Structures*, John Wiley & Sons: New York, 1957, 181 pp.

Hawkins, S.W., *A Brief History of Time* (From the Big Bang to Black Holes), Bantam Books: New York, 1988, 198 pp.

Hearst, J.E. and J.B. Ifft, *Contemporary Chemistry*, W.H. Freeman and Company: San Francisco, 1976, 753 pp.

Hecht, E. and A. Zajac, *Optics*, Addison-Wesley Publishing Company: Reading, MA, 1975, 565 pp.

Helmholtz, H.L.F., *On the Sensations of Tone as a Physiological Basis for the Theory of Music*, Dover Publications: New York, 1954, 576 pp. (Translation of the 4th German Edition of 1877).

Herman, R.A., *A Treatise on Geometrical Optics*, Cambridge University Press, 1900, 344 pp.

Hill, T.L., *Statistical Mechanics*, McGraw-Hill Book Company: New York, 1956, 432 pp.

Hoyle, Fred, *The Intelligent Universe*, 1983.

Holden, A., *The Nature of Solids*, Dover Publications: New York, 1992, 241 pp.

Huang, K., *Statistical Mechanics*, John Wiley & Sons: New York, 1963, 470 pp.

Itzyksson, C. and J.-B. Zuber, *Quantum Field Theory*, McGraw-Hill Book Company: New York, 1980, 705 pp.

Jackson, J.D., *Classical Electrodynamics*, John Wiley & Sons: New York, 1975, 848 pp.

Jose, J.V. and E.J., Saletan *Classical dynamics*, Cambridge University Press, 2000, 670 pp.

Kak, A.C. and M. Slaney, *Principles of Computerized Tomographic Imaging*, IEEE Press: New York, 1988, 327 pp.

Kaku, M., *Parallel Worlds*, Anchor Books: New York, 2006, 428 pp.

Karamcheti, K., *Principles of Ideal-Fluid Aerodynamics*, John Wiley & Sons: New York, 1966, 635 pp.

Kittel, C., *Thermal Physics*, John Wiley & Sons: New York, 1969, 418 pp.

Kittel, C., *Introduction to Solid State Physics*, John Wiley & Sons: New York, 1986, 646 pp.

Lal, B., *Mathematical Theory of Electromagnetism*, Asia Publishing House: London, 1965, 510 pp.

Lanczos, C., *The Variational Principles of Mechanics*, University of Toronto Press: Toronto, 1964, 367 pp.

Landau, L.D. and E.M. Lifshitz, *The Classical Theory of Fields*, Pergamon Press: Oxford, 1962, 404 pp.

Leighton, R.B., *Principles of Modern Physics*, McGraw-Hill Book Company: New York, 1959, 795 pp.

Leimanis, E., *The General Problem of the Motion of Coupled Rigid Bodies About a Fixed Point*, Springer-Verlag: New York, 1965.

Li, W.-H. and S.-H. Lam, *Principles of Fluid Mechanics*, Addison-Wesley Publishing Company: Reading, MA, 1964, 374 pp.

Livesey, D.L., *Atomic and Nuclear Physics*, Blaisdell Publishing Company: Waltham, MA, 1966, 525 pp.

Livinstone, James D., *Driving Force*, Harvard University Press, 1996, 311 pp.

Lizuka, K., *Engineering Optics*, Springer-Verlag: New York, 1987, 489 pp.

Love, A.E.H., *A Treatise on the Mathematical Theory of Elasticity*, Dover Publications: New York, 1944, 643 pp.

Mach, E., *The Science of Mechanics*, Open Court, 1989, 634 pp.

Marion, J.B. and W.F. Hornyak, *Physics* (For Science and Engineering), Saunders College Publishing: New York, 1982, 1274 pp.

Matthews, P.T., *Introduction to Quantum Mechanics*, McGraw-Hill Book Company: New York, 1974, 208 pp.

Matveev, A.N., *Molecular Physics*, Mir Publishers: Moscow, 1985, 448 pp.

Matveev, A.N., *Optics*, Mir Publishers: Moscow, 1988, 446 pp.

McMahon, D., *String Theory DeMYSTiFied*, McGraw Hill, 2009, 306 pp.

Milne-Thomson, L.M., *Theoretical Aerodynamics*, Dover Publications: New York, 1958, 430 pp.

Minnaert, M.G.J., *Light and Color in the Outdoors*, Springer Verlag: Berlin, 1992, 417 pp.

Morris, P.G., *Nuclear Magnetic Resonance Imaging in Medicine and Biology*, Clarendon Press: Oxford, 1986, 388 pp.

Novikov, I.D. and V.P. Frolov, *Physics of Black Holes*, Kluwer Academic Publishers, 1991.

Nicolis, G. and I. Prigogine, *Self-Organization in Nonequilibrium Systems*, John Wiley & Sons: New York, 1977, 491 pp.

Ohanian, H.C., *Gravitation and Spacetime*, W.W. Norton and Company: New York, 1976, 461 pp.

O'Neill, E.L., *Introduction to Statistical Optics*, Dover, 1992, 179 pp.

Pagels, H.R., *The Cosmic Code*, Bantam Books: New York, 1983, 333 pp.

Pauli, W., *Theory of Relativity*, Dover Publications: New York, 1958, 241 pp.

Penrose, R., *The Emperor's New Mind*, Oxford University Press: New York, 1989, 466 pp.

Pierce, A.D., *Acoustics*, McGraw-Hill Book Company: New York, 1981.

Polkinghorne, J.C., *The Quantum World*, Princeton University Press: Princeton, NJ, 1985, 100 pp.

Ponomarev, L.I., *The Quantum Dice*, Institute of Physics Publishing: Bristol and Philadelphia, 1993, 255 pp.

Power, E.A., *Introductory Quantum Electrodynamics*, Longmans: London, 1964, 147 pp.

Prigogine, I., *From Being to Becoming* (Time and Complexity in the Physical Sciences), W.H. Freeman and Company: New York, 1980, 272 pp.

Prigogine, I. and I. Stengers, *Order out of Chaos* (Man's New Dialogue with Nature), Bantam Books: New York, 1984, 349 pp.

Rae, A., *Quantum Physics: Illusion or Reality?* Cambridge University Press: Cambridge, 1986, 123 pp.

Rao, K.N.S., *The Rotation and Lorentz Groups and their Representations for Physicists*, John Wiley & Sons: New York, 1988, 358 pp.

Rees, Martin, *Just Six Numbers*, Basic Books: New York, 2000.

Rindler, W., *Essential Relativity* (Special, General, and Cosmological), Springer-Verlag: New York, 1977, 284 pp.

Rocard, Y., *Thermodynamics*, Isaac Pitman & Sons: London, 1961, 681 pp.

Rogers, E.M., *Physics for the Inquiring Mind* (The Methods, Nature, and Philosophy of Physical Science), Princeton University Press: Princeton, NJ, 1960, 778 pp.

Rohlf, J.W., *Modern Physics from α to Z^0*, John Wiley and Sons, Inc: New York, 1994, 646 pp.

Ronchi, V., *The Nature of Light, an Historical Survey*, William Hinemann: London, 1970.

Sandin, T.R., *Essentials of Modern Physics*, Addison-Wesley Publishing Company: Reading, MA, 1989, 575 pp.

Sciama, D.W., *Modern Cosmology*, Cambridge University Press: Cambridge, 1975, 214 pp.

Schiff, L.I., *Quantum Mechanics*, McGraw-Hill Book Company: New York, 1968, 544 pp.

Schrödinger, E., *Statistical Thermodynamics*, Cambridge University Press: Cambridge, 1962, 95 pp.

Serway, R.A., *Physics for Scientists and Engineers*, Saunders College Publishing: New York, 1986, 942 pp.

Shadowitz, A., *Special Relativity*, Dover Publications: New York, 1968, 203 pp.

Shapiro, S.L. and S.A. Teukolsky, *Black Holes, White Dwards, and Neutron Stars* (The Physics of Compact Objects), John Wiley & Sons: New York, 1983, 645 pp.

Silk, J., *The Big Bang* (The Creation and Evolution of the Universe), W.H. Freeman: San Francisco, 1980, 394 pp.

Singh, J., *Great Ideas and Theories of Modern Cosmology*, Dover Publications: New York, 1970, 416 pp.

Singh, *Big Bang*, Harper Perennial: New York, 2005, 532 pp.

Singh, S., *The Code Book*, Anchor Books: New York, 2000, 411 pp.

Smythe, W.R., *Static and Dynamic Electricity*, McGraw-Hill Book Company: New York, 1950, 616 pp.

Sommerfeld, A., *Optics* (Lectures on Theoretical Physics), Vol. 4, Academic Press: New York, 1964, 383 pp.

Sommerfeld, A., *Thermodynamics and Statistical Mechanics* (Lectures on Theoretical Physics), Vol. 5, Academic Press: New York, 1964, 401 pp.

Spiridonov, O.P., *Universal Physical Constants*, Mir Publications: Moscow, 1986, 204 pp.

Sullivan, W., *Black Holes: The Edge of Space, The End of Time*, Doubleday Anchor Books: New York, 1979, 303 pp.

Tabor, D., *Gases, Liquids and Solids*, Cambridge University Press, 3rd edition, 1979, 418 pp.

Takayasu, H., *Fractals in the Physical Sciences*, Manchester University Press: Manchester, 1990, 170 pp.

Tarasov, L., *This Amazingly Symmetrical World*, Mir Publishers: Moscow, 1986, 164 pp.

Tarasov, L., *The World is Built on Probability*, Mir Publishers: Moscow, 1988, 190 pp.

Taylor, J.C., *Hidden Unity in Nature's Laws*, Cambridge University Press: New York, 2002, 490 pp.

Tolman, R.C., *Relativity, Thermodynamics and Cosmology*, Oxford University Press: London, 1950, 502 pp.

Tinkham, M., *Group Theory and Quantum Mechanics*, McGraw-Hill Book Company: New York, 1964, 340 pp.

Unsöld, A., *The New Cosmos*, Springer-Verlag: New York, 1969, 373 pp.

Van Kampen, N.G., *Stochastic Processes in Physics and chemistry*, North Holland Publishing Company: Amsterdam, 1987, 419 pp.

Van Ness, H.C., *Understanding Thermodynamics*, Dover Publications: New York, 1969, 103 pp.

Volkenshtein, M.V., *Biophysics*, Mir Publishers: Moscow, 1983, 640 pp.

Von Karman, T., *Aerodynamics*, McGraw-Hill Book Company: New York, 1954, 203 pp.

Walecka, J.D., *Fundamentals of Statistical Mechanics*, Imperial College Press, London, 302 pp.

Walker, J., *The Flying Circus of Physics* (*With Answers*), John Wiley & Sons: New York, 1977, 295 pp.

Wax, N. (Editor), *Selected Papers on Noise and Stochastic Processes*, Dover Publications: New York, 1954, 337 pp.

Weinberg, S., *Dreams of a Final Theory*, Vintage Press: London, 1993, 260 pp.

Weinberg, S., *Gravitation and Cosmology* (Principles and Applications of the General Theory of Relativity), John Wiley & Sons: New York, 1972, 657 pp.

Weinberg, S., *The First Three Minutes* (A Modern View of the Origin of the Universe), Bantam Books: New York, 1977, 177 pp.

Westphal, W.H., *Physik*, Springer–Verlag: Berlin, 20/21 Edition, 1959, 746 pp.

Westphal, W.H., *A Short Textbook of Physics*, Springer-Verlag, New York, 1968, 347 pp.

Will, C.M., *Was Einstein Right* (Putting General Relativity to the Test), Basic Books: New York, 1986, 274 pp.

Williams, L.P., *The Origins of Field Theory*, Random House: New York, 1966, 148 pp.

Withrow, G.J., *The Nature of Time*, Penguin Books: England, 1972, 153 pp.

Yin, C.-S., *Fluid Mechanics*, West River Press: Ann Arbor, MI, 1979, 622 pp.

Yourgrau, W. and S. Mandelstam, *Variational Principles in Dynamics and Quantum Theory*, Dover Publications: New York, 1968, 201 pp.

Yourgrau, W. *et al.*, *Treatise on Irreversible and Statistical Thermophysics*, Dover Publications: New York, 1982, 268 pp.

Zeldovich, Ya.B., et al. *The Almighty Chance*, World Scientific: Singapore, 1990, 316 pp.

Zhi, F.L. and L.S. Xian, *Creation of the Universe*, World Scientific: Singapore, 1989, 180 pp.

Ziman, J.M., *Principles of the Theory of Solids*, Cambridge University Press, 2nd edition, 1972, 435 pp.

Zwiebach, B., *A First Course in String Theory* , Cambridge University Press, 2004, 558 pp.

5. Chemistry

Atkins, P. and J. de Paula, *Physical Chemistry*, 7th edition, Oxford University Press, 2002, 1149 pp.

Atkins, P.W., *Molecules*, W.H. Freeman and Company: New York, 1987, 197 pp.

Atkins, P., and L. Jones, *Chemistry – Molecules, Matter, and Change*, 3rd ed., H. Freeman and Company: New York, 1997, 886 pp.

Bishop, D.M., *Group Theory and Chemistry*, Dover Publications: New York, 1993, 300 pp.

Chang, R., *Physical Chemistry with Applications to Biological Sciences*, Collier Macmillan International: New York, 1977, 538 pp.

Hearst, J.E. and J.B. Ifft, *Contemporary Chemistry*, W.H. Freeman and Company: San Francisco, 1976, 753 pp.

Ihde, A.J., *The Development of Modern Chemistry*, Harper and Row: New York, 1964, 851 pp.

Leicester, H.M. and H.S. Klickstein, *A Source Book in Chemistry*, McGraw-Hill Book Company: New York, 1952.

Partington, J.R., *A Text-Book of Inorganic Chemistry*, 5th edition, Macmillan and Co.: London, 1947, 1062 pp.

Pauling, L.C., *General Chemistry*, Dover Publications: New York, 1988, 959 pp.

Peter, K., C. Vollhardt and N.E. Schore, *Organic Chemistry – Structure and Function*, 4th ed., W.H. Freeman and Company: New York, 1003, 1203 pp.

Petrucci, R.H., W.S. Harwood and F.G. Herring, *General Chemistry*, Prentice Hall: New Jersey, 1160 pp.

Schatz, G.C. and M.A. Ratner, *Quantum Mechanics in Chemistry*, Dover Publications: New York, 2002, 361 pp.

6. General Biology and Medicine

Campbell, N., *Biology*, Benjamin–Cummings, Publishing company, 2004

Fruton, J., *Proteins, Genes, Enzymes: The Interplay of Chemistry and Biology*, Yale University Press, New Haven, 1999

Johnson, G.B., *Biology, Visualizing Life*, Holt, Rinehart and Winston, 2005

Sneader, W., *Drug Discovery: The Evolution of Modern Medicines*, Wiley, 1985,

Solomon, E.P., Berg, L.R. and D.W. Martin, *Biology*, Thomson: Brooks/Cole, 2005.

Evolution

Darwin, C.R., *Origin of the Species*, 1859

Futuyama, J.F., *Evolution*, Sinauer Associates, 2005

Gregory, T.R. (editor) *The Evolution of the Genome*, Elsevier, 2005

Kimura, M., *The Neutral Theory of Molecular Evolution*, 1983

Lewontin R.C., *The Genetic Basis of Evolutionary Change*, Columbia University Press, 1974

Ridley, M., *Evolution*, Blackwell, 2003

Rose, M.R. and Mueller, *Evolution and Ecology of the Organism*, Prentice Hall, 2005

Biophysics

Morowitz, H.J., *Energy Flow in Biology*, Ox Bow Press, 1979, 179 pp.

Nölting, B., *Methods in Modern Biophysics*, Springer–Verlag, 2006

Volkenstein, M.V., *Biophysics*, Mir Publications, 1983

Mathematical Biology

Ewens, W.J., *Mathematical Population Genetics*, Springer–Verlag, 2004

Hoppenstadt, F.C., *An Introduction to the Mathematics of Neurons*, Cambridge University Press, 1986

Murray, J.D., *Mathematical Biology*, Springer-Verlag: Berlin, 1989, 767 pp.

Segel, L.A., *Modeling Dynamic Phenomena in Molecular and Cellular Biology*, Cambridge University Press, 1984

Neurobiology

Martin, A.R., Wallace, B.G., Fuchs, P.A. and J.G., Nichols, *From Neuron to Brain*, Sinauer Associates, 2001

Shepard, G.M., *Neurobiology*, Oxford University Press, 1994

Ecology

Real, A.L. and S.A., Levin, *Foundation of Ecology*, Chicago University Press, 1991

Bioinformatics

Barnes, M.R. and I.C., Gray (Editors), *Bioinformatics for Geneticists*, Wiley, 2003

Baxevanis, A.D. and B.F.F., Ouellette (Editors), *Bioinformatics: A Practical Guide to the Analysis of Genes and Proteins*, Wiley, 2005

Molecular Biology

Alberts, B. (Editor), *Molecular Biology of the Cell*, Garland Science, 2002

Buchanan, B.B., Gruissem, W. and R.L., Jones, *Biochemistry and Molecular Biology of Plants*, American Society of Plant Physiologists, 2000

Campbell, M.K., *Biochemistry*, Saunders College Publishing, 1995

Lodish, H. (Editor), *Molecular Cell Biology*, W.H. Freeman and Company, 2000

Population Biology

Babloyantz, A., *Molecules, Dynamics, and Life* (An Introduction to Self-Organization of Matter), John Wiley & Sons: New York, 1986, 345 pp.

Bartlett, M.S., *Stochastic Population Models*, Methuen: London, 1960, 87 pp.

Berg, H.C., *Random Walks in Biology*, Princeton University Press: Princeton, N.J., 1993, 152 pp.

Calder III, W.A., *Size, Function and Life History*, Dover Publications: New York, 1996, 431 pp.

Campbell, M.K., *Biochemistry*, Saunders College Publishing: New York, 1995, 657 pp.

Cavalli–Sforza, L.C., *The History and Geography of Human Genes*, Princeton University Press, 1994

D'Arcy, W. Thompson, *On Growth and Form* (Abridged Edition by J.T. Bonner), Cambridge University Press: Cambridge, 1961, 346 pp.

Dyson, F., *Origins of Life*, Cambridge University Press: Cambridge, 1985, 81 pp.

Fogel, S., *Life's Devices, The Physical World of Animals and Plants*, Princeton University Press: Princeton, NJ, 1988, 367 pp.

Goldsmith, D. and T.Owen, *The Search for Life in the Universe*, Addison-Wesley, Reading MA, 1992, 530 pp.

Hartl, D., *Primer of Population Genetics*, Sinauer, 2000

Hoppensteadt, F.C., *Mathematical Methods of Population Biology*, Cambridge University Press, 1982, 149 pp.

Jacob, F., *The Logic of Life* (A History of Heredity), Pantheon Books: New York, 1982, 348 pp.

Jacob, F., *The Possible and the Actual*, Pantheon Books: New York, 1982, 70 pp.

Klein, R.M., *The Green World*, Harper and Row Publishers: New York, 1979, 437 pp.

Monod, J., *Chance and Necessity*, Vintage Books: New York, 1971, 199 pp.

Parker, S. (Editor) *The Human Body Book*, DK Publishers: New York, 2007, 256 pp.

Prigogine, I. and I. Stengers, *Order out of Chaos* (Man's New Dialogue with Nature), Bantam Books: New York, 1984, 349 pp.

Prigogine, I., *From Being to Becoming* (Time and Complexity in the Physical Sciences), W.H. Freeman and Company: New York, 1980, 272 pp.

Smith, J. Maynard, *The Theory of Evolution*, 1997

Schrödinger, E., *What is Life?* (The Physical Aspect of the Living Cell), Cambridge University Press: Cambridge, 1951, 92 pp.

Svirezhev, Yu.M. and D.O., Logofet, *Stability of Biological Communities*, Mir: Moscow, 1983, 319 pp.

Von Frisch, K., *Animal Architecture*, Harcourt, Brace and Jovanovich: New York, 1974, 306 pp.

Von Neumann, John, *The Computer and the Brain*, Yale University Press: New Haven, CT, 1958, 82 pp.

Wells, S., *Deep Ancestry-Inside the Genographic Project*, National Geographic: Washington D.C., 2006, 247 pp.

Wilson, E.O. and W.H., Bossert, *A Primer of Population Biology*, Sinauer, 1971, 92 pp.

7. *Earth Sciences*

Adams, F.D., *Birth and Development of the Geological Sciences*, Dover Publications: New York, 1954, 606 pp.

Asimov, I., *Guide to Earth and Space*, Fawcett Crest: New York, 1991, 274 pp.

Båth, M., *Introduction to Seismology*, Birkhäuser Verlag: Basel, 1979, 428 pp.

Ben-Menahem, A. and S.J. Singh, *Seismic Waves and Sources*, Springer-Verlag: New York, 1981, 1108 pp.

Ben-Menahem, A. and S.J. Singh, *Seismic Waves and Sources*, Second Edition, Dover Publications: New York, 2000.

Berry, A., *A Short History of Astronomy*, John Murray: London, 1898.

Binney, J. and S. Termaine, *Galactic Dynamics*, Princeton University Press: Princeton, NJ, 1987, 733 pp.

Boyer, C.B., *The Rainbow*, Thomas Yoseloff: New York, 1959, 376 pp.

Brancazio, P.J. and A.G.W. Cameron (Editors), *The Origin and Evolution of Atmospheres and Oceans*, John Wiley & Sons: New York, 1964, 314 pp.

Broecker, W.S., *How to Build a Habitable Planet*, Eldigio Press: Palisades, NY, 1985, 291 pp.

Brooks, C.E.P., *Climate through the Ages*, Dover: New York, 1970.

Bullard, F.M., *Volcanoes*, University of Texas Press: Texas, 1962, 441 pp.

Burroughs, W.J., et al., *Guide to Weather*, Fog city Press: San Francisco, 2007, 288 pp.

Corliss, W.R., *Handbook of Unusual Natural Phenomena*, Anchor Books, New York, 1983, 423 pp.

Gould, R.T., *The Marine Chronometer*, J.D. Porter: London, 1923.

Graedel, T.E. and P.J. Crutzen *Atmosphere, Climate and Change*, Scientific American Library: New York, 1995, 196 pp.

Gribbin, J., *The Hole in the Sky* (Man's Threat to the Ozone Layer), Bantam Books: New York, 1988, 192 pp.

Hawkins, G.S. and J.B. White, *Stonehenge Decoded*, Dell Publishing Company, A Delta Book: New York, 1965, 202 pp.

Henson, R., *The Rough Guide to Climatic Change*, The Penguin Group: London, 2006, 341 pp.

Hoyle, F., *On Stonehenge*, W.H. Freeman and Company: San Francisco, 1977, 157 pp.

Humphreys, W.J., *Physics of the Air*, Dover Publications: New York, 1964.

Imbrie, J. and K. Palmer Imbrie, *Ice Ages: Solving the Mystery*, Enslow Publications: New Jersey, 1979, 224 pp.

Ince, M., *The Rough Guide to Earth*, Penguin Group: The London, 2007, 294 pp.

Lamb, H.H., *Climate: Present, Past and Future*, Methuen: London, 1977.

Lovelock, J.E., *Gaja: A New Look at Life on Earth*, Oxford University Press: Oxford, 1987, 157 pp.

Lucretius (Titus Lucretius Carus), *On the Nature of the Universe*, Translated by R.E. Lathom, Penguin Books: New York, 1951, 275 pp.

Lutgens, F.K. and E.J. Tarbuck, *The Atmosphere*, Prentice-Hall: New Jersey, 1982, 478 pp.

Macdonald, G.A., *Volcanoes*, Prentice-Hall: New Jersey, 1972, 510 pp.

Menard, H.W. (Editor), *Ocean Science*, Reading from Scientific American, W.H. Freeman and Company: San Francisco, 1965, 307 pp.

Menke, W. and D. Abbot, *Geophysical Theory*, Columbia University Press: New York, 1990, 458 pp.

Monroe, J.S. and R. Wicander *Physical Geology* 4th Edition. Brooks/Cole: United States, 2001, 712 pp.

Myers, N. (Editor), *Gaia – an Atlas of Planet Management* Anchor Press: Doubleday, 1984, 272 pp

Reingold, E.M. and N. Dershowitz *Calendrical Calculations*, Cambridge University Press: Cambridge, UK, 2001, 422 pp.

Ridley, B.K., *The Physical Environment*, Ellis Horwood: Chichester, England, 1979, 236 pp.

Ridley, B.K., *The Physical Environment*, John Wiley: New York, 1979, 236 pp.

Ryan, W. and W. Pitman, *Noah's Flood*, Simon and Schuster: New York, 1998, 319 pp.

Simkin, T. and L. Siebert, *Volcanoes of The World*, Geoscience Press: Tucson Arizona, 1994, 349 pp.

Tarling, D. and Tarling M., *Continental Drift*, Anchor Books: New York, 1975, 142 pp.

Taube, M., *Evolution of Matter and Energy on a Cosmic and Planetary Scale*, Springer-Verlag: Berlin, 1985, 289 pp.

Thompson, P.D., *Numerical Weather Analysis and Prediction*, Macmillan and Company: New York, 1961, 170 pp.

Verschuur, G.L., *Impact!*, Oxford University Press, 1996, 237 pp.

Volcano, Time–Life Books, 1982, 176 pp.

Von Arx, W.S., *An Introduction to Physical Oceanography*, Addison-Wesley Publishing Company: Reading, MA, 1962, 422 pp.

Weiner, J., *Planet Earth*, Bantam Books: New York, 1986, 370 pp.

Weyl, P.K., *Oceanography*, John Wiley & Sons: New York, 1970, 535 pp.

Yasso, W.E., *Oceanography*, Holt, Rinehart and Winston: New York, 1965, 176 pp.

8. Space Science, Astronomy and Astrophysics

Abel, G.O., *Exploration of the Universe*, Saunders College Publishing: Philadelphia, PA, 1982, 721 pp.

Asimov, I., *The Universe* (From Flat Earth to Quasar), Avon Books: New York, 1966, 315 pp.

Asimov, I., *Jupiter*, Ace Books: New York, 1975, 247 pp.

Berry, A., *A Short History of Astronomy*, John Murray: London, 1898.

Binney, J. and S. Termaine, *Galactic Dynamics*, Princeton University Press: Princeton, NJ, 1987, 733 pp.

Burnham, R., et al., *A Guide to Backyard Astronomy*, Fog City press: San Francisco, 1977, 288 pp.

Celnikier, L.M., *Basics of Space Flight*, Editions Frontieres: Gif-sur-Yvette Cedex: France, 1993, 356 pp.

Celnikier, L.M., *Basic Cosmic Structures*, Editions Frontieres: France, 1989, 408 pp.

Chandrasekhar, S., *Ellipsoidal Figures of Equilibrium*, Dover Publications: New York, 1969, 254 pp.

Coveney, P. and R. Highfield, *The Arrow of Time*, Fawcett Columbine: New York, 1990, 378 pp.

Flammarion, C. (Editor), *The Flammarion Book of Astronomy*, Simon and Schuster: New York, 1964, 670 pp.

Harrison, E., *Darkness at Night*, Harvard University Press: Cambridge, 1987, 293 pp.

Harwitt, M., *Astrophysical Concepts*, John Wiley & Sons: New York, 1973, 561 pp.

Illingworth, Valerie, *Dictionary of Astronomy Facts On File*, 1994, 520 pp.

Kaufmann, W.J., *Universe*, W.H. Freeman and Company: New York, 1985, 594 pp.

Lightman, A., *Time For The Stars*, Warner Books, 1992, 124 pp.

Longair, M.S., *The Origins of our Universe*, Cambridge University Press, 1991, 128 pp.

Morris, R., *Times Arrow*, Simon and Schuster: New York, 1985, 240 pp.

Moulton, F.R., *An Introduction to Celestial Mechanics*, Dover Publications: New York, 1970, 436 pp.

Payne-Gaposchkin, C., *Introduction to Astronomy*, Prentice-Hall: New York, 1954, 508 pp.

Rees, M. (Editor) *Universe*, DK Publications: New York, 2005, 510 pp.

Shapiro, S.L. and S.A. Teukolsky, *Black Holes, White Dwards, and Neutron Stars* (The Physics of Compact Objects), John Wiley & Sons: New York, 1983, 645 pp.

Shklovskii, I.S. and C. Sagan, *Intelligent Life in the Universe*, Halden Day: San Francisco, 1966, 509 pp.

Shu, F.H., *The Physical Universe, An Introduction to Astronomy*, University Science Books: Mill Valley, CA, 1982, 584 pp.

Silk, J., *A Short History of the Universe*, Scientific American Library: New York, 1994, 246 pp.

Sterne, T.E., *An Introduction to Celestial Mechanics*, Interscience Publishers: New York, 1960, 206 pp.

Struve, O. and V. Zelbergs, *Astronomy of the 20th Century*, Macmillan and Company: New York, 1962, 544 pp.

Waters, T.R., *Smithsonian Guide to the Planets*, Macmillan: New York, 256 pp.

9. Social Science

Craig, C. (Editor), *Dictionary of Social Science*, Oxford University Press: New York, 2002, 584 pp.

Economics

Henderson, D.R., *The Concise Encyclopedia of Economics*, Liberty Fund, 2007, 637 pp.

Klein, M.W. and W.W., Klein, *Mathematical Methods for Economics*, Addison Wesley, 2001, 600 pp.

Mokyr, J. (Editor), *The Oxford Encyclopedia of Economics*, Oxford University Press: New York, 2003, 2824 pp

Wheelan, C., *Naked Economics*, Norton, New York, 2002, 260 pp.

Sociology

Ritzer, G. (Editor), *The Blackwell Encyclopedia of Sociology* , Wiley–Blackwell, 2007, 6384 pp

Scott, J. and M., Gordon (Editors), *A Dictionary of Sociology* , Oxford University Press: New York, 2005, 7204 pp

Turner, B.S. (Editor), *The Cambridge Dictionary of Sociology* , Cambridge University Press: London, 2006, 708 pp

Psychology

Hunt, M., *The Story of Psychology*, Anchor: New York, 2007, 896 pp.

Kazdin, A.E. (Editor), *Encyclopedia of Psychology*, (8 vol.), Oxford University Press: New York, 2000, 4160 pp

Strickland, B.B., *The Gale Encyclopedia of Psychology*, Gale Gengage, 2000, 701 pp.

Anthropology

Birx, H.J., *Encyclopedia of Anthropology*, (5 vol.), Sage Publications: New York, 2005, 2500 pp

Cavalli–Sforza, L.L. and F., Cavalli–Sforza,*The Great Human Diasporas*, Helix Books, Addison-Wesley Publishing Company: New York, 1995, 300 pp

Eriksen, T.H. and F.S., Nielsen, *A History of Anthropology*, Pluto Press: New York, 2001, 216 pp

Etymology, Terminology and Linguistics

Ayto, J., *Dictionary of Word Origins*, Arcade Publishing: New York, 1993, 583 pp.

Bodmer, F., *The Loom of Language*, W.W. Norton and Company: New York, 1944, 692 pp.

Cary, M., et all (Editors), *The Oxford Classical Dictionary*, The Clarendon Press: Oxford, 1957, 971 pp.

Clifford, A. H., Perkins, S. R. and T., Barr, *Latin For Dummies*, Wiley Publishing, Inc., Hoboken N.J., 2002, 363 pp.

Ehrlich, E., *Amo, Amas, Amat and More*, Harper and Row: New York, 1985, 328 pp.

Elkoshi, G.,*Thesaurus Proverbiorum Et Idiomatum Latinorum*, The Magnes Press, The Hebrew University: Jerusalem, 1981, 555 pp.

Guterman, N., *Latin Quotations*

Hayakawa, S.I., (and the Funk and Wagnalls Dictionary Staff), *Modern Guide to Synonyms and Related Words*, Funk and Wagnalls: New York, 1968, 726 pp.

Henderson, I. F. and W. D., Henderson, *A Dictionary of Scientific Terms*, Van Nostrand: New York, 1952

Hough, J.N., *Scientific Terminology*, Holt, Reinhart and Winston, 1963

Lewis, C. T. and C., Short, *Latin Dictionary*, Oxford University Press: New York, 1957.

Liddell, H. G. and R., Scott, *A Greek – English Lexicon. Two volumes. Revised by H. S. Jones*, Oxford University Press, 1951.

Moore, B., and M. Moore, *Dictionary of Latin and Greek Origins*, Anchor Books, Doubleday: New York, 1966, 433 pp.

Morris, W. Ed, *The American Heritage Dictionary of the English Language*, Houghton Mifflin Company: New York, 1975, 1550 pp.

Morwood, J. and M., Warman, *Our Greek and Latin Roots*, Cambridge University Press, 2006, 56 pp

Mount, H., *Carpe Diem*, Hyperion: New York, 2007, 259 pp.

Nybakken, O.E., *Greek and Latin in Scientific Terminology* , Blackwell Publishing, 1985, 321 pp

Sack, A., *Shortcut to Word Power: Essential Latin and Greek Roots and Prefixes*, White Pine Printers, Ann Arbor: Michigan, 2003, 1138 pp

Shipley, J.T., *Dictionary of Word Origins*, Littlefield Adams Co.: Totowa NJ, 1974, 430 pp.

Simpson, D. P., Cassell's New Compact Latin – English and English – Latin Dictionary, Funk ad Wagnalls: New York, 1963, 379 pp.

Webster's Encyclopedic Unabridged Dictionary of the English Language, New Revised Edition, 1994, Random House, Inc., 1693 pp.

10. Quotations

Bartlett, J., *Bartlett's Familiar Quotations*, Little Brown and Com: London, 2002, 1472 pp.

Bernstein, L., *The Unanswered Questions-Six Talks at Harvard*, Harvard Univ. Press, London, 1976, 428 pp.

Bernstein, L., *The Infinite Variety of Music*, Amadeus Press, New York, 1993, 288 pp.

Cary, M., et al. (Editors), *The Oxford Classical Dictionary*, Oxford at the Clarendon Press, 1957, 971 pp.

Howatson, M.C. (Editor), *The Oxford Companion to Classical Literature*, Oxford University Press: New York, 1988, 6158 pp.

Knowels, E. (Editor), *The Oxford Dictionary of Quotations*, Oxford University Press: London, 2004, 1168 pp.

Mackay, A.L., *A Dictionary of Scientific Quotations*, Institute of Physics Publishings: Bristol and Philadelphia, 1991, 297 pp.

Runes, D.D. (Editor), *Reflections and Maxims by Baruch Spinoza* , Philosophical Library: New York, 1965, 92 pp.

Schmidt, A., *Shakespeare Lexicon and Quotation Dictionary*, Dover Publications: New York, 1971, 767 pp.

Shapira, F.R. (Editor), *The Yale Book of Quotations*, Yale University Press, 2006, 1104 pp.

GigaQuotes.com (WEB)

BrainyQuotes.com (WEB)

Name Index

A

C

D

E

F

G

H

I

J

K

L

M

N

O

P

Q

R

S

X

Y

Z

Printing and Binding: Stürtz GmbH, Würzburg